MODELING AND CONTROL IN VIBRATIONAL AND STRUCTURAL DYNAMICS

A Differential Geometric Approach

Published Titles

Advanced Differential Quadrature Methods, Zhi Zong and Yingyan Zhang
Computing with hp-ADAPTIVE FINITE ELEMENTS, Volume 1, One and Two Dimensional Elliptic and Maxwell Problems, Leszek Demkowicz
Computing with hp-ADAPTIVE FINITE ELEMENTS, Volume 2, Frontiers: Three Dimensional Elliptic and Maxwell Problems with Applications, Leszek Demkowicz, Jason Kurtz, David Pardo, Maciej Paszyński, Waldemar Rachowicz, and Adam Zdunek
CRC Standard Curves and Surfaces with Mathematica®: *Second Edition*, David H. von Seggern
Discovering Evolution Equations with Applications: Volume 1-Deterministic Equations, Mark A. McKibben
Discovering Evolution Equations with Applications: Volume 2-Stochastic Equations, Mark A. McKibben
Exact Solutions and Invariant Subspaces of Nonlinear Partial Differential Equations in Mechanics and Physics, Victor A. Galaktionov and Sergey R. Svirshchevskii
Fourier Series in Several Variables with Applications to Partial Differential Equations, Victor L. Shapiro
Geometric Sturmian Theory of Nonlinear Parabolic Equations and Applications, Victor A. Galaktionov
Green's Functions and Linear Differential Equations: Theory, Applications, and Computation, Prem K. Kythe
Introduction to Fuzzy Systems, Guanrong Chen and Trung Tat Pham
Introduction to non-Kerr Law Optical Solitons, Anjan Biswas and Swapan Konar
Introduction to Partial Differential Equations with MATLAB®, Matthew P. Coleman
Introduction to Quantum Control and Dynamics, Domenico D'Alessandro
Mathematical Methods in Physics and Engineering with Mathematica, Ferdinand F. Cap
Mathematical Theory of Quantum Computation, Goong Chen and Zijian Diao
Mathematics of Quantum Computation and Quantum Technology, Goong Chen, Louis Kauffman, and Samuel J. Lomonaco
Mixed Boundary Value Problems, Dean G. Duffy
Modeling and Control in Vibrational and Structual Dynamics, Peng-Fei Yao
Multi-Resolution Methods for Modeling and Control of Dynamical Systems, Puneet Singla and John L. Junkins
Optimal Estimation of Dynamic Systems, John L. Crassidis and John L. Junkins
Quantum Computing Devices: Principles, Designs, and Analysis, Goong Chen, David A. Church, Berthold-Georg Englert, Carsten Henkel, Bernd Rohwedder, Marlan O. Scully, and M. Suhail Zubairy
A Shock-Fitting Primer, Manuel D. Salas
Stochastic Partial Differential Equations, Pao-Liu Chow

CHAPMAN & HALL/CRC APPLIED MATHEMATICS
AND NONLINEAR SCIENCE SERIES

MODELING AND CONTROL IN VIBRATIONAL AND STRUCTURAL DYNAMICS

A Differential Geometric Approach

Peng-Fei Yao

Chinese Academy of Sciences

Beijing, China

CRC Press
Taylor & Francis Group
Boca Raton London New York

CRC Press is an imprint of the
Taylor & Francis Group, an **informa** business

CRC Press
Taylor & Francis Group
6000 Broken Sound Parkway NW, Suite 300
Boca Raton, FL 33487-2742

First issued in paperback 2017

© 2011 by Taylor & Francis Group, LLC
CRC Press is an imprint of Taylor & Francis Group, an Informa business

No claim to original U.S. Government works

Version Date: 20110506

ISBN-13: 978-1-4398-3455-8 (hbk)
ISBN-13: 978-1-138-11664-1 (pbk)

Visit the Taylor & Francis Web site at
http://www.taylorandfrancis.com

and the CRC Press Web site at
http://www.crcpress.com

To Professor John E. Lagnese

Contents

List of Symbols

$\mathbb{R}$	set of real numbers
(M, g)	Riemannian manifold with metric g
$C^\infty(M) = T^0(M)$	all C^∞ functions on M
$\mathcal{X}(M) = T(M) = \Lambda(M)$	all vector fields on M
$T^k(M)$	all tensor fields of rank k on M
D	Levi–Civita connection
$[\cdot, \cdot]$	Lie bracket
Γ^k_{ij}	Christoffel symbol
$\otimes$	tensor product
$D^2 f$	Hessian of f in metric g
$\text{tr}\, S$	trace of tensor S of rank 2
$\mathbf{R}_{XY}$	curvature operator
$\mathbf{R}(\cdot, \cdot, \cdot, \cdot)$	curvature tensor
Φ_*	differential of map Φ
$\kappa(\Xi)$	sectional curvature of subspace Ξ of dimension 2
N	normal vector field of hypersurface
$\Pi(\cdot, \cdot)$	second fundamental form
$\text{Ric}\,(\cdot, \cdot)$	Ricci tensor
$L(\gamma)$	length of curve γ
$d(x, y)$	distance between x and y in metric g
$\Sigma(x_0)$	interior of tangent cut locus of x_0
$\exp_{x_0} \Sigma(x_0)$	interior of cut locus of x_0
Δ	Laplacian in metric
Δ_g	Laplacian in metric g
$\boldsymbol{\Delta}$	Hodge–Laplacian
$\Lambda^k(M)$	all k-forms
$\wedge$	exterior product
$\text{i}(X)T$	interior product of tensor field T by vector field X
$\text{div}\, X$	divergence of vector field X
d	exterior derivative
$\mathbb{R}^n$	Euclidean space of dimension n
Ω	bounded open set in $\mathbb{R}^n$, or middle surface of plate or shell
$\Gamma = \partial\Omega$	boundary of Ω
$L^2(\Omega, T^k)$	all square integrable tensor fields of rank k on Ω
$L^2(\Omega, \Lambda^k)$	all square integrable k-forms on Ω
$H^k(\Omega, \Lambda)$	all 1-forms with all its i-th differential square integrable

	for $0 \leq i \leq k$ on Ω
Q	formal adjoint of covariant differential operator
$(\cdot, \cdot)$ and $\| \cdot \|$	inner product and norm of $L^2(\Omega)$, respectively
ν	outside normal of boundary in Euclidean metric
$\Upsilon = \dfrac{1}{2}(\overline{g} - g)$	strain tensor of middle surface
$\rho = \overline{\Pi} - \Pi$	change of curvature tensor of middle surface
μ	Poisson's coefficient
	$\mathcal{L}^2(\Omega) = (L^2(\Omega, \Lambda))^2 \times (L^2(\Omega))^2$
	$\mathcal{H}^1(\Omega) = (H^1(\Omega, \Lambda))^2 \times (H^1(\Omega))^2$
	$\mathcal{H}^1_{\hat{\Gamma}}(\Omega) = (H^1_{\hat{\Gamma}}(\Omega, \Lambda))^2 \times (H^1_{\hat{\Gamma}}(\Omega))^2$
$\mathrm{lo}\,(\zeta),\, L(\zeta)$	lower order terms
$\aleph$	canonical isomorphism
$\mathrm{B}^{\,m}_{D}(\Sigma_0)$	space of boundary control in Dirichlet action
$\mathrm{B}^{\,m}_{N}(\Sigma_0)$	space of boundary control in Neumann action

Preface

Control theory of partial differential equations is a many-faceted subject. Created to describe the control behavior of mechanical objects such as wave equations, plates, and shells, it has developed into a body of material that interacts with many branches of mathematics and mechanics, such as differential geometry and elasticity.

The differential geometrical approach was introduced more than a decade ago. Since then, many important advances in modeling and control in vibrational and structural dynamics have been made. This book is intended to present a systematic, updated account of this direction of research.

The original motivation for the differential geometrical approach may be said to have arisen from the need to cope with the following three situations:

(i) The case where the coefficients of the partial differential equations are variable in space (waves/plates);

(ii) The case where the partial differential equations themselves are defined on curved surfaces (shells);

(iii) The case where the systems have quasilinear principal parts.

Regarding the first need (i), the differential geometric energy methods described in this book may be viewed as far-reaching generalizations of the classical energy methods of the early/mid-1980s. The first reason this approach is necessary is that the curvature theory in Riemannian geometry provides us with checkable assumptions for controllability/stabilization for the systems with variable coefficients. The classical analysis was originally successful in dealing with the canonical (constant coefficients) wave/plate equations, but proved inadequate for treating variably coefficient cases mainly because controllability/stabilization is a global property while the classical case corresponds to the zero curvature. Although the curvature is locally defined at each point, it interprets global information on almost everything on manifolds, including controllability/stabilization. That is one of the reasons Riemannian geometry is called global geometry.

As to the second need (ii), it would seem self-explanatory that if an equation is defined on a manifold (as in the case of shells), the natural setting that is necessary for its analysis should be that of differential geometry. In particular, modeling of shells in the form of free coordinates is the key step in analysis and control. Classically, the topic of static shells is covered by many books. They all assume the middle surface of a shell to be described by one coordinate patch: this is the image in $\mathbb{R}^3$ of a smooth function defined on a connected domain of $\mathbb{R}^2$, rooted in classical differential geometry. This view has geometrical limitations, as it forces the exclusion of interesting objects such as a sphere. Moreover, the classical models use traditional geometry and end up with highly complicated resultant equations. In these, the explicit presence of the Christoffel symbols makes them unsuitable for energy method computations of the type needed for continuous observability/stabilization estimates. We view a middle surface of a shell as a Riemannian manifold of dimension 2 that is not subject to any particular coordinate patch. A basic computational technique, first used by Bochner, allows us to have mathematical models for shells in the form of free coordinates and to carry out multiplier schemes for controls. The basic ideas are as follows: In order to verify an identity or a pointwise estimate, it suffices to do so at each point p relative to a coordinate system or frame field that offers the greatest computational simplification. What prevents a computation from being as simple as the corresponding classical (i.e., Euclidean) situation is the presence of the Christoffel symbols. Thus, we use a coordinate system or a frame field relative to which the Christoffel symbols vanish at the given point p. Such a coordinate system varies with point p of the middle surface. Clearly, once the middle surface of a shell is fixed by one particular coordinate patch, this technique does not work. More importantly, it seems clear that, without this differential geometric tool, many of these more sophisticated theorems would probably not have been discovered or, at least, their discoveries would otherwise have been delayed.

As the third need (iii), the quasilinearity of models comes from nonlinear materials, and their controllability/stabilization depends not only on their variable coefficients but also on their equilibria. Similarly, as a necessary tool, the curvature theory yields checkable assumptions.

This book will cover the following topics:

(i) control of the wave equation with variable coefficients (Chapter 2);

(ii) control of plates with variable coefficients (Chapter 3);

(iii) modeling and control of shallow shells (Chapter 4);

(iv) modeling and control of Naghdi's shells (Chapter 5);

(v) modeling and control of Koiter's shells (Chapter 6);

(vi) control of the quasilinear wave equation (Chapter 7), with particular emphasis on research results using the differential geometrical approach.

In order to make this book self-contained, I have condensed the main content of the book, *Introduction to Riemannian Geometry* (in Chinese), by H. Wu, C. L. Shen, and Y. L. Yu, into Chapter 1. I hope this chapter will help readers who are not familiar with Riemannian geometry to obtain some necessary knowledge for our problems.

I thank Goong Chen for inviting me to write this book.

I have greatly profited from the comments and thoughtful suggestions of many of my colleagues, friends and students, in particular: X. Cao, M. M. Cavalcanti, S. G. Chai, G. Chen, M. C. Delfour, L. Deng, S. Feng, D. X. Feng, H. Gao, B. Z. Guo, Y. X. Guo, V. Komornik, M. Krstic, I. Lasiecka, T. T. Li,

S. J. Li, S. Li, B. Miara, B. P. Rao, Z. C. Shao, R. Triggiani, M. Tucsnak, G. Weiss, H. G. Wu, J. Q. Wu, B. Y. Zhang, and Z. F. Zhang.

I am extremely grateful for the suggestions and lists of mistakes from drafts of this book sent to me by S. G. Chai, L. Deng, Y. X. Guo, S. Li, Z. C. Shao, J. Q. Wu, and Z. F. Zhang.

I have been supported by the National Science Foundation of China during the writing, most recently under grants no. 60821091, no. 60334040, no. 60221301, no. 60774025, no. 10831007, and no. YB20098000101(the excellent PhD adviser program of Beijing).

Peng-Fei Yao

Key Laboratory of Systems and Control
Institute of Systems Science
Academy of Mathematics and Systems Science
Chinese Academy of Sciences
Beijing 100190, P. R. China
E-mail: pfyao@iss.ac.cn

Chapter 1

Preliminaries from Differential Geometry

This chapter will cover some necessary knowledge of Riemann geometry which will be needed in the subsequent chapters to make this book self-contained as far as possible.

The main reference is book [206] in Chinese.

1.1 Linear Connections, Differential of Tensor Fields and Curvature

Connections Let M be a C^∞ manifold of dimension n. We say that g is a *Riemannian metric* on M which means that for each tangential space M_x ($x \in M$), there is an inner product $g_x(\cdot, \cdot)$ (sometimes denoted by $\langle \cdot, \cdot \rangle$) on M_x and this relation is C^∞. Here the C^∞ means: For any local coordinate $(x_1, \cdots, x_n)$, if we let

$$g_{ij}(x) = g_x(\partial_{x_i}, \partial_{x_j}),$$

where $\partial_{x_i} = \frac{\partial}{\partial x_i}$, then they are C^∞ functions of the coordinate $(x_1, \cdots, x_n)$. Note that the inner products $g_x(\cdot, \cdot)$ are equal to that the matrices $(g_{ij}(x))_{n \times n}$ are positive, symmetrical. In the terminology of tensors, g is actually a C^∞ covariant tensor field of rank 2 such that, for any vector fields X, Y on M,

$$g(X, Y) = g(Y, X), \quad g(X, X) \geq 0,$$

and $g_x(X, X) = 0$ if and only if $X(x) = 0$.

In this chapter (M, g) will always denote a Riemannian manifold of dimension n with metric g.

1

The following examples are two of the most important Riemannian manifolds in our book.

Example 1.1 *Let $G(x)$ be a symmetrical, positive matrix for each $x \in \mathbb{R}^n$. We introduce an inner product on each $\mathbb{R}_x^n = \mathbb{R}^n$ by*

$$g(X,Y) = \langle G(x)X, Y \rangle \quad for \quad X, \, Y \in R_x^n,$$

where $\langle \cdot, \cdot \rangle$ is the Euclidean metric in $\mathbb{R}^n$. Then $(\mathbb{R}^n, g)$ is a Riemannian manifold. g is the Euclidean metric if and only if $G(x) = I$ is the unit matrix on $\mathbb{R}^n$.

Example 1.2 *Let M be a hypersurface in $\mathbb{R}^{n+1}$. For each $x \in M$, M_x is an n-dimensional tangential plane of M at x. We define an inner product on each M_x by*

$$g(X,Y) = \langle X, Y \rangle \quad for \quad X, \, Y \in M_x, \tag{1.1}$$

where $\langle \cdot, \cdot \rangle$ is the Euclidean metric in $\mathbb{R}^{n+1}$. This metric is called the induced metric from $\mathbb{R}^{n+1}$. For example, let

$$M = \{\, x = (x_1, \cdots, x_{n+1}) \in \mathbb{R}^{n+1} \mid \sum_{i=1}^{n+1} x_i^2 = 1 \,\},$$

be the unit sphere in $\mathbb{R}^{n+1}$. Let $x \in M$ be given. Then $\alpha \in M_x$ if and only if there is a curve $\gamma \colon (-\varepsilon, \varepsilon) \to M$ such that $\gamma(0) = x$ and $\dot{\gamma}(0) = \alpha$. Let $\gamma(t) = (\gamma_1(t), \cdots, \gamma_{n+1}(t))$. Then $\gamma(t) \in M$ implies that

$$\sum_{i=1}^{n+1} \gamma_i^2(t) = 1 \quad for \quad t \in (-\varepsilon, \varepsilon).$$

We differentiate the above identity in the variable t at $t = 0$ to obtain

$$M_x = \{\, \alpha \in \mathbb{R}^{n+1} \mid \langle x, \alpha \rangle = 0 \,\}.$$

If $x_{n+1} \neq 0$, then the induced metric g of (1.1) on M_x is

$$g(\alpha, \beta) = \langle \alpha, \beta \rangle = \sum_{i=1}^{n+1} \alpha_i \beta_i = \sum_{i=1}^{n} \alpha_i \beta_i + \frac{1}{x_{n+1}^2} \sum_{ij=1}^{n} \alpha_i \beta_j x_i x_j$$

$$= \sum_{ij} \alpha_i \beta_j (\delta_{ij} + x_i x_j / x_{n+1}^2)$$

for any $\alpha = (\alpha_1, \cdots, \alpha_{n+1})$, $\beta = (\beta_1, \cdots, \beta_{n+1}) \in M_x$.

Before any study of Riemannian metrics, let us introduce the concept of "connections." Roughly speaking, a "connection" is a means to differentiate tensor fields. Let us recall how we differentiate a vector field in $\mathbb{R}^n$. If v is a vector at $p \in \mathbb{R}^n$ and f is a function differentiable at p, then the directional derivative $D_v f$ can be defined, as a real number, by

$$D_v f = \lim_{t \to 0} \frac{f(p + tv) - f(p)}{t}.$$

Let X be a vector field at p. Then $X = (X_1, \cdots, X_n)$ where X_i are given by the equation

$$X = \sum_{i=1}^{n} X_i \partial_{x_i},$$

and $(x_1, \cdots, x_n)$ is the natural coordinates in $\mathbb{R}^n$. Then the derivative of the vector field X along the vector v is defined by

$$D_v X = (D_v X_1, \cdots, D_v X_n),$$

or

$$D_v X = \sum_{i=1}^{n} (D_v X_i) \partial_{x_i}.$$

Clearly, $D_v X$ satisfies the following properties:
   (a) $D_{\alpha v} X = \alpha D_v X$ for any $\alpha \in \mathbb{R}$;
   (b) $D_v(fX) = (D_v f)X + f D_v X$ for any function f;
   (c) $D_v(X_1 + X_2) = D_v X_1 + D_v X_2$ for any vector fields X_1 and X_2;
   (d) $D_{v_1 + v_2} X = D_{v_1} X + D_{v_2} X$ for any vectors v_1 and v_2.
The above properties are exactly ones of the directional derivatives. Let V be a vector field such that $V(p) = v$. We define a vector field $D_V X$ by

$$(D_V X)(p) = D_v X.$$

Then properties (a)-(d) become
   (C1) $D_{fV + gW} X = f D_V X + g D_W X$;
   (C2) $D_V(fX) = (D_V f)X + f D_V X$;
   (C3) $D_V(X + Y) = D_V X + D_V Y$,
where V, W, X, and Y are vector fields on $\mathbb{R}^n$ and f, g are functions on $\mathbb{R}^n$, respectively.

Let us summarize the points of the above arguments: Given vector fields V and X in $\mathbb{R}^n$, we define a directional derivative $D_V X$ which satisfies the properties (C1)-(C3).

Now we come back to manifolds. Denote by $C^\infty(M)$ and $\mathcal{X}(M)$ all C^∞ functions and vector fields, respectively, on manifold M. For $X \in \mathcal{X}(M)$ and $f \in C^\infty(M)$, $V(f)$ denotes the directional derivative of f along the vector field V. Then $V(f) \in C^\infty(M)$.

Definition 1.1 *Let M be a C^∞ manifold. A connection is a map $D: \mathcal{X}(M) \times \mathcal{X}(M) \to \mathcal{X}(M)$ which satisfies properties (C1)−(C3). In (C1)−(C3), $D_V f = V(f)$.*

Remark 1.1 *Given a connection, $D_V X$ is called the covariant derivative of the vector field X along the vector field V. Then D is also called a covariant differential. Traditionally, "covariant" means that something varies with local coordinates. In the classical differential geometry, everything was done by local coordinates. However, what we can do is actually something which is independent of local coordinates. Here the definition of a connection is free of local coordinates. We do not use local coordinates except when they are necessary.*

Remark 1.2 *Property (C1) implies that, if V and W are vector fields on M such that for some $x \in M$, $V(x) = W(x)$, then*

$$(D_V X)(x) = (D_W X)(x) \tag{1.2}$$

(Exercise 1.1.1).

Let D be a connection. By Remark 1.2, for any $v \in M_x$, we may define $D_v X$ as follows: Taking any vector field $V \in \mathcal{X}(M)$ with $V(x) = v$, let $D_v X = (D_V X)(x)$.

Remark 1.3 *Unlike that in Remark 1.2, for $v \in M_x$ and $V \in \mathcal{X}(M)$, we are not able to define $D_V v$ by extending v to a vector field (Exercise 1.1.2).*

Nevertheless, the following things are true and very useful: Let D be a connection on manifold M. Let $v \in M_x$ and let $\gamma \colon (-\varepsilon, \varepsilon) \to M$ be a curve with $\dot\gamma(0) = v$. We further assume that $X, Y \in \mathcal{X}(M)$ are such that

$$X(\gamma(t)) = Y(\gamma(t)) \quad for \quad t \in (-\varepsilon, \varepsilon).$$

Then

$$D_v X = D_v Y \tag{1.3}$$

(Exercise 1.1.3).

There is a consequence of the relation (1.3): If $\gamma \colon [0, a] \to M$ is a curve and $X \in \mathcal{X}(M)$ is a vector field, then $D_{\dot\gamma(0)} X$ is determined by $X \circ \gamma$ completely where $X \circ \gamma(t) = X(\gamma(t))$.

Remark 1.4 *There are many connections on a manifold. For example, we can construct one for each local coordinate and then put them together by a partition of unity.*

The significant one among all connections is the so-called *Levi-Civita connection* which is the one uniquely given by the metric in Theorem 1.1 below.

Theorem 1.1 *Let M be a Riemannian manifold with the metric $g = \langle \cdot, \cdot \rangle$. Then there is a unique connection D that is determined by the following formulas: For vector fields X, Y and Z,*

(L1) $X\langle Y, Z \rangle = \langle D_X Y, Z \rangle + \langle Y, D_X Z \rangle$;

(L2) $D_X Y = D_Y X + [X, Y]$ *where* $[X, Y] = XY - YX$ *is the Lie bracket.*

Proof. First, we prove the uniqueness. Consider a local coordinate $(x_1, \cdots, x_n)$. Let functions Γ_{ij}^k be given by the following formulas

$$D_{\partial_{x_i}} \partial_{x_j} = \sum_{k=1}^{n} \Gamma_{ij}^k \partial_{x_k} \quad \text{for} \quad 1 \leq i, j \leq n. \tag{1.4}$$

For our purposes, it suffices to prove that functions Γ_{ij}^k are uniquely determined by the metric g. As usual, let $g_{ij} = \langle \partial_{x_i}, \partial_{x_j} \rangle$. Then the formula (L1) implies that

$$
\begin{aligned}
g_{jkx_i} &= \partial_{x_i} g_{jk} = \partial_{x_i} \langle \partial_{x_j}, \partial_{x_k} \rangle = \langle D_{\partial_{x_i}} \partial_{x_j}, \partial_{x_k} \rangle + \langle \partial_{x_j}, D_{\partial_{x_i}} \partial_{x_k} \rangle \\
&= \sum_{l=1}^{n} (\Gamma_{ij}^l g_{lk} + \Gamma_{ik}^l g_{jl}).
\end{aligned}
\tag{1.5}
$$

On the other hand, it follows from the relations (L2) and (1.4) that

$$0 = D_{\partial_{x_i}} \partial_{x_j} - D_{\partial_{x_j}} \partial_{x_i} - [\partial_{x_i}, \partial_{x_j}] = \sum_{k=1}^{n} [\Gamma_{ij}^k - \Gamma_{ji}^k] \partial_{x_k},$$

that is,

$$\Gamma_{ij}^k = \Gamma_{ji}^k \quad \text{for} \quad 1 \leq i, j, k \leq n. \tag{1.6}$$

We take indexes i, j, and k in the formula (1.5) in turn to give

$$g_{kix_j} = \sum_{l=1}^{n} (\Gamma_{jk}^l g_{li} + \Gamma_{ji}^l g_{kl}), \tag{1.7}$$

$$g_{ijx_k} = \sum_{l=1}^{n} (\Gamma_{ki}^l g_{lj} + \Gamma_{kj}^l g_{il}). \tag{1.8}$$

Summing up the formulas (1.5) and (1.7) minus (1.8) and using the formula (1.6), we obtain

$$2 \sum_{l=1}^{n} \Gamma_{ij}^l g_{kl} = g_{jkx_i} + g_{kix_j} - g_{ijx_k},$$

or

$$\Gamma_{ij}^k = \frac{1}{2} \sum_{l=1}^{n} g^{kl} (g_{jlx_i} + g_{lix_j} - g_{ijx_l}) \quad \text{for} \quad 1 \leq i, j, k \leq n, \tag{1.9}$$

where $(g^{ij}) = (g_{ij})^{-1}$, which show that Γ_{ij}^k are uniquely determined by the metric g.

To complete the proof, it remains to show that there exists a connection D which satisfies the relations (L1) and (L2). We do it as follows. Let $\mathcal{U}$ be any chart with a coordinate function $(x_1, \cdots, x_n)$. Let Γ_{ij}^k be given on $\mathcal{U}$ by the formulas (1.9). Let X and Y be vector fields on M which are given by

$$X = \sum_{i=1}^{n} f_i \partial_{x_i}, \quad Y = \sum_{i=1}^{n} h_i \partial_{x_i} \quad \text{on} \quad \mathcal{U}.$$

We let

$$D_X Y = \sum_{ij=1}^{n} f_i (h_{jx_i} + \sum_{l=1}^{n} h_l \Gamma_{il}^j) \partial_{x_j} \quad \text{on} \quad \mathcal{U}.$$

It is easy to check, as an exercise, that
  (a) If $\mathcal{V}$ is another chart, then $D_X Y$ is the same on $\mathcal{V} \cap \mathcal{U}$;
  (b) D is a connection satisfying the relations (L1) and (L2).   □

Definition 1.2 *The connection D, determined by Theorem 1.1 is called the Levi-Civita connection of M.*

In the sequel, D will always denote the Levi-Civita connection.

Definition 1.3 *In the formulas (1.9), Γ_{ij}^k are called the Christoffel symbols.*

The formula (1.9) can be written as a form of coordinates free. For any vector fields X, Y, and Z,

$$\langle D_X Y, Z \rangle = \frac{1}{2} \{ X \langle Y, Z \rangle + Y \langle Z, X \rangle - Z \langle X, Y \rangle + \langle Z, [X, Y] \rangle$$
$$+ \langle Y, [Z, X] \rangle - \langle X, [Y, Z] \rangle \}. \tag{1.10}$$

If we take X, Y, and Z as ∂_{x_i}, ∂_{x_j}, and ∂_{x_k}, respectively, in the formula (1.10), then we have the formula (1.9). Conversely, it is easy to verify (1.10) by (1.9) (Exercise 1.1.5).
  A useful means for computation is the following:

Definition 1.4 *n fields $E_1, \cdots, E_n$, defined on some open set $\mathcal{O}$ of M, is said to be an orthonormal frame (basis) if $\langle E_i, E_j \rangle = \delta_{ij}$ for $x \in \mathcal{O}$ and $1 \leq i, j \leq n$.*

Let $(x_1, \cdots, x_n)$ be a local coordinate. We orthogonalize $\partial_{x_1}, \cdots, \partial_{x_n}$ to have

Theorem 1.2 *Let M be a Riemannian manifold of dimension n. Let $\mathcal{U}$ be a chart. Then there exists an orthonormal frame on $\mathcal{U}$.*

Let $E_1, \cdots, E_n$ be an orthonormal frame on M. It follows from the formula (1.10) that

$$\langle D_{E_i} E_j, E_k \rangle = \frac{1}{2} \{ \langle E_k, [E_i, E_j] \rangle + \langle E_j, [E_k, E_i] \rangle - \langle X_i, [E_j, E_k] \rangle \}.$$

The above equations determine the Levi-Civita connection D completely.
  Parallel Transport To understand what a connection connects, we need the concept of parallel transports. Let $\gamma: [a, b] \to M$ be an embedded curve. Let X be a vector field along γ, that is, for each $t \in [a, b]$, $X(t) \in M_{\gamma(t)}$ such

that $X(t)$ is C^∞ with respect to the variable t. Since γ is embedded, X can be extended to be a vector field $\tilde{X}$ on M, that is,

$$\tilde{X}(\gamma(t)) = X(t) \quad \text{for} \quad t \in [a, b].$$

By Remark 1.3, $D_{\dot{\gamma}(t)}\tilde{X}$ is independent of the extension $\tilde{X}$, so we write as $D_{\dot{\gamma}(t)}X$ but not mentioning the extension.

Question When does $D_{\dot{\gamma}(t)}X = 0$ for all $t \in [a, b]$?

Let us write $D_{\dot{\gamma}(t)}X = 0$ in detail. Denote $\dot{\gamma}(t) = \sum_{i=1}^{n} \dot{\gamma}_i(t)\partial_{x_i}(\gamma(t))$ and $X(t) = \sum_{i=1}^{n} X_i(t)\partial_{x_i}(\gamma(t))$. Using the formula (1.4), we have

$$
\begin{aligned}
D_{\dot{\gamma}(t)}X &= \sum_j D_{\dot{\gamma}(t)}(X_j(t)\partial_{x_j}(\gamma(t))) \\
&= \sum_j \dot{X}_j\partial_{x_j}(\gamma(t)) + \sum_j X_j(t)D_{\dot{\gamma}(t)}\partial_{x_j} \\
&= \sum_k [\dot{X}_k(t) + \sum_{ij} \Gamma_{ij}^k \dot{\gamma}_i(t)X_j]\partial_{x_k}.
\end{aligned}
$$

Then $D_{\dot{\gamma}(t)}X = 0$ is equivalent to

$$\dot{X}_k(t) + \sum_{ij} \Gamma_{ij}^k \dot{\gamma}_i(t)X_j = 0 \quad \text{for} \quad 1 \le k \le n, \tag{1.11}$$

which is a system of linear ordinary differential equations. Its unknown is $X = (X_1(t), \cdots, X_n(t))^\tau$ where the superscript τ denotes the transpose of a vector. Thus, for any $v \in M_{\gamma(a)}$ given, there is a unique solution $X(t)$ such that

$$X(a) = v, \quad D_{\dot{\gamma}(t)}X = 0.$$

Remark 1.5 *The linearity of the problem* (1.11) *guarantees its solutions to exist on the whole interval* $[a, b]$. *Otherwise, we only know that its solutions exist on a small interval* $[a, a + \varepsilon)$.

Definition 1.5 *A vector field X is said to be parallel along γ if $D_{\dot{\gamma}(t)}X = 0$. A vector $w \in M_{\gamma(b)}$ is called the parallel transport of a vector $v \in M_{\gamma(a)}$ if there is a parallel vector field X along γ such that $X(a) = v$ and $X(b) = w$.*

Remark 1.6 *In the above argument, we assumed that γ is contained in a local chart. Clearly, it is not necessary. We could divide the interval $[a, b]$ into many small intervals $[a, a_1], \cdots,$ and $[a_k, b]$ such that each is in one chart. For any $v \in M_{\gamma(a)}$, we parallelize it to $X(a_1) \in M_{\gamma(a_1)}$ along $[a, a_1]$, and then to $X(a_2) \in M_{\gamma(a_2)}$ along $[a_1, a_2]$. Repeating this procedure, we get $X(b) = w$.*

We have proved

Theorem 1.3 *Let $\gamma(t)$: $[a, b] \to M$ be an embedded curve. For each $v \in M_{\gamma(a)}$, there is a unique $w \in M_{\gamma(b)}$ such that it is the parallel transport of v along γ.*

From Theorem 1.3, for each embedded curve $\gamma: [a, b] \to M$, we are able to define a map $\mathbf{P}^\gamma: M_{\gamma(a)} \to M_{\gamma(b)}$ such that $\mathbf{P}^\gamma v$ is the parallel transport of v along γ. Then $\mathbf{P}^\gamma$ is a linear isomorphism from $M_{\gamma(a)}$ to $M_{\gamma(b)}$.

Definition 1.6 $\mathbf{P}^\gamma$ *is called a parallel isomorphism, given by the connection* D.

Now we explain what a connection connects. Let D be a connection of M. If x, $y \in M$ are connected by a curve γ, then D yields a parallel isomorphism $\mathbf{P}^\gamma: M_x \to M_y$. In this sense, D connects any two tangential spaces M_x and M_y if there is a curve which connects x with y.

For a curve, if its tangential vector field is parallel along itself, it will be especially interesting, that is,

Definition 1.7 *A curve* γ *is called a geodesic of the connection* D *if* $D_{\dot{\gamma}(t)}\dot{\gamma} = 0$.

A geodesic exists at least locally. Let $(x_1, \cdots, x_n)$ be a local coordinate. Denote $\gamma = (\gamma_1, \cdots, \gamma_n)$ where $\gamma_i = x_i \circ \gamma$. Then

$$\dot{\gamma}(t) = \sum_i \dot{\gamma}_i(t) \partial_{x_i}(\gamma(t)),$$

where $\dot{\gamma}_i(t) = \frac{d\gamma_i}{dt}(t)$. It follows from $X = \dot{\gamma}$ and the formulas (1.11) that γ is a geodesic if and only if

$$\ddot{\gamma}_k + \sum_{ij}(\Gamma_{ij}^k \circ \gamma)\dot{\gamma}_i\dot{\gamma}_j = 0 \quad \text{for} \quad 1 \le k \le n. \tag{1.12}$$

This is a system of nonlinear ordinary differential equations of rank 2. We only have a unique local solution. Since $(\gamma_1(0), \cdots, \gamma_n(0))$ and $(\dot{\gamma}_1(0), \cdots, \dot{\gamma}_n(0))$ can be given arbitrarily, we have

Theorem 1.4 *Given* $v \in M_x$ $(x \in M)$ *and a connection* D *of* M, *there exists a unique geodesic* γ *such that*

$$\gamma(0) = x, \quad \dot{\gamma}(0) = v.$$

Differential of Tensor Field Let W be a linear space with an inner product $\langle \cdot, \cdot \rangle$ and let $k \ge 0$ be an integer. A k-linear functional α on W is a functional from $W^k = W \otimes \cdots \otimes W \to \mathbb{R}$ such that, for any $X_1, \cdots,$ $X_{i-1}, X_{i+1}, \cdots, X_k$ fixed in W, $\alpha(X_1, \cdots, X_{i-1}, X_i, X_{i+1}, \cdots, X_k)$ is a linear functional with respect to the variable $X_i \in W$. In particular, α is a 0-linear functional on W if and only if α is a constant; α is a 1-linear functional on W if and only if $\alpha \in W$ and

$$\alpha(X) = \langle \alpha, X \rangle \quad \text{for} \quad X \in W.$$

Definition 1.8 *Let M be a Riemannian manifold. A C^∞ function is called a 0-rank tensor field on M; a vector field is called a 1-rank tensor field on M. Let $k \geq 2$ be an integer. T is said to be a tensor field of rank k if, for each $x \in M$, T is a k-linear functional on M_x such that $T(X_1, \cdots, X_k) \in C^\infty(M)$ where $X_1, \cdots, X_k$ are k vector fields on M.*

Remark 1.7 *Let W be a linear space. In the literature tensors are defined as multiple linear functionals on $W \otimes \cdots \otimes W \otimes W^* \otimes \cdots \otimes W^*$ where W^* is the adjoint space of W. For convenience, we here assume that W has an inner product so that $W = W^*$ in the duality sense.*

We denote by $T^k(M)$ all k-rank tensor fields on M. Then $T^0(M) = C^\infty(M)$ and $T^1(M) = \mathcal{X}(M)$. Let $v \in M_x$ and $T \in T^k(M)$ be a tensor field. In order to define $D_v T$, we consider parallel isometries. Let $\gamma; [0, a] \to M$ be a curve such that $\gamma(0) = x$ and $\dot{\gamma}(0) = v$. Let

$$\mathbf{P}(t): \ M_x \to M_{\gamma(t)} \quad \text{for} \quad t \in [0, a]$$

be the parallel isometry of the connection D along γ. We define $D_v T$ as follows. For $v_1, \cdots, v_k \in M_x$,

$$D_v T(v_1, \cdots, v_k) = \frac{d}{dt}[T(\gamma(t))(\mathbf{P}(t)v_1, \cdots, \mathbf{P}(t)v_k)]_{t=0}. \tag{1.13}$$

Then $D_v T$ is again a k-linear functional on M_x. It is easy to check that the formula (1.13) is independent of the choice of the curve γ (Exercise 1.1.8). $D_v T$ is called the covariant derivative of T in the direction v. Since $D_v T$ is still linear in $v \in M_x$, we have

Definition 1.9 *Let M be a Riemannian manifold with the Levi-Civita connection D and let $k \geq 0$ be an integer. Let $T \in T^k(M)$ be a tensor field of rank k. The differential of the tensor field T is a tensor field of rank $k+1$, denoted by DT, to be defined by the formula*

$$DT(X_1, \cdots, X_k, X) = D_X T(X_1, \cdots, X_k) \tag{1.14}$$

for $X, X_1, \cdots, X_k \in \mathcal{X}(M)$, where, for each $x \in M$, the right hand side of (1.14) is defined by the formula (1.13).

We have

Theorem 1.5 (1) *Let T_1, T_2 be tensor fields. Then*

$$D(T_1 \otimes T_2) = DT_1 \otimes T_2 + T_1 \otimes DT_2.$$

(2) *Let $X_1, \cdots, X_{k+1} \in \mathcal{X}(M)$ be vector fields and let $T \in T^k(M)$ be a tensor field. Then*

$$DT(X_1, \cdots, X_{k+1}) = X_{k+1}(T(X_1, \cdots, X_k))$$
$$- \sum_{i=1}^{k} T(X_1, \cdots, D_{X_{k+1}} X_i, \cdots, X_k). \tag{1.15}$$

Proof. Let $x \in M$ be given. We assume that $T_1 \in T^k(M)$ and $T_2 \in T^l(M)$. Let $X \in \mathcal{X}(M)$ and let $\gamma \colon [0, a] \to M$ be a curve such that

$$\gamma(0) = x, \quad \dot{\gamma}(0) = X(x).$$

For vector fields $X_1, \cdots, X_k, Y_1, \cdots, Y_l \in \mathcal{X}(M)$, we have, via the formula (1.13) at x,

$$
\begin{aligned}
&D(T_1 \otimes T_2)(X_1, \cdots, X_k, Y_1, \cdots, Y_l, X) \\
&= D_X(T_1 \otimes T_2)(X_1, \cdots, X_k, Y_1, \cdots, Y_l) \\
&= \frac{d}{dt}[T_1(\mathbf{P}(t)X_1(x), \cdots, \mathbf{P}(t)X_k(x))T_2(\mathbf{P}(t)Y_1(x), \cdots, \mathbf{P}(t)Y_l(x))]|_{t=0} \\
&= [D_{X(x)}T_1(X_1, \cdots, X_k)]T_2(Y_1, \cdots, Y_l) \\
&\quad + T_1(X_1, \cdots, X_k)D_{X(x)}T_2(Y_1, \cdots, Y_l) \\
&= [(DT_1) \otimes T_2](X_1, \cdots, X_k, X, Y_1, \cdots, Y_l) \\
&\quad + T_1 \otimes (DT_2)(X_1, \cdots, X_k, Y_1, \cdots, Y_l, X).
\end{aligned}
$$

We assume $k = 2$. Similar arguments will prove the formula (2) for the other k. Let $E_1, \cdots, E_n$ be a local basis on M. Then $E_i \otimes E_j$ $(1 \le i, j \le n)$ forms a local basis in $T^2(M)$. Let

$$T = \sum_{ij=1}^{n} T_{ij} E_i \otimes E_j.$$

It follows from the formulas (1) and (L1) that

$$D_X T = \sum_{ij}[X(T_{ij})E_i \otimes E_j + T_{ij}D_X E_i \otimes E_j + T_{ij}E_i \otimes D_X E_j].$$

We obtain

$$
\begin{aligned}
X(T(X_1, X_2)) &= X\Big[\sum_{ij} T_{ij}\langle X_1, E_i\rangle\langle X_2, E_j\rangle\Big] \\
&= \sum_{ij}[X(T_{ij})\langle X_1, E_i\rangle\langle X_2, E_j\rangle + T_{ij}\langle X_1, D_X E_i\rangle\langle X_2, E_j\rangle \\
&\quad + T_{ij}\langle X_1, E_i\rangle\langle X_2, D_X E_j\rangle + T_{ij}\langle D_X X_1, E_i\rangle\langle X_2, E_j\rangle \\
&\quad + T_{ij}\langle X_1, E_i\rangle\langle D_X X_2, E_j\rangle] \\
&= D_X T(X_1, X_2) + T(D_X X_1, X_2) + T(X_1, D_X X_2),
\end{aligned}
$$

that is, the formula (1.15) is true for $k = 2$. $\qquad\square$

Let $f \in C^\infty(M)$ be a C^∞ function. The differential Df is a vector field which is also called *the gradient of the connection D*. Sometimes we denote Df by df. For $X \in \mathcal{X}(M)$,

$$Df(X) = X(f) = \langle Df, X\rangle \quad \text{for} \quad x \in M.$$

The differential $D^2 f$ of Df is a tensor field of rank 2. For a tensor field T, we denote

$$D^2 T = D(DT), \quad D^3 T = D(D^2 T), \quad \cdots.$$

Note that in general

$$D^2 T(\cdots, X, Y) \neq D^2 T(\cdots, Y, X) \tag{1.16}$$

where "$\cdots$" denotes the number of the variables.

Definition 1.10 *Let $f \in C^\infty(M)$. $D^2 f$ is called the Hessian of f in the connection D.*

In general we cannot change the order of the differential on a tensor field as in (1.16) but if T is a 0 rank tensor field, we have

Theorem 1.6 *Let $f \in C^\infty(M)$. Then*

$$D^2 f(X, Y) = D^2 f(Y, X) \tag{1.17}$$

for any $X, Y \in \mathcal{X}(M)$.

Proof. Using the formula (1.15) and (L2), we have

$$D^2 f(X, Y) = Y(Df(X)) - Df(D_Y X) = YX(f) - D_Y X(f)$$
$$= YX(f) - D_X Y(f) - [Y, X]f = XY(f) - D_X Y(f) = D^2 f(Y, X).$$

$\square$

If S is a symmetric 2-rank tensor field, we define its *trace* $\operatorname{tr} S$ to be a function on M by

$$\operatorname{tr} S(x) = \sum_{i=1}^n S(e_i, e_i) \quad \text{for} \quad x \in M, \tag{1.18}$$

where $e_1, \cdots, e_n$ is an orthonormal basis of M_x. It is easy to check that the above definition of $\operatorname{tr} S$ is independent of choices of the orthonormal basis (Exercise 1.1.11).

For $f \in C^\infty(M)$, the formula (1.17) means that $D^2 f$ is a symmetrical tensor field of rank 2. We define

$$\Delta f = \operatorname{tr} D^2 f,$$

and call Δ the Laplacian acting on functions. It is also called the Laplace-Beltrami operator. If $\mathbb{R}^n$ has the Euclidean metric, a simple computation yields $\Delta f = \sum_i f_{x_i x_i}$, the classical Laplacian. Exercise 1.1.12 presents some formulas for Δ in a general metric g.

The second operator which is important to us is *the curvature operator* that is defined as follows.

Definition 1.11 *Let (M, g) be a Riemannian manifold. Let $X, Y \in \mathcal{X}(M)$ be vector fields and let $k \geq 0$ be an integer. We define a map*

$$\mathbf{R}_{XY} : \ T^k(M) \to T^k(M)$$

by

$$\mathbf{R}_{XY} = -D_X D_Y + D_Y D_X + D_{[X,Y]}.$$

$\mathbf{R}_{XY}$ *is called the curvature operator.*

As an exercise (Exercise 1.1.13) for the readers, we have

Theorem 1.7 $\mathbf{R}_{XY}$ *has the following properties:*
  (1) *For $T_1 \in T^k(M)$ and $T_2 \in T^l(M)$,*

$$\mathbf{R}_{XY}(T_1 \otimes T_2) = \mathbf{R}_{XY}T_1 \otimes T_2 + T_1 \otimes \mathbf{R}_{XY}T_2;$$

  (2) *For any $f \in C^\infty(M)$ and $T \in T^k(M)$,*

$$\mathbf{R}_{(fX)Y}T = \mathbf{R}_{X(fY)}T = \mathbf{R}_{XY}(fT) = f\mathbf{R}_{XY}T;$$

  (3) *For any $f \in C^\infty(M)$, $\mathbf{R}_{XY}f = 0$.*

Let $Z, W \in \mathcal{X}(M)$ be vector fields. Then $\mathbf{R}_{XY}Z$ is a vector field. By (2), $\mathbf{R}_{XY}Z$ is linear in each variable X, Y, and Z, respectively. We define

$$\mathbf{R}(X, Y, Z, W) = \langle R_{XY}Z, W \rangle \tag{1.19}$$

for vector fields $X, Y, Z, W \in \mathcal{X}(M)$. It is a tensor field of rank 4.

Definition 1.12 *The tensor field $\mathbf{R}$ of 4 rank, given by (1.19), is called the curvature tensor.*

Remark 1.8 *Although it looks not to be very complicated, the curvature tensor controls almost everything in (M, g); see comments in $[190]$, II. This tensor field is determined by the second derivatives of the metric g. In a local coordinate, we have the formulas $([190], II, 4D - 7$ page)*

$$\mathbf{R}(\partial_{x_i}, \partial_{x_j}, \partial_{x_k}, \partial_{x_l}) \ = \ \frac{1}{2}(g_{il x_j x_k} + g_{jk x_i x_l} - g_{ik x_j x_l} - g_{jl x_i x_k})$$
$$+ \sum_{rs}(g_{rs}\Gamma_{jk}^r\Gamma_{il}^s + g_{rs}\Gamma_{jl}^r\Gamma_{ik}^s) \tag{1.20}$$

where $g_{ij} = g(\partial_{x_i}, \partial_{x_j})$ and Γ_{ij}^k are the Christoffel symbols.

The curvature tensor has the following properties (Exercise 16).

Theorem 1.8 *For any vector fields X, Y, Z, and W, we have*
  (1) $\mathbf{R}_{XY} = -\mathbf{R}_{YX}$;
  (2) $\mathbf{R}_{XY}Z + \mathbf{R}_{YZ}X + \mathbf{R}_{ZX}Y = 0$ *(the first Bianchi identity);*
  (3) $\mathbf{R}(X, Y, Z, W) = -\mathbf{R}(X, Y, W, Z)$;
  (4) $\mathbf{R}(X, Y, Z, W) = \mathbf{R}(Z, W, X, Y)$.

We introduce a quadratic form on $M_x \otimes M_x$ by

$$Q(X,Y) = \mathbf{R}(X,Y,X,Y),$$

which determines the curvature tensor completely.

Theorem 1.9 *Let $\mathbf{R}$ and $\mathbf{R}'$ be tensor fields which satisfy the properties (1)-(4) in Theorem 1.8. If $Q = Q'$ on $M_x \otimes M_x$, then $\mathbf{R} = \mathbf{R}'$.*

Proof. Exercise 17. ☐

Now we introduce the concept of sectional curvatures.

Definition 1.13 *Let Ξ be a 2-dimensional linear subspace of M_x. For any basis v_1, v_2 of Ξ, we define the sectional curvature of Ξ by*

$$\kappa(\Xi) = \frac{\mathbf{R}(v_1, v_2, v_1, v_2)}{|v_1 \wedge v_2|^2} \tag{1.21}$$

where $|v_1 \wedge v_2|^2 = |v_1|^2 |v_2|^2 - \langle v_1, v_2 \rangle^2$.

It follows from (1)-(4) in Theorem 1.8 that the right hand side of (1.21) is independent of choice of the basis of Π (Exercise 1.1.16). If e_1, e_2 is an orthonormal basis of Ξ, then

$$\kappa(\Xi) = \mathbf{R}(e_1, e_2, e_1, e_2).$$

Remark 1.9 *κ is a function which is defined on all 2-dimensional tangential spaces. It follows from Theorem 1.9 that the sectional curvature determines the whole curvature tensor.*

Remark 1.10 *Although sectional curvatures are defined locally at point $x \in M$ (we may compute it by the formula (1.20)), they will yield all the global information about the Riemannian manifold (M, g). We will explain this point more later.*

We have noted that in general it is not possible to change the order of the differentials, see (1.16). However, the following theorem, named *the Ricci identity*, tells us that their difference is just a curvature term.

Theorem 1.10 *Let T be a tensor field and let X, Y be vector fields. Then*

$$D^2 T(\cdots, X, Y) = D^2 T(\cdots, Y, X) + (\mathbf{R}_{XY} T)(\cdots). \tag{1.22}$$

Proof. By the formula (1.15), we have

$$\begin{aligned}
D^2 T(\cdots, X, Y) &= D_Y(DT)(\cdots, X) = D_Y D_X T(\cdots) - D_{D_Y X} T(\cdots) \\
&= D_X D_Y T(\cdots) - D_{D_X Y} T(\cdots) - D_X D_Y T(\cdots) \\
&\quad + D_Y D_X T(\cdots) - D_{D_Y X} T(\cdots) + D_{D_X Y} T(\cdots) \\
&= D^2 T(\cdots, Y, X) + (\mathbf{R}_{XY} T)(\cdots),
\end{aligned}$$

since $D_X Y = D_Y X + [X, Y]$.   □

Finally, we introduce the Ricci tensor and the Ricci curvature function. The *Ricci tensor* is a 2 rank tensor field which is defined by

$$\text{Ric}\,(X, Y) = \sum_{i=1}^{n} \mathbf{R}(X, e_i, Y, e_i) \quad \text{at each} \quad x \in M$$

where X and Y are vector fields and $e_1, \cdots, e_n$ is an orthonormal basis of M_x. It is easy to check that Ric is independent of choices of bases. By (4) in Theorem 1.8, Ric is symmetric, that is, $\text{Ric}\,(X, Y) = \text{Ric}\,(Y, X)$.

Let $x \in M$ and $X \in M_x$ with $|X| = 1$. We define $\text{Ric}\,(X, X)$ to be *the Ricci curvature* of the vector X. Let $e_1, \cdots, e_n$ be an orthonormal basis of M_x such that $e_1 = X$. Then

$$\text{Ric}\,(X, X) = \sum_{i=2}^{n} \mathbf{R}(e_1, e_i, e_1, e_i),$$

since $\mathbf{R}(e_1, e_1, e_1, e_1) = 0$ by (3) in Theorem 1.8. Then the Ricci curvature is the sum of $n - 1$ sectional curvatures.

Isometry Map Let M and N be two Riemannian manifolds. Let Φ: $M \to N$ be a C^∞ map. Let $x \in M$ be given. We define a linear map Φ_*: $M_x \to N_{\Phi(x)}$ as follows. Let $v \in M_x$. Let $\gamma: (-\varepsilon, \varepsilon) \to M$ be a curve such that

$$\gamma(0) = x, \quad \dot{\gamma}(0) = v.$$

We define $\Phi_* v \in N_{\Phi(x)}$ by

$$\Phi_* v = \dot{\beta}(0)$$

where

$$\beta(t) = \Phi(\gamma(t)) \quad \text{for} \quad t \in (-\varepsilon, \varepsilon)$$

is a curve on N with $\beta(0) = \Phi(x)$. Sometimes, we denote $d\Phi = \Phi_*$ and call it *the differential of the map* Φ. Then for any $f \in C^\infty(N)$,

$$(\Phi_* v)(f) = [f \circ \Phi \circ \gamma(t)]'_{t=0} = v(f \circ \Phi).$$

Definition 1.14 *A C^∞ map $\Phi: M \to N$ is called a local isometry between two Riemannian manifolds M and N, if, for each $x \in M$, Φ is a local isomorphism at x such that $\Phi_*: M_x \to N_{\Phi(x)}$ is an isometry between the two inner product spaces. Further, Φ is said to be an isometry from M to N, if Φ is a local isometry and, at the same time, a diffeomorphism.*

Remark 1.11 *If there is a local isometry between M and N, then the two manifolds have the same dimension.*

A local isometry keeps the connections the same in the following sense.

Theorem 1.11 *Let M and N be two Riemannian manifolds. Let D and D' be the Levi-Civita connections of M and N, respectively. Let $\Phi: M \to N$ be a local isometry. Let X, Y be vector fields on M. Then*

$$\Phi_* D_X Y = D'_{\Phi_* X} \Phi_* Y.$$

Proof. Let $p \in M$ be given and let $q = \Phi(p)$. Let $\varphi = (x_1, \cdots, x_n)$ be a local coordinate system near p on M. Since Φ is a local isomorphism, $y = \varphi \circ \Phi^{-1}$ is a local coordinate system around q on N. For $f \in C^\infty(N)$, we have

$$\partial_{y_i}(f) = \frac{\partial}{\partial y_i}[f \circ (\varphi \circ \Phi^{-1})^{-1}(y)] = \frac{\partial}{\partial y_i}[f \circ \Phi \circ \varphi^{-1}(y)]$$
$$= \partial_{x_i}(f \circ \Phi) = \Phi_* \partial_{x_i}(f),$$

that is,

$$\partial_{y_i} = \Phi_* \partial_{x_i} \quad \text{for all} \quad i. \tag{1.23}$$

Let

$$g_{ij} = \langle \partial_{x_i}, \partial_{x_j} \rangle_M, \quad g'_{ij} = \langle \partial_{y_i}, \partial_{y_j} \rangle_N \quad \text{for all} \quad i, j. \tag{1.24}$$

Let Γ^k_{ij} and Γ'^k_{ij} be the Christoffel symbols, respectively for D and D'. Since Φ is a local isometry, the relations (1.23) imply

$$g_{ij} = g'_{ij} \circ \Phi \quad \text{for all} \quad i, \tag{1.25}$$

$$\Gamma^k_{ij} = \Gamma'^k_{ij} \circ \Phi. \quad \text{for all} \quad i, j, k. \tag{1.26}$$

Let

$$X = \sum_i X_i \partial_{x_i}, \quad Y = \sum_i Y_i \partial_{x_i}.$$

Then by (1.23)

$$\Phi_* X = \sum_i (X_i \circ \Phi^{-1}) \partial_{y_i}, \quad \Phi_* Y = \sum_i (Y_i \circ \Phi^{-1}) \partial_{y_i}.$$

Since $X(Y_i) = X(Y_i \circ \Phi^{-1} \circ \Phi) = \Phi_* X(Y_i \circ \Phi^{-1})$ for all i, we obtain, via (1.23)-(1.26),

$$\Phi_* D_X Y = \Phi_* [\sum_{k=1}^n (X(Y_k) + \sum_{ij=1}^n X_i Y_j \Gamma^k_{ij}) \partial_{x_k}]$$
$$= \sum_{k=1}^n [\Phi_* X(Y_k \circ \Phi^{-1}) + \sum_{ij=1}^n (X_i \circ \Phi^{-1})(Y_j \circ \Phi^{-1}) \Gamma'^k_{ij}] \partial_{y_k}$$
$$= D'_{\Phi_* X} \Phi_* Y.$$

$\square$

Theorem 1.12 *Let* $\Phi\colon M \to N$ *be a local isometry between the two Riemannian manifolds. Let* $\mathbf{R}$ *and* $\mathbf{R}'$ *be the curvature tensors on* M *and* N, *respectively. Then, for any vector fields* $X_i \in \mathcal{X}(M)$ *for* $1 \le i \le 4$, *we have*

$$\mathbf{R}(X_1, X_2, X_3, X_4) = \mathbf{R}'(\Phi_* X_1, \Phi_* X_2, \Phi_* X_3, \Phi_* X_4) \circ \Phi. \qquad (1.27)$$

Proof. Let g_{ij} and g'_{ij} be defined in (1.24) for all i, j. It follows from (1.23) and (1.26) that

$$g_{ijx_k} = \partial_{x_k}(g'_{ij} \circ \Phi) = \partial_{y_k} g'_{ij} \circ \Phi = g'_{ijy_k} \circ \Phi,$$

$$g_{ijx_k x_l} = \partial_{x_l}(g'_{ijy_k} \circ \Phi) = \partial_{y_l}(g'_{ijy_k}) \circ \Phi = g'_{ijy_k y_l} \circ \Phi, \qquad (1.28)$$

for all i, j, k, l.

From the relations (1.20), (1.25), (1.26), and (1.28), we have the formula (1.27). $\qquad \Box$

Second Fundamental Form of Hypersurface In general it is not easy to compute sectional curvatures. We will study *the second fundamental forms of hypersurfaces* which will provide us with a useful tool to compute sectional curvatures.

Let $\tilde{M}$ be an $(n+1)$-dimensional Riemannian manifold. An n-dimensional submanifold of $\tilde{M}$, denoted by M, is called *a hypersurface*. A vector field $N \in \mathcal{X}(\tilde{M})$ is called *the normal vector field of* M if for each $p \in M$, $N \perp M_p$ and $|N| = 1$.

We introduce a Riemannian metric on a hypersurface M as follows, which is called *the induced metric from* $\tilde{M}$:

$$g(X, Y) = \langle X, Y \rangle \quad \text{for} \quad X, Y \in \mathcal{X}(M), \qquad (1.29)$$

where $\langle \cdot, \cdot \rangle$ is the metric of $\tilde{M}$. Let $\tilde{D}$ and D be the Levi-Civita connections of $\tilde{M}$ and M. Then

Lemma 1.1 *For any* X, $Y \in \mathcal{X}(M)$,

$$D_X Y = \tilde{D}_X Y - \langle \tilde{D}_X Y, N \rangle N. \qquad (1.30)$$

Proof. We define a map $P\colon \mathcal{X}(M) \times \mathcal{X}(M) \to \mathcal{X}(M)$ by

$$P(X, Y) = \tilde{D}_X Y - \langle \tilde{D}_X Y, N \rangle N \quad \text{for} \quad X, Y \in \mathcal{X}(M).$$

Clearly P satisfies (C1)-(C3). Then P is a connection on M. It suffices to show that P satisfies (L1) and (L2) in Theorem 1.1. Indeed, for X, Y, $Z \in \mathcal{X}(M)$, since $N \perp \mathcal{X}(M)$,

$$X\langle Y, Z \rangle = \langle \tilde{D}_X Y, Z \rangle + \langle Y, \tilde{D}_X Z \rangle = \langle P(X, Y), Z \rangle + \langle Y, P(X, Z) \rangle,$$

$$P(X, Y) - P(Y, X) = \tilde{D}_X Y - \tilde{D}_Y X = [X, Y].$$

Then $D_X Y = P(X, Y)$. $\qquad \Box$

Definition 1.15 *The second fundamental form of a hypersurface M is a 2-rank tensor field, given by*

$$\Pi(X,Y) = \langle \tilde{D}_X N, Y \rangle \quad for \quad X,\ Y \in \mathcal{X}(M). \tag{1.31}$$

Lemma 1.2 *The second fundamental form Π of M is symmetric, that is,*

$$\Pi(X,Y) = \Pi(Y,X) \quad for \quad X,\ Y \in \mathcal{X}(M). \tag{1.32}$$

Proof. Since $[X,Y] \in \mathcal{X}(M)$, we have

$$\begin{aligned}
\Pi(X,Y) &= \langle \tilde{D}_X N, Y \rangle = X\langle N,Y \rangle - \langle N, \tilde{D}_X Y \rangle = -\langle N, \tilde{D}_X Y \rangle \\
&= -\langle N, \tilde{D}_Y X + [X,Y] \rangle = -\langle N, \tilde{D}_Y X \rangle = \Pi(Y,X).
\end{aligned}$$

$\square$

Remark 1.12 *From Lemmas 1.1 and 1.2, the following formula is true:*

$$D_X Y = \tilde{D}_X Y + \Pi(X,Y)N. \tag{1.33}$$

Theorem 1.13 *For $X,\ Y \in \mathcal{X}(M)$,*

$$\tilde{\mathbf{R}}(X,Y,X,Y) = \mathbf{R}(X,Y,X,Y) - \Pi(X,X)\Pi(Y,Y) + \Pi^2(X,Y) \tag{1.34}$$

where $\tilde{\mathbf{R}}$ and $\mathbf{R}$ are, the curvature tensors of $\tilde{M}$ and M, respectively.

Proof. The proof consists of a straight computation:

$$\tilde{\mathbf{R}}(X,Y,X,Y) = -\langle \tilde{D}_X \tilde{D}_Y X, Y \rangle + \langle \tilde{D}_Y \tilde{D}_X X, Y \rangle + \langle \tilde{D}_{[X,Y]} X, Y \rangle,$$

$$\begin{aligned}
-\langle \tilde{D}_X \tilde{D}_Y X, Y \rangle &= -X\langle \tilde{D}_Y X, Y \rangle + \langle \tilde{D}_Y X, \tilde{D}_X Y \rangle \\
&= -X\langle D_Y X, Y \rangle + \langle D_Y X, D_X Y \rangle + \Pi^2(X,Y) \text{(by (1.33))} \\
&= -\langle D_X D_Y X, Y \rangle + \Pi^2(X,Y),
\end{aligned}$$

$$\begin{aligned}
\langle \tilde{D}_Y \tilde{D}_X X, Y \rangle &= Y\langle \tilde{D}_X X, Y \rangle - \langle \tilde{D}_X X, \tilde{D}_Y Y \rangle \\
&= \langle D_Y D_X X, Y \rangle - \Pi(X,X)\Pi(Y,Y) \text{(by (1.33))},
\end{aligned}$$

$$\langle \tilde{D}_{[X,Y]} X, Y \rangle = \langle D_{[X,Y]} X, Y \rangle \text{(by (1.33))}.$$

Summing up the above yields the formula (1.34). $\square$

Codazzi's Equation We introduce *the Codazzi equation* which will be needed in the modeling of shells in this book.

Theorem 1.14 *Let $\tilde{\mathbf{R}}$ be the curvature tensor of $\tilde{M}$. Then*

$$D\Pi(Z,X,Y) = D\Pi(Z,Y,X) + \tilde{\mathbf{R}}(X,Y,N,Z) \tag{1.35}$$

for all $X,\ Y,\ Z \in \mathcal{X}(M)$.

Proof. For any $X \in \mathcal{X}(M)$,

$$\langle \tilde{D}_X N, N \rangle = X(|N|^2)/2 = 0. \tag{1.36}$$

From (1.32), (1.33), and (1.36), we have

$$\begin{aligned}
Y(\Pi(Z,X)) - X(\Pi(Z,Y)) &= Y \langle \tilde{D}_X N, Z \rangle - X \langle \tilde{D}_Y N, Z \rangle \\
&= \langle \tilde{D}_Y \tilde{D}_X N, Z \rangle - \langle \tilde{D}_X \tilde{D}_Y N, Z \rangle + \langle \tilde{D}_X N, \tilde{D}_Y Z \rangle - \langle \tilde{D}_Y N, \tilde{D}_X Z \rangle \\
&= \langle \tilde{\mathbf{R}}_{XY} N, Z \rangle - \langle \tilde{D}_{[X,Y]} N, Z \rangle + \langle \tilde{D}_X N, D_Y Z \rangle - \langle \tilde{D}_Y N, D_X Z \rangle \\
&= \tilde{\mathbf{R}}(X,Y,N,Z) - \Pi([X,Y],Z) + \Pi(X,D_Y Z) - \Pi(Y,D_X Z) \tag{1.37}
\end{aligned}$$

for all X, Y, $Z \in \mathcal{X}(M)$. On the other hand,

$$\begin{aligned}
\text{the left hand side of (1.37)} &= D\Pi(Z,X,Y) - D\Pi(Z,Y,X) \\
&\quad + \Pi(D_Y Z, X) + \Pi(Z, D_Y X) - \Pi(D_X Z, Y) - \Pi(Z, D_X Y) \\
&= D\Pi(Z,X,Y) - D\Pi(Z,Y,X) - \Pi(Z,[X,Y]) \\
&\quad + \Pi(D_Y Z, X) - \Pi(D_X Z, Y). \tag{1.38}
\end{aligned}$$

Then the formula (1.35) follows from (1.37) and (1.38). $\square$

Let $\tilde{M} = \mathbb{R}^{n+1}$. Then M is a hypersurface of R^{n+1}. Let X, Y, $Z \in \mathcal{X}(\tilde{M})$ with $Z = (Z_1, \cdots, Z_{n+1})$. Then

$$\tilde{D}_X \tilde{D}_Y Z = \sum_{i=1}^{n+1} XY(Z_i) \partial_{x_i}$$

which yields

$$\tilde{\mathbf{R}}_{XY} Z = 0. \quad \text{In particular,} \quad \tilde{\mathbf{R}}(X,Y,N,Z) = 0.$$

It follows from Theorem 1.14 that

Corollary 1.1 *Let M be a hypersurface of $\mathbb{R}^{n+1}$ and let Π be its second fundamental form. Then $D\Pi$ is symmetric in its variables.*

Definition 1.16 *Let M be a hypersurface of $\mathbb{R}^{n+1}$. The induced metric g of M is called the first fundamental form of M.*

Remark 1.13 *Let M and M' be two hypersurfaces of $\mathbb{R}^{n+1}$. If there is an isometry $\Phi \colon M \to M'$, then*

$$g(X,Y) = g'(\Phi_* X, \Phi_* Y) \tag{1.39}$$

for all $X, Y \in \mathcal{X}(M)$, where g and g' are the induced metrics of M and M', respectively. We say that two isometrical hypersurfaces have the same first fundamental form in the sense (1.39).

In general, two isometrical hypersurfaces may have different shapes. In order that they have the same shape, the two isometrical hypersurfaces must have the same second fundamental form in the following sense.

Theorem 1.15 *Let M and M' be two hypersurfaces of $\mathbb{R}^{n+1}$. Let N and N' be the normals of M and M', respectively. Let $\phi\colon M \to M'$ be an isometry such that*

$$\mathrm{II}(X,Y) = \mathrm{II}'(\phi_* X, \phi_* Y)$$

for all X, $Y \in \mathcal{X}(M)$, where II and II' are the second fundamental forms of M and M', respectively. Then there is an isometry $\Phi\colon \mathbb{R}^{n+1} \to \mathbb{R}^{n+1}$ such that $\Phi|_M = \phi$.

For a proof of the above theorem, see [190], IV.

Let us give an example to end this section.

Example 1.3 *Let*

$$M = \{\, x = (x_1, \cdots, x_{n+1}) \in \mathbb{R}^{n+1} \mid \sum_{i=1}^{n+1} x_i^2 = 1 \,\}$$

be the n-dimensional unit sphere in $\mathbb{R}^{n+1}$. Then (M, g) is a Riemannian manifold where g is the induced metric from $\mathbb{R}^{n+1}$.

The normal field of M is $N = x$. Let $X = (X_1, \cdots, X_{n+1})$, $Y = (Y_1, \cdots, Y_{n+1}) \in \mathcal{X}(M)$. Then the second fundamental form of M is given by

$$\mathrm{II}(X,Y) = \langle \tilde{D}_X N, Y \rangle = \langle \sum_{i=1}^{n+1} X(x_i)\partial_{x_i}, Y \rangle = \sum_{i=1}^{n+1} X_i Y_i = \langle X, Y \rangle,$$

that is,

$$\mathrm{II} = g. \tag{1.40}$$

Let $x \in M$ be given. Let e_1, e_2 be an orthonormal basis of a 2-dimensional subspace Ξ of M_x. Since $\tilde{\mathbf{R}}(e_1, e_2, e_1, e_2) = 0$, by the formulas (1.34) and (1.40), we obtain

the sectional curvature of $\Xi = \mathbf{R}(e_1, e_2, e_1, e_2) = \mathrm{II}(e_1, e_1)\mathrm{II}(e_2, e_2)$
$$-\mathrm{II}^2(e_1, e_2) = 1.$$

□

Exercise 1.1

1.1.1 *Prove the relation* (1.2).

1.1.2 *Find out vector fields V, T_1 and T_2 on $\mathbb{R}^n$ such that, for some $p \in \mathbb{R}^n$, $T_1(p) = T_2(p)$ but $(D_V T_1)(p) \neq (D_V T_2)(p)$.*

1.1.3 *Prove the relation* (1.3).

1.1.4 *Prove the relations* (a) *and* (b) *in the proof of Theorem 1.1.*

1.1.5 *Let M be a Riemannian manifold and let D be the Levi-Civita connection. Prove the formula* (1.10).

1.1.6 *Let M be a Riemannian manifold and let D be its Levi-Civita connection. Then D satisfies the relation* (L1) *in Theorem 1.1 if and only if all parallel transports, given by D, are isometries between the tangential spaces.*

1.1.7 *Let M be a Riemannian manifold and let D be its Levi-Civita connection. Let $\gamma \colon [a, b] \to M$ be a geodesic. We define the length of γ by*

$$L(\gamma) = \int_a^b |\dot{\gamma}(t)| dt$$

where $|\dot{\gamma}(t)| = \langle \dot{\gamma}(t), \dot{\gamma}(t) \rangle^{1/2}$. Prove $L(\gamma) = (b - a)|\dot{\gamma}(0)|$.

1.1.8 *Prove that the formula* (1.13) *is independent of the choice of the curve γ.*

1.1.9 *Let M be a Riemannian manifold and let D be its Levi-Civita connection. A tensor field T is called parallel if $DT = 0$. Let $A \colon M_x \to M_y$ be a linear operator and α be a k-linear functional on M_y. We define $A_* \alpha$ to be a k-linear functional on M_x by*

$$A_* \alpha(v_1, \cdots, v_k) = \alpha(Av_1, \cdots, Av_k) \quad \text{for} \quad v_i \in M_x \qquad (1.41)$$

with $1 \leq i \leq k$. Prove that T is parallel if and only if, for each curve $\gamma \colon [0, 1] \to M$,

$$\mathbf{P}_*(1)T = T|_{\gamma(0)}.$$

1.1.10 *Let M be a Riemannian manifold with the metric g and let D be its Levi-Civita connection. Then g is parallel.*

1.1.11 *Let S be a tensor field of rank 2. Prove that the definition of $\operatorname{tr} S$ by* (1.18) *is free from choices of the orthonormal basis.*

1.1.12 *Let M be a Riemannian manifold with the metric g. Let $(x_1, \cdots, x_n)$ be a local coordinate. Let $(g^{ij}) = (g_{ij})^{-1}$ where $g_{ij} = g(\partial_{x_i}, \partial_{x_j})$ for $1 \le i, j \le n$. Let $f \in C^\infty(M)$. Then*

   (1) $\Delta f = \sum_{ij} g^{ij} D^2 f(\partial_{x_i}, \partial_{x_j})$;

   (2) $\Delta f = \frac{1}{\sqrt{G}} \sum_{ij} (g^{ij} \sqrt{G} f_{x_i})_{x_j}$ *where $G = \det(g_{ij})$.*

1.1.13 *Prove Theorem 1.7.*

1.1.14 *Prove Theorem 1.8([190], II).*

1.1.15 *Prove Theorem 1.9 (The proof of Lemma 1.6 later; [72]; [101], I, page 199).*

1.1.16 *Prove that the definition of (1.21) is independent of choice of bases of Ξ.*

1.2 Distance Functions

Distance functions will play an important role in the application of Riemannian geometry.

Let (M, g) be a complete Riemannian manifold. Let $\gamma\colon [a, b] \to M$ be a curve. The *length of γ* is defined as

$$L(\gamma) = \int_a^b |\dot{\gamma}(t)| dt.$$

If γ is contained in a local coordinate $(x_1, \cdots, x_n)$, then

$$L(\gamma) = \int_a^b \sqrt{\sum_{ij=1}^n g_{ij}(\gamma(t)) \dot{\gamma}_i(t) \dot{\gamma}_j(t)} \, dt,$$

where $g_{ij} = g(\partial_{x_i}, \partial_{x_j})$ and $\dot{\gamma}(t) = \sum_i \dot{\gamma}_i(t) \partial_{x_i}(\gamma(t))$. We remark that the length of a (continuous and) piecewise smooth curve may be defined as the sum of the lengths of the smooth pieces.

For $x, y \in M$, set

$$\text{curve}(x, y) = \{ \gamma \mid \gamma \colon [a, b] \to M \text{ piecewise smooth curve}$$
$$\text{with } \gamma(a) = x, \ \gamma(b) = y \}.$$

The *distance* between two points $x, y \in M$ can be defined:

$$d(x, y) = \inf_{\gamma \in \text{curve}(x, y)} L(\gamma).$$

Remark 1.14 *It is easy to show that any two points x, $y \in M$ can be connected by a piecewise smooth curve and therefore, $d(x, y)$ is always defined (Exercise 1.2.1). The distance function satisfies the usual axioms as in Exercise 1.2.2.*

We consider whether a geodesic $\gamma \in \text{curve}\,(x, y)$ satisfies that $d(x, y) = L(\gamma)$.

Formula for the first variation of arc length Let $\gamma \colon [a, b] \to M$ be a curve. A C^∞ map $\alpha(\cdot, \cdot) \colon [a, b] \times [0, \varepsilon] \to M$ is called *a variation of the base curve γ* if

$$\alpha(t, 0) = \gamma(t) \quad \text{for} \quad t \in [a, b].$$

If, in addition, $\alpha(a, s) = \gamma(a)$ and $\alpha(b, s) = \gamma(b)$ for all $s \in [0, \varepsilon]$, then α is called *a proper variation of the base curve γ*.

Let $L(s) = L(\alpha(\cdot, s))$ for $s \in [0, \varepsilon]$. We will compute $L'(0)$. For convenience, we assume that γ is parameterized by arc length, that is, $|\dot{\gamma}(t)| = 1$ for all $t \in [a, b]$.

For $s \in [0, \varepsilon]$, let $T = \dot{\alpha}(t, s)$ be *the tangential vector field of the one parameter family α of curves.* Since $L(s) = \int_a^b |T(t, s)| dt$, we have

$$L'(s) = \int_a^b \frac{d}{ds} |T(t, s)| dt.$$

Let us compute $\frac{d}{ds}(|T|^2)$. For each t fixed, $\alpha(t, \cdot) \colon [0, \varepsilon] \to M$ is a curve in the parameter s. Let $U = \alpha_s(t, s)$ be the tangential vector field of the curve $\alpha(t, \cdot)$, which is called *the transversal vector field of the one parameter family α of curves.* Then

$$\frac{d}{ds}(|T|^2) = [\langle T, T\rangle(\alpha(t, s))]_s = U\langle T, T\rangle = 2\langle D_U T, T\rangle.$$

In addition, it is easy to check that (Exercise 1.2.5)

$$[T, U] = 0.$$

Then $D_U T = D_T U$ and we have

$$L'(s) = \int_a^b \frac{\langle T, D_T U\rangle}{|T|} dt. \tag{1.42}$$

In particular, at $s = 0$,

$$\frac{\langle T, D_T U\rangle}{|T|}\Big|_{s=0} = \frac{\langle T, D_T U\rangle}{|\dot\gamma|} = T\langle T, U\rangle - \langle D_T T, U\rangle$$

$$= \frac{d}{dt}\langle T, U\rangle - \langle D_{\dot\gamma(t)}\dot\gamma, U\rangle.$$

We have the *formula for the first variation of arc length*

$$L'(0) = \langle \dot\gamma, U\rangle\big|_a^b - \int_a^b \langle D_{\dot\gamma(t)}\dot\gamma, U\rangle dt, \tag{1.43}$$

where $U(t) = U|_{s=0}$.

As an application of the above formula, we have

Theorem 1.16 *Let x, $y \in M$ be given. Let $\gamma \colon [a, b] \to M$ be a curve with the arc length parameter such that $\gamma(a) = x$, $\gamma(b) = y$, and*

$$d(x, y) = L(\gamma).$$

Then γ is a geodesic.

Proof. Let $\alpha \colon [a, b] \times [0, \varepsilon] \to M$ be a proper variation of the base curve γ. Then $\alpha(a, s) = x$ and $\alpha(b, s) = y$ imply that $U(a) = U(b) = 0$. Then $L'(0) = 0$ and the formula (1.43) yield $D_{\dot\gamma}\dot\gamma = 0$ for all $t \in [a, b]$. □

The above theorem tells us that, if a curve γ is shortest in length, it must be a geodesic. Conversely, this is not true, that is, a geodesic may not be shortest in length globally.

Formula for the second variation of arc length A curve is called *a normal geodesic* if it is a geodesic with the arc length parameter. Let γ be a normal geodesic connecting x with y. Then $L'(0) = 0$ for a proper variation of the base curve γ. From the formula (1.42), we obtain

$$L''(s) = \int_a^b \frac{d}{ds} \frac{\langle T, D_T U \rangle}{|T|} dt, \tag{1.44}$$

where the integrand is

$$
\begin{aligned}
U\left(\frac{\langle T, D_T U \rangle}{|T|}\right) &= -\frac{1}{|T|^3} \langle T, D_T U \rangle^2 + \frac{1}{|T|}(\langle D_T U, D_T U \rangle + \langle T, D_U D_T U \rangle) \\
&= -\frac{1}{|T|^3}(T\langle T, U \rangle - \langle D_T T, U \rangle)^2 + \frac{|D_T U|^2}{|T|} \\
&\quad + \frac{1}{|T|}(-\langle T, \mathbf{R}_{UT} U \rangle + \langle T, D_T D_U U \rangle),
\end{aligned}
$$

where the relation $[T, U] = 0$ is used again. Substitute it into the formula (1.44) and compute at $s = 0$. Noting that

$$D_T T = D_{\dot\gamma}\dot\gamma = 0, \quad |T| = |\dot\gamma| = 1, \quad \langle T, D_T D_U U \rangle = T\langle T, D_U U \rangle,$$

at $s = 0$, we obtain

$$
\begin{aligned}
L''(0) &= \langle \dot\gamma, D_U U \rangle|_a^b \\
&\quad + \int_a^b [|\dot U(t)|^2 - \langle \dot\gamma, \mathbf{R}_{U\dot\gamma} U \rangle - (\langle \dot\gamma, U(t) \rangle')^2] dt, \tag{1.45}
\end{aligned}
$$

which is called the *formula for the second variation of arc length*.

Remark 1.15 *In the formula (1.45) the crucial term is the curvature one which relates $L''(0)$ to the sectional curvatures. $\langle \dot{\gamma}, D_U U \rangle|_a^b$ are called the boundary terms. If α is a proper variation, then $\langle \dot{\gamma}, D_U U \rangle|_a^b = 0$.*

We decompose

$$U(t) = U^\perp(t) + \langle \dot{\gamma}, U \rangle \dot{\gamma} \quad \text{for} \quad t \in [a, b],$$

where $\langle U^\perp(t), \dot{\gamma}(t) \rangle = 0$. Since $\dot{U}(t) = D_{\dot{\gamma}}[U^\perp(t) + \langle \dot{\gamma}, U \rangle \dot{\gamma}] = \dot{U}^\perp(t) + \langle \dot{\gamma}(t), U(t) \rangle' \dot{\gamma}(t)$, it follows from the formula (1.45) that

$$L''(0) = \langle \dot{\gamma}, D_U U \rangle|_a^b + \int_a^b [\dot{U}^\perp(t)|^2 - \langle \dot{\gamma}, \mathbf{R}_{U^\perp \dot{\gamma}} U^\perp \rangle] dt, \qquad (1.46)$$

The above formula will be needed when we compute the Hessian of the distance later.

Exponential Map In order to show that a geodesic is locally shortest in length, we need the concept of the exponential map. Let (M, g) be a complete Riemannian manifold. Let $x \in M$ be given. We define a map $\exp_x: M_x \to M$ as follows: For any $v \in M_x$, let γ be a geodesic starting from x with the initial tangential vector v. Take the point y on γ such that the arc length of γ between x and y is $|v|$. Then we set $\exp_x v = y$.

Definition 1.17 *The map $\exp_x: M_x \to M$ is called the exponential map of M at x.*

Remark 1.16 *By the theory of ordinary differential equations and from the equations (1.12), the exponential map is locally well-defined: Given $x \in M$, there are $\varepsilon > 0$ and a neighborhood $\mathcal{U}$ of x on M such that, for any $y \in \mathcal{U}$ and $v \in M_y$, if $|v| < \varepsilon$, then there is a unique geodesic γ satisfying $\gamma(0) = y$ and $\dot{\gamma}(0) = v$. Here we assume that M is complete so that $\exp_x$ has the domain M_x.*

For any $v \in M_x$, the exponential map maps a straight line tv in M_x into a geodesic $\gamma(t) = \exp_x tv$ satisfying $\gamma(0) = x$ and $\dot{\gamma}(0) = v$. Then the exponential map is a local diffeomorphism at the origin of M_x, that is,

Lemma 1.3 *Let $x \in M$ be given. Then there is $\varepsilon > 0$ such that $\exp_x: B_x(\varepsilon) \to M$ is a diffeomorphism, where $B_x(\varepsilon) = \{ v \in M_x \, | \, |v| < \varepsilon \}$.*

Proof. Since M_x is an n-dimensional linear space, its tangential spaces are itself. It will suffice to prove that $(d \exp_x)|_0: M_x \to M_x$ is invertible, where 0 is the origin of M_x. For any $v \in M_x$, $c(t) = tv$ is a curve in M_x such that $c(0) = 0$ and $\dot{c}(0) = v$. By the definition of d, $(d \exp_x)|_0 v = [\exp_x c(t)]'_{t=0} = v$, that is, $(d \exp_x)|_0 = I$, the identical map from M_x to itself. $\qquad \square$

Fix $\varepsilon > 0$ small such that $\exp_x$ on $B_x(\varepsilon) = \{ v \in M_x \, | \, |v| < \varepsilon \}$ is a diffeomorphism. Let $B(x, \varepsilon) = \exp_x B_x(\varepsilon)$, which is called *the geodesic ball centered at x with radius ε*. We define a map by

$$F(t, v) = \exp_x tv \quad \text{for} \quad t \in \mathbb{R}, \ v \in M_x. \tag{1.47}$$

Then $\dot{F}(t, v) = d\exp_x v \in M_{\exp_x tv}$ is called *the radial vector field* on M starting from x. Since, for any $v \in M_x$ fixed, $F(t, v)$ is a geodesic, we have

$$|d\exp_x v|_{\exp_x tv} = |\dot{F}(t, v)|_{F(t,v)} = |\dot{F}(0, v)|_x = |v| \tag{1.48}$$

for all $t \in \mathbb{R}$ and $v \in M_x$.

For $0 < s < \varepsilon$, let $S_x(s) = \{v \in M_x \,|\, |v| = s\}$ and $S(x, s) = \exp_x S_x(s)$. $S(x, s)$ is called *the geodesic sphere centered at x with radius s*. The following is called *the Gaussian lemma* which states that the radial vector field orthogonalizes the tangential space of the sphere on Riemannian manifolds.

Lemma 1.4 *We have*

$$\langle d\exp_x v, X\rangle(\exp_x v) = 0 \quad \text{for all} \quad X \in [S(x, |v|)]_{\exp_x v}.$$

Proof. Let $y = \exp_x v \in S(x, |v|)$ and let $X \in [S(x, |v|)]_y$ be given. Since $\exp_x$ is a diffeomorphism, there is a curve $\tilde{\sigma}$ on $S_x(|v|)$ such that $\tilde{\sigma}(0) = v$ and $\dot{\tilde{\sigma}}(0) = (d\exp_x)^{-1}X$. In addition

$$|\tilde{\sigma}(s)| = |v| \quad \text{for all} \quad s. \tag{1.49}$$

Then the curve $\sigma(t) = \exp_x \tilde{\sigma}(t)$ on $S(x, |v|)$ satisfies that $\sigma(0) = y$ and $\dot{\sigma}(0) = d\exp_x \dot{\tilde{\sigma}}(0) = X$.

Consider one parameter family of curves

$$\alpha(t, s) = \exp_x t\tilde{\sigma}(s) \quad \text{for} \quad (t, s) \in [0, 1] \times [0, a]$$

with the base curve $\gamma(t) = \alpha(t, 0) = \exp_x tv$ for $t \in [0, 1]$. Let $T(t, s) = \dot{\alpha}(t, s) = d\exp_x \tilde{\sigma}(s)$ be the tangential vector field and $U(t, s) = \alpha_s(t, s) = td\exp_x \dot{\tilde{\sigma}}(s)$ be the transversal vector field of $\alpha(t, s)$, respectively. Then

$$[T, U] = 0 \quad \text{for} \quad (t, s) \in [0, 1] \times [0, a],$$

$T(1, 0) = \dot{\gamma}(1) = d\exp_x v \in M_y$, and $U(1, 0) = X \in [S(x, |v|)]_y$. Moreover, it follows from (1.49) that

$$|T(t, s)| = |\dot{\alpha}(t, s)| = |\tilde{\sigma}(s)| = |v| \quad \text{for} \quad (t, s) \in [0, 1] \times [0, a]. \tag{1.50}$$

In addition, $U(0, 0) = 0$, the origin of M_x.

We need to prove $\langle\dot{\gamma}(1), U(1, 0)\rangle(y) = 0$. Denote $\dot{U}(t, 0) = D_{\dot{\gamma}(t)}U$. By the formula (L2) in Theorem 1.1, $\dot{U}(t, 0) = D_{U(t,0)}T + [T, U](t, 0) = D_U T$. Let

$f(t) = \langle \dot{\gamma}(t), U(t,0) \rangle$ for $t \in [0,1]$. Since γ is a geodesic, we have, via the formula (1.50),

$$\dot{f}(t) = \dot{\gamma}\langle \dot{\gamma}, U(t,0) \rangle = \langle D_{\dot{\gamma}(t)}\dot{\gamma}, U(t,0) \rangle + \langle \dot{\gamma}, \dot{U}(t,0) \rangle$$
$$= \langle T, D_U T \rangle = \frac{1}{2}U(|T|^2) = \frac{1}{2}U(|v|^2) = 0 \quad \text{for} \quad t \in [0,1],$$

which yields

$$\langle d\exp_x v, X \rangle = \langle \dot{\gamma}(1), U(1,0) \rangle = \langle \dot{\gamma}(0), U(0,0) \rangle = \langle v, 0 \rangle = 0.$$

$\square$

The following theorem shows that a geodesic is locally shortest in length.

Theorem 1.17 *Let $\varepsilon > 0$ be such that $\exp_x \colon B_x(\varepsilon) \to B(x,\varepsilon)$ is a diffeomorphism. Let $v \in B_x(\varepsilon)$. Let $\gamma(t) = \exp_x tv$ for $t \in [0,1]$. Then $d(x,y) = L(\gamma) = |v|$ where $y = \gamma(1)$.*

Proof. Let $\sigma\colon [0,1] \to M$ be a piecewise smooth curve such that $\sigma(0) = x$ and $\sigma(1) = y$. We will prove $L(\sigma) \geq |v|$. We assume that σ is C^∞; otherwise, we confine it to several small segments.

Case 1 Let $\sigma(s) \in B(x,\varepsilon)$ for all $s \in [0,1]$. Let $\tilde{\sigma}(s) = \exp_x^{-1}\sigma(s)$. Then $\tilde{\sigma}(s)$ is a curve on M_x such that $\tilde{\sigma}(0) = 0$ and $\tilde{\sigma}(1) = v$.

Consider one parameter family of curves

$$\alpha(t,s) = \exp_x t\frac{\tilde{\sigma}(s)}{|\tilde{\sigma}(s)|}.$$

Then

$$\sigma(s) = \alpha(|\tilde{\sigma}(s)|, s).$$

Let $T = \dot{\alpha}(t,s)$ and $U = \alpha_s(t,s)$. By the formula (1.48), $|T| = 1$ for all $(t,s) \in [0,a] \times (0,1]$. By the Gaussian lemma,

$$\langle T, U \rangle = 0 \quad \text{for} \quad (t,s) \in [0,a] \times (0,1].$$

Thus, we obtain

$$|\dot{\sigma}(s)| = |(\frac{d}{ds}|\tilde{\sigma}(s)|)T + U| \geq |\frac{d}{ds}|\tilde{\sigma}(s)|| \geq \frac{d}{ds}|\tilde{\sigma}(s)|$$

which gives

$$L(\sigma) = \int_0^1 |\dot{\sigma}(s)|ds \geq \int_0^1 \frac{d}{ds}|\tilde{\sigma}(s)|ds = |\tilde{\sigma}(1)| = |v|.$$

Case 2 We assume that σ starts from $\sigma(0) = x$, goes out of $B(x,\varepsilon)$, and then comes back to arrive at $\sigma(1) = y$. Take $|v| < \varepsilon_0 < \varepsilon$. Then the curve σ must intersect $S(x,\varepsilon_0)$, for example as the first time, at some point

$y_0 \in S(x, \varepsilon_0)$ at $s = s_0$ after it starts from x. Since $\sigma: [0, s_0] \to B(x, \varepsilon)$, Case 1 implies that

$$L(\sigma) \geq L(\sigma|_{[0,s_0]}) \geq |\exp_x^{-1} y_0| = \varepsilon_0 > |v|.$$

□

How large ε in Theorem 1.17 can be taken is closely related to sectional curvatures. The following result, named as the Cartan-Hadamard theorem, is of globalness. In 1898, Hadamard proved that, for a complete, simply connected surface M in $\mathbb{R}^3$, if its Gaussian curvature is nonpositive, then for any $x \in M$, $\exp_x: M_x \to M$ is a diffeomorphism. In 1925, E. Cartan generalized this result to the n-dimensional case. Its proof involves much more knowledge in geometry and we omit it because of the limited extent of this chapter. For a proof, we refer to [38].

Theorem 1.18 (*Cartan-Hadamard*) *Let (M, g) be a complete, simply connected Riemannian manifold. If (M, g) has non-positive sectional curvatures, then for any $x \in M$, $\exp_x: M_x \to M$ is a diffeomorphism.*

It follows from Theorems 1.18 and 1.17 that

Corollary 1.2 *Let (M, g) be a complete, simply connected Riemannian manifold with non-positive sectional curvatures. Then all geodesics on (M, g) are shortest in length.*

Theorem 1.17 tells us that a geodesic is locally shortest in length without curvature information. However, it is actually shortest in a relatively large region which is confined by *a cut point*. Let $\gamma = \exp_{x_0} tv$ with $v \in M_{x_0}$ and $|v| = 1$. Then that γ is shortest in the interval $[0, t_0]$ is equivalent to

$$d(\gamma(0), \gamma(t_0)) = t_0 = L(\gamma|_{[0,t_0]}).$$

Moreover, if $\gamma|_{[0,t_0]}$ is shortest, then for any $t < t_0$, $\gamma|_{[0,t]}$ does. By Theorem 1.17, there is $t_0 > 0$ (or $t_0 = \infty$) such that, for any $t < t_0$, $\gamma|_{[0,t]}$ is shortest, but for $t > t_0$, $\gamma|_{[0,t]}$ is not. $\gamma(t_0)$ *is called the cut point of γ with regard to* $\gamma(0) = x_0$. Then $t_0 v \in M_{x_0}$ *is called the tangent cut point.* We define $\tau:$ $S_{x_0}(1) \to \mathbb{R}$ by

$$\tau(v) = t_0 \quad \text{for} \quad v \in S_{x_0}(1).$$

Let $C(x_0) = \{\tau(v)v \mid v \in S_{x_0}(1)\}$. $C(x_0)$ *is called the tangent cut locus of x_0.* The set $\exp_{x_0} C(x_0) \subset M$ *is called the cut locus of x_0.*

Let M be a complete Riemannian manifold. Let $x_0 \in M$. Let

$$\Sigma(x_0) = \{tv \mid v \in S_{x_0}(1), \ 0 \leq t < \tau(v)\}, \tag{1.51}$$

which is called *the interior of the tangent cut locus of x_0.* Then $\exp_{x_0}: \Sigma(x_0) \to \exp_{x_0} \Sigma(x_0)$ is a diffeomorphism ([72]).

Definition 1.18 $\exp_{x_0} \Sigma(x_0)$ *is called the interior of the cut locus of* x_0.

Clearly, any geodesic $\gamma(t) = \exp_{x_0} tv$ is shortest on $\exp_{x_0} \Sigma(x_0)$, that is, if $\gamma(b) \in \exp_{x_0} \Sigma(x_0)$, then

$$\rho(\gamma(b)) = b|v|,$$

where $\rho(x) = d(x, x_0)$ is the distance function. Furthermore, $\exp_{x_0} \Sigma(x_0)$ is a large region on M in the following sense. We have

$$M = \exp_{x_0} \Sigma(x_0) \cup \exp_{x_0} C(x_0). \tag{1.52}$$

It is easy to check that $C(x_0)$ is a zero measure set on M_{x_0}. Then $\exp_{x_0} C(x_0)$ is a zero measure set on M since it is the image of the zero measure set $C(x_0)$, that is, $\exp_{x_0} \Sigma(x_0)$ is M minus a zero measure set.

Corollary 1.3 *Let M be a complete Riemannian manifold and let $x_0 \in M$ be given. Let $\rho(x) = d(x, x_0)$ be the distance function from $x \in M$ to x_0 in the metric g. Let $\gamma(t) = \exp_{x_0} tv$ for $t \in \mathbb{R}$ where $v \in M_{x_0}$ with $|v| = 1$. Then for $x = \gamma(\rho(x)) \in \exp_{x_0} \Sigma(x_0)$*

$$D\rho|_{\gamma(t)} = \dot{\gamma}(t) \quad for \quad t \in [0, \rho(x)]. \tag{1.53}$$

Proof. That $x \in \exp_{x_0} \Sigma(x_0)$ implies that $\rho(\gamma(t)) = t$ for $t \in [0, \rho(x)]$. Then

$$\langle D\rho, \dot{\gamma}(t) \rangle = 1 \quad for \quad t \in [0, \rho(x)]. \tag{1.54}$$

In addition, since $D\rho \perp [S(x_0, t)]_{\gamma(t)}$, the relation (1.53) follows from (1.54). $\square$

In order to compute the Hessian $D^2\rho$ of the distance function ρ, we introduce the concept of Jacobi field.

Definition 1.19 *Let $\gamma \colon [a, b] \to M$ be a geodesic. A vector field J along γ is called a Jacobi field if it satisfies the equation*

$$\ddot{J}(t) + \mathbf{R}_{\dot{\gamma}J}\dot{\gamma} = 0 \quad for \quad t \in [a, b], \tag{1.55}$$

where $\ddot{J} = D_{\dot{\gamma}}D_{\dot{\gamma}}J$ and $\mathbf{R}_{\dot{\gamma}J}$ is the curvature operator. A Jacobi field J is called normal if $\langle \dot{\gamma}(t), J(t) \rangle = 0$ for all $t \in [a, b]$.

The following lemma plays a basic role in the structure of $D^2\rho$.

Lemma 1.5 *Let M be a complete Riemannian manifold and let $x_0 \in M$ be given. Let $\rho(x) = d(x, x_0)$ be the distance function from $x \in M$ to x_0 in the metric g. Let $\gamma(t) = \exp_{x_0} tv$ for $t \in \mathbb{R}$ where $v \in M_{x_0}$ with $|v| = 1$. Let $x = \exp_{x_0} bv \in \exp_{x_0} \Sigma(x_0)$. Then, for any $X \in [S(x_0, b)]_x$, there is a normal Jacobi field J along γ such that*

$$J(0) = 0, \quad J(b) = X; \tag{1.56}$$

$$D^2\rho(X, X) = \langle \dot{J}(b), J(b) \rangle. \tag{1.57}$$

Proof. Let $\sigma\colon [0, \varepsilon] \to M$ be a geodesic such that $\sigma(0) = x$ and $\dot{\sigma}(0) = X$. Denote $\tilde{\sigma}(s) = \exp_{x_0}^{-1} \sigma(s)$ for $s \in [0, \varepsilon]$. Set

$$\alpha(t, s) = \exp_{x_0} t\frac{\tilde{\sigma}(s)}{b} \quad \text{for} \quad (t, s) \in [0, b] \times [0, \varepsilon]. \tag{1.58}$$

Then the base curve of the one parameter family $\alpha(\cdot, s)$ of curves is $\alpha(t, 0) = \gamma(t)$. Let

$$T(t, s) = \dot{\alpha}(t, s), \quad U(t, s) = \alpha_s(t, s)$$

be the tangential vector field and the transversal vector field of α, respectively. Then

$$[T, U] = 0 \quad \text{for} \quad (t, s) \in [0, b] \times [0, \varepsilon]. \tag{1.59}$$

Since $\alpha(\cdot, s)$ is a geodesic for each $s \in [0, \varepsilon]$ fixed,

$$D_T T = 0 \quad \text{for} \quad (t, s) \in [0, b] \times [0, \varepsilon]. \tag{1.60}$$

Let $J(t) = U(t, 0)$. We will prove that J is a Jacobi field along γ to satisfy the relations (1.56) and (1.57). The relation (1.59) implies $D_T U = D_J T$. Then it follows from the relations (1.59) and (1.60) that

$$\begin{aligned} \ddot{J}(t) &= D_{\dot{\gamma}} D_{\dot{\gamma}} J = D_T D_T U = D_T D_U T \\ &= -[-D_T D_U T + D_U D_T T + D_{[T,U]} T] \\ &= -\mathbf{R}_{TU} T = -\mathbf{R}_{\dot{\gamma} J} \dot{\gamma} \quad \text{for} \quad t \in [0, b]. \end{aligned} \tag{1.61}$$

In addition, since $\alpha(0, s) = x_0$ and $\alpha(b, s) = \sigma(s)$, we have

$$J(b) = U(b, 0) = \dot{\sigma}(0) = X, \quad J(0) = U(0, 0) = 0, \tag{1.62}$$

$$D_U U|_{(0,s)} = 0, \quad D_U U|_{(b,s)} = D_{\dot{\sigma}(s)} \dot{\sigma} = 0.$$

Moreover, using the relations (1.61) and (3) of Theorem 1.8 , we have

$$\frac{d^2}{dt^2} \langle \dot{\gamma}(t), J(t) \rangle = \langle \dot{\gamma}(t), \ddot{J}(t) \rangle = -\mathbf{R}(\dot{\gamma}(t), J(t), \dot{\gamma}(t), \dot{\gamma}(t)) = 0,$$

and, therefore, we obtain $\langle \dot{\gamma}(t), J(t) \rangle = \sigma_0 t + \sigma_1$ and, via (1.62),

$$\langle \dot{\gamma}(t), J(t) \rangle = \langle \dot{\gamma}(0), J(0) \rangle = 0 \quad \text{for} \quad t \in [0, b],$$

that is, J is a normal Jacobi field with the relation (1.56).

Note that $\exp_{x_0} \Sigma(x_0)$ is star-shaped open set. Then that $x \in \exp_{x_0} \Sigma(x_0)$ implies that, for $\varepsilon > 0$ small, $\alpha(t, s) \in \exp_{x_0} \Sigma(x_0)$ for $(t, s) \in [0, b] \times [0, \varepsilon]$, that is,

$$\rho(\alpha(b, s)) = L(\alpha(\cdot, s)|_{[0,b]}) \quad \text{for} \quad s \in [0, \varepsilon]. \tag{1.63}$$

On one hand

$$\frac{d}{ds} \rho(\alpha(b, s)) = \langle D\rho, U(b, s) \rangle.$$

Then

$$\frac{d^2}{ds^2}\rho(\alpha(b,s))\Big|_{s=0} = D^2\rho(J(b), J(b)) + D\rho(D_U U)|_{(b,0)} = D^2\rho(X, X).$$

On the other hand, by the formulas (1.46), (1.63) and (1.55), we obtain

$$\frac{d^2}{ds^2}\rho(\alpha(b,s))\Big|_{s=0} = L''(0) = \int_0^b [|\dot{J}(t)|^2 - \mathbf{R}(\dot{\gamma}(t), J(t), \dot{\gamma}(t), J(t)]dt$$

$$= \int_0^b \frac{d}{dt}\langle \dot{J}(t), J(t)\rangle dt = \langle \dot{J}(b), J(b)\rangle.$$

$$\square$$

Remark 1.17 *On* $\exp_{x_0} \Sigma(x_0)$ *a Jacobi field satisfying the conditions* (1.56) *is unique* ([206]).

Let $x_0 \in M$ be given. Let $\rho(x) = d(x, x_0)$ be *the distance function* from $x \in M$ to x_0 in the metric g. Denote by $CD\rho^2$ all the points x on M such that

$$D^2\rho^2(X, X) > 0 \quad \text{for all} \quad X \in M_x, \quad X \neq 0.$$

The set $CD\rho^2$ will play an important role in application; see the following chapters in this book. Consider the Euclidean space $\mathbb{R}^n$, the distance function is $\rho(x) = |x - x_0|$. It is easy to check that (Exercise 1.2.6)

$$D^2\rho^2(X, X) = 2|X|^2 \quad \text{for all} \quad X \in \mathbb{R}^n_x, \; x \in \mathbb{R}^n$$

where D is the classical connection, given by the Euclidean metric. Then $CD\rho^2 = \mathbb{R}^n$. For a general metric without any information on curvature, we will only know that $CD\rho^2$ is just a neighborhood of x_0, see Theorem 1.20 later. However, the sectional curvatures will yield the global information on the set $CD\rho^2$, see Theorem 1.19 below.

As an application of Lemma 1.5 and Theorem 1.18, we will prove

Theorem 1.19 *Let* (M, g) *be a simply connected, complete Riemannian manifold with non-positive sectional curvatures. Then* $CD\rho^2 = M$.

Proof. By Theorem 1.18, $\exp_{x_0} \Sigma(x_0) = M$. Let $x \in M$ be given. Let $x = \exp_{x_0} v$. Let $\gamma(t) = \exp_{x_0} tv/|v|$ be the normal geodesic connecting x_0 and x. Let $X \in M_x$ be given with $|X| \neq 0$. We will prove $D^2\rho^2(X, X) > 0$.

We decompose X as

$$X = \langle X, \dot{\gamma}(|v|)\rangle\dot{\gamma}(|v|) + X_0$$

where $\langle\dot{\gamma}(|v|), X_0\rangle = 0$. It follows from the formula (1.53) that

$$D^2\rho^2(X, X) = 2(D\rho \otimes D\rho + \rho D^2\rho)(X, X)$$
$$= 2\langle X, \dot{\gamma}(|v|)\rangle^2 + 2|v|D^2\rho(X_0, X_0) \qquad (1.64)$$

where the relation $D^2\rho(X_0, \dot\gamma) = \langle D_{\dot\gamma}\dot\gamma, X_0 \rangle = 0$ is used.

If $X_0 = 0$, then by (1.64), $D^2\rho^2(X, X) = 2|X|^2 > 0$. Let $X_0 \neq 0$. By Lemma 1.5, there is a normal Jacobi field J along γ such that

$$J(0) = 0, \quad J(|v|) = X_0,$$

and

$$D^2\rho(X_0, X_0) = \langle \dot J(|v|), J(|v|) \rangle = \int_0^{|v|} \frac{d}{dt}\langle \dot J(t), J(t) \rangle dt$$

$$= \int_0^{|v|} [|\dot J(t)|^2 - \mathbf{R}(\dot\gamma, J, \dot\gamma, J)]dt \geq \int_0^{|v|} |\dot J(t)|^2 dt > 0$$

because $\mathbf{R}(\dot\gamma, J, \dot\gamma, J) \leq 0$ and $\dot J(t)$ is not identically zero. Indeed, if $\dot J(t) = 0$ for all $t \in [0, |v|]$, then J is parallel along γ and therefore, $|X_0| = |J(|v|)| = |J(0)| = 0$, which is a contradiction. $\qquad\Box$

Without curvature information, CD$\,\rho^2$ is just a neighborhood of x_0, that is,

Theorem 1.20 *Let M be a Riemannian manifold. Let $x_0 \in M$ and let $\rho(x) = d(x, x_0)$ be the distance function. Then $D^2\rho^2$ is positive in a neighborhood of x_0.*

Proof. Let $\varepsilon > 0$ be such that $\exp_{x_0}: B_{x_0}(\varepsilon) \to B(x_0, \varepsilon)$ is a diffeomorphism. For any $v \in M_{x_0}$, let $\gamma(t) = \exp_{x_0} tv/|v|$. Then

$$\rho^2(\gamma(t)) = t^2 \quad \text{for} \quad t \in [0, \varepsilon).$$

We differentiate the above identity twice in t at $t = 0$ to have

$$D^2\rho^2(v, v) = 2|v|^2 \tag{1.65}$$

since $\dot\gamma(0) = v/|v|$. Then there is a neighborhood of x_0 where $D^2\rho^2 > 0$. $\quad\Box$

In Theorem 1.20, we have checked the convexity of the function ρ^2. We introduce

Definition 1.20 *A differentiable function $f: M \to \mathbb{R}$ is called (strictly) convex if its Hessian $D^2 f$ is positive semidefinite (definite).*

Proposition 1.1 *Let $G(x) = (g_{ij})$ be a symmetric, positive, and smooth matrix for each $x \in \mathbb{R}^n$. Consider $(\mathbb{R}^n, g)$ as a Riemannian manifold with the metric $g = G(x)$. Let $\Omega \subset \mathbb{R}^n$ be an open set. Then f is strictly convex function on $\overline\Omega$ if and only if f satisfies that*

$$\sum_{ijkl} [2a_{il}(a_{kj}f_{x_k})_{x_l} - a_{ijx_l}a_{kl}f_{x_k}]\xi_i\xi_j \geq c\sum_{ij} a_{ij}\xi_i\xi_j \tag{1.66}$$

for all $(x, \xi) \in \overline\Omega \times \mathbb{R}^n$, where $c > 0$ and $(a_{ij}(x)) = G^{-1}(x)$ for $x \in \mathbb{R}^n$.

Proof. Let $x = (x_1, \cdots, x_n)$ be the natural coordinate system. Then

$$g_{ij}(x) = \langle \partial_{x_i}, \partial_{x_j} \rangle_g = \langle G(x)\partial_{x_i}, \partial_{x_j} \rangle \quad \text{for} \quad x \in \mathbb{R}^n$$

for all $1 \leq i, j \leq n$ where $\langle \cdot, \cdot \rangle$ denotes the Euclidean metric of $\mathbb{R}^n$. The Christoffel symbols are given by

$$\begin{aligned}
2\Gamma^k_{pq} &= \sum_l a_{kl}(g_{plx_q} + g_{qlx_p} - g_{pqx_l}) \\
&= \sum_l [(a_{kl}g_{pl})_{x_q} + (a_{kl}g_{ql})_{x_p} - a_{kl}g_{pqx_l} - a_{klx_q}g_{pl} - a_{klx_p}g_{ql}] \\
&= -\sum_l [a_{kl}g_{pqx_l} + a_{klx_q}g_{pl} + a_{klx_p}g_{ql}]
\end{aligned} \tag{1.67}$$

for all $1 \leq p, q, k \leq n$. It follows from (1.67) that

$$2\sum_{pq} a_{ip}a_{jq}\Gamma^k_{pq} = \sum_l a_{kl}a_{ijx_l} - \sum_l a_{ikx_l}a_{jl} - \sum_l a_{jkx_l}a_{il} \tag{1.68}$$

for all $1 \leq i, j, k \leq n$.

For $\xi = (\xi_1, \cdots, \xi_n) \in \mathbb{R}^n$ given, let

$$X = \sum_i \xi_i \partial_{x_i} \quad \text{for} \quad x \in \overline{\Omega}.$$

Let D be the Levi-Civita connection of the metric g. It follows from (1.68) that

$$\begin{aligned}
2D^2 f(G^{-1}X, G^{-1}X) &= 2\sum_{ijpq} a_{ip}a_{jq}\xi_i\xi_j D^2 f(\partial_{x_p}, \partial_{x_q}) \\
&= 2\sum_{ij}\xi_i\xi_j \sum_{pq} a_{ip}a_{jq}[f_{x_px_q} - D_{\partial_{x_p}}\partial_{x_q}f] \\
&= \sum_{ij}[2\sum_{kl} a_{ik}a_{jl}f_{x_kx_l} - 2\sum_{pqk} a_{ip}a_{jq}\Gamma^k_{pq}f_{x_k}]\xi_i\xi_j \\
&= \sum_{ij}[2\sum_{kl} a_{ik}a_{jl}f_{x_kx_l} - \sum_{kl} a_{kl}a_{ijx_l}f_{x_k}]\xi_i\xi_j \\
&\quad + \sum_{kl}(\sum_i \xi_i a_{ik})_{x_l}(\sum_j \xi_j a_{jl})f_{x_k} + \sum_{kl}(\sum_j \xi_j a_{jk})_{x_l}(\sum_i \xi_i a_{il})f_{x_k} \\
&= \sum_{ijkl}[2a_{il}(a_{kj}f_{x_k})_{x_l} - a_{ijx_l}a_{kl}f_{x_k}]\xi_i\xi_j.
\end{aligned} \tag{1.69}$$

Then the inequality (1.66) is equivalent to

$$2D^2 f(G^{-1}X, G^{-1}X) \geq c\langle G^{-1}X, X \rangle$$

for all $X \in \mathbb{R}^n_x$ with $x \in \overline{\Omega}$, that is,

$$D^2 f(X, X) \geq \frac{c}{2} |X|^2_g$$

for all $X \in \mathbb{R}^n_x$ with $x \in \overline{\Omega}$ where $|X|_g = \langle G(x)X, X \rangle^{1/2}$ is the norm of X in the metric g. $\qquad\square$

$D^2 \rho$ in **Spaces of Constant Curvature** Let M be a Riemannian manifold. M is called *a space of constant curvature* κ, *or a space form* if all its sectional curvatures are the same number κ.

Lemma 1.6 *Let M be a space of constant curvature κ. Let $x \in M$ and let* **R** *be the curvature tensor. Then*

$$\mathbf{R}(X_1, X_2, X_3, X_4) = \kappa[\langle X_1, X_3 \rangle \langle X_2, X_4 \rangle - \langle X_2, X_3 \rangle \langle X_1, X_4 \rangle] \qquad (1.70)$$

for $X_i \in M_x$ with $1 \leq i \leq 4$.

Proof. Let $X_1 \in M_x$ be given. Consider a bilinear form on M_x by

$$b(X_2, X_4) = \mathbf{R}(X_1, X_2, X_1, X_4) \quad \text{for} \quad X_2, \ X_4 \in M_x.$$

By (1.21),

$$b(X_2, X_2) = \kappa(|X_1|^2 |X_2|^2 - \langle X_1, X_2 \rangle^2).$$

By (4) of Theorem 1.8, $b(X_2, X_4) = b(X_4, X_2)$. Then

$$
\begin{aligned}
4\mathbf{R}(X_1, X_2, X_1, X_4) &= 4b(X_2, X_4) \\
&= b(X_2 + X_4, X_2 + X_4) - b(X_2 - X_4, X_2 - X_4) \\
&= \kappa(|X_1|^2 |X_2 + X_4|^2 - \langle X_1, X_2 + X_4 \rangle^2) \\
&\quad -\kappa(|X_1|^2 |X_2 - X_4|^2 - \langle X_1, X_2 - X_4 \rangle^2) \\
&= 4\kappa(|X_1|^2 \langle X_2, X_4 \rangle - \langle X_1, X_2 \rangle \langle X_1, X_4 \rangle),
\end{aligned}
$$

that is,

$$\mathbf{R}(X_1, X_2, X_1, X_4) = \kappa(|X_1|^2 \langle X_2, X_4 \rangle - \langle X_1, X_2 \rangle \langle X_1, X_4 \rangle) \qquad (1.71)$$

for all $X_1, \ X_2, \ X_4 \in M_x$.

For $X_i \in M_x$ with $1 \leq i \leq 4$, set

$$
\begin{aligned}
P(X_1, X_2, X_3, X_4) &= \mathbf{R}(X_1, X_2, X_3, X_4) \\
&\quad -\kappa[\langle X_1, X_3 \rangle \langle X_2, X_4 \rangle - \langle X_2, X_3 \rangle \langle X_1, X_4 \rangle].
\end{aligned}
$$

Then $P \in T^4(M_x)$. We will prove that $P(X_1, X_2, X_3, X_4) = 0$ for all $X_i \in M_x$ with $1 \leq i \leq 4$.

It follows from (1.71) that

$$P(X_1, X_2, X_1, X_3) = 0 \qquad (1.72)$$

for all $X_i \in M_x$ with $1 \leq i \leq 3$.

By (1)-(4) of Theorem 1.8 it is easy to check that P satisfies the following properties:

$$P(X_1, X_2, X_3, X_4) = -P(X_2, X_1, X_3, X_4); \qquad (1.73)$$

$$P(X_1, X_2, X_3, X_4) + P(X_2, X_3, X_1, X_4) + P(X_3, X_1, X_2, X_4) = 0; \quad (1.74)$$

$$P(X_1, X_2, X_3, X_4) = -P(X_1, X_2, X_4, X_3); \qquad (1.75)$$

$$P(X_1, X_2, X_3, X_4) = P(X_3, X_4, X_1, X_2). \qquad (1.76)$$

It follows from (1.73), (1.75), and (1.72) that

$$P(X_1, X_2, X_3, X_2) = P(X_2, X_1, X_2, X_3) = 0. \qquad (1.77)$$

Using (1.72) and (1.73), we obtain

$$
\begin{aligned}
0 &= P(X_1 + X_2, X_3, X_1 + X_2, X_4) \\
&= P(X_1, X_3, X_1 + X_2, X_4) + P(X_2, X_3, X_1 + X_2, X_4) \\
&= P(X_1, X_3, X_1, X_4) + P(X_1, X_3, X_2, X_4) + P(X_2, X_3, X_1, X_4) \\
&\quad + P(X_2, X_3, X_2, X_4) \\
&= P(X_1, X_3, X_2, X_4) + P(X_2, X_3, X_1, X_4) \\
&= -P(X_3, X_1, X_2, X_4) + P(X_2, X_3, X_1, X_4).
\end{aligned} \qquad (1.78)
$$

Similarly, we have

$$
\begin{aligned}
0 &= P(X_1, X_2 + X_4, X_3, X_2 + X_4) = P(X_1, X_2, X_3, X_4) \\
&\quad - P(X_2, X_3, X_1, X_4).
\end{aligned} \qquad (1.79)
$$

Summing up (1.78) with (1.79) yields

$$P(X_1, X_2, X_3, X_4) = P(X_3, X_1, X_2, X_4). \qquad (1.80)$$

In addition, the relation $0 = P(X_1 + X_3, X_2, X_1 + X_3, X_4)$ implies that

$$P(X_1, X_2, X_3, X_4) = P(X_2, X_3, X_1, X_4). \qquad (1.81)$$

Finally, from (1.80), (1.81), and (1.74), we obtain

$$
\begin{aligned}
3P(X_1, X_2, X_3, X_4) &= P(X_1, X_2, X_3, X_4) + P(X_2, X_3, X_1, X_4) \\
&\quad + P(X_3, X_1, X_2, X_4) = 0.
\end{aligned}
$$

$\square$

Remark 1.18 *If $X_i \in \mathcal{X}(M)$ are vector fields for $1 \leq i \leq 3$, it follows from (1.70) that*

$$\mathbf{R}_{X_1 X_2} X_3 = \kappa[\langle X_1, X_3 \rangle X_2 - \langle X_2, X_3 \rangle X_1]. \qquad (1.82)$$

Let M be a space of constant curvature. Let $x_0 \in M$ be given. We will compute the Hessian $D^2\rho$ of the distance function ρ on the interior of the cut locus of x_0, $\exp_{x_0} \Sigma(x_0)$. For this purpose, we need to compute Jacobi fields.

Lemma 1.7 *Let M be a space of constant curvature κ. Let $x \in \exp_{x_0} \Sigma(x_0)$ be given. Let $\gamma(t) = \exp_{x_0} tv$ with $|v| = 1$ such that $x = \gamma(b)$. Let $X \in [S(x_0, b)]_x$. Let $e \in M_{x_0}$ be such that the parallel transport $e(t) \in M_{\gamma(t)}$ of e along γ satisfies*

$$e(b) = X.$$

Let J be a Jacobi field along γ such that $J(0) = 0$ and $J(b) = X$. Then

$$J(t) = f_0(t)e(t) \quad for \quad t \in [0, b] \tag{1.83}$$

where

$$f_0(t) = \begin{cases} \frac{1}{\sin \sqrt{\kappa}b} \sin \sqrt{\kappa}t & \kappa > 0; \\ \frac{1}{b}t, & \kappa = 0; \\ \frac{1}{\sinh \sqrt{-\kappa}b} \sinh \sqrt{-\kappa}t, & \kappa < 0, \end{cases} \quad for \quad t \in [0, b].$$

Proof. Since $X = e(b)$ is the parallel transport of e along γ, $|e(t)| = |X|$ for all $t \in [0, b]$. Let $e_1, e_2, \cdots, e_n$ be an orthonormal basis of M_{x_0} with $e_1 = \dot{\gamma}(0)$ and $e_2 = e/|X|$. For $1 \le i \le n$, we parallelize e_i to $e_i(t) \in M_{\gamma(t)}$ along $\gamma|_{[0,t]}$ for $0 < t \le b$. Then for each $t \in [0, b]$, $e_1(t), \cdots, e_n(t)$ is an orthonormal basis of $M_{\gamma(t)}$ such that $\dot{e}_i(t) = D_{\dot{\gamma}(t)}e_i = 0$ for $1 \le i \le n$ where $e_1(t) = \dot{\gamma}(t)$ and $e_2(t) = e(t)/|X|$.

Let

$$J(t) = \sum_{i=1}^{n} J_i(t)e_i(t) \quad for \quad t \in [0, b].$$

Using the formula (1.70), we have

$$\mathbf{R}_{\dot{\gamma}J\dot{\gamma}} = \sum_{i=1}^{n} J_i(t)\mathbf{R}_{\dot{\gamma}(t)e_i(t)}\dot{\gamma} = \sum_{ij=1}^{n} J_i(t)\mathbf{R}(\dot{\gamma}(t), e_i(t), \dot{\gamma}, e_j(t))e_j(t)$$

$$= \sum_{ij=2}^{n} J_i(t)\mathbf{R}(\dot{\gamma}(t), e_i(t), \dot{\gamma}, e_j(t))e_j(t)$$

$$= \kappa \sum_{i=1}^{n} J_i(t)e_i(t) \quad for \quad t \in [0, b].$$

Then J, being a Jacobi field, is equivalent to

$$\ddot{J}_i(t) + \kappa J_i(t) = 0 \quad for \quad t \in [0, b], \quad 1 \le i \le n. \tag{1.84}$$

In addition, the conditions $J(0) = 0$ and $J(b) = X$ yield

$$J_1(0) = 0, \quad J_1(b) = |X|; \quad J_i(0) = J_i(b) = 0 \tag{1.85}$$

for $2 \leq i \leq n$.

We solve the problem (1.84)-(1.85) to give the formula (1.83). □

Combining Lemmas 1.7 with 1.5, we have

Theorem 1.21 *Let M be a space of constant curvature κ. Let $x_0 \in M$ be given. Let $\rho(x) = d(x, x_0)$ be the distance function from $x \in M$ to x_0 in the metric g. Then for $x \in \exp_{x_0} \Sigma(x_0)$*

$$D^2\rho(X, X) = f(\rho)|X|^2 \quad for \quad X \in M_x, \ \langle D\rho, X \rangle = 0, \qquad (1.86)$$

where

$$f(\rho) = \begin{cases} \sqrt{\kappa} \cot \sqrt{\kappa}\rho & \kappa > 0; \\ \dfrac{1}{\rho}, & \kappa = 0; \\ \sqrt{-\kappa} \coth \sqrt{-\kappa}\rho, & \kappa < 0. \end{cases} \qquad (1.87)$$

Hessian Comparison Theorem For a general Riemannian manifold with variable curvature, we can study the Hessian of the distance by comparing it with the one in a space of constant curvature.

Let M be a Riemannian manifold and let $\gamma(t)$ be a normal geodesic on M. Let $\mathbf{R}$ be the curvature tensor. Then for any $X \in M_{\gamma(t)}$ with $\langle X, \dot{\gamma}(t) \rangle = 0$ and $|X| = 1$, the sectional curvature $\mathbf{R}(\dot{\gamma}(t), X, \dot{\gamma}(t), X)$ is called *a radial curvature at* $\gamma(t)$.

Let M_1 and M_2 be complete Riemannian manifolds. Let $x_0^i \in M_i$ be given for $i = 1, 2$. Denote by ∂_{M_i} the radial vector fields on M_i starting from x_0^i, respectively, for $i = 1, 2$. Let ρ_i be the distance functions on M_i starting from x_0^i, respectively, for $i = 1, 2$. If μ_i are symmetric tensors of rank 2 on $M_i / \{x_0^i\}$, respectively, for $i = 1, 2$, then for $x^i \in M_i$,

$$\mu_1(x^1) \prec \mu_2(x^2) \qquad (1.88)$$

means: for all $X_i \in M_{x^i}$ with $|X_1| = |X_2|$ and $\langle X_1, \partial_{M_1} \rangle = \langle X_2, \partial_{M_2} \rangle$, the inequality $\mu_1(X_1, X_1) \leq \mu_2(X_2, X_2)$ is valid.

We introduce the following theorem without proofs. For a proof, see [71].

Theorem 1.22 *(Hessian comparison theorem) Let $\gamma_i \colon [0, b] \to M_i$ be normal geodesics with $\gamma_i(0) = x_0^i$ and $\gamma_i(b) \in \exp_{x_0^i} \Sigma(x_0^i)$, respectively, for $i = 1$ and 2. If*

each radial curvature at $\gamma_2(t) \geq$ every radial curvature at $\gamma_1(t)$,

then

$$D^2\rho_2 \prec D^2\rho_1$$

for every $t \in [0, b]$.

It follows from Theorems 1.21 and 1.22 that

Corollary 1.4 *Let $x_0 \in M$ be given. Let $\gamma\colon [0,b] \to M$ be a normal geodesic with $\gamma(0) = x_0$ and $\gamma(b) \in \exp_{x_0} \Sigma(x_0)$. Set*

$$\kappa_0 = \inf_{t \in [0,b]} \{all\ radial\ curvatures\ at\ \gamma(t)\},$$

$$\kappa_1 = \sup_{t \in [0,b]} \{all\ radial\ curvatures\ at\ \gamma(t)\}.$$

Then

$$f_0(t)|X|^2 \geq D^2\rho(X,X) \geq f_1(t)|X|^2 \tag{1.89}$$

for all $X \in M_{\gamma(t)}$ with $\langle X, \dot{\gamma}(t) \rangle = 0$ for all $t \in [0,b]$, where $\rho = d(x_0, x)$ and $f_0(\rho) = f(\rho)$ and $f_1(\rho) = f(\rho)$ are given by the formula (1.87) with κ replaced by κ_0 and by κ_1, respectively.

Exercise 1.2

1.2.1 *Let (M,g) be a complete Riemannian manifold. Then for any two points x, $y \in M$, there is a piecewise smooth curve on M that connects x with y.*

1.2.2 *The distance function on the Riemannian manifold (M,g) satisfies the usual axioms:*
  (a) *$d(x,y) \geq 0$ for all x, y, and $d(x,y) > 0$ for all $x \neq y$;*
  (b) *$d(x,y) = d(y,x)$;*
  (c) *$d(x,y) \leq d(x,z) + d(z,y)$ (triangle inequality) for all x, y, $z \in M$.*

1.2.3 *Let (M,g) be a Riemannian manifold. Then the topology on M induced by the distance function d coincides with the original manifold topology of M.*

1.2.4 *If $\gamma\colon [a,b] \to M$ is a smooth curve, and $\psi\colon [\alpha,\beta] \to [a,b]$ is a change of parameter, then*
$$L(\gamma \circ \psi) = L(\gamma).$$

1.2.5 *Consider a C^∞ map $\alpha(\cdot,\cdot)\colon [a,b] \times [0,\varepsilon] \to M$. Let $T = \dot{\alpha}(\cdot,s)$ and $U = \alpha_s(t,\cdot)$ be the tangential vector field and the transversal vector field of one parameter curve family $\alpha(t,\cdot)$, respectively. Then*

$$[T,U] = 0.$$

1.2.6 *Consider the Euclidean space $\mathbb{R}^n$ with the classical dot metric. Let $\rho(x) = |x - x_0|$ be the distance function. Prove $D^2\rho^2 = 2I$ for all $x \in \mathbb{R}^n$.*

1.2.7 *Let $x_0 \in M$ be given. Let $\rho(x) = d(x, x_0)$ be the distance function in the metric g. For $x \in M$ given, let $\gamma\colon [0, \rho(x)] \to M$ be a geodesic such that $\gamma(0) = x_0$ and $\gamma(\rho(x)) = x$. Then γ is a normal geodesic.*

1.2.8 *Let $x_0 \in M$ be given. Let $\rho(x) = d(x, x_0)$ be the distance function in the metric g. Let $x \in \exp_{x_0} \Sigma(x_0)$. Then $x = \exp_{x_0} \rho(x)v$ for some $v \in M_{x_0}$ with $|v| = 1$, and*

$$D^2 \rho(\dot{\gamma}(\rho(x)), X) = 0 \quad for \quad X \in M_x$$

where $\gamma(t) = \exp_{x_0} tv$.

1.2.9 *Let M be a Riemannian manifold. A differentiable function $f\colon M \to \mathbb{R}$ is (strictly) convex if and only if, for any geodesic γ, $((f \circ \gamma)'' > 0)$ $(f \circ \gamma)'' \geq 0$.*

1.3   A Basic Computational Technique

This section presents a basic computational technique that is useful not only in differential geometry (particularly in the Bochner technique) but in many other mathematical problems as well. We will need it as a basic necessary computational tool to obtain models of shells and to carry out multiplier schemes for control problems in this book. The basic ideas are as follows: In order to verify an identity or a pointwise estimate on a Riemannian manifold, it suffices to do so at each point p relative to a coordinate system or frame field that offers the greatest computational simplification. What prevents a computation from being as simple as the corresponding classical (i.e., Euclidean) situation is the presence of the Christoffel symbols Γ_{ij}^k. Thus we hope to find a coordinate system or a frame field relative to which the Γ_{ij}^k's vanish at the given point p. Such a coordinate system or frame field is then generically referred to as *being normal at p*. The possibility of finding such a coordinate system or frame field under various circumstances is the content of the present section.

Definition 1.21 *Let M be a Riemannian manifold and let $p \in M$ be given. Let $(x_1, \cdots, x_n)$ be a coordinate system around p. If*

$$\begin{cases} g_{ij}(p) = \delta_{ij}, \\ D_{\partial_{x_i}} \partial_{x_j}(p) = 0 \end{cases} \tag{1.90}$$

for all $i, j = 1, \cdots, n$, then the coordinate system $\{x_i\}$ is said to be normal at p.

A well known fact is that a coordinate system normal at p can be obtained from *geodesic coordinate systems*: Let $\exp_p$ denote the exponential map defined on an open neighborhood of 0 in the tangent space M_p. Let $e_1, \cdots, e_n$ be an orthonormal basis of M_p. We define a coordinate system $\varphi = (x_1, \cdots, x_n)$ at p by

$$\varphi(q) = (x_1, \cdots, x_n) \quad \text{for} \quad q = \exp_p \sum_{j=1}^{n} x_j e_j \in M. \tag{1.91}$$

The above coordinate system is called *a geodesic coordinate system*.

Theorem 1.23 *Let M be a Riemannian manifold and let $p \in M$ be given. Then a geodesic coordinate system is normal at p.*

Proof. For any $f \in C^\infty(M)$, by definition

$$\partial_{x_i} f = \frac{\partial}{\partial x_i} f(\exp_p \sum_{j=1}^{n} x_j e_j) = \langle Df, d\exp_p e_i \rangle,$$

that is ,

$$\partial_{x_i} = d\exp_p e_i \quad \text{for} \quad 1 \le i \le n.$$

Since $d\exp_p$ is the identical map from M_p to itself at p,

$$g_{ij}(p) = \langle e_i, e_j \rangle(p) = \delta_{ij}.$$

We define a bilinear map $\eta \colon M_p \otimes M_p \to M_p$ by

$$\eta(v, w) = \sum_{ij=1}^{n} \alpha_i \beta_j (D_{\partial_{x_i}} \partial_{x_j})(p)$$

where $v = \sum_i \alpha_i e_i$ and $w = \sum_i \beta_i e_i$. For any $v = \sum_i \alpha_i e_i \in M_p$ fixed, let $\gamma(t) = \exp_p tv$. Then $\varphi(\gamma(t)) = t(\alpha_1, \cdots, \alpha_n)$ which yields

$$\dot\gamma(t) = \sum_{i=1}^{n} \alpha_i \partial_{x_i}(\gamma(t)).$$

Then that γ is a geodesic implies that

$$\eta(v, v) = D_{\dot\gamma(0)} \dot\gamma = 0.$$

By the bilinearity of η, we have $\eta(v, w) = 0$ for any v, $w \in M_p$. In particular, the second system of equations in (1.90) hold true. $\square$

Definition 1.22 *Let M be a Riemannian manifold and let $p \in M$ be given. Let there be a frame field $\{E_1, \cdots, E_n\}$ near p, i.e., $E_1, \cdots, E_n$ are locally defined vector fields around p which satisfy $\langle E_i, E_j \rangle = \delta_{ij}$ for all i, j. The frame field $\{E_i\}$ is said to be normal at p if*

$$(D_{E_i} E_j)(p) = 0 \quad \text{for all} \quad i, \ j. \tag{1.92}$$

Theorem 1.24 *Given $p \in M$ and an orthonormal basis $\{e_1, \cdots, e_n\}$ of M_p, there exists a frame field $\{E_1, \cdots, E_n\}$ normal at p such that $E_i(p) = e_i$ for $1 \le i \le n$.*

Proof. Let $\varepsilon > 0$ be given such that $\exp_p \colon B_p(\varepsilon) \subset M_p \to M$ is a diffeomorphism. Let $U = \exp_p B_p(\varepsilon)$ which is a neighborhood of p. Then a frame field $\{E_1, \cdots, E_n\}$ can be defined in U by: If $q \in U$, $E_i(q) =$ the parallel transport of e_i from p to q along the unique minimizing geodesic in U jointing p and q. Consider the geodesic $\gamma_i(t) = \exp_x te_i$. Since E_j is a parallel transport along $\gamma(t)$ for $0 < t < \varepsilon$, we have

$$D_{E_i} E_j = D_{\dot{\gamma}(0)} E_j = 0 \quad \text{at} \quad x$$

for all i, j. $\qquad\qquad\square$

Remark 1.19 *For a frame field normal at p*

$$[E_i, E_j](p) = 0 \tag{1.93}$$

for all i, j because

$$[E_i, E_j](p) = (D_{E_i} E_j)(p) - (D_{E_j} E_i)(p) = 0.$$

As the first example to show how the computational principle works, we have

Lemma 1.8 *Let $\{x_i\}$ be a local coordinate system on M normal at $p \in M$. Then for $f \in C^2(M)$*

$$\Delta f(p) = \sum_{i=1}^{n} \frac{\partial^2 f}{\partial x_i^2}(p). \tag{1.94}$$

If $E_1, \cdots, E_n$ is a frame field normal at p, then

$$\Delta f(p) = \sum_{i=1}^{n} E_i E_i(f). \tag{1.95}$$

Proof. The conditions $g_{ij}(p) = \langle \partial_{x_i}, \partial_{x_j} \rangle(p) = \delta_{ij}$ mean that $\partial_{x_1}, \cdots, \partial_{x_n}$ is an orthonormal basis of M_p. Then

$$\Delta f = \operatorname{tr} D^2 f = \sum_{i=1}^{n} D^2 f(\partial_{x_i}, \partial_{x_i}) = \sum_i \frac{\partial^2}{\partial x_i^2} f - \sum_i D_{\partial_{x_i}} \partial_{x_i} f$$

which yields the formula (1.94) since $(D_{\partial_{x_i}} \partial_{x_i})(p) = 0$ for all i.

Let $\{E_i\}$ be a frame field normal at p. Then the relation $(D_{E_i} E_i)(p) = 0$ yields, at p,

$$\Delta f = \sum_i D^2 f(E_i, E_i) = \sum_i [E_i E_i(f) - D_{E_i} E_i(f)] = \sum_i E_i E_i(f).$$

$\qquad\qquad\square$

In order to compute *the Hodge-Laplacian* $\mathbf{\Delta}$, we need

Definition 1.23 *A vector field on M is called a 1-form. Let $k \geq 2$. A k-rank tensor field $\omega \in T^k(M)$ is called a k-form if ω is antisymmetric, i.e., for X_1, $\cdots$, $X_k \in \mathcal{X}(M)$,*

$$\omega(\cdots, X_i, \cdots, X_j, \cdots) = -\omega(\cdots, X_j, \cdots, X_i, \cdots)$$

for all i, j where the variables in the position "$\cdots$" are kept the same. All the k-forms on M is denoted by $\Lambda^k(M)$. In particular, $\Lambda(M) = \mathcal{X}(M) = T(M)$.

We have two important operations: First, *the exterior product* by a vector field $X \in \mathcal{X}(M)$: $\Lambda^k(M) \to \Lambda^{k+1}(M)$,

$$\omega \longmapsto e(X)\omega = X \wedge \omega$$

where $X \wedge \omega$ is defined by

$$X \wedge \omega(X_1, \cdots, X_{k+1})$$
$$= \sum_{i=1}^{k+1}(-1)^{i+1}\langle X, X_i\rangle\omega(X_1, \cdots, \hat{X}_i, \cdots, X_{k+1}), \tag{1.96}$$

for $X_1, \cdots, X_{k+1} \in \mathcal{X}(M)$, where $\hat{X}_i$ denotes that the variable X_i does not appear. Second, *the interior product* by a vector field $X \in \mathcal{X}(M)$: $T^{k+1}(M) \to T^k(M)$,

$$T \longmapsto i(X)T,$$

$$[i(X)T](X_1, \cdots, X_k) = T(X, X_1, \cdots, X_k)$$

for all $X_1, \cdots, X_k \in \mathcal{X}(M)$. In particular, if $X, Y \in \mathcal{X}(M)$ are vector fields, then

$$i(X)Y = \langle X, Y\rangle \tag{1.97}$$

is a function on M.

Let M be an oriented n-dimensional Riemannian manifold. An n-form ω_0 on M is called *the volume element* if for any frame field $\{E_1, \cdots, E_n\}$, $|\omega_0(E_1, \cdots, E_n)| = 1$. Let $X \in \mathcal{X}(M)$ be a vector field and $\omega_0 \in \Lambda^n(M)$ be a volume element. Then there is a unique function on M, denoted by $\mathrm{div}\,X$, such that

$$(\mathrm{div}\,X)\omega_0 = d(i(X)\omega_0)$$

where d is *the exterior derivative*. The function $\mathrm{div}\,X$ is called *the divergence* of X.

Lemma 1.9 *Let $E_1, \cdots, E_n$ be a frame field. Then*

$$d = \sum_{i=1}^{n} E_i \wedge D_{E_i}. \tag{1.98}$$

Proof. Let $d_0 = \sum_{i=1}^{n} E_i \wedge D_{E_i}$. We must show $d = d_0$. Note that d_0 is independent of the choice of $\{E_i\}$: Let $Y_1, \cdots, Y_n$ be another frame field with the same domain of definition as $\{E_i\}$ and let $Y_i = \sum_{j=1}^{n} \alpha_{ij} E_j$. Then (α_{ij}) is an orthogonal matrix and $\sum_{k=1}^{n} \alpha_{ki} \alpha_{kj} = \delta_{ij}$ for all i, j. Thus

$$\sum_{k=1}^{n} Y_k \wedge D_{Y_k} = \sum_{ij=1}^{n} \sum_{k=1}^{n} \alpha_{ki} \alpha_{kj} E_i \wedge D_{E_j} = \sum_{i=1}^{n} E_i \wedge D_{E_i}.$$

Let $f \in C^\infty(M)$. Clearly, $d_0 f = \sum_{i=1}^{n} E_i(f) E_i = Df = df$.

Now fix a point $p \in M$ and let a coordinate system $(x_1, \cdots, x_n)$ be normal at p. It suffices to show that relative to this particular choice of $\{E_i\}$, $d = d_0$ holds at the point p. Since both d and d_0 are linear operators, by renumbering the indices if necessary, it suffices to show that, at p, $d = d_0$ on the k-form $f dx_1 \wedge \cdots \wedge dx_k$, where f is a C^∞ function defined near p. Since $\{x_i\}$ is normal at p, $(D_X \partial_{x_i})(p) = 0$ for all $X \in \mathcal{X}(M)$. Then the fact that $dx_i(\partial_{x_j}) = \delta_{ij}$ implies that $0 = E_k[dx_i(\partial_{x_j})] = (D_{E_k} dx_i)(\partial_{x_j}) + dx_i(D_{E_k} \partial_{x_j}) = (D_{E_k} dx_i)(\partial_{x_j})$ at p for all i, j, k. Thus

$$d_0(f dx_1 \wedge \cdots \wedge dx_k) = d_0 f \wedge dx_1 \wedge \cdots \wedge dx_k$$

$$+ f \sum_{j=1}^{n} E_j \wedge (\sum_{i=1}^{k} dx_1 \wedge \cdots \wedge D_{E_j} dx_i \wedge \cdots \wedge dx_k)$$

$$= d(f dx_1 \wedge \cdots \wedge dx_k) \tag{1.99}$$

at p. $\qquad\square$

Lemma 1.10 *Let M be an oriented n-dimensional Riemannian manifold. Let $X \in \mathcal{X}(M)$. Then*

$$\operatorname{div} X = \operatorname{tr} DX. \tag{1.100}$$

Let $p \in M$ be given. If $\{E_i\}$ is a frame field normal at p, then

$$\operatorname{div} X = \sum_{i=1}^{n} E_i \langle X, E_i \rangle \quad \text{at } p. \tag{1.101}$$

Proof. Let $p \in M$ be given. Let ω_0 be the volume element. Let $E_1, \cdots, E_n$ be a frame field normal at p with the positive orientation. Then $\langle E_i, E_j \rangle = \delta_{ij}$ around p, $(D_{E_i} E_j)(p) = 0$ for all i, j, and $\omega_0 = E_1 \wedge \cdots \wedge E_n$. It follows from (1.98) that

$$dE_k = \sum_i E_i \wedge D_{E_i} E_k = 0 \quad \text{at} \quad p \tag{1.102}$$

for all k.

Let $X = \sum_i X_i E_i$. We have

$$i(X)\omega_0(E_1, \cdots, \hat{E}_i, \cdots, E_n) = \omega_0(X, E_1, \cdots, \hat{E}_i, \cdots, E_n)$$

$$= (-1)^{i-1} \omega_0(E_1, \cdots, X, \cdots, E_n) = (-1)^{i-1} X_i$$

for all i. Then

$$\mathrm{i}\,(X)\omega_0 = \sum_{i=1}^{n}(-1)^{i-1}X_iE_1 \wedge \cdots \wedge \hat{E}_i \wedge \cdots \wedge E_n.$$

Using the formula (1.102), we obtain, at p,

$$d[\,\mathrm{i}\,(X)\omega_0] = \sum_{i=1}^{n}(-1)^{i-1}d(X_i) \wedge E_1 \wedge \cdots \wedge \hat{E}_i \wedge \cdots \wedge E_n$$

$$= \sum_{ij=1}^{n}(-1)^{i-1}E_j(X_i)E_j \wedge E_1 \wedge \cdots \wedge \hat{E}_i \wedge \cdots \wedge E_n = \sum_{i=1}^{n}E_i(X_i)\omega_0$$

which gives

$$\mathrm{div}\,X = \sum_{i=1}^{n}E_i(X_i) = \sum_{i}E_i\langle X, E_i\rangle = \sum_{i}\langle D_{E_i}X, E_i\rangle = \mathrm{tr}\,DX.$$

In particular, (1.101) holds true. $\qquad\qquad\square$

Remark 1.20 *To define the divergence for a vector field it is not necessary to assume the Riemannian manifold to be orientable. The right hand side of the formula (1.100) can be regarded as the definition of the divergence of the vector field X for a Riemannian manifold without an orientation.*

Let $p \in M$ be given. Let $T^k(M_p)$ be all k-rank tensors on M_p. We will introduce an inner product, still denoted by $\langle\cdot,\cdot\rangle$, on $T^k(M_p)$ as follows. For any $\alpha, \beta \in T^k(M_p)$,

$$\langle\alpha,\beta\rangle = \sum_{i_1\cdots i_k=1}^{n}\alpha(e_{i_1},\cdots,e_{i_k})\beta(e_{i_1},\cdots,e_{i_k}) \qquad (1.103)$$

where $\{e_i\}$ is an orthonormal basis. It is easy to check that the definition (1.103) is independent of the choice of $\{e_i\}$. Let $\Lambda^k(M_p)$ be all skew-symmetric tensors of rank k on M_p. Then a similar inner product is defined by

$$\langle\alpha,\beta\rangle = \sum_{i_1<\cdots<i_k}\alpha(e_{i_1},\cdots,e_{i_k})\beta(e_{i_1},\cdots,e_{i_k}) \qquad (1.104)$$

for $\alpha, \beta \in \Lambda^k(M)$. In particular, for $k = 1$, $T(M_p) = \Lambda(M_p) = M_p$ and $\langle\cdot,\cdot\rangle = g$.

If $T_1, T_2 \in T^k(M)$ are tensor fields on M, then

$$\langle T_1, T_2\rangle$$

is a function on M, which is defined by the formula (1.103) at each $p \in M$. Similarly, if $\alpha, \beta \in \Lambda^k(M)$ are k-forms, then the function $\langle\alpha,\beta\rangle$ is well defined at each point $p \in M$ by the formula (1.104).

Let $p \in M$ be given. Let $e_1, \cdots, e_n$ be an orthonormal basis of M_p. Then

$$e_{i_1} \wedge \cdots \wedge e_{i_k} \quad \text{with} \quad 1 \le i_1 < \cdots < i_k \le n \tag{1.105}$$

constitutes an orthonormal basis of $\Lambda^k(M_p)$ under the inner product (1.104).

Let M_p carry an orientation. We define the linear star operator $*$: $\Lambda^k(M_p) \to \Lambda^{n-k}(M_p)$ $(0 \le k \le n)$ by

$$*(e_{i_1} \wedge \cdots \wedge e_{i_k}) = e_{j_1} \wedge \cdots \wedge e_{j_{n-k}} \tag{1.106}$$

where $j_1, \cdots, j_{n-k}$ is selected such that $e_{i_1}, \cdots, e_{i_k}, e_{j_1}, \cdots, e_{j_{n-k}}$ is a positive basis of M_p. Since the star operator is supposed to be linear, it is determined by its values (1.106) on the basis (1.105).

In particular,

$$*(1) = e_1 \wedge \cdots \wedge e_n, \tag{1.107}$$

$$*(e_1 \wedge \cdots \wedge e_n) = 1, \tag{1.108}$$

if $e_1, \cdots, e_n$ is positive basis.

The Hodge-Laplacian is $\mathbf{\Delta} = d\delta + \delta d$, where δ acting on k-forms is given in terms of the star operator by

$$\delta = (-1)^{nk+n+1} * d *. \tag{1.109}$$

We need

Theorem 1.25 (a) $d^2 = 0$.
(b) $\delta^2 = 0$.

Proof. (a) Let $\omega \in \Lambda^k(M)$. Let $\{x_i\}$ be a local coordinate system. We may assume that $\omega = f dx_{i_1} \wedge \cdots \wedge dx_{i_k}$. Then

$$
\begin{aligned}
d^2\omega &= d(\sum_{i=1}^{n} f_{x_i} dx_i \wedge dx_{i_1} \wedge \cdots \wedge dx_{i_k}) \\
&= \sum_{ij=1}^{n} f_{x_i x_j} dx_j \wedge dx_i \wedge dx_{i_1} \wedge \cdots \wedge dx_{i_k} \\
&= \sum_{j<i} (f_{x_i x_j} - f_{x_j x_i}) dx_j \wedge dx_i \wedge dx_{i_1} \wedge \cdots \wedge dx_{i_k} \\
&= 0.
\end{aligned}
\tag{1.110}
$$

(b) Let $*$ be the star operator. By (1.106), $** = (-1)^{k(n-k)} I$ where I denote the identical operator. From (a) and (1.109),

$$\delta^2 = (-1)^n * d * *d* = (-1)^{n+k(n-k)} * d^2 * = 0.$$

$\square$

Lemma 1.11 *Let $E_1, \cdots, E_n$ be a frame field. Then*

$$\delta = -\sum_{j=1}^{n} \mathrm{i}\,(E_j)D_{E_j}. \tag{1.111}$$

Proof. Let $\delta_0 = -\sum_{j=1}^{n} \mathrm{i}\,(E_j)D_{E_j}$. Observe that the definition of δ_0 is independent of the particular frame field $\{E_i\}$ chosen. Indeed, if $\{Y_i\}$ is another frame field, letting $Y_i = \sum_j \alpha_{ij}E_j$ and noting that (α_{ij}) is an orthogonal matrix at each point, we have

$$-\sum_{i=1}^{n} \mathrm{i}\,(Y_i)D_{Y_i} = -\sum_{kl}\sum_{i}\alpha_{ik}\alpha_{il}\,\mathrm{i}\,(E_k)D_{E_l} = -\sum_{k}\mathrm{i}\,(E_k)D_{E_k}.$$

So to prove $\delta = \delta_0$, we may check it at a fixed $p \in M$ and choose $\{E_i\}$ to be a frame field normal at p. Let $e_1, \cdots, e_n$ be a positive basis of M_P. By Theorem 1.24, there is a frame field $\{E_i\}$ normal at p such that

$$E_i(p) = e_i \quad \text{for} \quad 1 \le i \le n.$$

Then

$$E_{i_1} \wedge \cdots \wedge E_{i_k} \quad \text{with} \quad 1 \le i_1 < \cdots < i_k \le n \tag{1.112}$$

constitutes an orthonormal basis of $\Lambda^k(M_q)$ for q in a neighborhood of p. It suffices to prove $\delta\omega = \delta_0\omega$ at p where $\omega = fE_{i_1} \wedge \cdots \wedge E_{i_k}$.

Let $j_1, \cdots, j_{n-k}$ be selected such that $E_{i_1}, \cdots, E_{i_k}, E_{j_1}, \cdots, E_{j_{n-k}}$ is a positive basis of M_q with q in a neighborhood of p. Using the formulas (1.109) and (1.102), we obtain, at p,

$$\begin{aligned}
\delta\omega &= (-1)^{nk+n+1} * d * (fE_{i_1} \wedge \cdots \wedge E_{i_k}) \\
&= (-1)^{nk+n+1} * d(fE_{j_1} \wedge \cdots \wedge E_{j_{n-k}}) \\
&= (-1)^{nk+n+1} * (df \wedge E_{j_1} \wedge \cdots \wedge E_{j_{n-k}}) \\
&= (-1)^{nk+n+1} \sum_{l=1}^{k} *[E_{i_l}(f)E_{i_l} \wedge E_{j_1} \wedge \cdots \wedge E_{j_{n-k}}] \\
&= \sum_{l=1}^{k}(-1)^{l}E_{i_l}(f)(E_{i_1} \wedge \cdots \wedge \hat{E}_{i_l} \wedge \cdots \wedge E_{i_k})
\end{aligned} \tag{1.113}$$

where $\hat{E}_{i_l}$ means that E_{i_l} does not appear in the formula.

Then at p, via the formulas (1.125) in Exercise 1.3.3 in the end of this section and (1.113),

$$
\begin{aligned}
\delta_0\omega &= -\sum_j \mathrm{i}\,(E_j)D_{E_j}\omega = -\sum_j \mathrm{i}\,(E_j)[E_j(f)E_{i_1} \wedge \cdots \wedge E_{i_k}] \\
&= \sum_{j=1}^{n}\sum_{l=1}^{k}(-1)^l E_j(f)\langle E_{i_l}, E_j\rangle E_{i_1} \wedge \cdots \wedge \hat{E}_{i_l} \wedge \cdots \wedge E_{i_k} \\
&= \sum_{l=1}^{k}(-1)^l E_{i_l}(f)E_{i_1} \wedge \cdots \wedge \hat{E}_{i_l} \wedge \cdots \wedge E_{i_k} = \delta\omega.
\end{aligned}
$$

$\square$

Theorem 1.26 *On an oriented Riemannian manifold M of dimension n, if $\{E_i\}$ is a locally defined frame field, then*

$$
\boldsymbol{\Delta} = -\sum_{i=1}^{n} D^2_{E_i E_i} + \sum_{ij=1}^{n} E_i \wedge \mathrm{i}\,(E_j)\mathbf{R}_{E_i E_j} \tag{1.114}
$$

where $D^2_{XY} = D_X D_Y - D_{D_X Y}$ is the second order covariant derivative and $\mathbf{R}_{E_i E_j}$ is the curvature operator.

Proof. Since the right hand side of the formula for $\boldsymbol{\Delta}$ is independent of the choice of the frame field $\{E_i\}$, it suffices to check this formula at a point $p \in M$ with $\{E_i\}$ chosen to be normal at p. Note that at p, the following simplifications take place (cf. (1.92), (1.93)):

$$
D^2_{E_i E_j} = D_{E_i} D_{E_j}; \tag{1.115}
$$

$$
\mathbf{R}_{E_i E_j} = -D_{E_i} D_{E_j} + D_{E_j} D_{E_i} \tag{1.116}
$$

for all i, j. Now with all the subsequent computations understood to hold only at p, we have

$$
\begin{aligned}
\delta d &= -\sum_{ij} \mathrm{i}\,(E_j)D_{E_j}[E_i \wedge D_{E_i}] = -\sum_{ij} \mathrm{i}\,(E_j)[E_i \wedge D_{E_j}D_{E_i}]\text{(by (1.92))} \\
&= -\sum_{i} D_{E_i}D_{E_i} + \sum_{ij} E_i \wedge \mathrm{i}\,(E_j)D_{E_j}D_{E_i}\text{(by (1.124))} \\
&= -\sum_{i} D^2_{E_i E_i} + \sum_{ij} E_i \wedge \mathrm{i}\,(E_j)D_{E_j}D_{E_i} \tag{1.117}
\end{aligned}
$$

where we have used (1.115).

Using the formula (1.125) in Exercise 1.3.3 later and the formula (1.92), we have that, at p, the identity

$$
\mathrm{i}\,(E_k)D_{E_j} = D_{E_j}\mathrm{i}\,(E_k) \tag{1.118}
$$

is valid on forms for all k, j. It follows from (1.118) that, at p,

$$d\delta = -\sum_{ij} E_i \wedge D_{E_i}[\mathrm{i}\,(E_j)D_{E_j}] = -\sum_{ij} E_i \wedge \mathrm{i}\,(E_j)D_{E_i}D_{E_j}. \qquad (1.119)$$

So,

$$\boldsymbol{\Delta} = \delta d + d\delta = -\sum_i D^2_{E_i E_i} + \sum_{ij} E_i \wedge \mathrm{i}\,(E_j)\mathbf{R}_{E_i E_j}$$

by (1.116). $\qquad\qquad\qquad\qquad\qquad\qquad\qquad\qquad\qquad\qquad\quad\square$

We get the minus Laplacian $-\Delta f = \delta df = \boldsymbol{\Delta} f$ if we apply the Hodge-Laplacian to functions.

We will need the following formula in Chapter 3 later, which is called *the Weitzenböck formula.*

Theorem 1.27 *Let X, Y be vector fields, or equivalently 1-forms. Then*

$$\langle \boldsymbol{\Delta} X, Y \rangle + \langle X, \boldsymbol{\Delta} Y \rangle + \Delta \langle X, Y \rangle = 2\langle DX, DY \rangle + 2\,\mathrm{Ric}\,(X, Y) \qquad (1.120)$$

where Δ is the Laplacian on M and $\boldsymbol{\Delta}$ is the Hodge-Laplacian and Ric is the Ricci tensor. In the above formula, $\langle DX, DY \rangle$ is defined at each $p \in M$ as the inner product of two tensors of rank 2 on M_p by the formula (1.103).

Proof. Let $p \in M$ be given. Let $\{E_i\}$ be a frame field normal at p. By (1.114) and (1.97), we have

$$\begin{aligned}
\boldsymbol{\Delta} X &= -\sum_i D^2_{E_i E_i} X + \sum_{ij} \mathbf{R}(E_i, E_j, X, E_j)E_i \\
&= -\sum_i D^2_{E_i E_i} X + \sum_i \mathrm{Ric}\,(E_i, X)E_i. \qquad (1.121)
\end{aligned}$$

Let us compute $\langle D^2_{E_i E_i} X, Y \rangle$ at p for each i. The relation $(D_{E_i} E_i)(p) = 0$ implies that, at p,

$$\begin{aligned}
\langle D^2_{E_i E_i} X, Y \rangle &= \langle D_{E_i} D_{E_i} X, Y \rangle = E_i \langle D_{E_i} X, Y \rangle - \langle D_{E_i} X, D_{E_i} Y \rangle \\
&= E_i[E_i \langle X, Y \rangle - \langle X, D_{E_i} Y \rangle] - \langle D_{E_i} X, D_{E_i} Y \rangle \\
&= E_i E_i \langle X, Y \rangle - 2\langle D_{E_i} X, D_{E_i} Y \rangle - \langle X, D_{E_i} D_{E_i} Y \rangle \\
&= E_i E_i \langle X, Y \rangle - 2\langle D_{E_i} X, D_{E_i} Y \rangle - \langle X, D^2_{E_i E_i} Y \rangle \qquad (1.122)
\end{aligned}$$

for all i. On the other hand, by (1.103) with $k = 2$,

$$\begin{aligned}
\sum_i \langle D_{E_i} X, D_{E_i} Y \rangle &= \sum_{ij} \langle D_{E_i} X, E_j \rangle \langle E_j, D_{E_i} Y \rangle \\
&= \sum_{ij} DX(E_j, E_i)DY(E_j, E_i) = \langle DX, DY \rangle. \qquad (1.123)
\end{aligned}$$

It follows from (1.121), (1.122), (1.123) and (1.95) that

$$\langle \mathbf{\Delta} X, Y \rangle = -\sum_i \langle D^2_{E_i E_i} X, Y \rangle + \mathrm{Ric}\,(X, Y)$$

$$= -\sum_i E_i E_i \langle X, Y \rangle + 2\langle DX, DY \rangle + \langle X, \sum_i D^2_{E_i E_i} Y \rangle + \mathrm{Ric}\,(X, Y)$$

$$= -\Delta \langle X, Y \rangle + 2\langle DX, DY \rangle + 2\,\mathrm{Ric}\,(X, Y) - \langle X, \mathbf{\Delta} Y \rangle$$

at p where the formula (1.121) is again used with X replaced by Y.   □

Exercise 1.3

1.3.1 *Let $X_i \in X(M)$ be vector fields, or equivalently, 1-forms for $1 \le i \le k$. Then $X_1 \wedge \cdots \wedge X_k \in \Lambda^k(M)$ is given by*

$$X_1 \wedge \cdots \wedge X_k(Y_1, \cdots, Y_k) = \det(\langle X_i, Y_j \rangle)$$

for any $Y_1, \cdots, Y_k \in \mathcal{X}(M)$.

1.3.2 *Let X, Y be vector fields and let ω be a form. Then*

$$\mathrm{i}(X)[Y \wedge \omega] = \langle X, Y \rangle \omega - Y \wedge \mathrm{i}(X)\omega. \qquad (1.124)$$

1.3.3 *Let $X, X_1, \cdots, X_k \in \mathcal{X}(M)$ be $k+1$ vector fields. Then*

$$\mathrm{i}(X)(X_1 \wedge \cdots \wedge X_k) = \sum_{j=1}^k (-1)^{j+1} \langle X, X_j \rangle X_1 \wedge \cdots \wedge \hat{X}_j \wedge \cdots \wedge X_k \quad (1.125)$$

where $\hat{X}_j$ means that X_j does not appear in the formula.

1.3.4 *Let X, Y be vector fields and let ω be a form. Then*

$$\mathrm{i}(X)D_Y \omega = D_Y \mathrm{i}(X)\omega - \mathrm{i}(D_Y X)\omega. \qquad (1.126)$$

1.3.5 *Let $\{E_1, \cdots, E_n\}$ be a frame field. Then*
(a) $Df = \sum_i E_i(f)E_i$ where $f \in C^1(M)$ is a function;
(b) $\mathrm{div}\, X = \sum_i \langle D_{E_i} X, E_i \rangle$ where $X \in \mathcal{X}(M)$ is a vector field.

1.3.6 *Prove that the right hand side of the formula (1.114) is independent of the choice of the frame field $\{E_i\}$.*

1.4 Sobolev Spaces of Tensor Field and Some Basic Differential Operators

Sobolev spaces of tensor fields are necessary to apply. To deal with them, we need the concept of *the integral in the metric*.

Integral in Metric Let (M, g) be a Riemannian manifold. Let $\mathcal{U} \subset M$ be a coordinate neighborhood with a coordinate system $\varphi = \{x_i\}: \mathcal{U} \to \mathbb{R}^n$. Then the metric g can be expressed as

$$g = \sum_{ij=1}^{n} g_{ij} dx_i \otimes dx_j.$$

Let $G = \det(g_{ij})$. Then $G(q) > 0$ for $q \in \mathcal{U}$. Let $f \in C(M)$ be such that $\operatorname{supp} f \subset \mathcal{U}$. We define

$$\int_{\mathcal{U}} f dg = \int_{\mathbb{R}^n} f \circ \varphi^{-1}(x) \sqrt{G \circ \varphi^{-1}(x)} dx_1 \cdots dx_n \qquad (1.127)$$

with the right hand side defined as normal n-integrals in $\mathbb{R}^n$.

Let $\Omega \subset M$ be an open set. We will use (1.127) and *a partition of unity* to define $\int_{\Omega} f dg$. Let $\{(\mathcal{U}_i, \varphi_i)\}$ be a locally finite coordinate covering of Ω. Let $\{\phi_i\}$ be a partition of unity subordinate to $\{\mathcal{U}_i\}$. We define

$$\int_{\Omega} f dg = \sum_i \int_{\Omega \cap \mathcal{U}_i} f \phi_i dg. \qquad (1.128)$$

It is easy to check that the above integral is well defined (Exercise 1.4.1).

Remark 1.21 *For the integral* (1.128) *to make sense, it is not necessary for the Riemannian manifold M to be orientable.*

For a Riemannian manifold with an orientation, we have

Lemma 1.12 *Let M be an oriented Riemannian manifold and let ω_0 be a volume element. Then for $f \in C(M)$*

$$\int_M f dg = \pm \int_M f \omega_0. \qquad (1.129)$$

Proof. Clearly, it suffices to prove (1.129) holds for any coordinate neighborhood $(\mathcal{U}, \{x_i\})$ and $\operatorname{supp} f \subset \mathcal{U}$. There is a function h on $\mathcal{U}$ such that

$$\omega_0 = h dx_1 \wedge \cdots \wedge dx_n \quad \text{on} \quad \mathcal{U} \qquad (1.130)$$

since $\Lambda^n(M)$ is a space of dimension 1. On the other hand, since ω_0 is of unit,

$$\omega_0 = \pm \frac{\partial_{x_1} \wedge \cdots \wedge \partial_{x_n}}{|\partial_{x_1} \wedge \cdots \wedge \partial_{x_n}|} \qquad (1.131)$$

where $\pm$ correspond to if $\{x_i\}$ is positively orientated or not and $|\partial_{x_1} \wedge \cdots \wedge \partial_{x_n}| = \sqrt{\det(\langle \partial_{x_i}, \partial_{x_j} \rangle)}$. It follows from (1.130) and (1.131) that

$$h = \omega_0(\partial_{x_1}, \cdots, \partial_{x_n}) = \pm |\partial_{x_1} \wedge \cdots \wedge \partial_{x_n}| = \pm\sqrt{G}.$$

Then

$$\int_{\mathcal{U}} f\omega_0 = \int_{\mathcal{U}} fh dx_1 \cdots dx_n = \pm \int_{\mathcal{U}} f\sqrt{G} dx_1 \cdots dx_n = \pm \int_{\mathcal{U}} f dg.$$

$\square$

Theorem 1.28 (*Green's formula*) *Let (M, g) be an orientated Riemannian manifold. Let $\Omega \subset M$ be a bounded, open region with a regular boundary Γ or without boundary (when Γ is empty). Let ν be the outside normal vector field along Γ. Then for a vector field $X \in \mathcal{X}(M)$*

$$\int_{\Omega} \operatorname{div} X \, dg = \int_{\Gamma} \langle X, \nu \rangle d\Gamma \qquad (1.132)$$

where $d\Gamma$ denotes the volume element of Γ with an induced metric from (M, g).

Proof. Let ω_0 be the volume element on M. Let $p \in \Gamma$ be given and let $e_1, \cdots, e_n$ be a positive orthonormal basis of M_p such that $e_1 = \nu$. Then

$$\omega_0 = e_1 \wedge \cdots \wedge e_n.$$

The volume element of Γ at p is given by

$$\omega_\Gamma = e_2 \wedge \cdots \wedge e_n.$$

Noting that, by (1.125) at p,

$$\begin{aligned}
i(X)\omega_0 &= i(X)(e_1 \wedge \cdots \wedge e_n) \\
&= \langle X, e_1 \rangle e_2 \wedge \cdots \wedge e_n + \text{terms including } e_1 \\
&= \langle X, \nu \rangle \omega_\Gamma,
\end{aligned}$$

we have, by Stokes' Theorem,

$$\int_{\Omega} \operatorname{div} X dg = \int_{\Omega} \operatorname{div} X \omega_0 = \int_{\Omega} d(i(X)\omega_0) = \int_{\Gamma} i(X)\omega_0 = \int_{\Gamma} \langle X, \nu \rangle d\Gamma.$$

$\square$

Sobolev Spaces of Tensor Field Let (M, g) be a Riemannian manifold. Let $\Omega \subset M$ be a bounded, open region with a regular boundary Γ or without boundary (when Γ is empty). We introduce an inner product on $T^k(\Omega)$ by

$$(T_1, T_2)_{T^k(\Omega)} = \int_{\Omega} \langle T_1, T_2 \rangle \, dg, \quad \text{for} \quad T_1, T_2 \in T^k(\Omega), \qquad (1.133)$$

where $\langle T_1, T_2 \rangle$ is defined by (1.103) for each $p \in M$. The completion of $T^k(\Omega)$ in the inner product (1.133) is denoted by $L^2(\Omega, T^k)$. Let $L^2(\Omega)$ be the completion of $C^\infty(\Omega)$ in the following inner product

$$(f, h) = \int_\Omega f(x)h(x)\, dg \quad f,\, h \in C^\infty(\Omega). \tag{1.134}$$

Similarly, for $k \geq 2$, we introduce an inner product on $\Lambda^k(\Omega)$ by

$$(\varphi, \phi)_{\Lambda^k(\Omega)} = \int_\Omega \langle \varphi, \phi \rangle\, dg \quad \text{for} \quad \varphi,\, \phi \in \Lambda^k(\Omega), \tag{1.135}$$

where $\langle \varphi, \phi \rangle$ is defined by (1.104) for each $p \in M$. Its completion is denoted by $L^2(\Omega, \Lambda^k)$. We have

$$L^2(\Omega, \Lambda) = L^2(\Omega, T).$$

Let the Sobolev space $H^k(\Omega)$ be the completion of $C^\infty(\Omega)$ with respect to the norm

$$\|f\|_{H^k(\Omega)}^2 = \sum_{i=1}^k \|D^i f\|_{L^2(\Omega, T^i)}^2 + \|f\|^2 \quad \text{for} \quad f \in C^\infty(\Omega),$$

where $D^i f$ is the i-th covariant differential of f in the metric g. Let $\|\cdot\|_{L^2(\Omega, T^k)}$ and $\|\cdot\|_{L^2(\Omega)}$ be the induced norms in the inner product (1.133) and the inner product (1.134), respectively. For details on Sobolev spaces on Riemannian manifolds, we refer to [85] or [194].

Another important Sobolev space for us is $H^k(\Omega, \Lambda)$, to be defined by

$$H^k(\Omega, \Lambda) = \{\, U \,|\, U \in L^2(\Omega, \Lambda),\ D^i U \in L^2(\Omega, T^{i+1}),\ 1 \leq i \leq k \,\}$$

with an inner product

$$(U, V)_{H^k(\Omega, \Lambda)} = \sum_{i=0}^k (D^i U, D^i V)_{L^2(\Omega, T^{i+1})}, \quad \forall\, U,\, V \in H^k(\Omega, \Lambda).$$

For details, see [204]. In particular, $H^0(\Omega, \Lambda) = L^2(\Omega, \Lambda)$.

Theorem 1.29 *Let M be an orientated Riemannian manifold. Then*

$$(d\alpha, \beta)_{L^2(\Omega, \Lambda^2)} = (\alpha, \delta\beta)_{L^2(\Omega, \Lambda)} + \int_\Gamma \langle \nu \wedge \alpha, \beta \rangle\, d\Gamma, \tag{1.136}$$

for $\alpha \in \Lambda(\Omega)$, $\beta \in \Lambda^2(\Omega)$. Moreover,

$$(\alpha, d\beta)_{L^2(\Omega, \Lambda)} = (\delta\alpha, \beta) + \int_\Gamma \beta \langle \alpha, \nu \rangle\, d\Gamma, \tag{1.137}$$

for $\alpha \in \Lambda(\Omega)$, $\beta \in C^\infty(\Omega)$.

Proof. Let $\alpha \in \Lambda(\Omega)$ and $\beta \in \Lambda^2(\Omega)$. Let $p \in \Omega$ be given. Let $E_1, \cdots, E_n$ be a frame field normal at p. Then

$$(D_{E_i} E_j)(p) = 0 \quad \text{for all } i,j. \tag{1.138}$$

Let $\alpha = \sum_i a_i E_i$ and $\beta = \sum_{i<j} b_{ij} E_i \wedge E_j$. We compute at p. By (1.98), (1.138), at p

$$
\begin{aligned}
d\alpha &= \sum_{ij} E_i \wedge D_{E_i}(a_j E_j) = \sum_{ij} E_i(a_j) E_i \wedge E_j \\
&= \sum_{i<j} [E_i(a_j) - E_j(a_i)] E_i \wedge E_j,
\end{aligned}
$$

which yields, by (1.104) at p

$$
\begin{aligned}
\langle d\alpha, \beta \rangle &= \sum_{i<j} [E_i(a_j) - E_j(a_i)] b_{ij} = \sum_{i<j} [E_i(a_j b_{ij}) - E_j(a_i b_{ij})] \\
&\quad + \sum_{i<j} [a_i E_j(b_{ij}) - a_j E_i(b_{ij})]. \tag{1.139}
\end{aligned}
$$

On the other hand, at p, by (1.111), (1.138) and (1.124),

$$\delta\beta = -\sum_{i<j}\sum_k E_k(b_{ij})\, \mathrm{i}\,(E_k) E_i \wedge E_j = \sum_{i<j}[E_j(b_{ij})E_i - E_i(b_{ij})E_j]$$

which gives, by (1.139),

$$\langle \alpha, \delta\beta \rangle = \sum_{i<j}[a_i E_j(b_{ij}) - a_j E_i(b_{ij})] = \langle d\alpha, \beta \rangle - \lambda \tag{1.140}$$

where

$$\lambda = \sum_{i<j}[E_i(a_j b_{ij}) - E_j(a_i b_{ij})].$$

Noting that $b_{ij} = \beta(E_i, E_j) = -\beta(E_j, E_i) = -b_{ji}$, we have

$$
\begin{aligned}
\sum_{i\geq j}[E_i(a_j b_{ij}) - E_j(a_i b_{ij})] &= \sum_{i<j}[E_j(a_i b_{ji}) - E_i(a_j b_{ji})] \\
&= \sum_{i<j}[E_i(a_j b_{ij}) - E_j(a_i b_{ij})] = \lambda.
\end{aligned}
$$

Thus, we obtain, at p by (1.101),

$$
\begin{aligned}
2\lambda &= \sum_{ij=1}^n [E_i(a_j b_{ij}) - E_j(a_i b_{ij})] = \sum_i [E_i(\beta(E_i, \alpha)) - E_i(\beta(\alpha, E_i))] \\
&= -2\sum_i E_i(\beta(\alpha, E_i)) = -2\,\mathrm{div}\,\mathrm{i}\,(\alpha)\beta. \tag{1.141}
\end{aligned}
$$

We use (1.141) in (1.140) and integrate it over Ω to have, by (1.132),

$$(d\alpha, \beta)_{L^2(\Omega, \Lambda^2)} = (\alpha, \delta\beta)_{L^2(\Omega, \Lambda)} + \int_\Gamma \beta(\nu, \alpha)d\Gamma. \tag{1.142}$$

For $p \in \Gamma$, let $\{e_i\}$ be an orthonormal basis of M_p such that $e_1 = \nu$. Then on Γ

$$\begin{aligned}
\beta(\nu, \alpha) &= \sum_{i<j} \langle \beta, e_i \wedge e_j \rangle e_i \wedge e_j(\nu, \alpha) = \sum_{1<j} \langle \beta, \nu \wedge e_j \rangle \nu \wedge e_j(\nu, \alpha) \\
&= \sum_{1<j} \langle \beta, \nu \wedge e_j \rangle \langle \alpha, e_j \rangle = \langle \beta, \nu \wedge \alpha \rangle.
\end{aligned} \tag{1.143}$$

Then the formula (1.136) follows from (1.142) and (1.143).

A similar argument shows that the formula (1.137) holds. We left its proof to the reader (Exercise 1.4.2). $\qquad\square$

Let $T \in T^2(M)$ be a tensor field of rank 2. We define $T^* \in T^2(M)$ by

$$T^*(X, Y) = T(Y, X) \quad \text{for} \quad X, Y \in \mathcal{X}(M).$$

T^* is called *the transpose of tensor field T of 2 rank.*

We define an operator $\mathcal{Q}: T^2(\Omega) \to \mathcal{X}(\Omega)$ as follows. For any $T \in T^2(\Omega)$, since $\operatorname{tr} i(X)DT$ is a linear functional in the variable X, there is a unique $\mathcal{Q}T \in \mathcal{X}(\Omega)$ such that

$$\langle X, \mathcal{Q}T \rangle = \operatorname{tr} i(X)DT \quad \text{for} \quad X \in \mathcal{X}(\Omega). \tag{1.144}$$

The following result shows that the operator $\mathcal{Q}$ is *the formal adjoint of the covariant differential operator* $D: \mathcal{X}(\Omega) \to T^2(\Omega)$.

Theorem 1.30 *For $T \in T^2(\Omega)$ and $U \in \mathcal{X}(\Omega)$,*

$$(T, DU)_{L^2(\Omega, T^2)} = (-\mathcal{Q}T, U)_{L^2(\Omega, \Lambda)} + \int_\Gamma \langle i(\nu)T^*, U \rangle \, d\Gamma, \tag{1.145}$$

where T^* is the transpose of T and $i(\nu)T^*$ is the interior product of T^* by ν which is a vector field on Γ, given by

$$\langle i(\nu)T^*, Z \rangle = T^*(\nu, Z) \quad \text{for all} \quad Z \in \mathcal{X}(\Gamma). \tag{1.146}$$

Proof. We define a vector field Y on $\overline{\Omega}$ as follows. Let $p \in \overline{\Omega}$ be given. Let $E_1, \cdots, E_n$ be a frame field in a neighborhood of p. Set

$$Y = \sum_{j=1}^n \langle i(E_j)T^*, U \rangle E_j. \tag{1.147}$$

It is easily checked that the right hand side of (1.147) is independent of the choice of $\{E_i\}$. So we can assume that $\{E_i\}$ is normal at p, that is,

$$(D_{E_i}E_j)(p) = 0 \quad \text{for all } i, j. \tag{1.148}$$

Using the formulas (1.103), (1.148) and (1.100), we compute, at p,

$$
\begin{aligned}
\langle T, DU \rangle &= \sum_{ij=1}^n T(E_i, E_j)DU(E_i, E_j) = \sum_{ij=1}^n T(E_i, E_j)E_j(U(E_i)) \\
&= \sum_{ij=1}^n E_j(T(E_i, E_j)U(E_i)) - \sum_{ij=1}^n DT(E_i, E_j, E_j)U(E_i) \\
&= \sum_{j=1}^n E_j \langle \mathrm{i}\,(E_j)T^*, U \rangle - \sum_{i=1}^n U(E_i)\,\mathrm{tr}\,\mathrm{i}\,(E_i)DT \\
&= \mathrm{div}\,Y - \mathrm{tr}\,\mathrm{i}\,(U)DT. \tag{1.149}
\end{aligned}
$$

From the formulas (1.149), (1.132), (1.144), and (1.147), we obtain

$$
\begin{aligned}
(T, DU)_{L^2(\Omega, T^2)} &= \int_\Omega \langle T, DU \rangle \, dx = \int_\Gamma \langle Y, \nu \rangle \, d\Gamma - \int_\Omega \mathrm{tr}\,\mathrm{i}\,(U)DT \, dx \\
&= (-QT, U)_{L^2(\Omega, \Lambda)} + \int_\Gamma \langle \mathrm{i}\,(\nu)T^*, U \rangle \, d\Gamma.
\end{aligned}
$$

$\square$

We have

Theorem 1.31 *Let X be a vector field (or equivalently 1-form). Then*

$$QDX = -\mathbf{\Delta}X + \mathrm{i}\,(X)\,\mathrm{Ric}, \tag{1.150}$$

$$QD^*X = -d\delta X + \mathrm{i}\,(X)\,\mathrm{Ric}, \tag{1.151}$$

where D^*X denotes the transpose of DX and Ric is the Ricci tensor.

Proof. Let $p \in \Omega$ be given. Let $E_1, \cdots, E_n$ be a frame field normal at p. Let $\mathbf{R}_{XY}$ be the curvature operator. Then

$$
\begin{aligned}
\sum_{kl} E_k \wedge \mathrm{i}\,(E_l)\mathbf{R}_{E_k E_l}X &= \sum_{kl} E_k \wedge \langle \mathbf{R}_{E_k E_l}X, E_l \rangle \\
&= \sum_{kl} \mathbf{R}(E_k, E_l, X, E_l)E_k = \sum_k \mathrm{Ric}\,(E_k, X)E_k = \mathrm{i}\,(X)\,\mathrm{Ric} \tag{1.152}
\end{aligned}
$$

since Ric is symmetric. It follows from formulas (1.144), (1.148), (1.114), and

(1.152) that, at p,

$$
\begin{aligned}
\langle QDX, E_i \rangle &= \operatorname{tr} \mathrm{i}(E_i) D^2 X = \sum_{j=1}^{n} D^2 X(E_i, E_j, E_j) = \sum_{j=1}^{n} E_j (DX(E_i, E_j)) \\
&= \sum_{j=1}^{n} E_j (D_{E_j} X(E_i)) = \sum_{j=1}^{n} D_{E_j} D_{E_j} X(E_i) = \sum_{j} (D^2_{E_j E_j} X)(E_i) \\
&= -(\mathbf{\Delta} X)(E_i) + [\sum_{kl} E_k \wedge \mathrm{i}(E_l) \mathbf{R}_{E_k E_l} X](E_i) \\
&= -(\mathbf{\Delta} X)(E_i) + [\mathrm{i}(X) \operatorname{Ric}](E_i)
\end{aligned}
$$

for all i, that is, the identity (1.150).

To obtain (1.151), we compute, at p, via Theorem 1.10,

$$
\begin{aligned}
\langle D_{E_i} D_{E_j} X, E_j \rangle &= E_i (DX(E_j, E_j)) = D^2 X(E_j, E_j, E_i) \\
&= D^2 X(E_j, E_i, E_j) + \mathbf{R}_{E_j E_i} X(E_j) \\
&= D^2 X(E_j, E_i, E_j) - \mathbf{R}(E_i, E_j, X, E_j) \qquad (1.153)
\end{aligned}
$$

for all i, j, where the formulas (1.148) have been used. It follows from (1.119) and (1.153) that, at p,

$$
\begin{aligned}
d\delta X &= -\sum_{ij=1}^{n} E_i \wedge \mathrm{i}(E_j) D_{E_i} D_{E_j} X = -\sum_{ij=1}^{n} \langle D_{E_i} D_{E_j} X, E_j \rangle E_i \\
&= -\sum_{ij} D^2 X(E_j, E_i, E_j) E_i + \mathrm{i}(X) \operatorname{Ric}. \qquad (1.154)
\end{aligned}
$$

Using (1.144), (1.148), and (1.154), we obtain, at p,

$$
\begin{aligned}
QD^* X &= \sum_{j=1}^{n} (\operatorname{tr} \mathrm{i}(E_j) DD^* X) E_j = \sum_{ij=1}^{n} DD^* X(E_j, E_i, E_i) E_j \\
&= \sum_{ij=1}^{n} E_i (DX(E_i, E_j)) E_j = \sum_{ij=1}^{n} D^2 X(E_i, E_j, E_i) E_j \\
&= -d\delta X + \mathrm{i}(X) \operatorname{Ric}.
\end{aligned}
$$

$\square$

Theorem 1.32 *For $V, U \in \Lambda(\Omega)$, we have*

$$
(DV, DU)_{L^2(\Omega, T^2)} = (\mathbf{\Delta} V - \mathrm{i}(V) \operatorname{Ric}, U)_{L^2(\Omega, \Lambda)} + \int_{\Gamma} \langle D_\nu V, U \rangle \, d\Gamma, \quad (1.155)
$$

$$
(D^* V, DU)_{L^2(\Omega, T^2)} = (d\delta V - \mathrm{i}(V) \operatorname{Ric}, U)_{L^2(\Omega, \Lambda)} + \int_{\Gamma} \langle D_U V, \nu \rangle \, d\Gamma. \quad (1.156)
$$

Proof. Observe that on Γ

$$\langle i(\nu)D^*V, U \rangle = DV(U, \nu) = \langle D_\nu V, U \rangle,$$

we obtain the formula (1.155) from the formula (1.150) after we take $T = DV$ in the formula (1.145).

Next, letting $T = D^*V$ in the formula (1.145), we have (1.156) by (1.151) since $\langle i(\nu)DV, U \rangle = \langle D_U V, \nu \rangle$ on Γ. $\square$

Exercise 1.4

1.4.1 *Prove that the definition (1.128) is independent of the choice of the coordinate covering of Ω and the partition of unity.*

1.4.2 *Prove the formula (1.137).*

Chapter 2

Control of the Wave Equation with Variable Coefficients in Space

First we present a brief comparison between several methods which have been used in the literature to show why Riemannian geometry is necessary for control of the wave equation with variable coefficients. Then we generalize the classical multipliers ([39], [40], [41], [86], [99], [104], [105], [117]-[122], [146], and many others) in a version of the Riemannian geometry setting with help of the Bochner technique to find out verifiable, geometric assumptions for control of the wave equation with variable coefficients in space. Controllability/stabilization of the wave equation with variable coefficients comes down to *an escape vector field for the Riemannian metric*, given by the variable coefficients.

2.1 How to Understand Riemannian Geometry as a Necessary Tool for Control of the Wave Equation with Variable Coefficients

The exact controllability of the wave equation with variable coefficients has been a difficult topic for almost 50 years. There were many papers which changed the controllability into some uncheckable assumptions to claim the problem solved. This emphasis seems to us misleading, as checkability is the most important criterion in judging the quality of a result. The reason that these assumptions are uncheckable is because controllability is a global property and the classical analysis works well for local problems only and is insufficient to cope with global problems. Here, we will briefly compare the Riemannian geometrical approach with some other main methods to show why it is a necessary tool to overcome the above defects.

Exact Controllability Let $A(x)$ be a symmetric, positive matrix for each $x \in \mathbb{R}^n$ with smooth elements. Let $\Omega \subset \mathbb{R}^n$ be an open, bounded set with a smooth boundary $\Gamma = \Gamma_0 \cup \Gamma_1$. Let $T > 0$ be given. Consider the problem

$$\begin{cases} u_{tt} = \operatorname{div} A(x)\nabla u & \text{in} \quad (0,T) \times \Omega, \\ u = 0 \quad \text{on} \quad (0,T) \times \Gamma_1, \quad u = \varphi \quad \text{on} \quad (0,T) \times \Gamma_0, \\ u(0,x) = u_0(x), \quad u_t(0,x) = u_1(x) \quad \text{on} \quad \Omega. \end{cases} \quad (2.1)$$

We say the problem (2.1) is *exactly controllable* on the time interval $[0,T]$ if we can find $T > 0$ such that, for any (u_0, u_1), $(\hat{u}_0, \hat{u}_1)$ given, there is a boundary function φ on $(0,T) \times \Gamma_0$ such that the solution u to the problem (2.1) satisfies

$$u(T,x) = \hat{u}_0(x), \quad u_t(T,x) = \hat{u}_1(x) \quad \text{for} \quad x \in \Omega. \quad (2.2)$$

The problem (2.1) is called *of constant coefficient* if $A(x)$ is a constant matrix for all $x \in \mathbb{R}^n$; otherwise, it is said to be *a variable coefficient problem*.

Extension Method Russell [180], [181], and [182] obtained exact controllability in the case of constant coefficients by extending the problem (2.1) to be a Cauchy problem of the wave equation on the whole space $\mathbb{R}^n$. This method was also used in [148], [149]. They constructed boundary controls in terms of the solution formula of the Cauchy problem. However, this method does not work for the variable coefficient problem because the solution formula of the Cauchy problem is no longer useful as in the case of constant coefficients.

Duality Method The concept of duality between control and observation was first noted by Kalman who gave a duality theorem in [92] (also see [102]) for linear systems and was generalized to distributed parameter systems by [58] and the references there: *Exact controllability is equivalent to an observability inequality in an operator theory form* (see Theorem 2.10). However, they did not apply it to the particular problem (2.1)(now we can apply their theorems to obtain the particular observability estimate of the problem (2.1); see Theorem 2.11).

Lions [146], [147] clarified the problem (2.1), worked out the particular observability inequality, and obtained constructive boundary controls (see Theorem 2.14). Since then, instead of looking for a boundary control directly, one has struggled on proving the observability estimate. However, it was not easy to prove. It was first proved by [86] that the observability inequality is true in the case of constant coefficients by the multiplier methods. The multiplier methods have been used by [39], [40], [41], [104], [105], [99],[117]-[122], [146], [155], [156], and many others. But in the classical analysis the multiplier methods do not work for the variable coefficient problems. At that time the observability inequality for the wave equation with variable coefficients was not checkable and, thus, Lions [146] put forward his open question: *Is the observability estimate true for the variable coefficient problem?*

Geometric Optics Method A systematic attempt to understand the wave equation controllability and non-controllability properties can be found in [174], [175] where the author proved that to have exact controllability the so-called Geometric Control Condition (GCC) needs to be satisfied. Later on it was proved by [8] that in fact GCC is also sufficient to assure exact controllability. GCC says that *the problem* (2.1) *is exactly controllable if and only if any geodesic initiated from the domain must hit the control portion* Γ_0 *after a finite time.* This condition provided us with a visible description of controllability.

Although it is useful to help us understand the problem, whether GCC is an answer to our question depends on whether it is checkable. In the case of constant coefficients, a geodesic is a straight line. Since the domain is bounded, a straight line from the domain must hit Γ_0 if Γ_0 is choosen appropriately large so that GCC is true. But in the variable coefficient case, *a geodesic is a solution to a second order system of nonlinear ordinary differential equations* (see (1.12)). It is almost impossible to verify GCC from its definition. Then in this sense GCC just changed a problem of linear PDEs into the one of *nonlinear* ODEs. Thus Lions' question was still open after GCC.

Operator Theory Method [207] introduced an assumption for the stabilization of the wave equation with variable coefficients. By Russell's principle this assumption is sufficient to guarantee the exact controllability. However, there was no way to check this assumption until [64] proved that it is equivalent to the one given by [208].

Pseudodifferential Multiplier Method General pseudodifferential multipliers were first derived from pseudoconvex functions [87] where these techniques were applied to solutions with compact support. However, the pseudodifferential multiplier methods in [87] did not account for boundary traces (which are critical for continuous observability inequalities). The techniques with pseudodifferential Carleman multipliers which did account for the boundary traces were first proposed in [192]. However, [192] requires the existence of a pseudoconvex function, a property which essentially can be verified mostly, if not exclusively, in the case of constant coefficients.

Differential Geometric Method Differential Geometric Methods were introduced by [208] where the original motivation was to give *checkable conditions* to the exact controllability of the wave equation with variable coefficients to answer Lions' open question [146]. They were extended by [25], [30], [130], [131], [201], [209]-[211], [216], [215], [218], and many others.

There are two great ingredients in Riemannian geometry that play the key role in extending the multipliers in [86] to the case of variable coefficients: One is the Bochner technique(Section 1.3) which provides us with a great computational tool to carry out the multiplier scheme. The other one is the curvature theory which yields the global information for our assumptions.

The basic ideas of Bochner technique(Section 1.3) are as follows: In order to verify an identity or a pointwise estimate, it suffices to do so at each point p relative to a coordinate system or frame field that offers the greatest computational simplification. What prevents a computation from being as simple as the corresponding classical (i.e., Euclidean) situation is the presence of the Christoffel symbols Γ_{ij}^{k}. Thus we use a coordinate system or a frame field relative to which the Γ_{ij}^{k} vanish at the given point p.

The second great ingredient, the curvature theory, yields the global information for our assumptions: The multiplier techniques lead to exact controllability to assumption (**A**): There exists *an escape vector field on the domain for the metric* $g = A^{-1}(x)$. One of the useful consequences is that a convex function yields an escape vector field. Now the key question is whether this assumption is checkable. The classical analysis can only tell us that assumption (**A**) is locally true whenever the domain of the wave equation is small enough. However, the Riemannian curvature theory provides us with an effective tool to check assumption (**A**)(see Theorem 2.6 and examples in Section 2.3). Although the curvature is locally defined at each point, it interprets global information on almost everything in the metric, including assumption (**A**). That is one of the reasons why Riemannian geometry is called global geometry. Moreover, as far as the author is aware, up to now, the curvature theory has been the only way to check exact controllability for the variable coefficient problems.

Recently, observability estimates were derived by [60] under an assumption (**H**): There are a function d and a constant $c > 0$ such that

$$\sum_{ijkl}[2a_{il}(a_{kj}d_{x_k})_{x_l} - a_{ijx_l}a_{kl}d_{x_k}]\xi_i\xi_j \geq c\sum_{ij}a_{ij}\xi_i\xi_j \quad \text{for} \quad (x,\xi) \in \overline{\Omega} \times \mathbb{R}^n,$$

where $A(x) = (a_{ij}(x))$ is the coefficient matrix. One would expect that assumption (**H**) represented a new assumption. However, it was not the case because this assumption is equivalent to that d is a convex function(see Proposition 1.1). Therefore it is a consequence of assumption (**A**) for which, again, only the curvature theory is capable of checkability (Section 2.3).

2.2 Geometric Multiplier Identities

Let $\Omega \subset \mathbb{R}^n$ be a bounded, open set with smooth boundary Γ. We consider u as a regular solution to the problem

$$u_{tt} = \text{div}\, A(x)\nabla u + f \quad \text{in} \quad (0,\infty) \times \Omega \tag{2.3}$$

to establish some multiplier identities where $A(x)$ are symmetric, positive, smooth $n \times n$ matrices for all $x \in \mathbb{R}^n$ which are given by materials in application. The main multipliers are

$$2H(u) \quad \text{and} \quad 2hu$$

where H and h are a vector field and a function on Ω, respectively.

We introduce

$$g = A^{-1}(x) \quad \text{for} \quad x \in \mathbb{R}^n \tag{2.4}$$

as a Riemannian metric on $\mathbb{R}^n$ and consider the couple $(\mathbb{R}^n, g)$ as a Riemannian manifold. We denote by $g = \langle \cdot, \cdot \rangle_g$ the inner product and by D the covariant differential of the metric g, respectively. Denote by $\langle \cdot, \cdot \rangle = \cdot$ the Euclidean product of $\mathbb{R}^n$. Then

$$\langle X, Y \rangle_g = \langle A^{-1}(x)X, Y \rangle \quad \text{for} \quad X, Y \in \mathbb{R}^n_x, \ x \in \mathbb{R}^n. \tag{2.5}$$

If applied to functions, sometimes, we denote by ∇_g and ∇ the gradients of the metric g and the Euclidean metric, respectively. Similarly, div $_g$ and div are the divergences of the metric g and the Euclidean metric, respectively.

With two metrics on $\mathbb{R}^n$ in mind, one the Euclidean metric and the other the Riemannian metric g, we have to deal with various notation carefully. First, let us present some basic relations between the two metrics on $\mathbb{R}^n$.

It follows from the relation (2.5) that

Lemma 2.1 *Let $x = (x_1, x_2, \cdots, x_n)$ be the natural coordinate system on $\mathbb{R}^n$. Let $H, X \in \mathcal{X}(\mathbb{R}^n)$ be vector fields and let h, f be functions. Then*

i) $\langle A(x)H(x), X(x) \rangle_g = \langle H(x), X(x) \rangle = H(x) \cdot X(x) \quad \text{for} \quad x \in \mathbb{R}^n;$
ii) $\nabla_g f = A(x) \nabla f \quad \text{for} \quad x \in \mathbb{R}^n;$
iii) $\langle \nabla_g f, \nabla_g h \rangle_g = \nabla_g f(h) = \langle A(x) \nabla f, \nabla h \rangle \quad \text{for} \quad x \in \mathbb{R}^n \text{ where } A(x) \text{ is the coefficient matrix, given in (2.3).}$

The following identity plays a crucial role in establishing multiplier identities in a version of the metric g.

Lemma 2.2 *Let f, h be functions on $\mathbb{R}^n$ and let H be a vector field on $\mathbb{R}^n$. Then*

$$\langle \nabla_g f, \nabla_g \big(H(h) \big) \rangle_g + \langle \nabla_g h, \nabla_g \big(H(f) \big) \rangle_g$$
$$= DH(\nabla_g h, \nabla_g f) + DH(\nabla_g f, \nabla_g h) + \text{div} \left(\langle \nabla_g f, \nabla_g h \rangle_g H \right)$$
$$- \langle \nabla_g f, \nabla_g h \rangle_g \, \text{div} \, H, \quad x \in \mathbb{R}^n \tag{2.6}$$

where div H is the divergence of the vector field H in the Euclidean metric.

Proof. We compute the identity (2.6) at each point by the Bochner technique. Let $x \in \mathbb{R}^n$ be given. Let $E_1, \cdots, E_n$ be a frame field normal at x in the metric g. Then

$$\langle E_i, E_j \rangle_g = \delta_{ij} \quad \text{in a neighborhood of } x, \tag{2.7}$$

$$D_{E_i} E_j = 0 \quad \text{at the point } x. \tag{2.8}$$

Let $H = \sum_{i=1}^{n} h_i E_i$. The relations (2.7) and (2.8) imply that

$$\nabla_g h = \sum_{i=1}^{n} E_i(h) E_i \quad \text{in a neighborhood of } x, \tag{2.9}$$

$$DH(\nabla_g h, \nabla_g f) = \sum_{ij=1}^{n} E_i(h) E_j(h_i) E_j(f), \tag{2.10}$$

$$E_j E_i(f) = D^2 f(E_i, E_j) = E_i E_j(f) \quad \text{at } x, \tag{2.11}$$

where $D^2 f$ denotes the Hessian of f in the metric g. Using the relations (2.9)-(2.11), we obtain

$$\langle \nabla_g f, \nabla_g \big(H(h) \big) \rangle_g = \sum_{j=1}^{n} E_j(f) E_j \big(H(h) \big)$$

$$= \sum_{ij=1}^{n} E_j(f) \Big[E_j(h_i) E_i(h) + h_i E_j E_i(h) \Big]$$

$$= DH(\nabla_g h, \nabla_g f) + \sum_{j=1}^{n} H\big(E_j(h) \big) E_j(f) \tag{2.12}$$

at x. We change f and h into each other in the formula (2.12) and add them together. The identity (2.6) follows from the following formula

$$\sum_{j=1}^{n} H\big(E_j(h) \big) E_j(f) + \sum_{j=1}^{n} H\big(E_j(f) \big) E_j(h) = H(\langle \nabla_g f, \nabla_g h \rangle_g)$$

$$= \text{div} \Big(\langle \nabla_g f, \nabla_g h \rangle_g H \Big) - \langle \nabla_g f, \nabla_g h \rangle_g \text{ div } H.$$

$$\square$$

Now we are ready to present our main multiplier identities.

Theorem 2.1 *Let u solve the problem (2.3). Suppose that H is a vector field. Then*

$$\text{div} \{ 2H(u) A(x) \nabla u - (|\nabla_g u|_g^2 - u_t^2) H \} + 2f H(u)$$
$$= 2[u_t H(u)]_t + 2DH(\nabla_g u, \nabla_g u) + (u_t^2 - |\nabla_g u|_g^2) \text{ div } H. \tag{2.13}$$

Proof. Multiply the equation (2.3) by $2H(u)$. Using the formulas (2.6) and iii) of Lemma 2.1 with $f = h = u$, we obtain

$$2f H(u) = 2(u_{tt} - \text{div } A(x) \nabla u) H(u)$$

$$= \Big[2u_t H(u) \Big]_t - \text{div} \Big[2H(u) A(x) \nabla u \Big] - H(u_t^2) + 2\langle A(x) \nabla u, \nabla \big(H(u) \big) \rangle$$

$$= 2\Big[u_t H(u) \Big]_t + \text{div} \Big[(|\nabla_g u|_g^2 - u_t^2) H - 2H(u) A(x) \nabla u \Big]$$

$$+ 2DH(\nabla_g u, \nabla_g u) + (u_t^2 - |\nabla_g u|_g^2) \text{ div } H,$$

that is, the identity (2.13). □

Theorem 2.2 *Let u solve the problem (2.3) and let p be a function on Ω. Then*

$$\text{div}\,[2puA(x)\nabla u - u^2 A(x)\nabla p] + 2fpu$$
$$= 2p(uu_t)_t + 2p(|\nabla_g u|_g^2 - u_t^2) - u^2 \,\text{div}\, A(x)\nabla p. \qquad (2.14)$$

Proof. Multiply the equation (2.3) by $2pu$ and we have, by the formulas ii) and iii) of Lemma 2.1,

$$\begin{aligned}
2fpu &= 2(u_{tt} - \text{div}\, A(x)\nabla u)pu \\
&= 2p(uu_t)_t - 2\,\text{div}\,[puA(x)\nabla u] - 2pu_t^2 + 2\langle A\nabla u, \nabla(pu)\rangle \\
&= 2p(uu_t)_t + \text{div}\,[u^2 A\nabla p - 2puA(x)\nabla u] \\
&\quad + 2p(|\nabla_g u|_g^2 - u_t^2) - u^2 \,\text{div}\, A\nabla p.
\end{aligned}$$

□

2.3   Escape Vector Fields and Escape Regions for Metrics

The duality method will transfer controllability/stabilization of the wave equation to an assumption which is:

Definition 2.1 *Let g be the metric given by (2.4) and let $\Omega \subset \mathbb{R}^n$ be a bounded, open set. A vector field H is said to be an escape vector field for the metric g on $\overline{\Omega}$ if the covariant differential DH of H in the metric g is a positive tensor field on $\overline{\Omega}$, i.e., there is a constant $\varrho_0 > 0$ such that*

$$DH(x) \geq \varrho_0 g(x) \quad \text{for all} \quad x \in \overline{\Omega}. \qquad (2.15)$$

Remark 2.1 *Escape vector fields were introduced by [208] as a checkable assumption for the exact controllability of the wave equation with variable coefficients. In [156] a convex function was called an escape function.*

Remark 2.2 *If $A(x) = I$, the unit matrix, then for any $x_0 \in \mathbb{R}^n$, $H = x - x_0$ is an escape vector field for the Euclidean metric on the whole space $\mathbb{R}^n$ since $DH = g$ for all $x \in \mathbb{R}^n$.*

Remark 2.3 *Roughly speaking, if there is an escape vector field, then all waves will go to the boundary along this vector field and boundary controllability is true.*

In general it is very hard to verify this assumption. That is one of the reasons why the Riemannian geometric approach is necessary to us. The goal of this section is to show the checkability of the above assumption by the Riemannian curvature theory: It is locally true (when Ω is small in some sense); the sectional curvature provides the global information on its existence; see Theorems 2.4-2.6 below.

First, we give a necessary condition for escape vector fields

Theorem 2.3 *A necessary condition for escape vector fields for the metric g on $\overline{\Omega}$ is: $\overline{\Omega}$ does not include any closed geodesic.*

Proof. By contradiction. Let there be a closed geodesic $\gamma(t) \in \overline{\Omega}$ for all $t \in [0, b]$. Then

$$\gamma(0) = \gamma(b) \quad \text{and} \quad \dot{\gamma}(0) = \dot{\gamma}(b). \tag{2.16}$$

Assume that H is an escape vector field on $\overline{\Omega}$. Let

$$f(t) = \langle H(\gamma(t)), \dot{\gamma}(t) \rangle_g \quad \text{for} \quad t \in [0, b].$$

By (2.16), we get

$$f(0) = f(b). \tag{2.17}$$

Since H is escaping, we have

$$\dot{f}(t) = DH(\dot{\gamma}(t), \dot{\gamma}(t)) > 0 \quad \text{for} \quad t \in [0, b]$$

which implies that $f(b) > f(0)$, that contradicts (2.17).   □

If h is a strictly convex function in the metric g on $\overline{\Omega}$, then the Hessian $D^2 h$ of h in the metric g is positive on $\mathbb{R}^n_x \times \mathbb{R}^n_x$ for all $x \in \overline{\Omega}$. Then the first useful criterion is

Theorem 2.4 *If h is a strictly convex function in the metric g on $\overline{\Omega}$, then $H = Dh$ is an escape vector field for the metric g on $\overline{\Omega}$.*

Remark 2.4 *The class of escape vector fields for the metric is larger than that which is given by all gradients of strictly convex functions. There is an escape vector field which is not a gradient of any strictly convex function; see Example 2.6 later.*

For $x \in \mathbb{R}^n$, let $\Xi \subset \mathbb{R}^n_x$ be a two-dimensional subspace. Denote the sectional curvature of Ξ by $\kappa(x, \Xi)$ in the metric g. Let $x_0 \in \mathbb{R}^n$ be given. Let $B_g(x_0, \gamma)$ be a geodesic ball in $(\mathbb{R}^n, g)$ centered at x_0 with radius $\gamma > 0$. Set

$$\kappa(\Omega) = \sup_{x \in \Omega, \, \Xi \subset \mathbb{R}^n_x} \kappa(x, \Xi). \tag{2.18}$$

Denote by $\rho(x) = d_g(x, x_0)$ the distance function of the metric g from $x \in \mathbb{R}^n$ to x_0.

Theorem 2.5 *If $\gamma > 0$ is such that $4\gamma^2\kappa(\Omega) < \pi^2$ and $\overline{\Omega} \subset B_g(x_0, \gamma)$, then $H = \rho D\rho$ is an escape vector field for the metric g on $\overline{\Omega}$.*

Proof. Since the inequality (2.15) is only required to be true for each point in $\overline{\Omega}$, it is independent of points in $B_g(x_0, \gamma)/\overline{\Omega}$. So we may change the metric $g(x)$ for $x \in B_g(x_0, \gamma)/\overline{\Omega}$ such that

$$4\gamma^2\kappa(B_g) < \pi^2 \tag{2.19}$$

where

$$\kappa(B_g) = \sup_{x \in B_g(x_0, \gamma), \, \Xi \subset \mathbb{R}_x^n} \kappa(x, \Xi).$$

Since $(\mathbb{R}^n, g)$ is simply connected, by Theorem 1.22, we have

$$D^2\rho(X, X) \geq f(\rho)|X|_g^2 \quad \text{for} \quad X \in \mathbb{R}_x^n, \quad \langle X, D\rho \rangle_g = 0 \tag{2.20}$$

for $x \in B_g(x_0, \gamma)$ where

$$f(\rho) = \begin{cases} 1/\rho, & \kappa(B_g) = 0; \\ \sqrt{\kappa(B_g)}\cot(\sqrt{\kappa(B_g)}\rho), & \kappa(B_g) > 0; \\ \sqrt{-\kappa(B_g)}\coth(\sqrt{-\kappa(B_g)}\rho), & \kappa(B_g) < 0. \end{cases} \tag{2.21}$$

It follows from (2.20) and (2.21) that

$$DH = \rho D^2\rho + D\rho \otimes D\rho \geq \varrho_0 g \quad \text{for} \quad x \in \overline{\Omega} \subset B_g(x_0, \gamma) \tag{2.22}$$

where $\varrho_0 = \min_{x \in \Omega}\{\rho f(\rho), 1\} > 0$. $\square$

Remark 2.5 *Theorem 2.5 shows that an escape vector field for the metric g locally exists on $\overline{\Omega}$: When $\overline{\Omega}$ is included in a neighborhood of some point in $\mathbb{R}^n$, it exists.*

Remark 2.6 *For convenience, we may change the geodesic ball $B_g(x_0, \gamma)$ in the metric g in Theorem 2.5 into the one in the Euclidean metric as in Corollary 2.1 below if $\kappa(\Omega) > 0$.*

Corollary 2.1 *Let $\kappa(\Omega) > 0$. Let*

$$\sigma_0 = \sup_{x \in \Omega, \, X \in \mathbb{R}_x^n, \, |X| = 1} \langle A^{-1}(x)X, X \rangle.$$

If there is $x_0 \in \mathbb{R}^n$ such that

$$\overline{\Omega} \subset B(x_0, \sigma_1),$$

then there is an escape vector field for the metric g on $\overline{\Omega}$, where $\sigma_1 = \pi/(2\sqrt{\sigma_0\kappa(\Omega)})$ and

$$B(x_0, \sigma_1) = \{ x \in \mathbb{R}^n \,|\, |x - x_0| < \sigma_1 \}.$$

Proof. By a similar reason as for (2.19), we assume that

$$\langle A^{-1}(x)X, X\rangle \le \sigma_0 \quad \text{for} \quad X \in \mathbb{R}^n_x, \ |X| = 1,$$

for all $x \in B_0(x_0, \sigma_1)$. Let $x \in B(x_0, \sigma_1)$ be given. Let $r(t) = tx_0 + (1-t)x$. Then $r(\cdot)$ is a curve connecting x_0 and x. Then

$$d_g(x_0, x) \le \int_0^1 |\dot{r}(t)|_g dt = \int_0^1 \langle A^{-1}(r(t))(x_0 - x), x_0 - x\rangle^{1/2} dt$$

$$\le \sigma_0^{1/2}|x - x_0| < \pi/(2\sqrt{\kappa(\Omega)}).$$

Then there is $0 < \gamma < \pi/(2\sqrt{\kappa(\Omega)})$ such that

$$\overline{\Omega} \subset B_g(x_0, \gamma).$$

The corollary follows from Theorem 2.5.   □

Theorem 2.6 *Let metric g be given by (2.4). Then*

(a) *If $(\mathbb{R}^n, g)$ has non-positive sectional curvature, then there exists an escape vector field for the metric g on the whole space $\mathbb{R}^n$.*

(b) *If $(\mathbb{R}^n, g)$ is noncompact, complete, and its sectional curvature is positive everywhere on $\mathbb{R}^n$, then there exists an escape vector field in the metric g on the whole space $\mathbb{R}^n$.*

Proof. (a) is a straight consequence of Theorem 2.5 and (b) is given by [71].   □

Examples We give some examples which verify existence or non-existence of escape vector fields for a metric.

Example 2.1 *Let a coefficient matrix $A(x) = \left(a_{ij}\right)$ be constant, symmetric, positive. Then $(\mathbb{R}^n, g)$ is of zero sectional curvature. By Theorem 2.6, (a), for any $\Omega \subset \mathbb{R}^n$ bounded, there is an escape vector field on $\overline{\Omega}$.*

Example 2.2 *Let $a_i > 0$ be constants for $1 \le i \le n$. Let $\alpha(x) = (a_1 x_1, \cdots, a_n x_n)^\tau$ and*

$$A(x) = \frac{1}{1 + |\alpha(x)|^2}[(1 + |\alpha(x)|^2)I - \alpha(x) \otimes \alpha(x)] \quad \text{for} \quad x \in \mathbb{R}^n,$$

where I is the $n \times n$ unit matrix. Then the metric on $\mathbb{R}^n$ is

$$g = A^{-1}(x) = I + \alpha(x) \otimes \alpha(x) \quad \text{for} \quad x \in \mathbb{R}^n$$

and $(\mathbb{R}^n, g)$ is a complete noncompact Riemannian manifold.

We claim: There is an escape vector field for the metric g on the whole space $\mathbb{R}^n$.

Proof. **Step 1** Let

$$M = \{\, (x_1, \cdots, x_n, x_{n+1}) \in \mathbb{R}^{n+1} \mid 2x_{n+1} = \sum_{i=1}^{n} a_i x_i^2 \,\}$$

be a hypersurface of $\mathbb{R}^{n+1}$. Then M is a Riemannian manifold with the induced metric from $\mathbb{R}^{n+1}$.

Let $p = (x_1, \cdots, x_{n+1}) \in M$. The tangential space of M at p is given by

$$M_p = \{\, \beta \in \mathbb{R}^{n+1} \mid \beta_{n+1} = \sum_{i=1}^{n} a_i x_i \beta_i \,\}.$$

Then the normal of M is

$$N(p) = \frac{1}{(1 + |\alpha(x)|^2)^{1/2}} (a_1 x_1, \cdots, a_n x_n, -1) \quad \text{for} \quad p \in M. \tag{2.23}$$

Denote by $\langle \cdot, \cdot \rangle$ the Euclidean metric of $\mathbb{R}^{n+1}$. Let D^0 be the connection of the metric $\langle \cdot, \cdot \rangle$ on $\mathbb{R}^n$. Then the second fundamental form of M is given by

$$\mathrm{II}(X, Y) = -\langle D_X^0 Y, N \rangle \quad \text{for} \quad X, \, Y \in \mathcal{X}(M). \tag{2.24}$$

Let

$$X_i = \frac{\partial}{\partial x_i} + a_i x_i \frac{\partial}{\partial x_{n+1}} \quad \text{for} \quad i = 1, \cdots, n.$$

Then $X_1, \cdots, X_n$ is a vector field basis of M. Since $D^0_{\frac{\partial}{\partial x_i}} \frac{\partial}{\partial x_j} = 0$ for $1 \leq i, j \leq n+1$, we obtain

$$D^0_{X_i} X_j = X_i(a_j x_j) \frac{\partial}{\partial x_{n+1}} = a_j \delta_{ij} \frac{\partial}{\partial x_{n+1}} \tag{2.25}$$

for $1 \leq i, j \leq n$.

Let

$$X = \sum_{i=1}^{n} \alpha_i X_i, \quad Y = \sum_{i=1}^{n} \beta_i X_i. \tag{2.26}$$

Let $\mathbf{R}$ be the curvature tensor. Then by Theorem 1.13 and (2.23)-(2.25), we have

$$\begin{aligned}
\mathbf{R}(X, Y, X, Y) &= \mathrm{II}(X, X)\mathrm{II}(Y, Y) - |\mathrm{II}(X, Y)|^2 \\
&= \frac{1}{1 + |\alpha(x)|^2} [(\sum_{i=1}^{n} a_i \alpha_i^2)(\sum_{i=1}^{n} a_i \beta_i^2) - (\sum_{i=1}^{n} a_i \alpha_i \beta_i)^2] > 0
\end{aligned}$$

if and only if X, Y are linearly independent. Thus M has positive sectional curvature everywhere.

Step 2 Consider a map $\Phi \colon M \to (\mathbb{R}^n, g)$ given by

$$\Phi(p) = x \quad \text{for} \quad p = (x, x_{n+1}) \in M.$$

Let $p = (x, x_{n+1}) \in M$ and $\beta = (\beta_1, \cdots, \beta_{n+1}) \in M_p$ be given. Assume that $\gamma \colon (-\varepsilon, \varepsilon) \to M$ be a curve such that $\gamma(0) = p$ and $\dot{\gamma}(0) = \beta$. Let

$$\gamma(t) = (\gamma_1(t), \cdots, \gamma_n(t), \gamma_{n+1}(t)).$$

Then $\Phi(\gamma(t)) = (\gamma_1(t), \cdots, \gamma_n(t))$ imply that $\Phi_*\beta = (\beta_1, \cdots, \beta_n)$. We obtain

$$g(\Phi_*\beta, \Phi_*\beta) = \langle A^{-1}(x)\Phi_*\beta, \Phi_*\beta \rangle = \sum_{i=1}^{n} \beta_i^2 + (\sum_{i=1}^{n} a_i x_i \beta_i)^2 = \hat{g}(\beta, \beta)$$

where $\hat{g}$ denotes the induced metric of M from $\mathbb{R}^{n+1}$. Thus Φ is an isometry between M and $(\mathbb{R}^n, g)$. Therefore $(\mathbb{R}^n, g)$ is of positive curvature everywhere, noncompact complete by Step 1. Then what we need follows from (b) in Theorem 2.6.   □

Example 2.3 *Set*

$$A(x) = \frac{1}{1 + x_1^2 + x_2^6} \begin{pmatrix} 1 + x_2^6 & x_1 x_2^3 \\ x_1 x_2^3 & 1 + x_1^2 \end{pmatrix} \quad for \quad x = (x_1, x_2) \in \mathbb{R}^2.$$

Then the metric is

$$g = A^{-1}(x) = \begin{pmatrix} 1 + x_1^2 & -x_1 x_2^3 \\ -x_1 x_2^3 & 1 + x_2^6 \end{pmatrix} \quad for \quad x = (x_1, x_2) \in \mathbb{R}^2.$$

It is easy to check that $(\mathbb{R}^2, g)$ is isometric to a surface M in $\mathbb{R}^3$, given by

$$M = \{ (x_1, x_2, x_3) \in \mathbb{R}^3 \,|\, 4x_3 = 2x_1^2 - x_2^4 \}.$$

We have the Gauss curvature of $(\mathbb{R}^2, g)$ at $x = (x_1, x_2)$:

$$\kappa(x) = \text{the Gauss curvature of } M \text{ at } p = (x, x_3)$$

$$= \frac{-3x_2^2}{(1 + x_1^2 + x_2^6)^2} \le 0.$$

By (a) in Theorem 2.6, there is an escape vector field for the metric g on the whole space $\mathbb{R}^2$.   □

For convenience to compute sectional curvature, we give a lemma.

Lemma 2.3 *Let $\mathbb{R}^2$ have a metric*

$$g = g_1 dx_1 dx_1 + g_2 dx_2 dx_2.$$

Then the Gaussian curvature is

$$\kappa = \frac{1}{4g_1^2 g_2^2} [g_2 g_{1x_1} g_{2x_1} + g_1 g_{1x_2} g_{2x_2} + g_1 g_{2x_1}^2 + g_2 g_{1x_2}^2 - 2g_1 g_2 (g_{1x_2x_2} + g_{2x_1x_1})].$$

In particular, if $g_1 = g_2 = h$, then

$$\kappa = -\frac{1}{2h} \Delta \log h \tag{2.27}$$

where Δ is the Laplacian in the Euclidean metric of $\mathbb{R}^2$.

Proof. Set

$$X_i = \frac{1}{\sqrt{g_i}} \partial_{x_i} \quad \text{for} \quad i = 1, 2.$$

Then X_1, X_2 forms an orthonormal frame on $(\mathbb{R}^2, g)$. We have

$$\kappa = \langle \mathbf{R}_{X_1 X_2} X_1, X_2 \rangle_g = \frac{1}{g_1 g_2} \langle \mathbf{R}_{\partial_{x_1} \partial_{x_2}} \partial_{x_1}, \partial_{x_2} \rangle_g \qquad (2.28)$$

where $\mathbf{R}_{\partial_{x_1} \partial_{x_2}} \partial_{x_1} = -D_{\partial_{x_1}} D_{\partial_{x_2}} \partial_{x_1} + D_{\partial_{x_2}} D_{\partial_{x_1}} \partial_{x_1}$ is the curvature operator.

Let Γ_{ij}^k be the coefficients of the connection D in the metric g. From the formula (1.9), we obtain

$$\Gamma_{12}^1 = \frac{1}{2g_1} g_{1x_2}, \quad \Gamma_{11}^2 = -\frac{1}{2g_2} g_{1x_2}, \quad \Gamma_{12}^2 = \frac{1}{2g_2} g_{2x_1},$$

$$\Gamma_{11}^1 = \frac{1}{2g_1} g_{1x_1}, \quad \Gamma_{22}^2 = \frac{1}{2g_2} g_{2x_2}.$$

It follows from the properties of the connection that

$$D_{\partial_{x_1}} D_{\partial_{x_2}} \partial_{x_1} = D_{\partial_{x_1}} (\Gamma_{21}^1 \partial_{x_1} + \Gamma_{21}^2 \partial_{x_2})$$
$$= (\cdots) \partial_{x_1} + [\Gamma_{12}^1 \Gamma_{11}^2 + (\Gamma_{12}^2)_{x_1} + (\Gamma_{12}^2)^2] \partial_{x_2}.$$

Thus

$$\langle D_{\partial_{x_1}} D_{\partial_{x_2}} \partial_{x_1}, \partial_{x_2} \rangle_g = [\Gamma_{12}^1 \Gamma_{11}^2 + (\Gamma_{12}^2)_{x_1} + (\Gamma_{12}^2)^2] g_2$$
$$= \frac{1}{2} g_{2x_1 x_1} - \frac{1}{4g_1} g_{1x_2}^2 - \frac{1}{4g_2} g_{2x_1}^2. \qquad (2.29)$$

A similar computation yields

$$\langle D_{\partial_{x_2}} D_{\partial_{x_1}} \partial_{x_1}, \partial_{x_2} \rangle_g = -\frac{1}{2} g_{1x_2 x_2} + \frac{1}{4g_1} g_{1x_1} g_{2x_1} + \frac{1}{4g_2} g_{1x_2} g_{2x_2}. \qquad (2.30)$$

From the relations (2.28), (2.29) and (2.30), we obtain the formula of κ.
$\square$

Example 2.4 *Let $a(x)$ be a smooth, positive function on $\mathbb{R}^2$. Set*

$$A(x) = \begin{pmatrix} a(x) & 0 \\ 0 & a(x) \end{pmatrix} \quad \text{for} \quad x \in \mathbb{R}^2.$$

Then the metric is

$$g = \begin{pmatrix} 1/a(x) & 0 \\ 0 & 1/a(x) \end{pmatrix} \quad \text{for} \quad x \in \mathbb{R}^2.$$

By Lemma 2.3, the Gauss curvature of $(\mathbb{R}^2, g)$ is

$$\kappa = \frac{a \Delta \log a}{2} \quad \text{for} \quad x \in \mathbb{R}^2$$

where Δ is the Laplacian of the Euclidean metric of $\mathbb{R}^2$. By (a) in Theorem 2.6, if $\Delta \log a \leq 0$ for $x \in \mathbb{R}^2$, then there is an escape vector field for the metric on the whole space $\mathbb{R}^2$.

In particular, let $a(x) = e^{x_1 + x_2}$ for $x \in \mathbb{R}^2$. Then $\kappa = 0$ for $x \in \mathbb{R}^2$. There is an escape vector field for this metric on the whole space $\mathbb{R}^2$.

Example 2.5 *Consider a coefficient matrix, given by*

$$A(x) = \begin{pmatrix} e^{x_1 + x_2} & 0 \\ 0 & e^{-x_1 - x_2} \end{pmatrix} \quad \text{for} \quad x \in \mathbb{R}^2.$$

The metric on $\mathbb{R}^2$ is

$$g = \begin{pmatrix} e^{-x_1 - x_2} & 0 \\ 0 & e^{x_1 + x_2} \end{pmatrix}.$$

By Lemma 2.3, we have

$$\kappa(x) = -\frac{1}{2}(e^{-x_1 - x_2} + e^{x_1 + x_2}) < 0 \quad \text{for} \quad x \in \mathbb{R}^2.$$

Then there is an escape vector field for the metric g on the whole space $\mathbb{R}^2$ from Theorem 2.6.

Example 2.6 *Set*

$$A(x) = e^{x_1^3 + x_2^3} I \quad \text{for} \quad x \in \mathbb{R}^2$$

where I is the 2×2 unit matrix. Then the metric is

$$g = hI \quad \text{where} \quad h = e^{-x_1^3 - x_2^3} \quad \text{for} \quad x \in \mathbb{R}^2.$$

Then the coefficients of the connection are

$$\Gamma_{11}^1 = \Gamma_{12}^2 = -\Gamma_{22}^1 = -\frac{3}{2}x_1^2, \quad \Gamma_{12}^1 = \Gamma_{22}^2 = -\Gamma_{11}^1 = -\frac{3}{2}x_2^2.$$

Let

$$H = -e^{-x_1}\partial_{x_1} - e^{-x_2}\partial_{x_2} \quad \text{for} \quad x \in \mathbb{R}^2. \tag{2.31}$$

Let $X = X_1 \partial_{x_1} + X_2 \partial_{x_2}$. We have

$$
\begin{aligned}
D_X H &= e^{-x_1} X_1 \partial_{x_1} - e^{-x_1} D_X \partial_{x_1} + e^{-x_2} X_2 \partial_{x_2} - e^{-x_2} D_X \partial_{x_2} \\
&= [(e^{-x_1} - e^{-x_1}\Gamma_{11}^1 - e^{-x_2}\Gamma_{12}^1)X_1 - (e^{-x_1}\Gamma_{12}^1 + e^{-x_2}\Gamma_{22}^1)X_2]\partial_{x_1} \\
&\quad + [-(e^{-x_1}\Gamma_{11}^2 + e^{-x_2}\Gamma_{12}^2)X_1 + (e^{-x_2} - e^{-x_1}\Gamma_{12}^2 - e^{-x_2}\Gamma_{22}^2)X_2]\partial_{x_2}
\end{aligned}
$$

which yields

$$DH(X, X) = \langle D_X H, X \rangle_g = \sum_{ij=1}^{2} h_{ij} X_i X_j \tag{2.32}$$

where

$$\left(h_{ij}\right) = h\left(\begin{matrix} e^{-x_1} + \frac{3}{2}(x_1^2 e^{-x_1} + x_2^2 e^{-x_2}) & \frac{3}{2}(x_1^2 e^{-x_2} - x_2^2 e^{-x_1}) \\ \frac{3}{2}(x_2^2 e^{-x_1} - x_1^2 e^{-x_2}) & e^{-x_2} + \frac{3}{2}(x_1^2 e^{-x_1} + x_2^2 e^{-x_2}) \end{matrix}\right).$$

(2.33)

It follows from (2.32) and (2.33) that

$$DH(X, X) \geq \varrho |X|_g^2 \quad for \quad x \in \mathbb{R}^2$$

where

$$\varrho = \min(e^{-x_1}, e^{-x_2}) > 0 \quad for \quad x \in \mathbb{R}^2,$$

that is, the vector field H, given by (2.31), is an escape vector field for the metric g on the whole space $\mathbb{R}^2$. However, H is not a gradient of any function because the matrix (h_{ij}) is not symmetric.

Next, we introduce geometric conditions on interior feedback regions. Let J be an integer. Let $\Omega_i \subseteq \Omega$ be open sets with C^∞ boundary $\partial\Omega_i$ for $1 \leq i \leq J$ such that

$$\Omega_i \cap \Omega_j = \emptyset \quad for \quad 1 \leq i < j \leq J. \tag{2.34}$$

Let H^i be vector fields on Ω_i for $1 \leq i \leq J$ such that

$$DH^i(X, X) \geq \varrho_0 |X|_g^2 \quad for \quad X \in R_x^n, \ x \in \Omega_i \tag{2.35}$$

for all i where $\varrho_0 > 0$.

Definition 2.2 *An subset $\omega \subset \overline{\Omega}$ is said to be an escape region for the metric g on $\overline{\Omega}$ if there is $\varepsilon > 0$ small such that*

$$\omega \supseteq \overline{\Omega} \cap \mathcal{N}_\varepsilon[\ \cup_{i=1}^J \Gamma_0^i \cup (\Omega \setminus \cup_{i=1}^J \Omega_i)\ ] \tag{2.36}$$

where

$$\mathcal{N}_\varepsilon(S) = \cup_{x \in S}\{\, y \in R^n \,|\, |y - x| < \varepsilon \,\}, \quad S \subset R^n,$$

$$\Gamma_0^i = \{\, x \in \partial\Omega_i \,|\, H^i(x) \cdot \nu^i(x) > 0 \,\},$$

and $\nu^i(x)$ is the unit normal of $\partial\Omega_i$ in the Euclidean of $\mathbb{R}^n$, pointing outside of Ω_i.

Remark 2.7 *Escape regions were introduced by [151] for interior feedback stabilization of the wave equation with constant coefficients. They were given in the above form by [62] for the variable coefficient problem.*

Now we consider the structure of an escape region for the metric. Since the set in the right-hand side of (7.207) is always a subset of $\overline{\Omega}$, $\omega = \overline{\Omega}$ is an escape region for the metric. However, we are particularly interested in the case $\omega \neq \overline{\Omega}$ where ω is small as far as possible in some sense.

If there is an escape vector field H on $\overline{\Omega}$, then the inequality (7.206) holds for all $x \in \Omega$. We can take $J = 1$ and $\Omega_1 = \Omega$. In this case, an escape region ω for the metric only needs to be supported in a neighborhood of Γ_0, where

$$\Gamma_0 = \{\, x \in \Gamma \mid H(x) \cdot \nu(x) > 0 \,\}. \tag{2.37}$$

This roughly means that we can dig out almost the whole Ω from the domain $\overline{\Omega}$ and that escape regions only need to be supported in a neighborhood of a subset of the boundary.

Suppose that $A(x)$ is the unit matrix for all $x \in \mathbb{R}^n$. Then the metric is the Euclidean metric of R^n, and $D = \nabla$ is the gradient of the standard metric. Let $x_0 \in R^n$ be given. Then the vector field $H(x) = x - x_0$ on $\overline{\Omega}$ meets condition (7.206) with $\rho_0 = 1$. Thus, escape regions for the metric needs only to be supported in a neighborhood of

$$\Gamma_0 = \{\, x \in \Gamma \mid (x - x_0) \cdot \nu(x) > 0 \,\}.$$

Such regions were used in [196].

In general, whether there exists one escape vector field H satisfying estimate (7.206) on the whole domain Ω largely depends on the sectional curvature of the Riemannian metric g; see Theorems 2.5 and 2.6.

Let $x_0 \in \mathbb{R}^n$ be given. Let $\rho(x) = \rho(x, x_0)$ be the distance function from $x \in \overline{\Omega}$ to x_0 in the metric g. We define a vector field $H(x, x_0)$ on $\overline{\Omega}$ by

$$H(x, x_0) = \rho(x) D\rho(x) \quad \text{for} \quad x \in \overline{\Omega} \tag{2.38}$$

If g is the Euclidean metric, then $D = \nabla$ and $H(x, x_0) = x - x_0$. It follows from Theorem 2.5 that

Theorem 2.7 *If $\gamma > 0$ is such that $4\gamma^2 \kappa(B) < \pi^2$ and $\overline{\Omega} \subset B(x_0, \gamma)$, an escape region for the metric on $\overline{\Omega}$ can be supported in a neighborhood of*

$$\Gamma_0 = \{\, x \in \Gamma \mid H(x, x_0) \cdot \nu(x) > 0 \,\} \tag{2.39}$$

where $\kappa(B)$ is given by (2.18) and $H(x, x_0)$ is given by (2.38).

Remark 2.8 *If $(\mathbb{R}^n, g)$ has nonpositive curvature, then $\kappa(B) \le 0$ for any $\gamma > 0$ and the results in Theorem 2.7 hold for any $\Omega \subset \mathbb{R}^n$ bounded.*

If $\kappa(B) > 0$, one escape vector field on the whole $\overline{\Omega}$ does not exist in general when Ω is large. A counterexample will be given by Example 2.7 later. But we can dig out a finite number of geodesic balls with radius $\le \pi/2\sqrt{\kappa}$ from $\overline{\Omega}$ and let an escape region be supported in a neighborhood of the remaining part of $\overline{\Omega}$. This is the following.

Theorem 2.8 *Let $\kappa > 0$. Suppose that there are points $x_i \in \overline{\Omega}$ for $1 \le i \le J$ such that*

$$\rho(x_i, x_j) > \frac{\pi}{\sqrt{\kappa}}, \quad 1 \le i, j \le J, \ i \ne j.$$

Let $B(x_i, r_0)$ and $S(x_i, r_0)$ be the geodesic ball and the geodesic sphere centered at x_i with radius r_0 in the metric g, respectively, where $r_0 < \pi/2\sqrt{\kappa}$. Let $\Gamma_0^i = \{\, x \mid x \in S(x_i, r_0),\ H(x, x_i) \cdot \nu(x) > 0 \,\}$. If

$$\omega \supseteq \overline{\Omega} \cap \mathcal{N}_\varepsilon[\, \cup_{i=1}^J \Gamma_0^i \cup (\Omega \setminus \cup_{i=1}^J B(x_i, r_0)) \,],$$

then ω is an escape region on $\overline{\Omega}$ for the metric.

Since we can dig out geodesic balls with radii small from Ω as many as we want, the following results are immediate.

Theorem 2.9 *For $\varepsilon > 0$ given, there is an escape region $\omega \subset \overline{\Omega}$ with*

$$\mathrm{meas}\,(\omega) < \varepsilon$$

where $\mathrm{meas}\,(\omega)$ is the n-dimensional Lebesgue measure of ω.

2.4 Exact Controllability. Dirichlet/Neumann Action

The concept of duality between control and observation was first noted by Kalman who gave a duality theorem in [92] (also see [102]) for linear systems and was generalized to distributed parameter systems by [58] and the references there.

Duality Method Let X and Y stand for Banach spaces of general type. Let A be an operator defined on a subspace of one of these spaces. Its domain will be denoted by $\mathcal{D}(A)$. Its range is denoted by $\mathcal{R}(A)$.

Consider an abstract linear system

$$A : \mathcal{D}(A) \subset X \longrightarrow Y \tag{2.40}$$

where A is a linear operator with dense domain. Along with (2.40), we consider a dual system

$$A^* : \mathcal{D}(A^*) \subset Y^* \longrightarrow X^* \tag{2.41}$$

where X^* and Y^* are the *conjugate spaces* of X and Y, respectively, and A^* is the adjoint operator for A.

Definition 2.3 *The system (2.40) is observable if there is $c > 0$ such that*

$$\|Ax\|_Y \geq c\|x\|_X \quad for \quad x \in \mathcal{D}(A). \tag{2.42}$$

Definition 2.4 *The system (2.41) is exactly controllable if*

$$X^* \subseteq \mathcal{R}(A^*). \tag{2.43}$$

The following is a duality theorem relating observability and controllability ([58] or [69]).

Theorem 2.10 *The system* (2.40) *is observable if and only if the system* (2.41) *is exactly controllable.*

Proof. Let (2.42) hold true. Let $x^* \in X^*$ be given. We define a linear functional y^* on the subspace $\mathcal{R}(A)$ of Y by

$$y^*(Ax) = x^*(x) \quad \text{for all} \quad Ax \in \mathcal{R}(A). \tag{2.44}$$

Then by (2.42)

$$|y^*(Ax)| \leq \|x^*\|_{X^*} \|x\|_X \leq \|x^*\|_{X^*} \|Ax\|_Y / c \quad \text{for all} \quad Ax \in \mathcal{R}(A).$$

By the Hahn-Banach Theorem, y^* can be extended from $\mathcal{R}(A)$ to Y to belong Y^*, still denoted by y^*. Then the relation (2.44) means that $y^* \in \mathcal{D}(A^*)$ and $x^* = A^* y^* \in \mathcal{R}(A^*)$.

Conversely, let (2.43) be true. For $x \in \mathcal{D}(A)$, $x \neq 0$, we define $f_x \in X^{**}$ by

$$f_x(x^*) = \frac{x^*(x)}{\|Ax\|_Y} \quad \text{for all} \quad x^* \in X^*.$$

Then for each $x^* \in X^*$ fixed, the condition (2.43) implies that there is $y* \in Y^*$ such that $x^* = A^* y^*$. For this x^*, we have

$$|f_x(x^*)| = |y^*(\frac{Ax}{\|Ax\|_Y})| \leq \|y^*\|_{Y^*} \quad \text{for all} \quad x \in \mathcal{D}(A), \ x \neq 0.$$

By the resonance theorem, we have

$$\sup_{x \in \mathcal{D}(A), \ \|x\|_X = 1} \frac{\|x\|_X}{\|Ax\|_Y} = \sup_{x \in \mathcal{D}(A), \ \|x\|_X = 1} \|f_x\|_{X^{**}} < \infty$$

that is equivalent to (2.42) being true. □

Control in Dirichlet Boundary Condition Let Γ_0 be a portion of the boundary Γ which is relatively open. Set $\Gamma_1 = \Gamma / \Gamma_0$. Let $T > 0$ be given. Set

$$Q = (0, T) \times \Omega, \quad \Sigma = (0, T) \times \Gamma, \quad \Sigma_i = (0, T) \times \Gamma_i, \quad \text{for} \quad i = 0, \ 1.$$

We consider the control problem

$$\begin{cases} u_{tt} = \text{div } A(x) \nabla u \quad \text{in} \quad Q, \\ u = 0 \quad \text{on} \quad \Sigma_1, \\ u = \varphi \quad \text{on} \quad \Sigma_0, \\ u(0) = u_0, \quad u_t(0) = u_1 \quad \text{on} \quad \Omega \end{cases} \tag{2.45}$$

with control φ on Σ_0.

Following [116], we have

Proposition 2.1 *Let $T > 0$ given. For any $(u_0, u_1) \in L^2(\Omega) \times H^{-1}(\Omega)$ and $\varphi \in L^2(\Sigma_0)$, the problem (2.45) has the unique solution $u \in C[0, T; L^2(\Omega)] \cap C^1[0, T; H^{-1}(\Omega)]$.*

Definition 2.5 *We say that the problem (2.45) is exactly $L^2(\Omega) \times H^{-1}(\Omega)$ controllable by $L^2(\Sigma_0)$ control at time T if, for any (u_0, u_1), $(\hat{u}_0, \hat{u}_1) \in L^2(\Omega) \times H^{-1}(\Omega)$, there is $\varphi \in L^2(\Sigma_0)$ such that the solution u to the problem (2.45) satisfies*

$$u(T) = \hat{u}_0, \quad u_t(T) = \hat{u}_1 \quad on \quad \Omega.$$

Now we formulize the controllability of the problem (2.45) in order to apply Theorem 2.10. Let

$$X = H_0^1(\Omega) \times L^2(\Omega), \quad Y = L^2(\Sigma_0).$$

Then

$$X^* = H^{-1}(\Omega) \times L^2(\Omega), \quad Y^* = L^2(\Sigma_0).$$

We define an operator $A\colon X \to Y$ as follows. For any $(w_0, w_1) \in X$, we solve the problem

$$\begin{cases} w_{tt} = \operatorname{div} A(x)\nabla w & \text{in} \quad Q, \\ w = 0 & \text{on} \quad \Sigma, \\ w(0) = w_0, \quad w_t(0) = w_1 & \text{on} \quad \Omega \end{cases} \tag{2.46}$$

to define

$$A(w_0, w_1) = w_{\nu_A} \quad \text{on} \quad \Sigma_0$$

where $w_{\nu_A} = \langle A(x)\nabla w, \nu \rangle$ and ν is the outside normal of Ω on Γ in the Euclidean metric of $\mathbb{R}^n$.

Next, we define an operator $B\colon Y^* \to X^*$ as follows. For any $\varphi \in L^2(\Sigma_0)$, we solve the problem

$$\begin{cases} u_{tt} = \operatorname{div} A(x)\nabla u & \text{in} \quad Q, \\ u(T) = u_t(T) = 0 & \text{on} \quad \Omega, \\ u = 0 & \text{on} \quad \Sigma_1, \\ u = \varphi & \text{on} \quad \Sigma_0 \end{cases} \tag{2.47}$$

to define

$$B\varphi = (u_t(0), -u(0)) \quad \text{on} \quad \Omega.$$

Proposition 2.1 implies that B is a bounded, linear operator. $\mathcal{R}(B)$ is all initial data which are exactly controllable to the rest. Since the system (2.45) is time-reversible, the exact controllability for the problem (2.45) on $X^* = H^{-1}(\Omega) \times L^2(\Omega)$ is equivalent to the relation $\mathcal{R}(B) = X^*$.

For any $\varphi \in Y^*$ and $(w_0, w_1) \in X$, using the systems (2.46) and (2.47), we obtain

$$(B\varphi, (w_0, w_1))_{L^2(\Omega) \times L^2(\Omega)} = (u_t(0), w(0)) - (u(0), w_t(0))$$

$$= \int_0^T [(u(\tau), w_t(\tau)) - (u_t(\tau), w(\tau))]_t d\tau$$

$$= \int_0^T [(u(\tau), w_{tt}(\tau)) - (u_{tt}(\tau), w(\tau))] d\tau$$

$$= (\varphi, A(w_0, w_1))_{L^2(\Sigma_0)} = A^*\varphi((w_0, w_1)) = (A^*\varphi, (w_0, w_1))_{L^2(\Omega) \times L^2(\Omega)}$$

where $(\cdot, \cdot)$ is the inner product of $L^2(\Omega)$, which implies that

$$B = A^*.$$

In particular, $A: X \to Y$ is bounded since B is. It follows from Theorem 2.10 that

Theorem 2.11 *The system* (2.45) *is exactly* $L^2(\Omega) \times H^{-1}(\Omega)$ *controllable at time* $T > 0$ *by* $L^2(\Sigma_0)$ *control if and only if there is* $c > 0$ *such that*

$$\|w_{\nu_A}\|^2_{L^2(\Sigma_0)} \geq c(\|w_0\|^2_{H^1_0(\Omega)} + \|w_1\|^2_{L^2(\Omega)}) \tag{2.48}$$

for all $(w_0, w_1) \in H^1_0(\Omega) \times L^2(\Omega)$.

Remark 2.9 *Proposition* 2.1 *implies that* $A: H^1_0(\Omega) \times L^2(\Omega) \to L^2(\Sigma_0)$ *is bounded. Then Theorem* 2.11 *means that* $\|(w_0, w_1)\|_\star = \|w_{\nu_A}\|_{L^2(\Sigma_0)}$ *defines an equivalent norm on* $H^1_0(\Omega) \times L^2(\Omega)$.

Definition 2.6 *The system* (2.46) *and the inequality* (2.48) *are said to be the duality system and the observability estimate of the problem* (2.45), *respectively.*

Lions' Construction [146], [147]   First, we introduce a theorem which is capable of defining a space which is exactly controllable, given by J. L. Lions by the dual method. Let H and L be Hilbert spaces with $H \subset L$ and H being dense in L such that

$$\|y\|_L \leq C\|y\|_H \quad \text{for all} \quad y \in H.$$

Let H^* denote the completion of the following space

$$\{\, x \in L \mid \sup_{\|y\|_H = 1} |(x, y)_L| < \infty \,\}$$

with a norm $\|x\|_{H^*} = \sup_{\|y\|_H = 1} |(x, y)_L|$. Clearly,

$$H \subset L \subset H^*.$$

Definition 2.7 H^* *is said to be the dual space of* H *with respect to* L.

For each $x \in H^*$, $(x, \cdot)_L$ is a linear, bounded functional on H. Then, by the Riesz representation theorem, there is a unique element in H, denoted by Px, such that

$$(x, y)_L = (Px, y)_H \quad \text{for all} \quad y \in H. \tag{2.49}$$

Then $P \colon H^* \to H$ is a linear operator.

The following results are needed when we construct boundary control for the Neumann action.

Theorem 2.12 *The operator $P \colon H^* \to H$ is an isometry.*

Proof. It follows from (2.49) that

$$\|Px\|_H = \sup_{\|y\|_H = 1} (Px, y)_H = \sup_{\|y\|_H = 1} (x, y)_L = \|x\|_{H^*}$$

for all $x \in H$. Then $\mathcal{R}(P)$ is a closed subspace of H. To complete the proof, it remains to show $\mathcal{R}(P) = H$. If it is not true, then there is $x_0 \in H$, $x_0 \neq 0$, such that $(Px, x_0)_H = 0$. The condition (2.49) yields

$$(x, x_0)_L = 0 \quad \text{for all} \quad x \in L,$$

that is, $x_0 = 0$. This contradiction shows $\mathcal{R}(P) = H$. $\qquad\square$

Definition 2.8 *The operator $\aleph = P^{-1} \colon H \to H^*$ is called the canonical isomorphism from H to H^*.*

Remark 2.10 *It follows from (2.49) that*

$$(x, y)_H = (\aleph x, y)_L \quad \text{for all} \quad x, \ y \in H. \tag{2.50}$$

Suppose that there is a linear operator $\Lambda \colon H \to L$ with dense domain $\mathcal{D}(\Lambda) \subset H$, such that

$$(\Lambda x, y)_L = (x, \Lambda y)_L \quad \text{for} \quad x, \ y \in \mathcal{D}(\Lambda).$$

We have

Theorem 2.13 *Let there be a constant $c > 0$ such that*

$$(\Lambda x, x)_L \geq c \|x\|_H^2 \quad \text{for all} \quad x \in H. \tag{2.51}$$

Let

$$\mathcal{R} = \{ \Lambda x \,|\, (\Lambda x, x)_L < \infty \}$$

with an inner product $(\Lambda x, \Lambda y)_{\mathcal{R}} = (\Lambda x, y)_L$. Then

$$\mathcal{R} \supset H^*.$$

If, in addition to the condition (2.51), there is a constant C such that

$$(\Lambda x, x)_L \leq C \|x\|_H^2 \quad \text{for all} \quad x \in H, \tag{2.52}$$

then

$$\mathcal{R} = H^*.$$

Proof. We introduce a space

$$H_1 = \{ x \mid (\Lambda x, x)_L < \infty \}$$

with an inner product $(x, y)_{H_1} = (\Lambda x, y)_L$. It is easy to check that $\mathcal{R} = H_1^*$.

Let the inequality (2.51) hold. The inequality (2.51) implies $H_1 \subset H$. Thus $\mathcal{R} \supset H^*$.

Moreover, let the inequality (2.52) be true. Then $H = H_1$. Thus $\mathcal{R} = H_1^* = H^*$. $\qquad\square$

Remark 2.11 *In application $\mathcal{R}$ represents a space which is exactly controllable. In addition, the condition $(\Lambda x, x)_L < \infty$ stands for the regularity of control action; see Remark 2.15 later.*

We are now back to the controllability problem (2.45) again. Let $(w_0, w_1) \in H_0^1(\Omega) \times L^2(\Omega)$. We solve the duality problem (2.46) to obtain the solution w. Then we solve the problem

$$
\begin{cases}
u_{tt} = \operatorname{div} A(x)\nabla u & \text{in} \quad Q, \\
u(T) = u_t(T) = 0 & \text{on} \quad \Omega, \\
u = 0 & \text{on} \quad \Sigma_1, \\
u = w_{\nu_A} & \text{on} \quad \Sigma_0
\end{cases}
\tag{2.53}
$$

where $\nu_A = A(x)\nu$. We define an operator $\Lambda \colon H_0^1(\Omega) \times L^2(\Omega) \to H^{-1}(\Omega) \times L^2(\Omega)$ by

$$\Lambda(w_0, w_1) = (u_t(0), -u(0)) \quad \text{on} \quad \Omega.$$

We take $H = H_0^1(\Omega) \times L^2(\Omega)$ and $L = L^2(\Omega) \times L^2(\Omega)$. Let w and $\hat{w}$ be the solutions to the problem (2.46) for the initial data (w_0, w_1) and $(\hat{w}_0, \hat{w}_1)$, respectively. Suppose that u and $\hat{u}$ are the solutions to the problem (2.53) with the boundary data w_{ν_A} and $\hat{w}_{\nu_A}$, respectively, on Σ_0. Using (2.46) and (2.53), we have

$$(\Lambda(w_0, w_1), (\hat{w}_0, \hat{w}_1))_{L^2(\Omega) \times L^2(\Omega)} = (u_t(0), \hat{w}(0)) - (u(0), \hat{w}_t(0))$$

$$= \int_0^T [(u, \hat{w}_t) - (u_t, \hat{w})]_t \, d\tau = (w_{\nu_A}, \hat{w}_{\nu_A})_{L^2(\Sigma_0)}. \tag{2.54}$$

Then H^* represents the space which is exactly controllable by $L^2(\Sigma_0)$ control. Applying Theorem 2.13, we have

Theorem 2.14 *The system (2.45) is exactly $L^2(\Omega) \times H^{-1}(\Omega)$ controllable at time $T > 0$ by $L^2(\Sigma_0)$ control if and only if $\|(w_0, w_1)\|_{\star} = \|w_{\nu_A}\|_{L^2(\Sigma_0)}$ defines an equivalent norm on $H_0^1(\Omega) \times L^2(\Omega)$.*

Remark 2.12 *In general the problem (2.53) only has distributed solutions whether the boundary control $u = w_{\nu_A}$ is smooth or not. To have smooth controls, we need to smooth the boundary control $u = w_{\nu_A}$ as in (2.100).*

Let w be a solution to the duality problem (2.46). We introduce its energy by

$$2E(t) = \int_\Omega (w_t^2 + \langle A(x)\nabla w, \nabla w \rangle)dx \quad \text{for} \quad t \geq 0 \tag{2.55}$$

where $\langle \cdot, \cdot \rangle$ is the Euclidean metric of $\mathbb{R}^n$. Then $E(t) = E(0)$ for all $t \geq 0$.

Then the exact controllability of the problem (2.45) on $L^2(\Omega) \times H^{-1}(\Omega)$ follows from Lemma 2.4 and Theorem 2.15 below.

Lemma 2.4 *For any $T > 0$, there is $C_T > 0$ such that*

$$\|w_{\nu_A}\|_{L^2(\Sigma)}^2 \leq C_T E(0) \tag{2.56}$$

for all solutions to the problem (2.46).

Proof. Let $\hat{H}$ be a vector field on $\overline{\Omega}$ such that

$$\hat{H} = \nu_A \quad \text{for} \quad x \in \Gamma$$

where $\nu_A(x) = A(x)\nu(x)$. For $X \in \mathcal{X}(\Gamma)$, by (i) in Lemma 2.1, we have

$$\langle \nabla_g w, X \rangle_g = \langle \nabla w, X \rangle = X(w) = 0 \quad \text{for} \quad x \in \Gamma$$

since $w|_\Gamma = 0$. Then

$$\nabla_g w = \langle \nabla_g w, \frac{\nu_A}{|\nu_A|_g} \rangle_g \frac{\nu_A}{|\nu_A|_g} = w_{\nu_A} \frac{\nu_A}{|\nu_A|_g^2}$$

which yields

$$|\nabla_g w|_g^2 = \nabla_g w(w) = w_{\nu_A}^2 / |\nu_A|_g^2 \quad \text{for} \quad x \in \Gamma. \tag{2.57}$$

After taking $H = \hat{H}$, $u = w$, and $f = 0$ in the identity (2.13), we integrate it over Q to have the estimate (2.56). $\qquad\square$

Next, we need a uniqueness result which is introduced from [5] or [170] without proofs.

Proposition 2.2 *Let X be a vector field and let q be a smooth function on $\overline{\Omega}$, respectively. Then the problem*

$$\text{div}\, A(x)\nabla v = X(v) + qv \quad \text{on} \quad \Omega, \quad v|_{\Gamma_0} = v_{\nu_A}|_{\Gamma_0} = 0 \tag{2.58}$$

has a unique zero solution.

We have

Theorem 2.15 *Let H be an escape vector field for the metric g on $\overline{\Omega}$, where g is given by (2.4). Then for any $T > T_0$, there is a constant $c_T > 0$ such that*

$$\|w_{\nu_A}\|^2_{L^2(\Sigma_0)} \geq c_T E(0) \tag{2.59}$$

for all solution w to the problem (2.46), where

$$T_0 = \frac{2}{\varrho_0} \sup_{x \in \Omega} |H|_g(x), \quad \Gamma_0 = \{\, x \in \Gamma \,|\, \langle H, \nu \rangle > 0 \,\}, \tag{2.60}$$

and $\varrho_0 > 0$ is given by (2.15).

Proof. We decompose H into the direct sum in $(\mathbb{R}^n_x, g(x))$ as

$$H = \langle H, \frac{\nu_A}{|\nu_A|_g} \rangle_g \frac{\nu_A}{|\nu_A|_g} + Z = \frac{\langle H, \nu \rangle}{|\nu_A|^2_g} \nu_A + Z \quad \text{for} \quad x \in \Gamma$$

where $\langle Z, \nu_A \rangle_g = \langle Z, \nu \rangle = 0$, which gives

$$H(w) = \langle H, \nu \rangle w_{\nu_A} / |\nu_A|^2_g \quad \text{for} \quad x \in \Gamma. \tag{2.61}$$

Taking $f = 0$ and $u = w$ in the identity (2.13) and using the assumption (2.15), we obtain

$$\operatorname{div} X_1 \geq 2\varrho_0 |\nabla_g w|^2_g + (w^2_t - |\nabla_g w|^2_g) \operatorname{div} H + 2[w_t H(w)]_t$$
$$= \varrho_0(w^2_t + |\nabla_g w|^2_g) + (w^2_t - |\nabla_g w|^2_g)(\operatorname{div} H - \varrho_0) + 2[w_t H(w)]_t \tag{2.62}$$

where

$$X_1 = 2H(w)A(x)\nabla w - (|\nabla_g w|^2_g - w^2_t)H.$$

Letting $f = 0$, $u = w$, and $p = (\operatorname{div} H - \varrho_0)/2$ in the identity (2.14), and then adding it to the inequality (2.62), we get

$$\operatorname{div} X_2 \geq \varrho_0(w^2_t + |\nabla_g w|^2_g) - w^2 \operatorname{div} A(x)\nabla p$$
$$+ 2p(ww_t)_t + 2[w_t H(w)]_t \tag{2.63}$$

where

$$X_2 = X_1 + 2pwA(x)\nabla w - w^2 A(x)\nabla p.$$

Moreover, it follows from the relations (2.61) and (2.57) and the boundary conditions in the problem (2.46) that

$$\langle X_2, \nu \rangle = \langle H, \nu \rangle w^2_{\nu_A} / |\nu_A|^2_g \quad \text{for} \quad x \in \Gamma. \tag{2.64}$$

We integrate the inequality (2.63) over Q and obtain, by (2.64),

$$c\|w_{\nu_A}\|^2_{L^2(\Sigma_0)} \geq 2(\varrho_0 T - \varepsilon - 2 \sup_{x \in \Omega} |H|_g)E(0) - C_\varepsilon L(w)$$

for $\varepsilon > 0$ small where

$$L(w) = (\|w\|^2_{L^2(Q)} + \|w(0)\|^2 + \|w(T)\|^2)$$

is the lower order term. Since $T > T_0$, there are constants $c_i > 0$ for $1 \le i \le 3$ such that

$$c_1\|w_{\nu_A}\|^2_{L^2(\Sigma_0)} + c_2 L(w) \ge c_3 E(0) \tag{2.65}$$

for all solutions w to the problem (2.46). Then the inequality (2.59) follows by Lemma 2.5 below. □

Remark 2.13 T_0 *in* (2.60) *represents a control time which can be shown to be the shortest in some cases of constant coefficients.*

Lemma 2.5 *Let the inequality* (2.65) *hold true for all solutions w to the problem* (2.46). *Then there is $c > 0$, independent of solutions, such that*

$$\|w_{\nu_A}\|^2_{L^2(\Sigma_0)} \ge cE(0). \tag{2.66}$$

Proof. **Step 1** Let

$$Y = \{\, w \in H^1(Q) \,|\, w \text{ is a solution to (2.46) satisfying } w_{\nu_A}|_{\Sigma_0} = 0 \,\}.$$

Then

$$Y = \{\, 0 \,\}. \tag{2.67}$$

Indeed, from the inequality (2.65), we have

$$c_2 L(w) \ge c_3 E(0) \quad \text{for all} \quad w \in Y$$

which implies that any bounded, closed set in $Y \cap H^1(Q)$ is compact in $H^1(Q)$. Then Y is a finitely dimensional, linear space.

For any $w \in Y$, $w_t \in Y$. Then $\partial_t : Y \to Y$ is a linear operator. Let $Y \ne \{\, 0 \,\}$. Then ∂_t has at least one eigenvalue λ. Assume that $v \in Y$, $v \ne 0$, is one of its eigenfunctions. Then $v_t = \lambda v$. Thus v is a nonzero solution to the problem (2.58) with $X = 0$ and $q = \lambda^2$. This contradiction shows that (2.67) is true.

Step 2 By contradiction. Suppose that the estimate (2.66) is not true. Then there are $(w_0^k, w_1^k) \in H_0^1(\Omega) \times L^2(\Omega)$, whose solutions are denoted by w^k, such that

$$E(w^k, 0) = 1, \quad \|w_{\nu_A}^k\|^2_{L^2(\Sigma_0)} \le \frac{1}{k} \quad \text{for} \quad k \ge 1. \tag{2.68}$$

Then $\|w^k\|^2_{H^1(Q)} = 2T$ for all $k \ge 1$. Thus there is a subsequence, still denoted by $\{\, w^k \,\}$, such that

$$\{\, w^k \,\} \text{ converges in } L^2(\Omega) \text{ for each } t \in [0, T] \text{ and} \tag{2.69}$$

$$\{\, w^k \,\} \text{ converges in } L^2(Q). \tag{2.70}$$

It follows from the relations (2.65), (2.68), (2.69), and (2.70) that $\{w^k\}$ converges in $H^1(Q)$. Then there is a solution w^0 to the problem (2.46) such that

$$w^k \to w^0 \quad \text{as } k \text{ goes to } \infty \text{ in } \quad H^1(Q).$$

Then

$$E(w^0, 0) = 1, \quad w^0_{\nu_A}|_{\Sigma_0} = 0,$$

which contradicts the relation (2.67). □

Remark 2.14 *The proof of the above lemma is referred to as a compactness-uniqueness argument in the literature.*

Control in Neumann Condition (Γ, g) is an $(n-1)$-dimensional Riemannian manifold where g is the induced metric from $(\mathbb{R}^n, g)$. Denote by ∇_{Γ_g} the gradient of (Γ, g). Then for any $v \in H^1(Q)$,

$$|\nabla_g v|^2_g = v^2_{\nu_A}/|\nu_A|^2_g + |\nabla_{\Gamma_g} v|^2_g \quad \text{for} \quad x \in \Gamma.$$

We consider the control problem

$$\begin{cases} u_{tt} = \operatorname{div} A(x)\nabla u & \text{in} \quad Q, \\ u_{\nu_A} = \varphi & \text{on} \quad \Sigma, \\ u(0) = u_0, \quad u_t(0) = u_1 & \text{on} \quad \Omega, \end{cases} \tag{2.71}$$

with a Neumann action φ on Σ. The duality problem is

$$\begin{cases} w_{tt} = \operatorname{div} A(x)\nabla w & \text{in} \quad Q, \\ w_{\nu_A} = 0 & \text{on} \quad \Sigma, \\ w(0) = w_0, \quad w_t(0) = w_1 & \text{on} \quad \Omega. \end{cases} \tag{2.72}$$

For $T > 0$, let

$$L_0 = L^2(\Sigma_0), \quad H_0 = H^1([0, T]; L^2(\Gamma_0)),$$

$$L_1 = L^2(\Sigma_1), \quad H_1 = L^2([0, T]; H^1(\Gamma_1)).$$

Denote by $\aleph_i \colon H_i \to H_i^*$ the canonical isomorphisms with respect to L_i for $i = 0, 1$. Then it follows from (2.49) that

$$(\aleph_0 \phi, \varphi)_{L^2(\Sigma_0)} = \int_{\Sigma_0} (\phi_t \varphi_t + \phi\varphi)d\Sigma \tag{2.73}$$

for all $\phi, \varphi \in H^1([0, T]; L^2(\Gamma_0))$, and

$$(\aleph_1 \phi, \varphi)_{L^2(\Sigma_1)} = \int_{\Sigma_1} (\langle \nabla_{\Gamma_g}\phi, \nabla_{\Gamma_g}\varphi\rangle_g + \phi\varphi)d\Sigma \tag{2.74}$$

for all $\phi, \varphi \in L^2([0, T]; H^1(\Gamma_1))$.

For $(w_0, w_1) \in H^1(\Omega) \times L^2(\Omega)$, we solve the problem (2.72) to obtain w. Then we solve the problem

$$\begin{cases} u_{tt} = \text{div } A(x)\nabla u & \text{in } Q, \\ u(T) = u_t(T) = 0 & \text{on } \Omega, \\ u_{\nu_A} = \begin{cases} -\aleph_0 w & \text{on } \Sigma_0 \\ -\aleph_1 w & \text{on } \Sigma_1. \end{cases} \end{cases} \tag{2.75}$$

We define an operator $\Lambda: H^1(\Omega) \times L^2(\Omega) \to (H^1(\Omega))^* \times L^2(\Omega)$ by

$$\Lambda(w_0, w_1) = (u_t(0), -u(0)) \quad \text{on } \Omega.$$

We take $H = H^1(\Omega) \times L^2(\Omega)$ and $L = L^2(\Omega) \times L^2(\Omega)$. Let w and $\hat{w}$ be the solutions to the problem (2.72) for the initial data (w_0, w_1) and $(\hat{w}_0, \hat{w}_1)$, respectively. Suppose that u and $\hat{u}$ are the solutions to the problem (2.75) with the corresponding boundary data from w and $\hat{w}$, respectively. Using (2.72), (2.75), (2.73) and (2.74), we have

$$(\Lambda(w_0, w_1), (\hat{w}_0, \hat{w}_1))_{L^2(\Omega) \times L^2(\Omega)} = (u_t(0), \hat{w}(0)) - (u(0), \hat{w}_t(0))$$

$$= \int_0^T [(u, \hat{w}_t) - (u_t, \hat{w})]_t d\tau = -(\hat{w}, u_{\nu_A})_{L^2(\Sigma)}$$

$$= \int_{\Sigma_0} (w_t \hat{w}_t + w\hat{w})d\Sigma + \int_{\Sigma_1} (\langle \nabla_{\Gamma_g} w, \nabla_{\Gamma_g} \hat{w}\rangle_g + w\hat{w})d\Sigma. \tag{2.76}$$

Remark 2.15 *If a solution w to the duality problem (2.72) such that*

$$\int_{\Sigma_0} (w_t^2 + w^2)d\Sigma + \int_{\Sigma_1} (|\nabla_{\Gamma_g} w|_g^2 + w^2)d\Sigma$$

$$= (\Lambda(w_0, w_1), (w_0, w_1))_{L^2(\Omega) \times L^2(\Omega)} < \infty,$$

that is, $w|_{\Sigma_0} \in H^1([0,T]; L^2(\Gamma_0))$ and $w|_{\Sigma_1} \in L^2([0,T]; H^1(\Gamma_1))$, then the boundary control in the problem (2.75) satisfies

$$\aleph_0 w \in [H^1([0,T]; L^2(\Gamma_0))]^*; \quad \aleph_1 w \in [L^2([0,T]; H^1(\Gamma_1))]^*.$$

Then in Theorem 2.13, $(\Lambda x, x)_L < \infty$ describes the regularity of control action.

Applying Theorem 2.13, we have

Theorem 2.16 *If there is $c_T > 0$ such that*

$$\int_{\Sigma_0} (w_t^2 + w^2)d\Sigma + \int_{\Sigma_1} (|\nabla_{\Gamma_g} w|_g^2 + w^2)d\Sigma \geq c_T E(0) \tag{2.77}$$

for all solutions w to the problem (2.72), then the problem (2.71) is exactly $L^2(\Omega) \times (H^1(\Omega))^*$ controllable by the Neumann action.

Next, we establish the observability inequality (2.77).

Theorem 2.17 *Let H be an escape vector field for the metric g on $\overline{\Omega}$. Then the inequality (2.77) holds true for any $T > T_0$ where T_0 and Γ_0 are given in (2.60).*

Proof. We take $u = w$ and $f = 0$ in the identities (2.13) and (2.14). Let $p = (\operatorname{div} H - \varrho_0)/2$ where ϱ_0 is given in (2.15). Then we add the identity (2.14) to the identity (2.13) and obtain, as in (2.63),

$$\operatorname{div} X \geq \varrho_0(w_t^2 + |\nabla_g w|_g^2) + 2[w_t H(w) + p w w_t]_t - w^2 \operatorname{div} A(x)\nabla p \quad (2.78)$$

where

$$X = 2H(w)A(x)\nabla w - (|\nabla_g w|_g^2 - w_t^2)H + 2pwA(x)\nabla w - w^2 A(x)\nabla p.$$

Using the boundary conditions $w_{\nu_A} = 0$, we have

$$\int_{\Sigma} \langle X, \nu \rangle d\Sigma = \int_{\Sigma} [(w_t^2 - |\nabla_g w|_g^2)\langle H, \nu \rangle + w^2] d\Sigma - \int_{\Sigma} w^2 (1 + p_{\nu_A}) d\Sigma$$

$$\leq C\mathcal{I}(w) + C\|w\|_{H^{1/2}(Q)}^2 \quad (2.79)$$

where $\mathcal{I}(w)$ denotes the left hand side of the inequality (2.77).

We integrate the inequality (2.78) over Q and obtain, by (2.79),

$$C\mathcal{I}(w) + C_\varepsilon L(w) \geq 2(\varrho_0 T - 2 \sup_{x \in \Omega} |H|_g - \varepsilon)E(0) \quad (2.80)$$

for $\varepsilon > 0$ small enough, where $L(w)$ are lower order terms with respect to the norm of $H^1(Q)$.

The inequality (2.77) follows from (2.80) by a compactness-uniqueness argument as for Lemma 2.5. $\qquad\square$

Remark 2.16 *In Theorem 2.16 controls are taken in "large" spaces where $\aleph_0 w \in (H^1([0,T]; L^2(\Gamma_0)))^*$ and $\aleph_1 w \in (L^2([0,T]; H^1(\Gamma_1)))^*$. In general, even if w is smooth, $\aleph_i(w)$ may be not in $L^2(\Sigma_i)$. However, if the boundary Γ satisfies the following assumption (2.81), then controls can be taken in $L^2(\Sigma)$, see Theorem 2.18 below.*

Let H be an escape vector field for the metric g such that

$$\langle H, \nu \rangle \geq 0 \quad \text{for all} \quad x \in \Gamma. \quad (2.81)$$

Let T_0 be given in (2.60). Then an argument as in the proof of Theorem 2.17 yields the estimate

$$\int_{\Sigma} (w_t^2 + w^2) d\Sigma \geq c_T E(0) \quad (2.82)$$

where $T > T_0$ is given and w are solutions to the problem (2.72).

We use a trick in [146]. Let w_1 satisfy

$$\int_{\Omega} w_1 dx = 0.$$

Assume that ω solves

$$\text{div } A(x)\nabla\omega = w_1 \quad \text{on} \quad \Omega, \quad \omega_{\nu_A} = 0 \quad \text{on} \quad \Gamma.$$

Let

$$v = \int_0^t w d\tau + \omega.$$

Then v solves the problem (2.72) with initial data $v(0) = \omega$ and $v_t(0) = w_0$. Applying v to the inequality (2.82) gives

$$\int_\Sigma w^2 d\Sigma \geq c_T(\|w_0\|^2 + \|w_1\|^2_{(H^1(\Omega))^*}). \tag{2.83}$$

It follows from Theorem 2.13 that

Theorem 2.18 *Let H be an escape vector field for the metric g on $\overline{\Omega}$ and let T_0 be given in (2.60). Then the problem (2.71) is exactly $H^1(\Omega) \times L^2(\Omega)$ controllable by $L^2(\Sigma)$ control by the Neumann action.*

Finally, we consider the exact controllability problem

$$\begin{cases} u_{tt} = \text{div } A(x)\nabla u & \text{in} \quad Q, \\ u|_{\Sigma_1} = 0, \quad u_{\nu_A}|_{\Sigma_0} = \varphi, \\ u(0) = u_0, \quad u_t(0) = u_1 & \text{on} \quad \Omega \end{cases} \tag{2.84}$$

where $\Gamma_0 \cup \Gamma_1 = \Gamma$. The duality problem of the problem (2.84) is

$$\begin{cases} w_{tt} = \text{div } A(x)\nabla w & \text{in} \quad Q, \\ w|_{\Sigma_1} = w_{\nu_A}|_{\Sigma_0} = 0, \\ w(0) = w_0, \quad w_t(0) = w_1 & \text{on} \quad \Omega. \end{cases} \tag{2.85}$$

It is easy to check that the exact controllability of the problem (2.84) on the space $L^2(\Omega) \times (H^1_{\Gamma_1}(\Omega))^*$ leads to the following observability estimate: There is a constant $c_T > 0$ such that

$$\|w_t\|^2_{L^2(\Sigma_0)} \geq c_T E(0) \tag{2.86}$$

for all solutions w to the problem (2.85).

We have

Theorem 2.19 *Let $\Gamma_1 \neq \emptyset$. Let H be an escape vector field for the metric g on $\overline{\Omega}$ such that*

$$\langle H, \nu \rangle \leq 0 \quad \text{for} \quad x \in \Gamma_1. \tag{2.87}$$

Then for $T > T_0$, the estimate (2.86) holds true.

Proof. Let $\Gamma_{0+} = \{ x \,|\, \langle H, \nu \rangle \geq 0, x \in \Gamma_0 \}$. Using the boundary conditions $w|_{\Sigma_1} = w_{\nu_A}|_{\Sigma_0} = 0$, we obtain, via (2.87),

$$\int_{\Sigma_1} \langle X, \nu \rangle d\Sigma = \int_{\Sigma_1} \langle H, \nu \rangle w^2_{\nu_A} / |\nu_A|^2_g d\Sigma \leq 0 \tag{2.88}$$

where X is given in (2.78). Similar arguments as in the proof of Theorem 2.17 complete the proof. $\square$

2.5   Smooth Controls

We consider smooth controls which are needed for the control of the quasi-linear wave equation in Chapter 7.

Control with the Dirichlet Action Let F and q be a vector field and a function on $\overline{\Omega}$, respectively. Let $T > 0$ be given. We consider the problem

$$
\begin{cases}
u_{tt} = \operatorname{div} A(x)\nabla u + F(u) + qu & \text{in} \quad Q, \\
u = 0 & \text{on} \quad \Sigma_1, \\
u = \varphi & \text{on} \quad \Sigma_0, \\
u(0) = u_0, \quad u_t(0) = u_1 & \text{on} \quad \Omega.
\end{cases}
\tag{2.89}
$$

The duality system is

$$
\begin{cases}
w_{tt} = \mathcal{A}w & \text{in} \quad Q \\
w = 0 & \text{on} \quad \Sigma \\
w(0) = w_0, \quad w_t(0) = w_1 & \text{on} \quad \Omega
\end{cases}
\tag{2.90}
$$

where the operator $\mathcal{A}$ is defined by

$$
\mathcal{A}w = \operatorname{div} A(x)\nabla w - F(w) + (q - \operatorname{div} F)w.
\tag{2.91}
$$

We have the following Green formula

$$
(v, \mathcal{A}u) = (\mathcal{A}^\star v, u) + \int_\Gamma [vu_{\nu_A} - uv_{\nu_A} - uv\langle F, \nu\rangle]\, d\Gamma
\tag{2.92}
$$

for $u, v \in H^2(\Omega)$, where

$$
\mathcal{A}^\star u = \operatorname{div} A(x)\nabla u + F(u) + qu.
\tag{2.93}
$$

Let $m \geq 0$ be a given integer. We shall study the exact controllability of the problem (2.89) on $[H^{m+1}(\Omega) \cap H^1_{\Gamma_1}(\Omega)] \times [H^m(\Omega) \cap H^1_{\Gamma_1}(\Omega)]$ by appropriate boundary controls where $H^1_{\Gamma_1}(\Omega) = \{v \in H^1(\Omega)\,|\,v|_{\Gamma_1} = 0\}$. To this end, we make some preparation as follows.

Space of Boundary Control Let $B^0_D(\Sigma_0) = L^2(\Sigma_0)$. For $m \geq 1$, let $B^m_D(\Sigma_0)$ consist of all the functions

$$
\varphi \in \cap_{k=0}^{m-1} C^k\left([0, T], H^{m-1/2-k}(\Gamma_0)\right), \quad \varphi^{(k)} \in H^1(\Sigma_0),
\tag{2.94}
$$

for $0 \leq k \leq m - 1$ with the norm

$$
\|\varphi\|^2_{B^m_D} = \sum_{k=0}^{m-1} \|\varphi^{(k)}\|^2_{C([0,T], H^{m-k-1/2}(\Gamma_0))} + \sum_{k=0}^{m-1} \|\varphi^{(k)}\|^2_{H^1(\Sigma_0)}
$$

where the superscript (k) denotes the k-th derivative with respect to time variable t.

Let $\Xi_0^0(\Omega) = L^2(\Omega)$. For $k \geq 1$ an integer, let $\Xi_0^k(\Omega)$ consist of the functions u in $H^k(\Omega)$ with the boundary conditions

$$\begin{cases} \mathcal{A}^i u|_\Gamma = 0, & 0 \leq i \leq l-1, \quad \text{if} \quad k = 2l; \\ \mathcal{A}^i u|_\Gamma = 0, & 0 \leq i \leq l \quad \text{if} \quad k = 2l+1 \end{cases} \tag{2.95}$$

and with the norms of $H^k(\Omega)$ where $\mathcal{A}$ is given by (2.91). Then

$$\Xi_0^1(\Omega) = H_0^1(\Omega), \quad \Xi_0^2(\Omega) = H^2(\Omega) \cap H_0^1(\Omega),$$

$$\Xi_0^3(\Omega) = \{\, w \mid w \in H^3(\Omega),\ w|_\Gamma = 0,\ \mathcal{A}w|_\Gamma = 0 \,\}.$$

We consider the operator $\mathcal{A}$ on $L^2(\Omega)$, given by (2.91) with a domain

$$\mathcal{D}(\mathcal{A}) = H^2(\Omega) \cap H_0^1(\Omega). \tag{2.96}$$

It follows from (2.91) and (2.92) that the adjoint operator of $\mathcal{A}$ is given by

$$\mathcal{A}^* w = \mathcal{A}^\star w, \quad \mathcal{D}(\mathcal{A}^*) = H^2(\Omega) \cap H_0^1(\Omega) \tag{2.97}$$

where $\mathcal{A}^\star$ is given by (2.93). We have

Lemma 2.6 *Let l be a positive integer. Consider the operator $\mathcal{A}^l \colon H^{2l}(\Omega) \cap \Xi_0^{2l}(\Omega) \to L^2(\Omega)$. Then*

$$L^2(\Omega) = \mathcal{R}(\mathcal{A}^l) \oplus \mathcal{N}(\mathcal{A}^{*l}) \tag{2.98}$$

where

$$\mathcal{N}(\mathcal{A}^{*l}) = \{\, w \mid \mathcal{A}^{*l} w = 0,\ w \in H^{2l}(\Omega) \cap \Xi_0^l(\Omega) \,\}$$

is a finitely dimensional subspace of $L^2(\Omega)$.

Proof. It will suffice to prove that $\mathcal{R}(\mathcal{A}^l)$ is a closed subspace of $L^2(\Omega)$ and $\mathcal{N}(\mathcal{A}^{*l})$ is finitely dimensional. Set

$$\sigma_0 > \max\{\sup_{x \in \Omega}(q - \operatorname{div} F),\ 0\}.$$

Then

$$\mathcal{A}w = \mathcal{A}_{\sigma_0} w + \sigma_0 w \quad \text{for} \quad w \in \mathcal{D}(\mathcal{A})$$

where the operator $\mathcal{A}_{\sigma_0}$ is defined by

$$\mathcal{A}_{\sigma_0} w = \operatorname{div} A(x)\nabla w - F(w) + (q - \operatorname{div} F - \sigma_0)w,$$

$$\mathcal{D}(\mathcal{A}_{\sigma_0}) = H^2(\Omega) \cap H_0^1(\Omega).$$

Since $q - \operatorname{div} F - \sigma_0 \leq 0$, by the regularity of elliptic operators (see Theorem

8.12 in [68]), 0 belongs to the resolvent set of $\mathcal{A}_{\sigma_0}$ and $\mathcal{A}_{\sigma_0}^{-1}\colon L^2(\Omega) \to H^2(\Omega) \cap H_0^1(\Omega)$ is a compact operator on $L^2(\Omega)$. Let

$$\mathcal{C} = (I + \sigma_0 \mathcal{A}_{\sigma_0}^{-1})^l - I$$

where I is the identity operator on $L^2(\Omega)$. Then $\mathcal{C}$ is a compact operator on $L^2(\Omega)$. Thus $\mathcal{R}(I + \mathcal{C})$ is closed in $L^2(\Omega)$. Then

$$\mathcal{R}(\mathcal{A}^l) = \mathcal{R}(\mathcal{A}_{\sigma_0}^l(I + \mathcal{C}))$$

is closed in $L^2(\Omega)$.

Similarly, we can take a constant $\sigma_1 > \max\{q, 0\}$. Then $q - \sigma_1 \leq 0$. Thus the operator $\mathcal{A}_{\sigma_1}^* = \mathcal{A}^* - \sigma_1 I$ has the compact inverse. Then

$$\mathcal{A}^{*l} = \mathcal{A}^{*l}_{\sigma_1} + \sum_{i=1}^l C_i^l \sigma_1^i \mathcal{A}^{*l-i}_{\sigma_1}.$$

Then $\mathcal{A}^{*l} w = 0$ if and only if

$$w + \sum_{i=1}^l C_i^l \sigma_1^i \mathcal{A}^{*-i}_{\sigma_1} w = 0.$$

Then $\mathcal{N}(\mathcal{A}^{*l})$ is finitely dimensional since the operator $\sum_{i=1}^l C_i^l \sigma_1^i \mathcal{A}^{*-i}_{\sigma_1}$ is compact. □

Let $T > T_1 > 0$ be given. We assume that $z \in C^\infty(-\infty, \infty)$ is such that $0 \leq z(t) \leq 1$ with

$$z(t) = \begin{cases} 0 & t \geq T, \\ 1 & t \leq T_1. \end{cases} \tag{2.99}$$

For $(w_0, w_1) \in \Xi_0^{m+1}(\Omega) \times \Xi_0^m(\Omega)$ given, we solve the duality problem (2.90) and, then, we solve the following problem

$$\begin{cases} u_{tt} = \mathcal{A}^* u & \text{in} \quad Q, \\ u(T) = u_t(T) = 0 & \text{on} \quad \Omega, \\ u|_{\Sigma_1} = 0, \quad u|_{\Sigma_0} = z w_{\nu_A} \end{cases} \tag{2.100}$$

where $\mathcal{A}^*$ is defined by (2.93).

An operator Λ on $\Xi_0^{m+1}(\Omega) \times \Xi_0^m(\Omega)$ is defined by

$$\Lambda(w_0, w_1) = (u_t(0), -u(0)). \tag{2.101}$$

It is easy to check that, for any $(w_0, w_1), (\hat{w}_0, \hat{w}_1) \in H_0^1(\Omega) \times L^2(\Omega)$,

$$(\Lambda(w_0, w_1), (\hat{w}_0, \hat{w}_1))_{L^2(\Omega) \times L^2(\Omega)} = \int_{\Sigma_0} z(t) w_{\nu_A} \hat{w}_{\nu_A} d\Sigma \tag{2.102}$$

where w and $\hat{w}$ are solutions to the duality problem (2.90) with initial data (w_0, w_1) and $(\hat{w}_0, \hat{w}_1)$, respectively.

A similar argument as for Lemma 2.4 yields

Lemma 2.7 *There is a constant $C > 0$ such that*

$$\|w_{\nu_A}\|^2_{L^2(\Sigma)} \le C\|(w_0, w_1)\|^2_{H^1_0(\Omega) \times L^2(\Omega)} \tag{2.103}$$

for all solutions w to the problem (2.90).

The formula (2.102) implies the regularity of the operator Λ, that is, Proposition 2.3 below. For convenience, we denote by $\|\cdot\|_{i,j}$ the norms of $H^i(\Omega) \times H^j(\Omega)$ for $i, j \ge -1$.

Proposition 2.3 *Let $m \ge 0$ be an integer. Then the operator Λ, given by (2.101), is bounded from $\Xi^{m+1}_0(\Omega) \times \Xi^m_0(\Omega)$ to $H^{m-1}(\Omega) \times H^m(\Omega)$.*

Proof. We use induction on m.

(1) Let $m = 0$. It follows from the relations (2.102) and (2.103) that

$$\|\Lambda(w_0, w_1)\|_{-1,0} = \sup_{\|(\hat{w}_0, \hat{w}_1)\|_{1,0}=1} |(zw_{\nu_A}, \hat{w}_{\nu_A})_{L^2(\Sigma)}| \le C\|(w_0, w_1)\|_{1,0},$$

that is, the claim is true for $m = 0$.

(2) Assume that the claim holds for some $m \ge 0$. We shall prove that the claim is true for $m + 1$. It will suffice if we establish the estimate

$$\|u_t(0)\|_m + \|u(0))\|_{m+1} \le C(\|w_0\|_{m+2} + \|w_1\|_{m+1}) \tag{2.104}$$

for all $(w_0, w_1) \in \Xi^{m+2}_0(\Omega) \times \Xi^{m+1}_0(\Omega)$.

Case 1 Let $m = 2l$ with $l \ge 0$. Then $m + 1 = 2l + 1$.

Let $\hat{w}$ be the solution to the duality problem (2.90) with the initial data $(\hat{w}_0, \hat{w}_1)$. Then $\hat{w}^{(2i)}$ and $\hat{w}^{(2i+1)}$ are the solutions to the problem (2.90) corresponding to the initial data $(\mathcal{A}^i \hat{w}_0, \mathcal{A}^i \hat{w}_1)$ and $(\mathcal{A}^i \hat{w}_1, \mathcal{A}^{i+1} \hat{w}_0)$, respectively, for $i \ge 0$, where $\mathcal{A}$ is given by (2.91).

Using initial data (w_0, w_1) and $(\mathcal{A}^{m+1} \hat{w}_0, \mathcal{A}^{m+1} \hat{w}_1)$ in the formula (2.102), we have

$$(u_t(0), \mathcal{A}^{m+1} \hat{w}_0) - (u(0), \mathcal{A}^{m+1} \hat{w}_1) = \int_{\Sigma_0} z(t) w_{\nu_A} \hat{w}^{(2m+2)}_{\nu_A} d\Sigma. \tag{2.105}$$

Noting that $z(0) = 1$ and $z^{(i-1)}(T) = z^{(i)}(0) = 0$ for $i \ge 1$, we obtain, by integration by parts with respect to the variable t over $[0, T]$,

$$\int_{\Sigma_0} z(t) w_{\nu_A} \hat{w}^{(2m+2)}_{\nu_A} d\Sigma = \sum_{j=1}^{m+1} \int_{\Gamma_0} (-1)^j w^{(j-1)}_{\nu_A}(0) \hat{w}^{(2m+2-j)}_{\nu_A}(0) d\Gamma$$

$$+ (-1)^{m+1} \int_{\Sigma_0} [z(t) w_{\nu_A}]^{(m+1)} \hat{w}^{(m+1)}_{\nu_A} d\Sigma$$

$$= \sum_{j=0}^{l-1} \int_{\Gamma_0} (\mathcal{A}^j w_1)_{\nu_A} (\mathcal{A}^{m-j} \hat{w}_0)_{\nu_A} d\Gamma - \sum_{j=0}^{l} \int_{\Gamma_0} (\mathcal{A}^j w_0)_{\nu_A} (\mathcal{A}^{m-j} \hat{w}_1)_{\nu_A} d\Gamma$$

$$+ (-1)^{m+1} \mathcal{I}(w, \hat{w}) \tag{2.106}$$

where

$$\mathcal{I}(w, \hat{w}) = \int_{\Sigma_0} z(t) w_{\nu_A}^{(m+1)} \hat{w}_{\nu_A}^{(m+1)} d\Sigma$$

$$+ \sum_{j=1}^{m+1} \int_{\Sigma_0} C_j^{m+1} z^{(j)}(t) w_{\nu_A}^{(m+1-j)} \hat{w}_{\nu_A}^{(m+1)} d\Sigma. \qquad (2.107)$$

We assume that $(\hat{w}_0, \hat{w}_1) \in C_0^\infty(\Omega) \times C_0^\infty(\Omega)$. Noting that $(\mathcal{A}^i \hat{w}_0, \mathcal{A}^j \hat{w}_1) \in C_0^\infty(\Omega) \times C_0^\infty(\Omega)$ for all i, j, using the formula (2.92), and the boundary in (2.95), we obtain

$$(u(0), \mathcal{A}^{m+1}\hat{w}_1) = (\mathcal{A}^{\star l} u(0), \mathcal{A}^{l+1}\hat{w}_1) + \sum_{j=0}^{l-1} \int_\Gamma \mathcal{A}^{\star j} u(0) (\mathcal{A}^{m-j}\hat{w}_1)_{\nu_A} d\Gamma$$

$$= -(A\nabla(\mathcal{A}^{\star l} u(0)), \nabla(\mathcal{A}^l \hat{w}_1)) + \sum_{j=0}^{l} \int_{\Gamma_0} u^{(2j)}(0) (\mathcal{A}^{m-j}\hat{w}_1)_{\nu_A} d\Gamma$$

$$-(\mathcal{A}^{\star l} u(0), F(\mathcal{A}^l \hat{w}_1) + \mathcal{A}^l \hat{w}_1(\operatorname{div} F - q)) \qquad (2.108)$$

since u solves the problem (2.100). Similarly, we have

$$(u_t(0), \mathcal{A}^{m+1}\hat{w}_0) = (\mathcal{A}^{\star l} u_t(0), \mathcal{A}^{l+1}\hat{w}_0)$$

$$+ \sum_{j=0}^{l-1} \int_{\Gamma_0} u^{(2j+1)}(0) (\mathcal{A}^{m-j}\hat{w}_0)_{\nu_A} d\Gamma. \qquad (2.109)$$

Since $u^{(2j+1)}(0) = w_{\nu_A}^{(2j+1)}(0) = (\mathcal{A}^j w_1)_{\nu_A}$ and $u^{(2j)}(0) = w_{\nu_A}^{(2j)}(0) = (\mathcal{A}^j w_0)_{\nu_A}$ on Γ_0, it follows from the relations (2.105)-(2.108) that

$$(\mathcal{A}^{\star l} u_t(0), \mathcal{A}^{l+1}\hat{w}_0) + (A\nabla(\mathcal{A}^{\star l} u(0)), \nabla(\mathcal{A}^l \hat{w}_1))$$

$$= -\mathcal{I}(w, \hat{w}) - (\mathcal{A}^{\star l} u(0), F(\mathcal{A}^l \hat{w}_1) + \mathcal{A}^l \hat{w}_1(\operatorname{div} F - q)) \qquad (2.110)$$

for all $(w_0, w_1) \in \Xi_0^{m+2}(\Omega) \times \Xi_0^{m+1}(\Omega)$ and $(\hat{w}_0, \hat{w}_1) \in C_0^\infty(\Omega) \times C_0^\infty(\Omega)$.

Letting $\hat{w}_0 = 0$ in the identity (2.110), we obtain

$$(\mathcal{A}_0(\mathcal{A}^{\star l} u(0)), \mathcal{A}^l \hat{w}_1)$$

$$= \mathcal{I}(w, \hat{w}) + (\mathcal{A}^l u(0), F(\mathcal{A}^l \hat{w}_1) + \mathcal{A}^l \hat{w}_1(\operatorname{div} F - q)) \qquad (2.111)$$

for all $\hat{w}_1 \in C_0^\infty(\Omega)$, where

$$\mathcal{A}_0 = \operatorname{div} A(x)\nabla.$$

From (2.107) and (2.103), we have

$$
|\mathcal{I}(w,\hat{w})| \leq \sum_{j=0}^{l} |\int_{\Sigma_0} C_{2j}^{m+1} z^{(2j)}(t) w_{\nu_{\mathcal{A}}}^{(2(l-j)+1)} \hat{w}_{\nu_{\mathcal{A}}}^{(m+1)} d\Sigma|
$$

$$
+ C\sum_{j=0}^{l} |\int_{\Sigma_0} C_{2j+1}^{m+1} z^{(2j+1)}(t) w_{\nu_{\mathcal{A}}}^{(2(l-j))} \hat{w}_{\nu_{\mathcal{A}}}^{(m+1)} d\Sigma|
$$

$$
\leq C\sum_{j=0}^{l} (\|w_{\nu_{\mathcal{A}}}^{(2(l-j)+1)}\|_{L^2(\Sigma_0)} + \|w_{\nu_{\mathcal{A}}}^{(2(l-j))}\|_{L^2(\Sigma_0)}) \|\hat{w}_{\nu_{\mathcal{A}}}^{(m+1)}\|_{L^2(\Sigma_0)}
$$

$$
\leq C\sum_{j=0}^{l} (\|(\mathcal{A}^{l-j}w_1, \mathcal{A}^{l-j+1}w_0)\|_{1,0}
$$

$$
+ \|(\mathcal{A}^{l-j}w_0, \mathcal{A}^{l-j}w_1)\|_{1,0}) \|(\mathcal{A}^l\hat{w}_1, 0)\|_{1,0}
$$

$$
\leq C(\|w_0\|_{H^{m+2}(\Omega)}^2 + \|w_1\|_{H^{m+1}(\Omega)}^2)^{1/2} \|\mathcal{A}^l\hat{w}_1\|_{H^1(\Omega)}. \tag{2.112}
$$

Next, we estimate

$$
\|\mathcal{A}_0(\mathcal{A}^{*l}u(0))\|_{H^{-1}(\Omega)}
$$

through the identity (2.111). For this purpose, Lemma 2.6 is needed. We decompose $\mathcal{A}_0(\mathcal{A}^{*l}u(0))$ as a direct sum

$$
\mathcal{A}_0(\mathcal{A}^{*l}u(0)) = [\mathcal{A}_0(\mathcal{A}^{*l}u(0))]_0 + [\mathcal{A}_0(\mathcal{A}^{*l}u(0))]_1
$$

where $[\mathcal{A}_0(\mathcal{A}^{*l}u(0))]_0 \in I\!N(\mathcal{A}^{*l})$. By Lemma 2.8 later, we have

$$
\|[\mathcal{A}_0(\mathcal{A}^{*l}u(0))]_0\| \leq C\|w_0\|_{H^{m+3/2}(\Omega)} + C\|u(0)\|_{H^m(\Omega)}. \tag{2.113}
$$

Moreover, using the relations (2.98) and (2.112) in (2.111), we obtain, via (2.113),

$$
\|\mathcal{A}_0(\mathcal{A}^{*l}u(0))\|_{H^{-1}(\Omega)} \leq C\|(w_0, w_1)\|_{m+2,m+1} + C\|u(0)\|_{H^m(\Omega)}. \tag{2.114}
$$

Moreover, on the boundary Γ the problem (2.100) implies that

$$
\|\mathcal{A}^{*i}u(0)\|_{H^{m+1/2-2i}(\Gamma)} = \|u^{(2i)}(0)\|_{H^{m+1/2-2i}(\Gamma_0)} = \|w_{\nu_{\mathcal{A}}}^{(2i)}(0)\|_{H^{m+1/2-2i}(\Gamma_0)}
$$

$$
= \|(\mathcal{A}^i w_0)_{\nu_{\mathcal{A}}}\|_{H^{m+1/2-2i}(\Gamma)} \leq c\|w_0\|_{H^{m+2}(\Omega)} \quad \text{for} \quad 0 \leq i \leq l. \tag{2.115}
$$

Now, using the relations (2.114) and (2.115) and the ellipticity of the operator $\mathcal{A}^*$ and $\mathcal{A}_0$, we obtain

$$
\|u(0)\|_{H^{m+1}(\Omega)} \leq c(\|\mathcal{A}^*u(0)\|_{H^{m-1}(\Omega)} + \|u(0)\|_{H^{m+1/2}(\Gamma)} + \|u(0)\|_{H^m(\Omega)})
$$

$$
\leq C(\|\mathcal{A}^{*l}u(0)\|_{H^1(\Omega)} + \|u(0)\|_{H^{m+1/2}(\Gamma)} + \|u(0)\|_{H^m(\Omega)})
$$

$$
\leq C(\|\mathcal{A}_0\mathcal{A}^{*l}u(0)\|_{H^{-1}(\Omega)} + \|\mathcal{A}^{*l}u(0)\|_{H^{1/2}(\Gamma)} + \|(w_0, w_1)\|_{m+2,m+1})
$$

$$
\leq C\|(w_0, w_1)\|_{m+2,m+1}
$$

where the induction assumption $\|u(0)\|_{H^m(\Omega)} \le c\|(w_0, w_1)\|_{m+1,m}$ is used.
A similar argument yields

$$\|u_t(0)\|_m \le C(\|w_0\|_{m+2} + \|w_1\|_{m+1}). \qquad (2.116)$$

Case 2 Let $m = 2l + 1$. Similar arguments as in Case 1 yield the estimate
(2.104). $\square$.

Lemma 2.8 *The estimate* (2.113) *holds true.*

Proof. Since $\mathbb{N}(\mathcal{A}^{*l})$ is finitely dimensional, it suffices to estimate
$(\mathcal{A}_0(\mathcal{A}^{*l}u(0)), \theta)$ for $\theta \in \mathbb{N}(\mathcal{A}^{*l})$ given. In fact, it follows from the Green
formula that

$$|(\mathcal{A}_0(\mathcal{A}^{*l}u(0)), \theta)| = |-\int_\Gamma \mathcal{A}^{*l}u(0)\theta_{\nu_\mathcal{A}}d\Gamma + (\mathcal{A}^{*l}u(0), \mathcal{A}_0\theta)|$$

$$\le C\|u^{(m)}(0)\|_{L^2(\Gamma_0)} + C\|\mathcal{A}^{*l}u(0)\|$$

$$\le C\|\mathcal{A}^l w_0\|_{H^{3/2}(\Omega)} + C\|u(0)\|_{H^m(\Omega)} \le C\|w_0\|_{H^{m+3/2}(\Omega)} + C\|u(0)\|_{H^m(\Omega)}.$$

$\square$

We need to indicate the control function space, that is,

Lemma 2.9 *Let* $m \ge 0$ *be given. Let* w *solve the problem* (2.90) *with initial
data* $(w_0, w_1) \in \Xi_0^{m+1}(\Omega) \times \Xi_0^m(\Omega)$. *Then* $w_{\nu_\mathcal{A}} \in \mathrm{B}_D^{\,m}(\Sigma_0)$.

Proof. We only prove the case $m = 1$. The proof may be completed by
induction but we omit it.
Let $(w_0, w_1) \in \Xi_0^2(\Omega) \times \Xi_0^1(\Omega)$. We shall prove

$$w_{\nu_\mathcal{A}} \in C([0, T], H^{1/2}(\Gamma_0)) \cap H^1(\Sigma_0). \qquad (2.117)$$

Clearly, for any $T > 0$ given, there is $c_T > 0$ such that

$$\|w(t)\|_{H^2(\Omega)}^2 \le c_T(\|w_0\|_{H^2(\Omega)}^2 + \|w_1\|_{H^1(\Omega)}^2) \quad \forall t \in [0, T],$$

which implies $w_{\nu_\mathcal{A}} \in C([0, T], H^{1/2}(\Gamma))$.
Since w_t is the solution of the problem (2.90) for the initial data
$(w_1, \mathcal{A}w_0) \in H_0^1(\Omega) \times L^2(\Omega)$, Lemma 2.7 implies $w_{t\nu_\mathcal{A}} \in L^2(\Sigma_0)$.
To complete the proof, it remains to show that $w_{\nu_\mathcal{A}} \in L^2\left((0, T), H^1(\Gamma_0)\right)$.
Let X be a vector field of the manifold Γ, that is, $X(x) \in \Gamma_x$ for each
$x \in \Gamma$. We extend X to the whole $\overline{\Omega}$ to be a vector field on the manifold $(\overline{\Omega}, g)$
where $g = A^{-1}(x)$.
Let

$$v = X(w) \quad \text{for} \quad (t, x) \in Q = (0, T) \times \Omega.$$

Then v solves the problem

$$\begin{cases} \ddot{v} = \mathcal{A}v + [X,\mathcal{A}]w \quad \text{in} \quad Q, \\ v|_\Gamma = 0, \quad t \in (0,T), \\ v(0) = X(w_0) \in H_0^1(\Omega), \quad \dot{v}(0) = X(w_1) \in L^2(\Omega), \end{cases}$$

where $[X,\mathcal{A}]w = X(\mathcal{A}w) - \mathcal{A}X(w)$ with the estimate

$$\|[X,\mathcal{A}]w(t)\|^2 \le c_T(\|w_0\|_{H^2(\Omega)}^2 + \|w_1\|_{H^1(\Omega)}^2) \quad \text{for} \quad t \in [0,T]. \tag{2.118}$$

Let H be a vector field on $\overline{\Omega}$ with

$$H(x) = \nu_A \quad \text{for} \quad x \in \Gamma.$$

Letting $u = v$ and $f = [X,\mathcal{A}]w - F(v) + v(q - \operatorname{div} F)$ in the identity (2.13), we integrate it over Q to obtain

$$\int_\Sigma v_{\nu_A}^2 d\Sigma = 2(\dot{v}, H(v))|_0^T + \int_\Sigma [2DH(\nabla_g v, \nabla_g v) + (\dot{v}^2 - |\nabla_g v|_g^2)\operatorname{div} H]dQ$$
$$+2(F(v) + v(\operatorname{div} F - q) - [X,\mathcal{A}]w, \; H(v))$$

which yields, via (2.118)

$$\int_\Sigma v_{\nu_A}^2 d\Sigma \le c_T(\|w_0\|_{H^2(\Omega)}^2 + \|w_1\|_{H^1(\Omega)}^2). \tag{2.119}$$

Since

$$v_{\nu_A} = \nu_A(X(w)) = X(w_{\nu_A}) + [\nu_A, X]w \quad \text{for} \quad x \in \Gamma,$$

by (2.119), we have

$$\int_\Sigma |X(w_{\nu_A})|^2 d\Sigma \le c_{T,X}\left(\|w_0\|_{H^2(\Omega)}^2 + \|w_1\|_{H^1(\Omega)}^2\right),$$

for any vector field X of the manifold Γ, that is, $w_{\nu_A} \in L^2\left((0,T), H^1(\Gamma_0)\right)$. $\square$

We shall show that the problem (2.100) provides smooth controls by the following theorem.

Theorem 2.20 *Let $m \ge 0$ be an integer and let $T > 0$ be given. Then the problem (2.89) is exactly $[H^m(\Omega) \cap H_{\Gamma_1}^1(\Omega)] \times [H^{m-1}(\Omega) \cap H_{\Gamma_1}^1(\Omega)]$ controllable by $B_D^m(\Sigma_0)$ controls through the relations (2.90), (2.100) and (2.101) if and only if there is $c_{1T} > 0$ and $c_{2T} > 0$ such that*

$$c_{1T}\|(w_0, w_1)\|_{m+1,m} \le \|\Lambda(w_0, w_1)\|_{m-1,m} \le c_{2T}\|(w_0, w_1)\|_{m+1,m} \tag{2.120}$$

for all $(w_0, w_1) \in \Xi_0^{m+1}(\Omega) \times \Xi_0^m(\Omega)$.

Proof. By Theorem 2.13 the claim is true for $m = 0$.

We consider the case of $m = 1$. If Λ is surjective, then the inverse operator theorem implies that the estimate (2.120) holds true for $m = 1$. Conversely, we assume that the inequality (2.120) is true. If Λ is not surjective, we derive a contradiction as follows. Let there be $(v_0, v_1) \in L^2(\Omega) \times H^1_{\Gamma_1}(\Omega)$, $(v_0, v_1) \neq 0$, such that

$$(u_t(0), v_0) - (u(0), v_1)_{H^1(\Omega)} = (\Lambda(w_0, w_1), (v_0, v_1))_{0,1} = 0 \qquad (2.121)$$

for all $(w_0, w_1) \in C_0^\infty(\Omega) \times C_0^\infty(\Omega)$. Noting that $u(0)|_{\Gamma_1} = 0$ and $u(0)|_{\Gamma_0} = (w_0)_{\nu_A}|_{\Gamma_0} = 0$, we have $(u(0), v_1)_{H^1(\Omega)} = (u(0), -\Delta v_1)$ where Δ is the Laplacion in the Euclidean metric. Using this formula in (2.121), we obtain, via (2.102),

$$0 = (w_{\nu_A}, \hat{w}_{\nu_A})_{L^2(\Sigma_0)} = ((w_0, w_1), \Lambda(v_0, -\Delta v_1))_{L^2(\Omega) \times L^2(\Omega)}$$

where $\hat{w}$ solves the problem (2.90) with the initial data $(v_0, -\Delta v_1)$, for all $(w_0, w_1) \in C_0^\infty(\Omega) \times C_0^\infty(\Omega)$, that is, $(v_0, -\Delta v_1) = 0$, which is a contradiction.

We repeat this procedure to complete the proof. $\Box$.

Next, we assume that the observability inequality is true for distribution controls to derive smooth controls .

Assumption (A): Let $T_1 > 0$ be given such that there is $c_{T_1} > 0$ such that

$$\|w_{\nu_A}\|^2_{L^2((0,T_1) \times \Gamma_0)} \geq c_{T_1} \|(w_0, w_1)\|^2_{1,0} \qquad (2.122)$$

for all solutions w to the problem (2.90).

Remark 2.17 *Let $F = 0$ and $q = 0$. If there is an escape vector field for the metric g on $\overline{\Omega}$, then Assumption (A) is true, see Theorem 2.15. For any F and q, the inequality (2.122) was established in [130] under the assumption that there is a strictly convex function on $\overline{\Omega}$ in the metric g. [154] proved that the assumption (A) is also true if there is an escape vector field for the metric g on $\overline{\Omega}$.*

We have

Theorem 2.21 *Let $m \geq 0$ be given and let $T > T_1$ be given where $T_1 > 0$ is such that the assumption (A) holds true. Then the problem (2.89) is exactly $[H^m(\Omega) \cap H^1_{\Gamma_1}(\Omega)] \times [H^{m-1}(\Omega) \cap H^1_{\Gamma_1}(\Omega)]$ controllable on $[0,T]$ by $B^m_D(\Sigma_0)$ controls.*

Proof. By Proposition 2.3, it will suffice to prove that the left hand side of (2.120) is true. We prove it by induction.

If $m = 0$, it is just the assumption (A). We assume that the claim is true for some $m \geq 0$. We will prove that it holds true for $m + 1$.

Let us assume $m = 2l$ for some $l \geq 0$. Similar arguments can show the case $m = 2l + 1$.

Let w solves the problem (2.90) with initial data $(w_0, w_1) \in \Xi_0^{m+2}(\Omega) \times \Xi_0^{m+1}(\Omega)$. Then $w^{(2i)}$ and $w_t^{(2i)}$ solve the same problem with initial data $(\mathcal{A}^i w_0, \mathcal{A}^i w_1)$ and $(\mathcal{A}^i w_1, \mathcal{A}^{i+1} w_0)$, respectively, for $1 \leq i \leq l$.

By taking $(\hat{w}_0, \hat{w}_1) = (w_0, w_1)$ in the relations (2.110) and (2.107), we obtain, via the inequalities (2.122) and (2.103),

$$
\begin{aligned}
&c\|\Lambda(w_0, w_1)\|_{m, m+1}\|(w_0, w_1)\|_{m+2, m+1} \\
&\geq |\mathcal{I}(w, w)| - c\|\Lambda(w_0, w_1)\|_{m-1, m}\|(w_0, w_1)\|_{m+2, m+1} \\
&\geq c_1\|(\mathcal{A}^l w_1, \mathcal{A}^{l+1} w_0)\|_{1,0}^2 - c\|(\mathcal{A}^l w_1, \mathcal{A}^{l+1} w_0)\|_{1,0}\|(w_0, w_1)\|_{m+1, m} \\
&\quad - c\|\Lambda(w_0, w_1)\|_{m-1, m}\|(w_0, w_1)\|_{m+2, m+1}. \tag{2.123}
\end{aligned}
$$

On the other hand, $(w_0, w_1) \in \Xi_0^{m+2}(\Omega) \times \Xi_0^{m+1}(\Omega)$ implies, by the ellipticity of the operator $\mathcal{A}$,

$$
\|(w_0, w_1)\|_{m+2, m+1} \leq c(\|(\mathcal{A}^l w_1, \mathcal{A}^{l+1} w_0)\|_{1,0} + \|(w_0, w_1)\|_{m+1, m}). \tag{2.124}
$$

Using (2.124) in (2.123), we obtain, by the induction assumption, that the claim holds true for $m + 1$. $\qquad\square$

Control with the Neumann Action We now turn to the smooth control problem

$$
\begin{cases}
u_{tt} = \operatorname{div} A(x)\nabla u & \text{in} \quad Q, \\
u = 0 & \text{on} \quad \Sigma_1, \\
u_{\nu_A} = \varphi & \text{on} \quad \Sigma_0, \\
u(0) = u_0, \quad u_t(0) = u_1 & \text{on} \quad \Omega.
\end{cases} \tag{2.125}
$$

The duality problem of the system (2.125) is

$$
\begin{cases}
w_{tt} = \operatorname{div} A(x)\nabla w & \text{in} \quad Q, \\
w|_{\Sigma_1} = w_{\nu_A}|_{\Sigma_0} = 0, \\
w(0) = w_0, \quad w_t(0) = w_1 & \text{on} \quad \Omega.
\end{cases} \tag{2.126}
$$

The problem (2.125) is exactly $(H_{\Gamma_1}(\Omega))^* \times L^2(\Omega)$ controllable if and only if the inequality (2.86) holds true. However, to find the smooth control, the one-side observability estimate (2.86) is insufficient. We have to seek boundary estimates of another type controlled by the initial energy on both sides from below and also from above, as in Theorem 2.14.

Let $\varepsilon > 0$ be given small. Let $\eta_\varepsilon \in C^\infty(\mathbb{R})$ be such that $0 \leq \eta_\varepsilon \leq 1$ and

$$
\eta_\varepsilon(t) = 1 \quad t \leq -\varepsilon; \quad \eta_\varepsilon(t) = 0 \quad t \geq 0.
$$

For any $T > \varepsilon$, let

$$
z_\varepsilon(t) = \eta_\varepsilon(t - T). \tag{2.127}
$$

Then

$$
z_\varepsilon(t) = 1 \quad 0 \leq t \leq T - \varepsilon; \quad z_\varepsilon(t) = 0 \quad t \geq T.
$$

Let H be a vector field. We introduce a bilinear form by

$$\Psi(w, \hat{w}) = \int_{\Sigma_0} z_\varepsilon(t)[w_t \hat{w}_t - \langle \nabla_{\Gamma_g} w, \nabla_{\Gamma_g} \hat{w} \rangle_g] \langle H, \nu \rangle d\Sigma \qquad (2.128)$$

where w and $\hat{w}$ are solutions to the problem (2.126) with initial data (w_0, w_1) and $(\hat{w}_0, \hat{w}_1)$, respectively.

Let H be an escape vector field for the metric g on $\overline{\Omega}$ with $\varrho_0 > 0$ such that the inequality (2.15) is true. For $\varepsilon > 0$ fixed small, set

$$T_0 = [(1 + \sup_{-\varepsilon \le t \le 0} \varepsilon |\eta_\varepsilon'(t)|) \sup_{x \in \Omega} |H|_g. \qquad (2.129)$$

We further assume that

$$\langle H, \nu \rangle \le 0 \quad \text{for} \quad x \in \Gamma_1. \qquad (2.130)$$

Another observability estimate we need is the following.

Theorem 2.22 *Let* $\Gamma_1 \ne \emptyset$ *and* $\overline{\Gamma}_0 \cap \overline{\Gamma}_1 = \emptyset$. *Let* H *be an escape vector field for the metric* g *on* $\overline{\Omega}$ *such that the assumption* (2.130) *holds. Let* T_0 *be given in* (2.129). *For any* $T > 0$, *there are constant* $c_0 > 0$ *and* $c_T > 0$ *such that*

$$c_T \|(w_0, w_1)\|_{1,0}^2 \ge \Psi(w, w) + c_0(z_\varepsilon w, w)_{L^2(\Sigma_0)} + c_0\|w_0\|^2 + c_0\|w\|_{L^2(Q)}^2$$

$$\ge [\varrho_0 \int_0^T z_\varepsilon(t)dt - T_0 - \varepsilon]\|(w_0, w_1)\|_{1,0}^2 \qquad (2.131)$$

for all solutions w *to the problem* (2.126).

Proof. We use the multipliers $2z_\varepsilon(t)H(w)$ and $2z_\varepsilon(t)pw$ for the equation in (2.126) as in Theorems 2.1 and 2.2, to obtain

$$\int_\Sigma z_\varepsilon[2H(w)w_{\nu_A} + (w_t^2 - |\nabla_g w|_g^2)\langle H, \nu \rangle]d\Sigma$$

$$= -2(w_1, H(w_0)) - 2\int_{T-\varepsilon}^T z_{\varepsilon t}(t)(w_t, H(w))dt$$

$$+ \int_Q z_\varepsilon[2D_g H(\nabla_g w, \nabla_g w) + (w_t^2 - |\nabla_g w|_g^2) \operatorname{div} H]dQ, \qquad (2.132)$$

and

$$2\int_Q z_\varepsilon(t)p(w_t^2 - |\nabla_g w|_g^2)dQ = -2(w_1, pw_0) - \int_Q z_\varepsilon w^2 \operatorname{div} A\nabla p \, dQ$$

$$-2\int_{T-\varepsilon}^T z_{\varepsilon t}(t)(w_t, pw)dt + \int_{\Sigma_0} z_\varepsilon w^2 p_{\nu_A} d\Sigma, \qquad (2.133)$$

respectively, where p is a function.

Let $\varrho_0 > 0$ be given in (2.15). We take $p = \operatorname{div} H - \varrho_0$ in the identity (2.133) and obtain the estimate

$$2 \int_Q z_\varepsilon p(w_t^2 - |\nabla_g w|_g^2) dQ \geq -\varepsilon \|(w_0, w_1)\|_{1,0}$$
$$-c_\varepsilon (\|w_0\|^2 + \|w\|_{L^2(Q)}^2) - c_0 \|w\|_{L^2(Q)}^2 - c_0 (z_\varepsilon w, w)_{L^2(\Sigma_0)}. \quad (2.134)$$

On Σ_1, the boundary conditions $w = 0$ implies that $H(w) = \langle H, \nu \rangle w_{\nu_A} / |\nu_A|_g^2$ and $|\nabla_g w|_g^2 = w_{\nu_A}^2 / |\nu_A|_g^2$. Then

$$\int_{\Sigma_1} z_\varepsilon [2H(w)w_{\nu_A} + (w_t^2 - |\nabla_g w|_g^2)\langle H, \nu \rangle] d\Sigma = \int_{\Sigma_1} z_\varepsilon w_{\nu_A}^2 \langle H, \nu \rangle / |\nu_A|_g^2 d\Sigma \leq 0.$$

On Σ_0, $w_{\nu_A} = 0$ implies $\nabla_g w = \nabla_{\Gamma_g} w$. Using those relations in the identity (2.132), we have, via (2.134),

$$\Psi(w, w) \geq \varrho_0 \int_Q z_\varepsilon (w_t^2 + |\nabla_g w|_g^2 dQ) + \int_Q z_\varepsilon (w_t^2 - |\nabla_g w|_g^2) p dQ$$
$$- (1 + \sup_{-\varepsilon \leq t \leq 0} \varepsilon |\eta_\varepsilon'(t)|) \sup_{x \in \Omega} |H|_g \|(w_0, w_1)\|_{1,0}^2$$
$$\geq \{\varrho_0 \int_0^T z_\varepsilon(t) dt - T_0 - \varepsilon\} \|(w_0, w_1)\|_{1,0}^2$$
$$- c_0 \|w\|_{L^2(Q)}^2 - c_0 (z_\varepsilon w, w)_{L^2(\Sigma_0)} - c_0 \|w_0\|^2$$

which proves that the right hand side of the inequality (2.131) holds.

On the other hand, since $\overline{\Gamma}_1 \cap \overline{\Gamma}_0 = \emptyset$, we take two open sets ω_0 and ω_1 in $\mathbb{R}^n$ such that $\omega_0 \cap \omega_1 = \emptyset$ and $\Gamma_i \subset \omega_i$ for $i = 0, 1$, respectively. Let $h \in C^\infty(\mathbb{R}^n)$ be such that

$$h(x) = 1 \quad x \in \omega_0; \quad h(x) = 0 \quad x \in \omega_1.$$

Replacing H with hH in the identity (2.132) yields

$$\Psi(w, w) \leq c_\varepsilon T \|(w_0, w_1)\|_{1,0}^2.$$

$\square$

Let all the assumptions in Theorem 2.22 hold. Let $1/2 > \varepsilon > 0$. Using a boundary trace estimate, Lemma 7.2 in [122], we have an estimate

$$\int_{(\varepsilon, T-\varepsilon) \times \Gamma_0} |\nabla_{\Gamma_g} w|_g^2 d\Sigma \leq c_{T\varepsilon} (\|w_t\|_{L^2(\Sigma_0)}^2 + \|w\|_{H^{1/2+\varepsilon}(Q)}^2) \quad (2.135)$$

for all solutions to the problem (2.126). Let $T > 2\varepsilon$. We use the right hand side of the inequality (2.131) where the interval $[0, T]$ is replaced with $[\varepsilon, T-\varepsilon]$ to have

$$c_T (z_\varepsilon w_t, w_t)_{L^2(\Sigma_0)} + c_T L(w) \geq [\varrho_0(T - 2\varepsilon) - T_0 - \varepsilon] \|(w_0, w_1)\|_{1,0}^2 \quad (2.136)$$

for all solutions w to the problem (2.126), where $L(w)$ are some lower order terms related to the norm $\|w\|^2_{H^1(Q)} = \|(w_0, w_1)\|^2_{1,0}$. A compactness-uniqueness argument, as in Lemma 2.5, shows that the lower order terms in (2.136) can be absorbed if the following uniqueness problem holds true: The problem

$$\begin{cases} w_{tt} = \text{div}\, A(x)\nabla w & \text{in}\quad Q, \\ w|_{\Sigma_1} = 0, \\ w_t|_{\Sigma_0} = w_{\nu_A}|_{\Sigma_0} = 0, \end{cases} \tag{2.137}$$

has the unique zero solution. However, the uniqueness problem (2.137) is true because the problem (2.137) implies that $v = w_t$ to solve the problem

$$\begin{cases} v_{tt} = \text{div}\, A(x)\nabla v & \text{in}\quad Q, \\ v|_{\Sigma_0} = v_{\nu_A}|_{\Sigma_0} = 0, \end{cases}$$

which yields $w_t = 0$ on Ω, that is, $w = 0$ on Q since $\Gamma_1 \neq \emptyset$.

We have

Corollary 2.2 *Let all the assumptions in Theorem 2.22 hold. Then for $T > (T_0 + \varepsilon)/\varrho_0 + 2\varepsilon$, there is $c_T > 0$ such that*

$$c_T(z_\varepsilon w_t, w_t)_{L^2(\Sigma_0)} \geq \|(w_0, w_1)\|^2_{1,0} \tag{2.138}$$

for all solutions w to the problem (2.126).

We introduce an operator by

$$\mathcal{A}_0 v = \text{div}\, A(x)\nabla v, \quad D(\mathcal{A}_0) = \{\, w \,|\, w \in H^2(\Omega),\ w|_{\Gamma_1} = 0,\ w_{\nu_A}|_{\Gamma_0} = 0 \,\}.$$

Let $(w_0, w_1) \in D(\mathcal{A}_0) \times H^1_{\Gamma_1}(\Omega)$. Then $(w_1, \mathcal{A}_0 w_0) \in H^1_{\Gamma_1}(\Omega) \times L^2(\Omega)$. Since w_t solves the problem (2.126) with the initial data $(w_1, \mathcal{A}_0 w_0)$, the inequality (2.131) implies

$$c_{2T}\|(w_1, \mathcal{A}_0 w_0)\|^2_{1,0} \geq \Psi(w_t, w_t) + c_0(z_\varepsilon w_t, w_t)_{L^2(\Sigma_0)} + c_0\|w_1\|^2 + c_0\|w_t\|^2_{L^2(Q)}$$
$$\geq c_{1T}\|(w_1, \mathcal{A}_0 w_0)\|^2_{1,0}, \tag{2.139}$$

for $T > (T_0 + \varepsilon)/\varrho_0 + 2\varepsilon$. Moreover, by the inequality (2.138), we have

$$\|w_1\|^2 \leq \|(w_0, w_1)\|^2_{1,0} \leq c_T(z_\varepsilon w_t, w_t)_{L^2(\Sigma_0)}, \tag{2.140}$$

$$\|w_t\|^2_{L^2(Q)} \leq T\|(w_0, w_1)\|^2_{1,0} \leq T c_T(z_\varepsilon w_t, w_t)_{L^2(\Sigma_0)}. \tag{2.141}$$

We introduce a bilinear form by

$$\Psi_\star(w, \hat{w}) = \Psi(w, \hat{w}) + c_0 c_T (1 + T)(z_\varepsilon w, \hat{w})_{L^2(\Sigma_0)} \tag{2.142}$$

where the positive constants c_0 and c_T are defined in (2.131) and (2.138), respectively.

Noting that $\int_0^T z_\varepsilon(t)dt \geq (T - \varepsilon)$, from the inequalities (2.131) and (2.141), we have

Lemma 2.10 *Let all the assumptions in Theorem 2.22 hold. Then for $T > (T_0 + \varepsilon)/\varrho_0 + 2\varepsilon$ given, there is constant $c_T > 0$ such that*

$$c_T \|(w_1, \mathcal{A}_0 w_0)\|_{1,0}^2 \geq \Psi_\star(w_t, w_t)$$
$$\geq [\varrho_0(T - 2\varepsilon) - T_0 - \varepsilon]\|(w_1, \mathcal{A}_0 w_0)\|_{1,0}^2, \qquad (2.143)$$

for all $(w_0, w_1) \in D(\mathcal{A}_0) \times H_{\Gamma_1}^1(\Omega)$.

Space of Boundary Control Let $T > 0$ be given. This time, we introduce a Banach space $\mathrm{B}_N^m(\Sigma_0)$ as follows. We consider the space $H^1([0, T], L^2(\Gamma_0))$ which is the completion of $C_0^\infty(\Sigma_0)$ with an inner product

$$(\varphi, \phi)_{H^1([0,T],L^2(\Gamma_0))} = \int_{\Sigma_0} (\varphi_t \phi_t + \varphi\phi) d\Sigma$$

and set

$$H_0^1([0, T], L^2(\Gamma_0)) = \{ \varphi \mid \varphi \in H^1([0, T], L^2(\Gamma_0)), \ \varphi(0) = 0 \}.$$

For $m = 0$, we define

$$\mathrm{B}_N^0(\Sigma_0) = (H_0^1([0, T], L^2(\Gamma_0)))^*$$

that is the duality space of $H_0^1([0, T], L^2(\Gamma_0))$ with respect to the space $L^2(\Sigma_0)$. For $m = 1$, let

$$\mathrm{B}_N^1(\Sigma_0) = L^2(\Sigma_0).$$

For $m \geq 2$, let $\mathrm{B}_N^m(\Sigma_0)$ consist of all the functions

$$\varphi \in \cap_{k=0}^{m-2} C^k([0, T], H^{m-k-3/2}(\Gamma_0)), \quad \varphi^{(k)} \in H^1(\Sigma_0), \qquad (2.144)$$

with the norm

$$\|\varphi\|_{\mathrm{B}_N^m(\Sigma_0)}^2 = \sum_{k=0}^{m-2} \|\varphi^{(k)}\|_{C([0,T],H^{m-k-3/2}(\Gamma_0))}^2 + \sum_{k=0}^{m-2} \|\varphi^{(k)}\|_{H^1(\Sigma_0)}^2.$$

Let $\Xi_{0N}^1(\Omega) = H_{\Gamma_1}^1(\Omega)$. For $m \geq 2$, let $\Xi_{0N}^m(\Omega)$ consist of functions u in $H^m(\Omega)$ with the boundary conditions

$$\begin{cases} \mathcal{A}_0^i u|_{\Gamma_1} = (\mathcal{A}_0^i u)_{\nu_A}|_{\Gamma_0} = 0, & 0 \leq i \leq l-1, \quad \text{if} \quad m = 2l; \\ \mathcal{A}_0^i u|_{\Gamma_1} = (\mathcal{A}_0^j u)_{\nu_A}|_{\Gamma_0} = 0, & 0 \leq i \leq l, \ 0 \leq j \leq l-1, \quad \text{if} \quad m = 2l+1, \end{cases}$$

with the norms of $H^m(\Omega)$. In particular, $\Xi_{0N}^2(\Omega) = D(\mathcal{A}_0)$. Clearly, if $\Gamma_1 \neq \emptyset$ and $\overline{\Gamma}_1 \cap \overline{\Gamma}_0 = \emptyset$, $\Xi_{0N}^m(\Omega)$ has an equivalent norm: $\|\mathcal{A}_0^l u\|$ for $m = 2l$ or $\|\mathcal{A}_0^l w\|_{H^1(\Omega)}$ for $m = 2l+1$.

We now go back to the problem (2.125). Let w be a solution to the problem (2.126). We solve the problem

$$\begin{cases} u_{tt} = \text{div}\, A(x)\nabla u & \text{in } Q, \\ u(T) = u_t(T) = 0 & \text{on } \Omega, \\ u|_{\Sigma_1} = 0, \quad u_{\nu_A}|_{\Sigma_0} = z_\varepsilon[(w^{(3)} - \Delta_{\Gamma_g} w_t)h_0 - \lambda_T w_t], \end{cases} \tag{2.145}$$

where $h_0 = \langle H, \nu \rangle$ and $\lambda_T = c_0 c_T (1 + T) + \sup_{x \in \Gamma_0} |\Delta_{\Gamma_g} h_0|/2$. We define Λ_N: $\Xi_{0N}^{m+1}(\Omega) \times \Xi_{0N}^m(\Omega) \to H^{m-2}(\Omega) \times H^{m-1}(\Omega)$ by

$$\Lambda_N(w_0, w_1) = (u_t(0), -u(0)). \tag{2.146}$$

It is easy to check that

$$(\Lambda_N(w_0, w_1), (\hat{w}_0, \hat{w}_1))_{L^2(\Omega) \times L^2(\Omega)} = -\int_{\Sigma_0} \hat{w} u_{\nu_A} d\Sigma \tag{2.147}$$

for all $(w_0, w_1), (\hat{w}_0, \hat{w}_1) \in H_{\Gamma_1}^1(\Omega) \times L^2(\Omega)$.

We have the following observability estimates for the Neumann action.

Theorem 2.23 *Let $\Gamma_1 \neq \emptyset$ and $\overline{\Gamma}_0 \cap \overline{\Gamma}_1 = \emptyset$. Let H be an escape vector field for the metric g on $\overline{\Omega}$ such that the assumption (2.130) holds. Let $m \geq 1$ be given. Then, for $T > 0$ suitable large, there are $c_1 > 0$ and $c_2 > 0$ satisfying*

$$c_1 \|(w_0, w_1)\|_{m+1,m} \leq \|\Lambda_N(w_0, w_1)\|_{m-2,m-1} \leq c_2 \|(w_0, w_1)\|_{m+1,m} \tag{2.148}$$

for all $(w_0, w_1) \in \Xi_{0N}^{m+1}(\Omega) \times \Xi_{0N}^m(\Omega)$, where $\|\cdot\|_{-1,0} = \|\cdot\|_{(H_{\Gamma_1}^1(\Omega))^* \times L^2(\Omega)}$.

Proof. By induction on m.

Step 1 Let $m = 1$. Let $(w_0, w_1) \in \Xi_{0N}^2(\Omega) \times \Xi_{0N}^1(\Omega)$ be given. For any $(\hat{w}_0, \hat{w}_1) \in \Xi_{0N}^2(\Omega) \times \Xi_{0N}^1(\Omega)$ given, suppose that $\hat{w}$ solves the problem (2.126) with the initial data $(\hat{w}_0, \hat{w}_1)$. Then $\hat{w}_t$ solves the problem (2.126) with the initial $(\hat{w}_1, \mathcal{A}_0 \hat{w}_0) \in H_{\Gamma_1}^1(\Omega) \times L^2(\Omega)$. By the formulas (2.147) and (2.145), we obtain

$$(\Lambda_N(w_0, w_1), (\hat{w}_1, \mathcal{A}_0 \hat{w}_0))_{L^2(\Omega) \times L^2(\Omega)} = -\int_{\Sigma_0} \hat{w}_t u_{\nu_A} d\Sigma$$

$$= \Psi_*(w_t, \hat{w}_t) + \int_{\Gamma_0} h_0 \hat{w}_1 \, \text{div}\, A(x)\nabla w_0 d\Gamma + \int_{T-\varepsilon}^T \int_{\Gamma_0} z_{\varepsilon t} h_0 \hat{w}_t w_{tt} d\Gamma dt$$

$$- \int_{\Sigma_0} z_\varepsilon \hat{w}_t \nabla_{\Gamma_g} h_0(w_t) d\Sigma + \frac{1}{2} \sup_{x \in \Gamma_0} |\Delta_{\Gamma_g} h_0| \int_{\Sigma_0} z_\varepsilon w_t \hat{w}_t d\Sigma \tag{2.149}$$

$$\leq \Psi_*^{1/2}(w_t, w_t) \Psi_*^{1/2}(\hat{w}_t, \hat{w}_t) + c\|(w_1, \mathcal{A}_0 w_0)\|_{1,0}\|(\hat{w}_1, \mathcal{A}_0 \hat{w}_0)\|_{1,0}$$

$$+ c \int_0^T \|\hat{w}_t\|_{H^{1/2}(\Gamma_0)} (\|\nabla_{\Gamma_g} w_t\|_{H^{-1/2}(\Gamma_0)} + \|w_t\|_{H^{1/2}(\Gamma_0)}) dt$$

$$\leq c_T \|(w_1, \mathcal{A}_0 w_0)\|_{1,0}\|(\hat{w}_1, \mathcal{A}_0 \hat{w}_0)\|_{1,0},$$

for all $(w_0, w_1) \in \Xi_{0N}^2(\Omega) \times \Xi_{0N}^1(\Omega)$, which yields

$$\|\Lambda_N(w_0, w_1)\|_{(H_{\Gamma_1}^1(\Omega))^* \times L^2(\Omega)} \le \hat{c}_T \|(w_1, \mathcal{A}_0 w_0)\|_{1,0},$$

for $(w_0, w_1) \in \Xi_{0N}^2(\Omega) \times \Xi_{0N}^1(\Omega)$.

Since $\overline{\Gamma}_0 \cap \overline{\Gamma}_1 = \emptyset$, Γ_0 is a closed surface in $\mathbb{R}^n$ and, then,

$$2 \int_{\Sigma_0} z_\varepsilon w_t \nabla_{\Gamma_g} h_0(w_t) d\Sigma = \int_{\Sigma_0} z_\varepsilon \nabla_{\Gamma_g} h_0(w_t^2) d\Sigma = - \int_{\Sigma_0} z_\varepsilon w_t^2 \Delta_{\Gamma_g} h_0 d\Sigma,$$

that is,

$$2 \int_{\Sigma_0} z_\varepsilon w_t \nabla_{\Gamma_g} h_0(w_t) d\Sigma + \sup_{x \in \Gamma_0} |\Delta_{\Gamma_g} h_0| \int_{\Sigma_0} z_\varepsilon w_t^2 d\Sigma \ge 0. \tag{2.150}$$

Using the relations (2.149), (2.143) and (2.150), we have

$$(\Lambda_N(w_0, w_1), (w_1, \mathcal{A}_0 w_0))_{L^2(\Omega) \times L^2(\Omega)}$$
$$\ge [\varrho_0(T - 2\varepsilon) - T_0 - T_1 c(\Gamma_0) - \varepsilon] \|(w_1, \mathcal{A}_0 w_0)\|_{1,0}^2 \tag{2.151}$$

for $\varepsilon > 0$ small, where

$$T_1 = [(1 + \sup_{-\varepsilon \le t \le 0} \varepsilon |\eta_\varepsilon'(t)|) \sup_{x \in \Omega} |h_0|$$

and a constant $c(\Gamma_0) > 0$ satisfies

$$|(u, v)_{L^2(\Gamma_0)}| \le c(\Gamma_0) \|(u, v)\|_{1,0} \quad \text{for all} \quad (u, v) \in H^1(\Omega) \times L^2(\Omega).$$

Then there is a constant $c_1 > 0$ such that

$$\|\Lambda_N(w_0, w_1)\|_{(H_{\Gamma_1}^1(\Omega))^* \times L^2(\Omega)} \ge c_1 \|(w_1, \mathcal{A}_0 w_0)\|_{1,0}$$

when $T > [T_0 + T_1 c(\Gamma_0) + \varepsilon]/\varrho_0 + 2\varepsilon$.

Then the inequality (2.148) is true for $m = 1$.

Step 2 Suppose that the inequality (2.148) is true for some $m \ge 1$. We shall prove that it is true for $m + 1$.

Case I Let $m = 2l$ for some $l \ge 1$. Firstly, we assume that

$$(w_0, w_1) \in C_0^\infty(\Omega) \times C_0^\infty(\Omega). \tag{2.152}$$

Suppose that $\hat{w}$ solves the problem (2.126) with an initial data $(\hat{w}_0, \hat{w}_1) \in C_0^\infty(\Omega) \times C_0^\infty(\Omega)$. Then $\hat{w}^{(2i)}$ and $\hat{w}^{(2j+1)}$ solve the problem (2.126) with the initial data $(\mathcal{A}_0^i \hat{w}_0, \mathcal{A}_0^i \hat{w}_1)$ and $(\mathcal{A}_0^j \hat{w}_1, \mathcal{A}_0^{j+1} \hat{w}_0)$, respectively, for $0 \le i \le l+1$ and $0 \le j \le l$.

Claim 1 The following identity is true.

$$-(A\nabla(\mathcal{A}_0^{l-1}u_t(0)), \nabla(\mathcal{A}_0^l\hat{w}_1)) - (\mathcal{A}_0^l u(0), \mathcal{A}_0^{l+1}\hat{w}_0)$$

$$= \Psi_*(w^{(m+1)}, \hat{w}^{(m+1)}) + \frac{1}{2}\sup_{x\in\Gamma_0}|\Delta_{\Gamma_g}h_0|\int_{\Sigma_0}z_\varepsilon w^{(m+1)}\hat{w}^{(m+1)}d\Sigma$$

$$-\int_{\Sigma_0}z_\varepsilon\hat{w}^{(m+1)}\nabla_{\Gamma_g}h_0(w^{(m+1)})d\Sigma$$

$$+\sum_{j=1}^{m-1}\int_{T-\varepsilon}^T\int_{\Gamma_0}C_j^{m-1}z_\varepsilon^{(j)}w^{(m+2-j)}\hat{w}^{(m+2)}h_0 d\Gamma dt$$

$$+\sum_{j=1}^m\int_{T-\varepsilon}^T\int_{\Gamma_0}C_j^m z_\varepsilon^{(j)}\hat{w}^{(m+1)}(\Delta_{\Gamma_g}w^{(m+1-j)}h_0 + \lambda_T w^{(m+1-j)})d\Gamma dt. \quad (2.153)$$

Proof of (2.153) Using $\hat{w}^{(2m+1)}$ in place of $\hat{w}$ in the formula (2.147), we obtain

$$(u_t(0), \mathcal{A}_0^m\hat{w}_1) - (u(0), \mathcal{A}_0^{m+1}\hat{w}_0) = -\int_{\Sigma_0}u_{\nu_A}\hat{w}^{(2m+1)}d\Sigma$$

$$= -\int_{\Sigma_0}z_\varepsilon h_0 w^{(3)}\hat{w}^{(2m+1)}d\Sigma + \int_{\Sigma_0}z_\varepsilon h_0\hat{w}^{(2m+1)}\Delta_{\Gamma_g}w_t d\Sigma$$

$$+\lambda_T\int_{\Sigma_0}z_\varepsilon w_t\hat{w}^{(2m+1)}d\Sigma$$

$$= \text{Term } 1 + \text{Term } 2 + \lambda_T\text{Term } 3. \quad (2.154)$$

We compute the terms in the right-hand side of (2.154) by integrating by parts over $\Sigma_0 = (0, T) \times \Gamma_0$, respectively, as follows.

$$\text{Term } 1 = \int_{\Sigma_0}z_\varepsilon w^{(m+2)}\hat{w}^{(m+2)}h_0 d\Sigma + \sum_{j=0}^{m-2}(-1)^j\int_{\Gamma_0}w^{(j+3)}(0)\hat{w}^{(2m-j)}(0)h_0 d\Gamma$$

$$+\sum_{j=1}^{m-1}\int_{T-\varepsilon}^T\int_{\Gamma_0}C_j^{m-1}z_\varepsilon^{(j)}w^{(m+2-j)}\hat{w}^{(m+2)}h_0 d\Gamma dt$$

$$= \int_{\Sigma_0}z_\varepsilon w^{(m+2)}\hat{w}^{(m+2)}h_0 d\Sigma + \sum_{j=1}^l\int_{\Gamma_0}\mathcal{A}_0^j w_1\mathcal{A}_0^{m+1-j}\hat{w}_0 h_0 d\Gamma$$

$$-\sum_{j=2}^l\int_{\Gamma_0}\mathcal{A}_0^j w_0\mathcal{A}_0^{m+1-j}\hat{w}_1 h_0 d\Gamma$$

$$+\sum_{j=1}^{m-1}\int_{T-\varepsilon}^T\int_{\Gamma_0}C_j^{m-1}z_\varepsilon^{(j)}w^{(m+2-j)}\hat{w}^{(m+2)}h_0 d\Gamma dt; \quad (2.155)$$

$$\text{Term } 2 = \int_{\Sigma_0} z_\varepsilon h_0 \hat{w}^{(m+1)} \Delta_{\Gamma_g} w^{(m+1)} d\Sigma$$

$$+ \sum_{j=1}^{m} \int_{T-\varepsilon}^{T} \int_{\Gamma_0} C_j^m z_\varepsilon^{(j)} \hat{w}^{(m+1)} \Delta_{\Gamma_g} w^{(m+1-j)} h_0 d\Gamma dt$$

$$+ \sum_{j=0}^{m-1} (-1)^{j+1} \int_{\Gamma_0} h_0 \hat{w}^{(2m-j)}(0) \Delta_{\Gamma_g} w^{(j+1)}(0) d\Gamma$$

$$= - \int_{\Sigma_0} z_\varepsilon \langle \nabla_{\Gamma_g} w^{(m+1)}, \nabla_{\Gamma_g} \hat{w}^{(m+1)} \rangle_g h_0 d\Sigma - \int_{\Sigma_0} z_\varepsilon \hat{w}^{(m+1)} \nabla_{\Gamma_g} h_0 (w^{(m+1)}) d\Sigma$$

$$+ \sum_{j=1}^{l} \int_{\Gamma_0} h_0 \mathcal{A}_0^{m-j} \hat{w}_1 \Delta_{\Gamma_g} \mathcal{A}_0^j w_0 d\Gamma - \sum_{j=0}^{l-1} \int_{\Gamma_0} h_0 \mathcal{A}_0^{m-j} \hat{w}_0 \Delta_{\Gamma_g} \mathcal{A}_0^j w_1 d\Gamma$$

$$+ \sum_{j=1}^{m} \int_{T-\varepsilon}^{T} \int_{\Gamma_0} C_j^m z_\varepsilon^{(j)} \hat{w}^{(m+1)} \Delta_{\Gamma_g} w^{(m+1-j)} h_0 d\Gamma dt; \tag{2.156}$$

$$\text{Term } 3 = \int_{\Sigma_0} z_\varepsilon w^{(m+1)} \hat{w}^{(m+1)} d\Sigma + \sum_{j=1}^{m} \int_{T-\varepsilon}^{T} \int_{\Gamma_0} C_j^m z_\varepsilon^{(j)} w^{(m+1-j)} \hat{w}^{(m+1)} d\Gamma dt$$

$$+ \sum_{j=1}^{l} \int_{\Gamma_0} \mathcal{A}_0^j w_0 \mathcal{A}_0^{m-j} \hat{w}_1 d\Gamma - \sum_{j=0}^{l-1} \int_{\Gamma_0} \mathcal{A}_0^j w_1 \mathcal{A}_0^{m-j} \hat{w}_0 d\Gamma. \tag{2.157}$$

Moreover, by the problem (2.145), we have, on Γ_0 for $j \geq 0$,

$$(\mathcal{A}_0^j u_t(0))_{\nu_A} = u_{\nu_A}^{(2j+1)}(0) = (\mathcal{A}_0^{j+2} w_0 - \Delta_{\Gamma_g} \mathcal{A}_0^{j+1} w_0) h_0 - \lambda_T \mathcal{A}_0^{j+1} w_0; \tag{2.158}$$

$$(\mathcal{A}_0^j u(0))_{\nu_A} = (\mathcal{A}_0^{j+1} w_1 - \Delta_{\Gamma_g} \mathcal{A}_0^j w_1) h_0 - \lambda_T \mathcal{A}_0^j w_1. \tag{2.159}$$

We substitute (2.155)-(2.159) into (2.154) to yield

$$(u_t(0), \mathcal{A}_0^m \hat{w}_1) - (u(0), \mathcal{A}_0^{m+1} \hat{w}_0)$$
= the right hand side of the identity (2.153)
$$+ \sum_{j=0}^{l-1} \int_{\Gamma_0} (\mathcal{A}_0^j u(0))_{\nu_A} \mathcal{A}_0^{m-j} \hat{w}_0 d\Gamma - \sum_{j=0}^{l-2} \int_{\Gamma_0} (\mathcal{A}_0^j u_t(0))_{\nu_A} \mathcal{A}_0^{m-1-j} \hat{w}_1 d\Gamma. \tag{2.160}$$

On the other hand, for $(w_0, w_1), (\hat{w}_0, \hat{w}_1) \in \Xi_{0N}^{m+2}(\Omega) \times \Xi_{0N}^{m+1}(\Omega)$, we obtain, via the Green formula,

$$(u_t(0), \mathcal{A}_0^m \hat{w}_1) = -(A\nabla(\mathcal{A}_0^{l-1} u_t(0)), \nabla(\mathcal{A}_0^l \hat{w}_1))$$

$$- \sum_{j=0}^{l-2} \int_{\Gamma_0} (\mathcal{A}_0^j u_t(0))_{\nu_A} \mathcal{A}_0^{m-1-j} \hat{w}_1 d\Gamma; \tag{2.161}$$

$$-(u(0), \mathcal{A}_0^{m+1}\hat{w}_0) = -(\mathcal{A}_0^l u(0), \mathcal{A}_0^{l+1}\hat{w}_0)$$

$$+\sum_{j=0}^{l-1}\int_{\Gamma_0}(\mathcal{A}_0^j u(0))_{\nu_A}\mathcal{A}_0^{m-j}\hat{w}_0 d\Gamma. \qquad (2.162)$$

After substituting (2.161) and (2.162) into the left-hand side of the identity (2.160) and eliminating the same terms from the both sides, we obtain the identity (2.153).

Claim 2 There is $T_m > 0$ such that, for $T > T_m$ given, there exists $c_1 > 0$ such that

$$\|\Lambda_N(w_0, w_1)\|_{m-1,m}^2 \ge c_1\|(w_0, w_1)\|_{m+2,m+1}^2 \qquad (2.163)$$

for all $(w_0, w_1) \in \Xi_{0N}^{m+2}(\Omega) \times \Xi_{0N}^{m+1}(\Omega)$.

Proof of (2.163) Replace w with $w^{(m)}$ in the inequality (2.143) and obtain

$$c_T\|(\mathcal{A}_0^l w_1, \mathcal{A}_0^{l+1}w_0)\|_{1,0}^2 \ge \Psi_*(w^{(m+1)}, w^{(m+1)})$$

$$\ge [\rho_0(T - 2\varepsilon) - c]\|(\mathcal{A}_0^l w_1, \mathcal{A}_0^{l+1}w_0)\|_{1,0}^2, \qquad (2.164)$$

for all $(w_0, w_1) \in \Xi_{0N}^{m+2}(\Omega) \times \Xi_{0N}^{m+1}(\Omega)$, where c is a constant, independent of $T > 0$.

We let $\hat{w} = w$ in the identity (2.153) and observe that

$$\int_{\Sigma_0} z_\varepsilon w^{(m+1)}\nabla_{\Gamma_g}h_0(w^{(m+1)})d\Sigma = -\frac{1}{2}\int_{\Sigma_0}z_\varepsilon[w^{(m+1)}]^2\Delta_{\Gamma_g}h_0 d\Sigma; \qquad (2.165)$$

$$\left|\sum_{j=1}^{m-1}\int_{T-\varepsilon}^T\int_{\Gamma_0}C_j^{m-1}z_\varepsilon^{(j)}w^{(m+2-j)}w^{(m+2)}h_0 d\Gamma dt\right|$$

$$\le c\sum_{j=1}^{m-1}\sup_{T-\varepsilon\le t\le T}\|w^{(m+2-j)}\|_{H^{1/2}(\Gamma_0)}\|w^{(m+2)}\|_{H^{-1/2}(\Gamma_0)}$$

$$\le c\|(\mathcal{A}_0^l w_1, \mathcal{A}_0^{l+1}w_0)\|_{1,0}^2; \qquad (2.166)$$

$$\left|\sum_{j=1}^m\int_{T-\varepsilon}^T\int_{\Gamma_0}C_j^m z_\varepsilon^{(j)}w^{(m+1)}(\Delta_{\Gamma_g}w^{(m+1-j)}h_0 + \lambda_T w^{(m+1-j)})d\Gamma dt\right|$$

$$\le c\sum_{j=1}^m\sup_{T-\varepsilon\le t\le T}\|w^{(m+1-j)}\|_{H^{3/2}(\Gamma_0)}\|w^{(m+1)}\|_{H^{1/2}(\Gamma_0)}$$

$$+\lambda_T c\sum_{j=1}^m\sup_{T-\varepsilon\le t\le T}\|w^{(m+1-j)}\|_{H^{1/2}(\Gamma_0)}\|w^{(m+1)}\|_{H^{-1/2}(\Gamma_0)}$$

$$\le c\|(\mathcal{A}_0^l w_1, \mathcal{A}_0^{l+1}w_0)\|_{1,0}^2. \qquad (2.167)$$

Setting $\hat{w} = w$ in the identity (2.153) and via (2.143) and (2.165)-(2.167)

$$\|A\nabla(\mathcal{A}_0^{-1}u_t(0))\|^2 + \|\mathcal{A}_0^l u(0)\|^2 \ge [\rho_0(T - 2\varepsilon) - c]\|(\mathcal{A}_0^l w_1, \mathcal{A}_0^{l+1}w_0)\|_{1,0}^2,$$

which yields Claim 2 by the ellipticity of the operator $\mathcal{A}_0$.

Claim 3 There is $c_2 > 0$ such that

$$\|u_t(0)\|^2_{H^{m-1}(\Omega)} + \|u(0)\|^2_{H^m(\Omega)} \leq c_2\|(w_0, w_1)\|^2_{m+2,m+1} \tag{2.168}$$

for all $(w_0, w_1) \in \Xi^{m+2}_{0N}(\Omega) \times \Xi^{m+1}_{0N}(\Omega)$.

Proof of (2.168)   We assume that $(w_0, w_1) \in \Xi^{m+2}_{0N}(\Omega) \times \Xi^{m+1}_{0N}(\Omega)$. We take $\hat{w}_1 = 0$ in the identity (2.153) and use the inequality (2.143). We obtain

$$|(\mathcal{A}^l_0 u(0), \mathcal{A}^{l+1}_0 \hat{w}_0)| \leq c\|(\mathcal{A}^l_0 w_1, \mathcal{A}^{l+1}_0 w_0)\|_{1,0}\|\mathcal{A}^{l+1}_0 \hat{w}_0\|,$$

that is,

$$\|\mathcal{A}^l_0 u(0)\| \leq c\|(\mathcal{A}^l_0 w_1, \mathcal{A}^{l+1}_0 w_0)\|_{1,0}. \tag{2.169}$$

Using the ellipticity of the operator $\mathcal{A}_0$ and the equation in (2.145), we have

$$\|u(0)\|^2_{H^m(\Omega)} \leq c\|\mathcal{A}_0 u(0)\|^2_{H^{m-2}(\Omega)} + c\|u_{\nu_A}(0)\|^2_{H^{m-3/2}(\Gamma_0)} + c\|u(0)\|^2_{H^{m-1}(\Omega)}$$

$$\leq c\|\mathcal{A}^2_0 u(0)\|^2_{H^{m-4}(\Omega)} + c\|u^{(2)}_{\nu_A}(0)\|^2_{H^{m-7/2}(\Gamma_0)}$$

$$+c\|u_{\nu_A}(0)\|^2_{H^{m-3/2}(\Gamma_0)} + c\|u(0)\|^2_{H^{m-1}(\Omega)}.$$

Repeating this process gives

$$\|u(0)\|^2_{H^m(\Omega)} \leq c\|\mathcal{A}^l_0 u(0)\|^2$$

$$+c\sum_{j=0}^{l-1} \|u^{(2j)}_{\nu_A}(0)\|^2_{H^{m-2j-3/2}(\Gamma_0)} + c\|u(0)\|^2_{H^{m-1}(\Omega)}. \tag{2.170}$$

We use the boundary control in (2.145) and the equation in the duality problem (2.126) to obtain

$$\sum_{j=0}^{l-1} \|u^{(2j)}_{\nu_A}(0)\|^2_{H^{m-2j-3/2}(\Gamma_0)} \leq c\sum_{j=0}^{l-1} \|w^{(2j+1)}(0)\|^2_{H^{m-2j-3/2}(\Gamma_0)}$$

$$+c\sum_{j=0}^{l-1}(\|w^{(2j+3)}(0)\|^2_{H^{m-2j-3/2}(\Gamma_0)} + \|\Delta_{\Gamma_g} w^{(2j+1)}(0)\|^2_{H^{m-2j-3/2}(\Gamma_0)})$$

$$\leq c\sum_{j=0}^{l-1}(\|w^{(2j+3)}(0)\|^2_{H^{m-2j-1}(\Omega)} + \|\mathcal{A}_0 w^{(2j+1)}(0)\|^2_{H^{m-2j-1}(\Omega)})$$

$$\leq c\|w^{(2l+1)}(0)\|^2_1 \leq c\|(\mathcal{A}^l_0 w_1, \mathcal{A}^{l+1}_0 w_0)\|_{1,0}. \tag{2.171}$$

We combine (2.170) and (2.171) and use the induction assumption

$$\|u(0)\|^2_{H^{m-1}(\Omega)} \leq c\|(w_0, w_1)\|^2_{H^{m+1}(\Omega) \times H^m(\Omega)},$$

to have

$$\|u(0)\|^2_{H^m(\Omega)} \leq c\|(w_0, w_1)\|^2_{H^{m+2}(\Omega) \times H^{m+1}(\Omega)} \tag{2.172}$$

for all $(w_0, w_1) \in \Xi_{0N}^{m+2}(\Omega) \times \Xi_{0N}^{m+1}(\Omega)$.

A similar argument establishes the estimate for $u_t(0)$.

Case II Let $m = 2l + 1$ for some $l \geq 0$. A similar argument shows that the inequality (2.148) holds with m replaced by $m + 1$ if it is true for m.

Finally, the theorem follows by induction. $\square$

For the exact controllability of the problem (2.125), we have

Theorem 2.24 *Let $\Gamma_1 \neq \emptyset$ and $\overline{\Gamma}_0 \cap \overline{\Gamma}_1 = \emptyset$. Let H be an escape vector field for the metric g on $\overline{\Omega}$ such that*

$$\langle H, \nu \rangle \leq 0 \quad for \quad x \in \Gamma_1.$$

Then, for $m \geq 0$, there is $T_m > 0$ such that, for $T > T_m$, the problem (2.125) is exactly $L^2(\Omega) \times (H_{\Gamma_1}^1(\Omega))^*$ controllable for the case $m=0$ and exactly $(H^m(\Omega) \cap H_{\Gamma_1}^1(\Omega)) \times (H^{m-1}(\Omega) \cap H_{\Gamma_1}^1(\Omega))$ controllable for the case $m \geq 1$ on $[0, T]$ by $B_N^m(\Sigma_0)$ controls.

Proof. **Step 1** Let the operator Λ_N be given by (2.146). We shall prove that Λ_N are isomorphisms from $\Xi_{0N}^{m+2}(\Omega) \times \Xi_{0N}^{m+1}(\Omega)$ onto $(H^{m-1}(\Omega) \cap H_{\Gamma_1}^1(\Omega)) \times (H^m(\Omega) \cap H_{\Gamma_1}^1(\Omega))$. By Theorem 2.23 it will suffice to prove that Λ_N is surjective.

Let $m = 0$. Let $\mathcal{R}(\Lambda_N) \neq (H_{\Gamma_1}^1(\Omega))^* \times L^2(\Omega)$. Then there is a nonzero point $(v_0, v_1) \in (H_{\Gamma_1}^1(\Omega))^* \times L^2(\Omega)$ such that

$$(u_t(0), v_0)_{(H_{\Gamma_1}^1(\Omega))^*} - (u(0), v_1) = 0 \tag{2.173}$$

for all $(w_0, w_1) \in \mathcal{D}(\mathcal{A}_0) \times H_{\Gamma_1}^1(\Omega)$. Recall that $-\mathcal{A}_0 : H_{\Gamma_1}^1(\Omega) \to (H_{\Gamma_1}^1(\Omega))^*$ is the canonical isomorphism and the inner product of $(H_{\Gamma_1}^1(\Omega))^*$ is defined by

$$(z_1, z_2)_{(H_{\Gamma_1}^1(\Omega))^*} = (\mathcal{A}_0^{-1} z_1, \mathcal{A}_0^{-1} z_2)_{H_{\Gamma_1}^1(\Omega)}.$$

Then, by (2.50),

$$(u_t(0), v_0)_{(H_{\Gamma_1}^1(\Omega))^*} = (\mathcal{A}_0^{-1} u_t(0), \mathcal{A}_0^{-1} v_0)_{H_{\Gamma_1}^1(\Omega)} = (u_t(0), -\mathcal{A}_0^{-1} v_0).$$

Then it follows from the relation (2.173) that

$$(\Lambda_N(w_0, w_1), (-\mathcal{A}_0^{-1} v_0, v_1))_{L^2(\Omega) \times L^2(\Omega)} = 0 \tag{2.174}$$

for all $(w_0, w_1) \in \mathcal{D}(\mathcal{A}_0) \times H_{\Gamma_1}^1(\Omega)$. In particular, we take $w_0 = \mathcal{A}_0^{-1} v_1$ and $w_1 = -\mathcal{A}_0^{-1} v_0$ in (2.174) and obtain, via the inequality (2.151), that $\|(-\mathcal{A}_0^{-1} v_0, v_1)\|_{1,0} = 0$, that is, $(v_0, v_1) = 0$. This contradiction shows that Λ_N is surjective.

Repeating the above procedure, we may prove that Λ_N are surjective when $m \geq 1$. We omit it here.

Step 2 We consider the regularity of the boundary control function in the problem (2.145). Since w_t is a lower order term in the the boundary control, $z_\varepsilon(w^{(3)} - \Delta_{\Gamma_g} w_t)h_0$ is the principal part of the control. It will suffice to prove that $z_\varepsilon(w^{(3)} - \Delta_{\Gamma_g} w_t)h_0 \in \mathrm{B}_N^m(\Sigma_0)$ for $(w_0, w_1) \in \Xi_{0N}^{m+2}(\Omega) \times \Xi_{0N}^{m+1}(\Omega)$ with $m \geq 0$.

Let $m = 0$. Let $(w_0, w_1) \in \Xi_{0N}^2(\Omega) \times \Xi_{0N}^1(\Omega)$. For any $\varphi \in H^1([0,T], L^2(\Gamma_0))$, by Lemma 2.11 below, we have

$$
\begin{aligned}
&(z_\varepsilon(w^{(3)} - \Delta_{\Gamma_g} w_t)h_0, \ \varphi)_{L^2(\Sigma_0)} \\
&= -(z_\varepsilon(w_{tt} - \Delta_{\Gamma_g} w)h_0, \ \varphi_t)_{L^2(\Sigma_0)} - (z_{\varepsilon t}(w_{tt} - \Delta_{\Gamma_g} w)h_0, \ \varphi)_{L^2(\Sigma_0)} \\
&\leq c\|w_{tt} - \Delta_{\Gamma_g} w\|_{L^2(\Sigma_0)} \|\varphi\|_{H^1([0,T], L^2(\Gamma_0))}
\end{aligned}
$$

for all $\varphi \in H_0^1([0,T], L^2(\Gamma_0))$, which implies that $z_\varepsilon(w^{(3)} - \Delta_{\Gamma_g} w_t)h_0 \in \mathrm{B}_N^0(\Sigma_0)$

Let $m = 1$ and $(w_0, w_1) \in \Xi_{0N}^3(\Omega) \times \Xi_{0N}^2(\Omega)$. Then $(w_t(0), w_{tt}(0)) \in \Xi_{0N}^2(\Omega) \times \Xi_{0N}^1(\Omega)$. By Lemma 2.11 below, $w^{(3)} - \Delta_{\Gamma_g} w_t \in L^2(\Sigma_0)$, that is,

$$
z_\varepsilon(w^{(3)} - \Delta_{\Gamma_g} w_t)h_0 \in \mathrm{B}_N^1(\Sigma_0).
$$

Let $m \geq 2$ and $(w_0, w_1) \in \Xi_{0N}^{m+2}(\Omega) \times \Xi_{0N}^{m+1}(\Omega)$. Then $(w^{(m-1)}(0), w^{(m)}(0)) \in \Xi_{0N}^3(\Omega) \times \Xi_{0N}^2(\Omega)$. Lemma 2.11 yields

$$
w^{(m+1)} - \Delta_{\Gamma_g} w^{(m-1)} \in C([0,T], H^{1/2}(\Gamma_0)) \cap H^1(\Sigma_0),
$$

that is,

$$
[z_\varepsilon(w^{(3)} - \Delta_{\Gamma_g} w_t)h_0]^{(m-2)} \in C([0,T], H^{1/2}(\Gamma_0)) \cap H^1(\Sigma_0).
$$

Then $z_\varepsilon(w^{(3)} - \Delta_{\Gamma_g} w_t)h_0 \in \mathrm{B}_N^m(\Sigma_0)$. $\square$

Remark 2.18 *Unlike the control with the Dirichlet action, here* Λ_N: $H^{m+1}(\Omega) \times H^m(\Omega) \to H^{m-2}(\Omega) \times H^{m-1}(\Omega)$. *This is because the Neumann action loses a regularity of* 1 *order (actually, order* 1/2 *). For this point, see* [117].

Lemma 2.11 *Let* w *solve the duality problem* (2.126) *with the initial data* $(w_0, w_1) \in D(\mathcal{A}_0) \times H_{\Gamma_1}^1(\Omega)$. *Then*

$$
w_{tt} - \Delta_{\Gamma_g} w \in L^2(\Sigma_0). \tag{2.175}
$$

Furthermore, if $(w_0, w_1) \in \Xi_{0N}^3(\Omega) \times \Xi_{0N}^2(\Omega)$, *then*

$$
w_{tt} - \Delta_{\Gamma_g} w \in C([0,T], H^{1/2}(\Gamma_0)) \cap H^1(\Sigma_0). \tag{2.176}
$$

Proof. Let the Riemann metric g be given by (2.4). Then

$$
\operatorname{div} A(x)\nabla w = \Delta_g w + F(w) \quad \text{for} \quad x \in \Omega, \tag{2.177}
$$

where Δ_g is the Laplacian of the metric g and F is a vector field on Ω give by

$$F = -\frac{1}{2 \det A(x)} A(x) \nabla \det A(x).$$

Using the boundary condition $w_{\nu_A} = 0$ on Σ_0 and the relations (2.126) and (2.177), we obtain

$$w_{tt} - \Delta_{\Gamma_g} w = \frac{1}{|\nu_A|_g^2} D^2 w(\nu_A, \nu_A) + \langle F, \nabla_{\Gamma_g} w \rangle_g \quad \text{on} \quad \Sigma_0$$

where $D^2 w(\cdot, \cdot)$ is the Hessian of w in the metric g. Since $\|\nabla_{\Gamma_g} w\|_{L^2(\Sigma_0)}^2 \leq c_T \|(w_1, \mathcal{A}_0 w_0)\|_{1,0}^2$, to get the relation (2.175) it will suffice to prove

$$D^2 w(\nu_A, \nu_A) \in L^2(\Sigma_0). \tag{2.178}$$

Let H be a vector field on $\overline{\Omega}$ such that

$$H = 0 \quad \text{for} \quad x \in \Gamma_1; \quad H = \nu_A \quad \text{for} \quad x \in \Gamma_0.$$

We set

$$v = H(w) \quad \text{for} \quad x \in \Omega.$$

It is easy to check that this v solves the problem with the Dirichlet boundary conditions

$$\begin{cases} v_{tt} = \operatorname{div} A \nabla v + [H, \operatorname{div} A \nabla] w & \text{in} \quad Q, \\ v = 0 & \text{on} \quad \Sigma, \\ v(0) = H(w_0), \quad v_t(0) = H(w_1). \end{cases} \tag{2.179}$$

In addition, $(w_0, w_1) \in D(\mathcal{A}_0) \times H_{\Gamma_1}^1(\Omega)$ implies $(v(0), v_t(0)) \in H_0^1(\Omega) \times L^2(\Omega)$.

We integrate the identity (2.13) where $u = v$ and $f = [H, \operatorname{div} A \nabla] w$ to obtain

$$v_{\nu_A} = D^2 w(\nu_A, \nu_A) + \langle \nabla_{\Gamma_g} w, \ D_{\nu_A} H \rangle_g \in L^2(\Sigma_0),$$

which gives the relation (2.178).

Next, we assume that $(w_0, w_1) \in \Xi_{0N}^3(\Omega) \times \Xi_{0N}^2(\Omega)$. Then $(v(0), v_t(0)) \in \left(H^2(\Omega) \cap H_0^1(\Omega) \right) \times H_0^1(\Omega)$, where v solves the problem (2.179). A similar argument, as for the relation (2.117), shows that

$$v_{\nu_A} \in C([0, T], H^{1/2}(\Gamma_0)) \cap H^1(\Sigma_0),$$

which implies that the relation (2.176) is true. $\qquad\square$

2.6 A Counterexample without Exact Controllability

A systematic attempt to understand the wave equation controllability and non-controllability properties can be found in the classical articles [174] and

[175] where the author proved that to have exact controllability the so-called geometric optics condition (GOC) needs to be satisfied. Later on it was proved by [8] that in fact GOC is also sufficient to assure exact controllability. Although it is hard to check as a sufficient condition, GOC is useful as a necessary one to construct counterexamples by the geometrical approach; see Theorem 2.25 later.

Given the differential operator

$$Pu = u_{tt} - \operatorname{div} A(x)\nabla u$$

acting on functions, one sets $f_\alpha = e^{i\alpha(\langle x,\xi\rangle + t\tau)}$, and

$$p(x,t,\xi,\tau) = \lim_{\alpha\to\infty} \overline{f}_\alpha P f_\alpha = \langle A(x)\xi,\xi\rangle - \tau^2$$

for $(x,t,\xi,\tau) \in \mathbb{R}^n \times \mathbb{R} \times \mathbb{R}^n \times \mathbb{R}$. $p(x,t,\xi,\tau)$ is called the principal symbol of the operator P.

Definition 2.9 *Given $(x_0,t_0,\xi_0) \in \mathbb{R}^n \times \mathbb{R} \times \mathbb{R}^n\backslash\{0\}$, $(x(s),t(s),\xi(s),\tau(s))$ is said to be a null bicharacteristic curve through (x_0,t_0,ξ_0) if it satisfies the Hamiltonian system of ordinary differential equations*

$$\dot{x}(s) = \frac{1}{2}\nabla_\xi p, \quad \dot{t}(s) = \frac{1}{2}\frac{\partial p}{\partial \tau}, \quad \dot{\xi}(s) = -\frac{1}{2}\nabla p, \quad \dot{\tau}(s) = -\frac{1}{2}\frac{\partial p}{\partial t}, \qquad (2.180)$$

with $(x(0),t(0),\xi(0)) = (x_0,t_0,\xi_0)$ and $\tau(0)$ chosen so that $p(x_0,t_0,\xi_0,\tau(0)) = 0$, where ∇_ξ denotes the gradient of the Euclidean metric with respect to the variable $\xi \in \mathbb{R}^n$. The projection, $(x(s),t(s))$, of a bicharacteristic curve on (x,t)-space is called a ray.

Note that there are two choices for $\tau(0)$ and $p(x(s),t(s),\xi(s),\tau(s)) = 0$ for all s since the matrix $A(x)$ is positive for $x \in \mathbb{R}^n$.

The following proposition shows that a ray is actually a geodesic in the metric $g = A^{-1}(x)$.

Proposition 2.4 *Let $x(t)$ be a geodesic on the Riemannian manifold (R^n, g) with $x(0) = x_0$ parameterized by arc length. Then*

$$(x(t), \pm t, \xi(t), \mp 1)$$

are bicharacteristic curves through $(x_0, 0, \xi_0)$, where $\xi(t) = A^{-1}(x(t))\dot{x}(t)$ for all $t \in \mathbb{R}$.

Proof. It will suffice to verify the relation (2.180). The relation $\xi(t) = A^{-1}(x(t))\dot{x}(t)$ yields

$$\dot{x}(t) = A(x(t))\xi(t) = \frac{1}{2}\nabla_\xi p \quad \text{for} \quad t \in \mathbb{R}. \qquad (2.181)$$

Let $A(x) = (a_{ij}(x))$ and $A^{-1}(x) = (g_{ij}(x))$. Then

$$\sum_{l=1}^{n} g_{lpx_i} a_{lk} = -\sum_{l=1}^{n} g_{lp} a_{lkx_i} \quad \text{for} \quad 1 \le i, p \le n. \tag{2.182}$$

Since $x(t)$ is a geodesic, we have

$$\ddot{x}_j(t) + \sum_{lp=1}^{n} \Gamma_{lp}^j \dot{x}_l(t) \dot{x}_p(t) = 0 \quad \text{for} \quad 1 \le j \le n \tag{2.183}$$

where Γ_{lp}^j are the coefficients of the Levi-Civita connection, given by

$$\Gamma_{lp}^j = \frac{1}{2} \sum_{k=1}^{n} a_{jk}(g_{klx_p} + g_{kpx_l} - g_{lpx_k}),$$

from which we further have

$$\sum_{j=1}^{n} g_{ij} \Gamma_{lp}^j = \frac{1}{2}(g_{ilx_p} + g_{ipx_l} - g_{lpx_i}). \tag{2.184}$$

Using the relations (2.183), (2.184), (2.181) and (2.182), we obtain

$$\dot{\xi}_i(t) = \left[\sum_{l=1}^{n} g_{il}(x(t))\dot{x}_l(t)\right]_t = \sum_{lp=1}^{n} g_{ilx_p}(x(t))\dot{x}_l(t)\dot{x}_p(t) + \sum_{j=1}^{n} g_{ij}(x(t))\ddot{x}_j(t)$$

$$= \frac{1}{2}\sum_{lp=1}^{n} g_{lpx_i}\dot{x}_l(t)\dot{x}_p(t) = \frac{1}{2}\sum_{lpkj=1}^{n} g_{lpx_i} a_{kl} a_{jp}\xi_k\xi_j$$

$$= -\frac{1}{2}\sum_{lpkj=1}^{n} g_{lp}a_{jp}a_{klx_i}\xi_k\xi_j = -\frac{1}{2}\frac{\partial p}{\partial x_i} \tag{2.185}$$

for $1 \le i \le j$.

Finally, it follows that

$$p(x(t), \pm t, \xi(t), \mp 1) = \sum_{ij=1}^{n} a_{ij}(x(t))\xi_i(t)\xi_j(t) - 1$$

$$= \sum_{ijlp} a_{ij}g_{il}g_{jp}\dot{x}_l(t)\dot{x}_p(t) - 1 = \sum_{il} g_{li}\dot{x}_l(t)\dot{x}_i(t) - 1 = \langle \dot{x}(t), \dot{x}(t)\rangle_g - 1 = 0$$

for all $t \in \mathbb{R}$, since $x(t)$ is parameterized by arc length. This completes our proof. $\square$

From [8] and Proposition 2.4, the following result is immediate.

Theorem 2.25 *If there is a closed geodesic that is contained in Ω, then for any $T > 0$, the problem (2.45) has no exact controllability where $\Gamma_0 = \Gamma$.*

We give a counterexample to end this section.

Example 2.7 *Let*

$$A(x) = \begin{pmatrix} (1+|x|^2)^2 & 0 \\ 0 & (1+|x|^2)^2 \end{pmatrix}, \quad x = (x_1, x_2) \in \mathbb{R}^2.$$

The metric is

$$g = \begin{pmatrix} (1+|x|^2)^{-2} & 0 \\ 0 & (1+|x|^2)^{-2} \end{pmatrix}, \quad x = (x_1, x_2) \in \mathbb{R}^2.$$

Let

$$B = \{ x \,|\, |x|^2 < 1 \}, \quad S = \{ x \,|\, |x| = 1 \}.$$

We have the following conclusions.

(i) If $\overline{\Omega} \subset B$, or $\overline{\Omega} \subset \mathbb{R}^2 \backslash B$, then the problem (2.45) is exactly controllable;

(ii) If $S \subset \Omega$, then the problem (2.45) has no exact controllability even when controls act on the whole boundary Γ.

Proof. Let $k(x)$ be the Gaussian curvature function. By Lemma 2.3, we have

$$k(x) = 4 \quad \text{for} \quad x \in \mathbb{R}^2.$$

Let M be the sphere of radius $1/2$ and centered at $(0, 0, 1/2)$ in $\mathbb{R}^3$, given by

$$M = \{ (x_1, x_2, x_3) \,|\, x_1^2 + x_2^2 + x_3^2 = x_3 \},$$

with the induced Riemannian metric from $\mathbb{R}^3$. Then $(\mathbb{R}^2 \cup \{\infty\}, g)$ is isometric to M with an isometry $\Phi \colon M \to (\mathbb{R}^2 \cup \{\infty\}, g)$, defined by

$$\Phi(x_1, x_2, x_3) = (\frac{x_1}{1 - x_3}, \frac{x_2}{1 - x_3}) \quad \text{for} \quad (x_1, x_2, x_3) \in M.$$

It is easy to check that

$$\sup_{x \in \mathbb{R}^2} d_g(0, x) = \pi/2, \quad \sup_{x \in S} d_g(0, x) = \pi/4,$$

where d_g is the distance function in the metric g.

If $\overline{\Omega} \subset \{ x \,|\, |x| < 1 \}$, or $\overline{\Omega} \subset \{ x \,|\, |x| > 1 \}$, the exact controllability follows from Theorem 2.5.

Let $S \subset \Omega$. Since the big circle $C = \{ (x_1 x_2, 1/2) \,|\, x_1^2 + x_2^2 = 1/4 \}$ is a closed geodesic on the sphere M, $S = \Phi(C)$ is a geodesic in $(\mathbb{R}^2, g)$. Then (ii) follows from Theorem 2.25. $\qquad\square$

2.7   Stabilization

We will study stabilization from boundary and interior, respectively.
Stabilization from Boundary Consider the stabilization problem

$$\begin{cases} u_{tt} = \operatorname{div} A(x)\nabla u & \text{in} \quad (0,\infty) \times \Omega, \\ u = 0 & \text{on} \quad (0,\infty) \times \Gamma_1 \\ u_{\nu_A} + au_t = 0 & \text{on} \quad (0,\infty) \times \Gamma_0, \\ u(0) = u_0, \quad u_t(0) = u_1 & \text{on} \quad \Omega, \end{cases} \tag{2.186}$$

where a is a positive function on $\overline{\Gamma}_0$. Let

$$2E(t) = \int_\Omega (u_t^2 + \langle A(x)\nabla u, \nabla u\rangle)dx$$

be the energy. It follows that

$$E(t) = E(0) - \int_0^t \int_{\Gamma_0} au_t^2 d\Gamma \quad \text{for} \quad t \geq 0. \tag{2.187}$$

The well-posedness of the problem (2.186) can be obtained by the same arguments as in the case of constant coefficients but with a few modifications, for example, by the theory of semigroups ([39]-[41]). We omit the details.

Our purposes are to seek geometrical assumptions on the metric $g = A^{-1}(x)$ such that the energy of the problem (2.186) decays exponentially.

Remark 2.19 *In the case of constant coefficients the exponential decay of energy for the problem (2.186) from boundary was obtained under certain geometric conditions by [39], [40], [41], [99], [100], [122], [207], and many other authors.*

Trace Estimate on Boundary In order to remove some unnecessary restrictions on the control portion Γ_0, a trace estimate of the wave equation has to be introduced from [122]. Its proof needs a pseudodifferential analysis and we omit it.

Let u solve the wave equation

$$\begin{cases} u_{tt} = \operatorname{div} A(x)\nabla u + X(u) + pu + qu_t & \text{in} \quad Q, \\ u = 0 & \text{on} \quad \Sigma_0, \end{cases} \tag{2.188}$$

where X is a vector field and p, q are functions, respectively, on Ω. Then

Lemma 2.12 ([122]) *Let $T > 0$ be given. let $T/2 > \varepsilon > 0$ be given small. Then there is $c_{T\varepsilon} > 0$ such that*

$$\int_\varepsilon^{T-\varepsilon} \int_{\Gamma_0} |\nabla_{\Gamma_g} u|_g^2 d\Gamma dt \leq c_{T\varepsilon} \int_{\Sigma_0} (u_{\nu_A}^2 + u_t^2)d\Sigma + c_{T\varepsilon}\|u\|_{H^{\varepsilon+1/2}(Q)}^2 \tag{2.189}$$

for all solutions u to the problem (2.188) where

$$\nabla_{\Gamma_g} u = \nabla_g u - \frac{u_{\nu_A}}{|A\nu|_g^2} A\nu \quad for \quad x \in \Gamma. \tag{2.190}$$

We have

Theorem 2.26 *Let H be an escape vector field on $\overline{\Omega}$ for the metric $g = A^{-1}(x)$ such that*

$$\langle H, \nu \rangle \leq 0 \quad for \quad x \in \Gamma_1. \tag{2.191}$$

Then there are $C > 0$ and $\sigma > 0$ such that

$$E(t) \leq Ce^{-\sigma t} \quad for \quad t \geq 0 \tag{2.192}$$

for all solutions u to the problem (2.186).

Proof. For convenience, we assume

$$DH(X, X) \geq |X|_g^2 \quad for \quad X \in \mathbb{R}_x^n, \ x \in \Omega. \tag{2.193}$$

Then

$$\operatorname{div} H \geq n \geq 2 \quad for \quad x \in \Omega. \tag{2.194}$$

Set

$$\Phi(t) = 2 \int_{\Omega} u_t P u \, dx, \quad P u = H(u) + pu, \quad p = (\operatorname{div} H - 1)/2.$$

Then

$$|\Phi(t)| \leq C(E(t) + \|u(t)\|^2) \quad for \quad t \geq 0. \tag{2.195}$$

Noting that $\langle H, A\nu \rangle_g = \langle H, \nu \rangle$ for $x \in \Gamma$, we have a direct decomposition

$$H = \frac{\langle H, \nu \rangle}{|A\nu|_g^2} A\nu + H_{\Gamma_g} \quad for \quad x \in \Gamma$$

where $\langle A\nu, H_{\Gamma_g} \rangle_g = 0$ for $x \in \Gamma$. It follows that

$$Pu|_{\Sigma_1} = \langle H, \nu \rangle u_{\nu_A}/|A\nu|_g^2, \tag{2.196}$$

$$Pu|_{\Sigma_0} = \langle H, \nu \rangle u_{\nu_A}/|A\nu|_g^2 + H_{\Gamma_g}(u) + pu. \tag{2.197}$$

Using the equation in (2.186) and the formula (2.6) and noting that $\operatorname{div} H - 2p = 1$, we have

$$
\begin{aligned}
2u_{tt}Pu &= 2Pu \operatorname{div} A(x)\nabla u = 2\operatorname{div}(PuA(x)\nabla u) - 2\langle A(x)\nabla u, \nabla Pu \rangle \\
&= \operatorname{div}(2PuA(x)\nabla u - |\nabla_g u|_g^2 H) - 2DH(\nabla_g u, \nabla_g u) \\
&\quad + |\nabla_g u|_g^2 \operatorname{div} H - 2p|\nabla_g u|_g^2 - A(x)\nabla p(u^2) \\
&= \operatorname{div}(2PuA(x)\nabla u - |\nabla_g u|_g^2 H - u^2 A\nabla p) \\
&\quad - 2DH(\nabla_g u, \nabla_g u) + |\nabla_g u|_g^2 + u^2 \operatorname{div} A(x)\nabla p
\end{aligned}
$$

which yields, via the formulas (2.196), (2.197), (2.193), and (2.194),

$$
\begin{aligned}
\Phi_t(t) &= 2 \int_\Omega [u_{tt} Pu + u_t Pu_t]\, dx \\
&= \int_\Gamma [2Puu_{\nu_A} + (u_t^2 - |\nabla_g u|_g^2)\langle H, \nu \rangle]\, d\Gamma \\
&\quad - \int_\Omega [2DH(\nabla_g u, \nabla_g u) + (u_t^2 - |\nabla_g u|_g^2) - u^2\, \mathrm{div}\, A(x)\nabla p]\, dx \\
&\leq \int_{\Gamma_1} \langle H, \nu \rangle u_{\nu_A}^2 / |A\nu|_g^2 d\Gamma + C \int_{\Gamma_0} (|\nabla_{\Gamma_g} u|_g^2 + u_{\nu_A}^2 + u_t^2)\, d\Gamma \\
&\quad - E(t) + C\|u\|_{L^2(\Omega)}^2 \\
&\leq -E(t) + C \int_{\Gamma_0} (|\nabla_{\Gamma_g} u|_g^2 + u_{\nu_A}^2 + u_t^2)\, d\Gamma + C\|u\|_{L^2(\Omega)}^2. \quad (2.198)
\end{aligned}
$$

Using (2.189), (2.198), and (2.195), we obtain

$$
\int_\varepsilon^{T-\varepsilon} E(\tau)\, d\tau \leq |\Phi(T-\varepsilon)| + |\Phi(\varepsilon)| + C\|u\|_{L^2(Q)}^2
$$

$$
+ C \int_\varepsilon^{T-\varepsilon} \int_{\Gamma_0} (|\nabla_{\Gamma_g} u|_g^2 + u_{\nu_A}^2 + u_t^2)\, d\Gamma dt
$$

$$
\leq C[E(T-\varepsilon) + E(\varepsilon) + \|u(T-\varepsilon)\|^2 + \|u(\varepsilon)\|^2] + C\|u\|_{L^2(Q)}^2
$$

$$
+ c_{T\varepsilon}(\|u_{\nu_A}\|_{L^2(\Sigma_0)}^2 + \|u_t\|_{L^2(\Sigma_0)}^2) + c_{T\varepsilon}\|u\|_{H^{\varepsilon+1/2}(\Omega)}^2 \quad (2.199)
$$

where $\| \cdot \|$ denotes the norm of $L^2(\Omega)$. Using the estimate $E(T) \leq E(t)$ for $\varepsilon \leq t \leq T - \varepsilon$ in the left hand side of (2.199), using the relation (2.187) in the right hand side of (2.199), and using $u_{\nu_A} = -au_t$ for $(t, x) \in \Sigma_0$, we have

$$
(T - 2\varepsilon - 2C)E(T) \leq c_{T\varepsilon}\|u_t\|_{L^2(\Sigma_0)}^2 + C_T L(u),
$$

which yields, after we take $T > 2\varepsilon + 2C$, a constant $c_T > 0$ such that

$$
E(T) \leq c_T\|u_t\|_{L^2(\Sigma_0)}^2 + c_T L(u) \quad (2.200)
$$

where $L(u)$ denote some lower order terms with respect to the norm of $H^1(Q)$. Then a compactness-uniqueness argument as in Lemma 2.5 gives a constant $c_T > 0$ such that

$$
E(T) \leq c_T\|u_t\|_{L^2(\Sigma_0)}^2. \quad (2.201)
$$

Let $a \geq a_0 > 0$ for $x \in \Gamma_0$. The relation (2.187) implies that

$$
a_0 \int_{\Sigma_0} u_t^2 d\Sigma \leq E(0) - E(T).
$$

Using the above inequality in (2.201), we obtain

$$
E(T) \leq \frac{c_T}{a_0 + c_T} E(0)
$$

for all solutions u to the problem (2.186) which yields the formula (2.192) (Exercise 2.7).   □

Stabilization from Interior Consider the stabilization problem

$$\begin{cases} u_{tt} - \operatorname{div} A(x)\nabla u + a(x)u_t = 0 & \text{in} \quad (0,\infty) \times \Omega, \\ u = 0 \quad \text{on} \quad (0,\infty) \times \Gamma, \\ u(0) = u_0, \quad u_t(0) = u_1 \quad \text{on} \quad \Omega \end{cases} \qquad (2.202)$$

where $a(x)$ is a nonnegative smooth function on Ω. Let $E(t)$ be the energy of (2.202), given by (2.55). A simple computation gives

$$E(T) = E(0) - \int_Q a u_t^2 dQ, \quad Q = (0,T) \times \Omega, \quad T > 0. \qquad (2.203)$$

Remark 2.20 *For the constant coefficient problem the earlier work concerning a local control was due to [103]; [151] used the piecewise multiplier method to study the local feedback; [196] studied the nonlinear feedback acting on a local region which was a neighborhood of a suitable subset of the boundary.*

We need

Lemma 2.13 *Let $\hat{\Omega} \subset \Omega$ be an open set and let $\hat{\nu}$ be the normal of $\partial\hat{\Omega}$ in the Euclidean metric of $\mathbb{R}^n$ pointing outside of $\hat{\Omega}$. Let H be a vector field on $\hat{\Omega}$. Let*

$$\hat{Q} = (0,T) \times \hat{\Omega}, \quad \hat{\Sigma} = (0,T) \times \partial\hat{\Omega}.$$

Then

$$2\int_{\hat{Q}} DH(\nabla_g u, \nabla_g u)dQ = -2\int_{\hat{\Omega}} u_t[H(u) + pu]dx\Big|_0^T$$

$$+ \int_{\hat{Q}} \Phi(H, \hat{\Omega})dQ + \int_{\hat{\Sigma}} \Phi(H, \partial\hat{\Omega})d\Sigma, \qquad (2.204)$$

where u is a solution to the problem (2.202), $2p = \operatorname{div} H$,

$$\Phi(H, \hat{\Omega}) = u^2 \operatorname{div} A(x)\nabla p - 2au_t[H(u) + pu], \qquad (2.205)$$

$$\Phi(H, \partial\hat{\Omega}) = 2H(u)u_{\hat{\nu}_A} + (u_t^2 - |\nabla_g u|_g^2)\langle H, \hat{\nu}\rangle$$

$$+ 2puu_{\hat{\nu}_A} - u^2 p_{\hat{\nu}_A}, \qquad (2.206)$$

and $\hat{\nu}_A = A\hat{\nu}$.

Proof. We take $f = -a(x)u_t$ in the identity (2.13) to have

$$2DH(\nabla_g u, \nabla_g u) = \operatorname{div}\{2H(u)A(x)\nabla u - (|\nabla_g u|_g^2 - u_t^2)H\}$$

$$- 2au_t H(u) + 2p(|\nabla_g u|_g^2 - u_t^2) - 2[u_t H(u)]_t. \qquad (2.207)$$

Moreover, letting $f = -au_t$ in the identity (2.14) yields

$$
\begin{aligned}
2p(|\nabla_g u|_g^2 - u_t^2) &= \text{div}\,[2puA(x)\nabla u - u^2 A(x)\nabla p] - 2pau_t u \\
&\quad + u^2\,\text{div}\,A(x)\nabla p - 2p(uu_t)_t.
\end{aligned}
\tag{2.208}
$$

Inserting (2.208) into (2.207), and integrating over $\hat{Q}$, we obtain the identity (2.204).   □

Let

$$
\omega = \{\,x \in \Omega\,|\,a(x) > 0\,\}.
\tag{2.209}
$$

We hope that the region ω can be small in some sense so this problem is also referred as to *locally distributed control*. We have

Theorem 2.27 *If ω is an escape region on $\overline{\Omega}$ for the metric $g = A^{-1}(x)$, then there are constants $C > 0$ and $\sigma > 0$ such that*

$$
E(t) \le Ce^{-\sigma t} E(0)
\tag{2.210}
$$

for all solutions u to the problem (2.202).

Proof. It suffices to prove that there are $T > 0$ and $c > 0$, independent of solutions to the problem (2.202), such that

$$
E(T) \le c\int_Q au_t^2\,dQ, \quad Q = (0, T) \times \Omega.
\tag{2.211}
$$

Indeed, if (2.211) is true, then it follows from (2.203) that

$$
E(T) \le \frac{c}{1 + c} E(0),
$$

which implies (2.210).

Next, we prove (2.211).

Let Ω_i and H^i be given in (2.34) and (7.206), respectively, for $1 \le i \le J$. Let $\varepsilon > 0$ be such that

$$
\omega \supseteq \overline{\Omega} \cap \mathcal{N}_\varepsilon[\cup_{i=1}^J \Gamma_0^i \cup (\Omega \setminus \cup_{i=1}^J \Omega_i)],
\tag{2.212}
$$

where

$$
\mathcal{N}_\varepsilon(S) = \cup_{x \in S}\{\,y \in R^n\,|\,|y - x| < \varepsilon\,\}, \quad S \subset R^n,
$$

$$
\Gamma_0^i = \{\,x \in \partial\Omega_i\,|\,H^i(x) \cdot \nu^i(x) > 0\,\},
$$

and $\nu^i(x)$ is the unit normal of $\partial\Omega_i$ in the Euclidean metric, pointing towards the exterior of Ω_i for all i.

For $0 < \varepsilon_2 < \varepsilon_1 < \varepsilon_0 < \varepsilon$, set

$$
V_j = \mathcal{N}_{\varepsilon_j}[\cup_{i=1}^J \Gamma_0^i \cup (\Omega \setminus \cup_{i=1}^J \Omega_i)] \quad \text{for} \quad j = 0, 1, 2.
\tag{2.213}
$$

Then

$$\Omega \backslash V_j \subset \cup_{i=1}^{J} \Omega_i \quad \text{for} \quad j = 0, 1, 2, \tag{2.214}$$

$$V_2 \subset V_1 \subset V_0 \subset \overline{V}_0, \quad \overline{V}_0 \cap \overline{\Omega} \subset \omega. \tag{2.215}$$

Clearly, there is $a_0 > 0$ such that

$$a(x) \geq a_0 \quad \text{for} \quad x \in \overline{V}_0 \cap \overline{\Omega}. \tag{2.216}$$

Let ϕ^i satisfy $\phi^i \in C_0^\infty(\mathbb{R}^n)$, $0 \leq \phi^i \leq 1$, and

$$\phi^i = \begin{cases} 1 & \text{on} \quad \overline{\Omega}_i \backslash V_1, \\ 0 & \text{on} \quad V_2, \end{cases} \tag{2.217}$$

for $1 \leq j \leq J$.

Step 1 We estimate the integral

$$\int_0^T \int_{\Omega \backslash V_1} |\nabla_g u|_g^2 dx dt.$$

We take $H = \phi^i H^i$, $2p_i = \text{div}\,(\phi^i H^i)$, and $\hat{\Omega} = \Omega_i$ in the identity (2.204), respectively, to have

$$2\int_{Q_i} D(\phi^i H^i)(\nabla_g u, \nabla_g u) dQ = -2\int_{\Omega_i} u_t [\phi^i H^i(u) + p_i u] dx|_0^T$$

$$+ \int_{Q_i} \Phi(\phi^i H^i, \Omega_i) dQ + \int_{\Sigma_i} \Phi(\phi^i H^i, \partial\Omega_i) d\Sigma \tag{2.218}$$

where $Q_i = (0, T) \times \Omega_i$, $\Sigma_i = (0, T) \times \partial\Omega_i$,

$$\Phi(\phi^i H^i, \Omega_i) = u^2\,\text{div}\,A(x)\nabla p_i - 2au_t[\phi^i H^i(u) + p_i u], \tag{2.219}$$

and

$$\Phi(\phi^i H^i, \partial\Omega_i) = 2\phi^i H^i(u) u_{\nu_{i_A}} + \phi^i (u_t^2 - |\nabla_g u|_g^2)\langle H^i, \nu_i \rangle$$

$$+ 2p_i u u_{\nu_{i_A}} - u^2 p_{i\nu_{i_A}} \tag{2.220}$$

for all i.

Since $\phi^i(x) = 1$ for $x \in \Omega_i \backslash V_1$, it follows from (2.217) that

$$D(\phi^i H^i)(\nabla_g u, \nabla_g u) \geq \rho_0 |\nabla_g u|_g^2 \quad \text{for} \quad x \in \Omega_i \backslash V_1 \tag{2.221}$$

for all i.

Since $\partial\Omega_i \backslash \Gamma \subset \Omega \backslash \cup_{j=1}^J \Omega_j \subset V_2$ and $\Gamma_0^i \subset V_2$, we have $\phi^i = 0$ for $x \in \Gamma_0^i \cup (\partial\Omega_i \backslash \Gamma)$ and

$$\Phi(\phi^i H^i, \partial\Omega_i) = 0 \quad \text{for} \quad x \in \Gamma_0^i \cup (\partial\Omega_i \backslash \Gamma)$$

for all i. In addition, if $x \in \partial\Omega_i \cap \Gamma$, then the boundary condition $u = 0$ implies

$$\Phi(\phi^i H^i, \partial\Omega_i) = \phi^i \langle H^i, \nu_i \rangle \frac{u_{\nu_{iA}}^2}{|A\nu_i|^2} \quad \text{for} \quad x \in \partial\Omega_i \cap \Gamma$$

for all i. Noting that $\partial\Omega_i = [\Gamma_0^i \cup (\partial\Omega_i \backslash \Gamma)] \cup [(\partial\Omega_i \backslash \Gamma_0^i) \cap \Gamma]$, we have

$$\Phi(\phi^i H^i, \partial\Omega_i) \leq 0 \quad \text{for all} \quad x \in \partial\Omega_i \qquad (2.222)$$

for all i.

Using the estimates (2.221) and (2.222) in the identity (2.218) yields

$$\int_0^T \int_{\Omega_i \backslash V_1} |\nabla_g u|_g^2 dx dt \leq C \int_0^T \int_{\Omega_i \cap V_1} |\nabla_g u|_g^2 dx dt$$

$$+ C[E(T) + E(0) + \|u(0)\|^2 + \|u(T)\|^2) + C_\beta \int_{Q_i} a u_t^2 dQ$$

$$+ \beta \int_{Q_i} |\nabla_g u|_g^2 dQ + C_\beta \|u\|_{L^2(Q)}^2$$

where $\beta > 0$ can be chosen small, which gives, via the relation $\Omega \subset (\cup_{i=1}^J \Omega_i) \cup V_1$, that

$$\int_0^T \int_{\Omega \backslash V_1} |\nabla_g u|_g^2 dx dt \leq \sum_i \int_0^T \int_{\Omega_i} |\nabla_g u|_g^2 dx dt$$

$$\leq C \int_0^T \int_{\Omega \cap V_1} |\nabla_g u|_g^2 dx dt + C \int_Q a u_t^2 dQ$$

$$+ C[E(T) + E(0) + L(u)]. \qquad (2.223)$$

Step 2 We estimate

$$\int_0^T \int_{\Omega \cap V_1} |\nabla_g u|_g^2 dx dt.$$

Let $\xi \in C_0^\infty(\mathbb{R}^n)$ be such that $0 \leq \xi \leq 1$ and

$$\xi = \begin{cases} 0 & x \in \mathbb{R}^n / V_0 \\ 1 & x \in V_1. \end{cases}$$

We take $p = \xi$ and $f = -a u_t$ in the identity (2.14), respectively, and integrate it over $Q = (0, T) \times \Omega$ to obtain

$$2 \int_Q \xi |\nabla_g u|_g^2 dQ = \int_Q [2\xi u_t^2 - 2a u_t u + u^2 \text{ div } A(x)\nabla\xi] dQ - 2 \int_\Omega \xi u u_t dx |_0^T$$

which yields, via (2.216),

$$2\int_0^T \int_{\Omega \cap V_1} |\nabla_g u|_g^2 dx dt \leq 2\int_0^T \int_{V_0 \cap \Omega} u_t^2 dx dt + \beta \int_Q u_t^2 dQ$$
$$+ C_\beta \|u\|_{L^2(Q)}^2 + E(T) + E(0)$$
$$\leq \frac{2}{a_0} \int_Q a u_t^2 dQ + \beta \int_Q u_t^2 dQ + C_\beta \|u\|_{L^2(Q)}^2 + E(T) + E(0) \quad (2.224)$$

where $\beta > 0$ can be chosen small.

Step 3 It follows from (2.223) and (2.224) that

$$\int_Q |\nabla_g u|_g^2 dQ \leq C \int_Q a u_t^2 d + \beta \int_Q u_t^2 dQ$$
$$+ C_\beta [E(T) + E(0) + L(u)] \quad (2.225)$$

where $\beta > 0$ can be chosen small.

On the other hand, we take $p = 1/2$ in the identity (2.14) and then integrate it over $Q = (0, T) \times \Omega$. We have

$$\int_Q (u_t^2 - |\nabla_g u|_g^2) dQ = \int_Q a u_t u dQ + \int_\Omega u u_t dx|_0^T$$
$$\leq \int_Q a u_t^2 dQ + C[E(T) + E(0) + L(u)]. \quad (2.226)$$

From (2.225) and (2.226), we obtain

$$\int_0^T E(t) dt = \int_Q |\nabla_g u|_g^2 dQ + \frac{1}{2} \int_Q (u_t^2 - |\nabla_g u|_g^2) dQ$$
$$\leq C \int_Q a u_t^2 dQ + \beta \int_0^T E(t) dt + C[E(T) + E(0) + L(u)]. \quad (2.227)$$

From (2.203), we have $E(t) \geq E(T)$ for all $t \in [0, T]$ and $E(0) = E(T) + \int_Q a u_t^2 dQ$. Using those facts and letting $\beta = 1/2$ in the inequality (2.227), we obtain

$$(T/2 - 2C) E(T) \leq 2C \int_Q a u_t^2 dQ + C L(u)$$

where constant $C > 0$ is given in the inequality (2.227). Then for $T > 4C$, there is a constant $c_T > 0$ such that

$$E(T) \leq c_T \int_Q a u_t^2 dQ + c_T L(u) \quad (2.228)$$

for all solutions u to the problem (2.202). By an argument as in Lemma 2.5, the lower order terms can be absorbed and the inequality (2.211) follows from (2.228).   □

2.8   Transmission Stabilization

We consider the stabilization of transmission of the wave equation with variable coefficients where the material is assumed to be composed of two portions with connecting conditions between them. Feedback controls will be acting on the boundary and the connecting area, respectively.

Let Ω be a bounded domain in $\mathbb{R}^n$ with smooth boundary Γ. Let Ω be divided into two regions, Ω_1 and Ω_2, by a smooth hypersurface Γ_3 on Ω. Set $\Gamma_i = \partial\Omega \cap \partial\Omega_i$ for $i = 1$ and 2. Denote by ν_i the normal of $\partial\Omega_i$ pointing outside Ω_i. Let ω be a domain near Γ_3 such that $\Gamma_3 \subset \omega \subset \Omega$. Let $m > 0$, $a_1 > 0$, and $a_2 > 0$ be constants.

Definition 2.10 *A function w on Ω is said to satisfy the connecting conditions for the wave equation if*

$$w_1 = w_2, \quad a_1 w_{1\nu_{1A}} + a_2 w_{2\nu_{2A}} = 0 \quad for \quad x \in \Gamma_3 \qquad (2.229)$$

where $w_i = w|_{\Omega_i}$ for $i = 1$, 2.

Let

$$a(x) = a_1 \quad for \quad x \in \Omega_1; \quad a(x) = a_2 \quad for \quad x \in \Omega_2.$$

We consider the following system

$$\begin{cases} u_{tt} - a(x)\operatorname{div} A(x)\nabla u + m\chi_\omega u_t = 0 \quad \text{in } (0,\infty) \times \Omega\backslash\Gamma_3 \\ u(x,0) = u^0(x), \quad u_t(x,0) = u^1(x) \quad \text{on} \quad \Omega, \\ u = 0 \quad \text{on} \quad (0,\infty) \times \Gamma_1, \\ u_{\nu_A} + l(x)u_t = 0 \quad \text{on} \quad (0,\infty) \times \Gamma_2 \\ u \text{ satisfies the connecting conditions} \quad \text{on} \quad (0,\infty) \times \Gamma_3 \end{cases} \qquad (2.230)$$

where χ_ω denotes the characteristic function of ω and $l \in L^\infty(\Gamma_2)$ is such that $l(x) \geq l_0 > 0$ for all $x \in \Gamma_2$.

Let

$$V = \{u \in H^1_{\Gamma_1}(\Omega)|\, u_k \in H^2(\Omega_k),\ k = 1, 2,\ a_1 u_{1\nu_{1A}} + a_2 u_{2\nu_{2A}} = 0 \text{ on } \Gamma_3 \}.$$

The well-posedness for the problem (2.230) can be established by the semigroup theory: For all given initial data $(u^0, u^1) \in H^1_{\Gamma_1}(\Omega) \times L^2(\Omega)$, the problem (2.230) admits a unique global weak solution

$$u \in C(R_+, H^1_{\Gamma_1}(\Omega)) \cap C^1(R_+, L^2(\Omega)).$$

Furthermore, if $(u^0, u^1) \in (V \cap H^1_{\Gamma_1}(\Omega)) \times H^1_{\Gamma_1}(\Omega)$, then

$$u \in C(R_+, V) \cap C^1(R_+, H^1_{\Gamma_1}(\Omega)).$$

Let u be a solution to the problem (2.230). We define the energy of the problem (2.230) by

$$2E(t) = \int_{\Omega} [u_t^2 + a(x)\langle A(x)\nabla u, \nabla u\rangle]dx. \tag{2.231}$$

Lemma 2.14 *For $T > 0$,*

$$E(T) = E(0) - a_2 \int_0^T \int_{\Gamma_2} l(x)|u_{2t}|^2 d\Gamma dt - m \int_0^T \int_\omega |u_t|^2 dx\, dt \tag{2.232}$$

where $u_2 = u|_{\Omega_2}$.

Proof. Let $u_i = u|_{\Omega_i}$ for $i = 1, 2$.
Differentiating $E(t)$ and using the connecting conditions, we have

$$
\begin{aligned}
\frac{d}{dt}E(t) &= \sum_{k=1}^2 \int_{\Omega_k} [u_{ktt}u_{kt} + a_k\langle A(x)\nabla u_k, \nabla u_{kt}\rangle]dx \\
&= \sum_{k=1}^2 \int_{\Gamma_k \cup \Gamma_3} a_k u_{kt} u_{k\nu_{k_A}} d\Gamma - m \int_\omega |u_t|^2 dx dt \\
&= \int_{\Gamma_3} u_{1t}(a_1 u_{1\nu_{1_A}} + a_2 u_{2\nu_{2_A}})d\Gamma dt + a_2 \int_{\Gamma_2} u_{2t} u_{2\nu_{2_A}} d\Gamma dt \\
&\quad - m \int_\omega |u_t|^2 dx dt \\
&= -a_2 \int_{\Gamma_2} l(x)|u_{2t}|^2 d\Gamma - m \int_\omega |u_t|^2 dx
\end{aligned}
$$

which yields the identity (2.232). $\qquad\qquad\square$

The stabilization of the problem (2.230) is

Theorem 2.28 *Let $\Gamma_1 \neq \emptyset$ and $\overline{\Gamma}_1 \cap \overline{\Gamma}_2 = \emptyset$. Let $\omega \supset \Gamma_3$ be such that*

$$\overline{\omega} \cap \Gamma_i = \emptyset \quad \text{for} \quad i = 1, 2. \tag{2.233}$$

If there is an escape vector field H on $\overline{\Omega}$ such that

$$\langle H, \nu_1\rangle \leq 0 \quad \text{for all} \quad x \in \Gamma_1, \tag{2.234}$$

then there are positive constants C and σ such that

$$E(t) \leq Ce^{-\sigma t}E(0) \quad \text{for} \quad t \geq 0 \tag{2.235}$$

for all solutions of the problem (2.230) with $(u^0, u^1) \in H^1_{\Gamma_1}(\Omega) \times L^2(\Omega)$.

Proof. It will suffice to prove that there are $T > 0$ and $C_T > 0$ such that

$$E(T) \leq C_T[\int_0^T \int_\omega m|u_t|^2 dx dt + C \int_{\Sigma_2} a_2 l(x)|u_{2t}|^2 d\Sigma] \tag{2.236}$$

for all solutions u to the problem (2.230).

Multiplying the wave equation in (2.232) by $2H(u_k)$ and integrating on $Q_k = (0,T) \times \Omega_k$, we obtain

$$\int_{\partial\Omega_k} [2a_k H(u_k)u_{k\nu_{kA}} + (u_{kt}^2 - a_k|\nabla_g u_k|_g^2)\langle H, \nu_k\rangle d\Sigma$$

$$= 2\int_{\Omega_k} u_{kt} H(u_k)dx|_0^T + 2m\int_0^T \int_{\omega_k} u_{kt} H(u_k)dx dt$$

$$+ \int_{Q_k} [2a_k DH(\nabla_g u_k, \nabla_g u_k) + (u_{kt}^2 - a_k|\nabla_g u_k|_g^2)\,\text{div}\,H dQ \tag{2.237}$$

where $\Sigma_k = (0,T) \times \Gamma_k$ and $\omega_k = \omega \cap \Omega_k$ for $k = 1, 2$. Summing up the identities (2.237) for $k = 1, 2$ yields, via the connecting conditions in (2.230),

$$\int_\Sigma [2a(x)H(u)u_{\nu_A} + (u_t^2 - a(x)|\nabla_g u|_g^2)\langle H, \nu\rangle] d\Sigma + F(u)$$

$$= 2\int_\Omega u_t H(u)dx|_0^T + 2m\int_0^T \int_\omega u_t H(u)dx dt$$

$$+ \int_Q [2a(x)DH(\nabla_g u, \nabla_g u) + \varrho_0(u_t^2 - a(x)|\nabla_g u|_g^2)]dQ$$

$$+ \int_Q (u_t^2 - a(x)|\nabla_g u|_g^2)(\,\text{div}\,H - \varrho_0)dQ \tag{2.238}$$

where

$$F(u) = \int_{\Sigma_3} \{2a_1[H(u_1) - H(u_2)]u_{1\nu_{1A}}$$

$$+ (a_2|\nabla_g u_2|_g^2 - a_1|\nabla_g u_1|_g^2)\langle H, \nu_1\rangle\}d\Sigma \tag{2.239}$$

where $\Sigma_3 = (0,T) \times \Gamma_3$ and $\varrho_0 > 0$ is given in (2.15).

By multiplying the wave equation in (2.232) by $2pu$, a similar computation gives

$$2\int_Q p[u_t^2 - a(x)|\nabla_g u|_g^2]dQ = 2\int_\Omega pu_t u dx|_0^T - 2\int_Q a(x)u^2\,\text{div}\,A\nabla p dQ$$

$$+ 2m\int_0^T \int_\omega pu_t u dx dt + \int_\Sigma a(x)[2u^2 p_{\nu_A} - 2puu_{\nu_A}]d\Sigma$$

$$+ \int_{\Sigma_3} 2(a_1 - a_2)u_1^2\langle A\nabla p, \nu_1\rangle d\Sigma \tag{2.240}$$

where p is a function on Ω.

Letting $p = (\operatorname{div} H - \varrho_0)/2$ in (2.240) and then inserting it into (2.238), we obtain, via the estimate (2.15),

$$\int_\Sigma [2a(x)H(u)u_{\nu_A} + (u_t^2 - a(x)|\nabla_g u|_g^2)\langle H, \nu \rangle]d\Sigma + F(u)$$

$$\geq \varrho_0 \int_0^T E(t)dt - 2\int_Q a(x)u^2 \operatorname{div} A(x)\nabla p\, dQ$$

$$+2\int_\Omega u_t[H(u) + pu]dx\Big|_0^T + 2m\int_0^T \int_\omega u_t[H(u) + pu]dx\, dt$$

$$+\int_\Sigma a(x)[2u^2 p_{\nu_A} - 2puu_{\nu_A}]d\Sigma + \int_{\Sigma_3} 2(a_1 - a_2)u_1^2\langle A\nabla p, \nu_1 \rangle d\Sigma. \quad (2.241)$$

As in (2.57) and (2.61), the boundary condition $u_1 = 0$ on Γ_1 implies that

$$H(u_1)u_{1\nu_{1A}} = \langle H, \nu_1 \rangle|\nabla_g u_1|_g^2 \quad \text{for} \quad x \in \Gamma_1.$$

Noting that $u_{1t} = 0$ and $\langle H, \nu_1 \rangle \leq 0$ on Γ_1 , we have

$$\int_{\Sigma_1}[2a_1 H(u_1)u_{1\nu_{1A}} + (u_{1t}^2 - a_1|\nabla_g u_1|_g^2)\langle H, \nu_1 \rangle]d\Sigma$$

$$= \int_{\Sigma_1} a_1|\nabla_g u_1|_g^2\langle H, \nu_1 \rangle d\Sigma \leq 0. \quad (2.242)$$

Using (2.242) in (2.241), we obtain

$$\varrho_0 \int_0^T E(t)dt \leq F(u) + C\int_{\Sigma_2}(|u_{2t}|^2 + |u_{2\nu_{2A}}|^2 + |\nabla_{\Gamma_g} u_2|_g^2)d\Sigma + C_\varepsilon L(u)$$

$$+C[E(T) + E(0)] + \varepsilon\int_0^T E(t)dt + C_\varepsilon m\int_0^T \int_\omega u_t^2 dx\, dt \quad (2.243)$$

where ∇_{Γ_g} is defined in (2.190), $\varepsilon > 0$ can be chosen small, and

$$L(u) = \|u(0)\|^2 + \|u(T)\|^2 + \|u\|_{L^2(Q)}^2 + \|u\|_{H^{1/2}(Q)}^2.$$

Moreover, the trace estimate in Lemma 2.12 and the feedback relation on Γ_2 yield, for $1/2 > \alpha > 0$ small,

$$\int_\alpha^{T-\alpha} \int_{\Gamma_2}(|u_{2t}|^2 + |u_{2\nu_{2A}}|^2 + |\nabla_{\Gamma_g} u_2|_g^2)d\Gamma dt$$

$$\leq C_{T\alpha}\int_{\Sigma_2} a_2 l(x)|u_{2t}|^2 d\Sigma + C_{T\alpha}\|u\|_{H^{\alpha+1/2}(Q)}^2. \quad (2.244)$$

Next, we estimate the transmission term $F(u)$ in (2.239). Let us choose open subsets θ_1, θ_2 and a vector field $\hat{H}$ to be such that

$$\hat{H} = H \quad \text{on } \Gamma_3, \quad \operatorname{supp}\hat{H} \subset \theta_1, \quad \text{and} \quad \Gamma_3 \subset \theta_1 \subset \bar{\theta}_1 \subset \theta_2 \subset \bar{\theta}_2 \subset \omega.$$

The conditions (2.233) implies $\hat{H} = 0$ for $x \in \Gamma$. Then it follows from (2.238) with H replaced by $\hat{H}$ that

$$F(u) \leq C[E(T) + E(0) + \int_0^T \int_{\theta_1} (m|u_t|^2 + a(x)|\nabla_g u|_g^2)dxdt. \qquad (2.245)$$

Let $\phi \in C_0^\infty(\Omega)$ be such that $0 \leq \phi \leq 1$, $\phi = 1$ for $x \in \theta_1$, and $\phi = 0$ for $x \notin \theta_2$. Letting $p = \phi$ in (2.240), we obtain

$$\int_0^T \int_{\theta_1} a(x)|\nabla_g u|_g^2 dxdt \leq C \int_0^T \int_{\theta_2} |u_t|^2 dxdt$$

$$+C[E(T) + E(0) + \int_0^T \int_\omega |u_t|^2 dxdt] + CL(u)$$

$$\leq C[E(T) + E(0) + \int_0^T \int_\omega |u_t|^2 dxdt] + CL(u) \text{ (by } \theta_2 \subset \omega). \qquad (2.246)$$

Using the estimates (2.246) and (2.245) in (2.243) and by taking $0 < \varepsilon < \varrho_0/2$, we obtain

$$\int_0^T E(t)dt \leq C \int_{\Sigma_2} (|u_{2t}|^2 + |u_{2\nu_{2A}}|^2 + |\nabla_{\Gamma_g} u_2|_g^2)d\Sigma$$

$$+C[E(T) + E(0)] + Cm \int_0^T \int_\omega u_t^2 dxdt + CL(u). \qquad (2.247)$$

Replacing the integral interval $[0, T]$ by $[\alpha, T - \alpha]$ in (2.247) gives, via (2.244),

$$\int_\alpha^{T-\alpha} E(t)dt \leq C \int_0^T \int_\omega m|u_t|^2 dxd + C \int_{\Sigma_2} a_2 l(x)|u_{2t}|^2 d\Sigma$$

$$+C[E(T - \alpha) + E(\alpha)] + CL(u) + \|u\|_{H^{\alpha+1/2}(Q)}^2. \qquad (2.248)$$

Furthermore, the relation (2.232) yields

$$\int_\alpha^{T-\alpha} E(t)dt \geq (T - 2\alpha)E(T),$$

$$E(T - \alpha) = E(T) + a_2 \int_{T-\alpha}^T \int_{\Gamma_2} l(x)|u_{2t}|^2 d\Gamma dt + m \int_{T-\alpha}^T \int_\omega |u_t|^2 dx\, dt,$$

$$E(\alpha) = E(T) + a_2 \int_\alpha^T \int_{\Gamma_2} l(x)|u_{2t}|^2 d\Gamma dt + m \int_\alpha^T \int_\omega |u_t|^2 dx\, dt.$$

Using those relations in (2.248), for $T > 2\alpha + 2C$, we obtain a constant $C_T > 0$ such that

$$E(T) \leq C_T[\int_0^T \int_\omega m|u_t|^2 dxd + C \int_{\Sigma_2} a_2 l(x)|u_{2t}|^2 d\Sigma] + C_T L(u) \qquad (2.249)$$

where $L(u)$ are lower order terms.

Then (2.236) follows from (2.249) after the lower order terms are absorbed by a compactness-uniqueness argument as in Lemma 2.5. $\square$

Exercises

2.1 *Prove the Green formula* (2.92).

2.2 *Let F and q be a vector field and a function on Ω, respectively. Consider an operator on $L^2(\Omega)$, given by*

$$\mathcal{A}u = \operatorname{div} A(x)\nabla u - F(u) + (q - \operatorname{div} F)u, \quad \mathcal{D}(\mathcal{A}) = H^2(\Omega) \cap H_0^1(\Omega).$$

Then the adjoint operator of $\mathcal{A}$ is given by

$$\mathcal{A}^*u = \operatorname{div} A(x)\nabla u + F(u) + qu, \quad \mathcal{D}(\mathcal{A}^*) = H^2(\Omega) \cap H_0^1(\Omega).$$

2.3 *Prove Lemma 2.7.*

2.4 *Let the operator Λ be defined by* (2.101). *Prove the estimate* (2.116).

2.5 *Let $L = L^2(\Omega)$ and $H = H_0^1(\Omega)$. Prove that $(H_0^1(\Omega))^* = H^{-1}(\Omega)$ and the canonical isomorphism $\aleph: H_0^1(\Omega) \to H^{-1}(\Omega)$ is given by*

$$\aleph = -\Delta$$

where Δ is the classical Laplacian in $\mathbb{R}^n$.

2.6 *Let the inequality* (2.200) *hold true. Prove the inequality* (2.201) *by the compactness-uniqueness argument as in Lemma 2.5.*

2.7 *Let u be a solution to the problem* (2.186) *and let $E(t)$ be its energy, defined by* (2.55) *where w is replaced with u. Let there be constants $T > 0$ and $0 < \lambda < 1$ such that*

$$E(T) \leq \lambda E(0)$$

for all solutions u to the problem (2.186). *Prove that there are constant $C > 0$ and $\sigma > 0$ such that*

$$E(t) \leq Ce^{-\sigma t}E(0) \quad for \quad t \geq 0$$

for all solutions u.

2.9   Notes and References

Sections 2.2, 2.3, 2.4, and 2.6 are from [208]; Section 2.5 is from [216]; Section 2.7 is from [62], [63], [65], [66], and [80]; Section 2.8 is from [27].

The differential geometrical approach was first introduced by [208] for the controllability of the wave equation with variable coefficients. It was extended to cope with lower order terms by [130] and [201]. Then this method was used on various problems of the wave equation and the Schrödinger with variable coefficients by [20], [21], [22], [23], [31], [55], [56], [79], [83], [84], [162], [163], [164], [178], and many others. Those results are not included in this chapter.

Chapter 3

Control of the Plate with Variable Coefficients in Space

We will establish Riemannian multiplier identities by the Bochner technique for the plate with variable coefficients to derive exact controllability/stabilization. Those multipliers are a geometric version of the classical ones for the plate with constant coefficients. Then we introduce a checkable assumption which guarantees exact controllability/stabilization, which is called *escape vector fields for plate*. It is different from the one for the wave equation. In particular, Section 3.2 is devoted to study the existence of *escape vector fields for plate* by curvature and many examples will be given. Finally, in Section 3.5, we will derive a plate model which is defined on a curved surface by the variational principle. In this model only the deformation along its normal is considered while the deformation in the tangential is neglected so it may serve as a curved plate. Stabilization by boundary feedbacks is presented for this model.

3.1 Multiplier Identities

Let $\Omega \subset \mathbb{R}^n$ be a bounded, open set with smooth boundary Γ. We shall consider regular solutions u to the problem

$$u_{tt} + \mathcal{A}^2 u = f \quad \text{in} \quad (0, \infty) \times \Omega \tag{3.1}$$

and establish some multiplier identities where function f is given. In (3.1) we have defined

$$\mathcal{A}u = \operatorname{div} A(x)\nabla u \tag{3.2}$$

where $A(x) = \Big(a_{ij}(x) \Big)$ is symmetrical, positive for each $x \in \mathbb{R}^n$.

We introduce

$$g = A^{-1}(x) \quad \text{for} \quad x \in \mathbb{R}^n \tag{3.3}$$

as a Riemannian metric on $\mathbb{R}^n$ and consider the couple $(\mathbb{R}^n, g)$ as a Riemannian manifold. We denote by $g = \langle \cdot, \cdot \rangle_g$ the inner product and by D the covariant differential of the metric g, respectively. Denote by $\cdot = \langle \cdot, \cdot \rangle$ the Euclidean product of $\mathbb{R}^n$. If applied to functions, we denote by ∇_g and ∇ the gradients of the metric g and the Euclidean metric, respectively. Similarly, div $_g$ and div are the divergences of the metric g and the Euclidean metric, respectively.

With two metrics on $\mathbb{R}^n$ in mind, one the Euclidean metric and the other the Riemannian metric g, we have to deal with various notations carefully.

Let $x \in \mathbb{R}^n$ be given. We recall that $E_1, \cdots, E_n$ is a *frame field normal* at x on $(\mathbb{R}^n, g)$ if and only if it is a local basis for vector fields such that

$$\langle E_i, E_j \rangle_g = \delta_{ij} \quad \text{in a neighborhood of } x,$$

$$D_{E_i} E_j(x) = 0 \quad \text{for all} \quad 1 \le i, j \le n.$$

Denote by T_x^2 all tensors of rank 2 on $\mathbb{R}_x^n$. Then T_x^2 is an inner product space of dimension n^2 with the inner product

$$\langle F, G \rangle_g = \sum_{ij=1}^n F(e_i, e_j)G(e_i, e_j) \quad \text{for} \quad F, G \in T_x^2 \tag{3.4}$$

where $e_1, \cdots, e_n$ is an orthonormal basis of $(\mathbb{R}^n, g(x))$. It is easy to check that $\langle \cdot, \cdot \rangle_g$ is independent of the choice of the orthnoromal base $\{e_i\}$ of $(\mathbb{R}_x^n, g(x))$.

Let Δ_g be the Laplacian in the metric g. Then in the natural coordinate systems $x = (x_1, \cdots, x_n)$

$$\Delta_g w = \sqrt{\det A(x)} \sum_{ij=1}^n ([\det A(x)]^{-1/2} a_{ij}(x) w_{x_i})_{x_j} = \mathcal{A}w - \langle Dw, Dv_g \rangle_g$$

for $w \in C^2(\mathbb{R}^n)$ (Exercise 1.1.12), that is,

$$\mathcal{A} = \Delta_g + Dv_g, \tag{3.5}$$

where

$$2v_g = \log \det A(x) \quad \text{for} \quad x \in \mathbb{R}^n. \tag{3.6}$$

We denote by $\mathcal{X}(\mathbb{R}^n)$ all vector fields on $\mathbb{R}^n$. Denote by $\mathbf{\Delta}_g \colon \mathcal{X}(\mathbb{R}^n) \to \mathcal{X}(\mathbb{R}^n)$ the Hodge-Laplace operator in the metric g. We define an operator $\mathcal{A} \colon \mathcal{X}(\mathbb{R}^n) \to \mathcal{X}(\mathbb{R}^n)$ by

$$\mathcal{A}H = -\mathbf{\Delta}_g H + D_{Dv_g} H \quad \text{for} \quad H \in \mathcal{X}(\mathbb{R}^n). \tag{3.7}$$

We have

Lemma 3.1 *Let H be a vector field and w be a function on* $\mathbb{R}^n$. *Then the following Weitzenbock formula holds true*

$$
\begin{aligned}
\Delta_g(H(w)) = & -(\Delta_g H)(w) + 2\langle DH, D^2 w\rangle_g \\
& + H(\Delta_g w) + 2\operatorname{Ric}(H, Dw) \quad for \quad x \in \mathbb{R}^n
\end{aligned} \tag{3.8}
$$

where Ric *is the Ricci tensor of the metric g.*

Proof. Since

$$
\Delta_g Dw = (\delta d + d\delta)dw = d\delta dw = -D(\Delta_g w),
$$

the formula (3.8) follows from Theorem 1.27 by letting $X = H$ and $Y = Dw$ in the formula (1.120) where $(M, g) = (\mathbb{R}^n, g)$. □

The main multipliers we need are

$$
hu, \quad 2h\mathcal{A}u, \quad 2H(u), \quad \text{and} \quad 2H(\mathcal{A}u)
$$

where h is a function.

Let v be the normal vector field of $\Gamma = \partial\Omega$, pointing outside of Ω. For $T > 0$, let

$$
Q = (0, T) \times \Omega, \quad \Sigma = (0, T) \times \Gamma.
$$

Theorem 3.1 *Let H be a vector field on* $\overline{\Omega}$ *and let* v_g *be given by (3.6). Suppose that u is a solution to the problem (3.1). Then*

$$
\int_\Sigma \{2\mathcal{A}u(H(u))_{\nu_A} - 2H(u)(\mathcal{A}u)_{\nu_A} + [u_t^2 - (\mathcal{A}u)^2]\langle H, \nu\rangle\}d\Sigma
$$

$$
= 2\Big(u_t, H(u)\Big)\big|_0^T + \int_Q \{[u_t^2 - (\mathcal{A}u)^2]\operatorname{div} H
$$

$$
+ 4\mathcal{A}u\langle DH, D^2 u\rangle_g + 2\mathcal{A}uP(Du) - 2fH(u)\}dQ \tag{3.9}
$$

where $P(Du) = (\mathcal{A}H)(u) + (2\operatorname{Ric} - D^2 v_g)(H, Du)$.

Proof. First, the divergence formula yields

$$
2u_{tt}H(u) = 2[u_t H(u)]_t - \operatorname{div}(u_t^2 H) + u_t^2 \operatorname{div} H. \tag{3.10}
$$

Next, we compute term $2\mathcal{A}^2 uH(u)$. To this end, we need some formulas further.

By (3.5), we obtain

$$
\begin{aligned}
H(\mathcal{A}u) &= H(\Delta_g u) + H(\langle Du, Dv_g\rangle_g) \\
&= H(\Delta_g u) + D^2 u(Dv_g, H) + D^2 v_g(Du, H).
\end{aligned} \tag{3.11}
$$

In addition, we have

$$\langle D\upsilon_g, D(H(u))\rangle_g = D\upsilon_g(\langle H, Du\rangle_g)$$
$$= DH(Du, D\upsilon_g) + D^2u(H, D\upsilon_g). \qquad (3.12)$$

It follows from the formulas (3.11), (3.12), (3.5), and (3.8) that

$$\mathcal{A}(H(u)) = \Delta_g(H(u)) + D\upsilon_g(\langle H, Du\rangle_g) = -(\pmb{\Delta}_gH)(u) + 2\langle DH, D^2u\rangle_g$$
$$+H(\Delta_g u) + 2\,\mathrm{Ric}\,(H, Du) + DH(Du, D\upsilon_g) + D^2u(H, D\upsilon_g)$$
$$= H(\mathcal{A}u) + 2\langle DH, D^2u\rangle_g + P(Du). \qquad (3.13)$$

From the formula (3.13), we have

$$2\mathcal{A}^2uH(u) = 2\,\mathrm{div}\,H(u)A(x)\nabla(\mathcal{A}u) - 2\langle D(\mathcal{A}u), D(H(u))\rangle_g$$
$$= 2\,\mathrm{div}\,[H(u)A(x)\nabla(\mathcal{A}u) - \mathcal{A}uA(x)\nabla(H(u))] + 2\mathcal{A}u\mathcal{A}(H(u))$$
$$= \mathrm{div}\,\{2[H(u)A(x)\nabla(\mathcal{A}u) - \mathcal{A}uA(x)\nabla(H(u))] + (\mathcal{A}u)^2H\}$$
$$+4\mathcal{A}u\langle DH, D^2u\rangle_g - (\mathcal{A}u)^2\,\mathrm{div}\,H + 2\mathcal{A}uP(Du). \qquad (3.14)$$

We multiply the equation (3.1) by $2H(u)$, integrate it over Q by parts, and obtain the formula (3.9) from the formulas (3.1), (3.10), and (3.14).   □

Remark 3.1 *In the right hand side of the identity (3.9), div H is the divergence of the vector field H in the Euclidean metric in $\mathbb{R}^n$, not in the metric g.*

Theorem 3.2 *Let H be a vector field on $\overline{\Omega}$ and let u be a solution to the problem (3.1). Set $p = \mathrm{div}\,H$. Then*

$$\int_\Sigma \{2[pu_t + H(u_t)](u_t)_{\nu_A} + 2H(\mathcal{A}u)(\mathcal{A}u)_{\nu_A}$$
$$-(2u_t\mathcal{A}u + |\nabla_g u_t|_g^2 + |\nabla_g(\mathcal{A}u)|_g^2)\langle H, \nu\rangle - u_t^2\langle \nabla_g p, \nu\rangle\}d\Sigma$$
$$= -2\Big(u_t, H(\mathcal{A}u)\Big)\big|_0^T + \int_Q [2DH(\nabla_g u_t, \nabla_g u_t) + 2DH(\nabla_g(\mathcal{A}u), \nabla_g(\mathcal{A}u))$$
$$+(|\nabla_g u_t|_g^2 - |\nabla_g(\mathcal{A}u)|_g^2)p - u_t^2\mathcal{A}p + 2fH(\mathcal{A}u)]dQ. \qquad (3.15)$$

Proof. We use the multiplier $2H(\mathcal{A}u)$.
Using the identity in Lemma 2.2, we have

$$2\mathcal{A}u_tH(u_t) = 2\,\mathrm{div}\,H(u_t)A(x)\nabla u_t - 2\langle\nabla_g u_t, \nabla_g(H(u_t))\rangle_g$$
$$= \mathrm{div}\,[2H(u_t)A(x)\nabla u_t - |\nabla_g u_t|_g^2 H]$$
$$-2DH(\nabla_g u_t, \nabla_g u_t) + |\nabla_g u_t|_g^2 p. \qquad (3.16)$$

In addition,

$$2pu_t\mathcal{A}u_t = \mathrm{div}\,[2pu_tA\nabla u_t - u_t^2A\nabla p] + u_t^2\mathcal{A}p - 2|\nabla_g u_t|_g^2 p. \qquad (3.17)$$

From the formulas (3.16) and (3.17), we have

$$
\begin{aligned}
2u_{tt}H(\mathcal{A}u) &= 2[u_t H(\mathcal{A}u)]_t - 2\operatorname{div}(u_t \mathcal{A}u_t)H + 2pu_t \mathcal{A}u_t + 2H(u_t)\mathcal{A}u_t \\
&= 2[u_t H(\mathcal{A}u)]_t + \operatorname{div}\{2[pu_t + H(u_t)]A\nabla u_t - u_t^2 A\nabla p \\
&\quad -(2u_t \mathcal{A}u_t + |\nabla_g u_t|_g^2)H\} + u_t^2 Ap - |\nabla_g u_t|_g^2 p \\
&\quad -2DH(\nabla_g u_t, \nabla_g u_t).
\end{aligned}
\tag{3.18}
$$

On the other hand, by the identity in Lemma 2.2 again,

$$
\begin{aligned}
2\mathcal{A}^2 u H(\mathcal{A}u) &= \operatorname{div}[2H(\mathcal{A}u)A(x)\nabla(\mathcal{A}u) - |\nabla_g(\mathcal{A}u)|_g^2 H] \\
&\quad +|\nabla_g(\mathcal{A}u)|_g^2 p - 2DH(\nabla_g(\mathcal{A}u), \nabla_g(\mathcal{A}u)).
\end{aligned}
\tag{3.19}
$$

We multiply the equation (3.1) by $2H(\mathcal{A}u)$, integrate it over Q, and obtain the identity (3.15) from (3.18) and (3.19). $\qquad\square$

Theorem 3.3 *Let p be a function on $\overline{\Omega}$. Suppose that u is a solution to the problem (3.1). Then*

$$
(i) \quad \int_Q p[u_t^2 - (\mathcal{A}u)^2]dQ = \int_Q \{\mathcal{A}u[2Dp(u) + u\mathcal{A}p] - fpu\}dQ
$$

$$
+(u_t, pu)|_0^T + \int_\Sigma [pu(\mathcal{A}u)_{\nu_A} - \mathcal{A}u(pu)_{\nu_A}]d\Sigma.
\tag{3.20}
$$

$$
(ii) \quad 2\int_Q p[|\nabla_g u_t|_g^2 - |\nabla_g(\mathcal{A}u)|_g^2]dQ
$$

$$
= \int_\Sigma \{2p[u_t(u_t)_{\nu_A} - \mathcal{A}u(\mathcal{A}u)_{\nu_A}] + [(\mathcal{A}u)^2 - u_t^2]p_{\nu_A}\}d\Sigma
$$

$$
-2(u_t, p\mathcal{A}u)|_0^T + \int_Q \{\mathcal{A}p[u_t^2 - (\mathcal{A}u)^2] + 2fp\mathcal{A}u\}dQ.
\tag{3.21}
$$

Proof. The divergence formula yields

$$
\begin{aligned}
\mathcal{A}(pu\mathcal{A}u) &= \mathcal{A}(pu)\mathcal{A}u + 2\langle\nabla_g(pu), \nabla_g(\mathcal{A}u)\rangle_g + pu\mathcal{A}^2 u \\
&= 2\operatorname{div} pu A\nabla(\mathcal{A}u) + \mathcal{A}(pu)\mathcal{A}u - pu\mathcal{A}^2 u.
\end{aligned}
\tag{3.22}
$$

Using the formula (3.22) and the equation (3.1), we compute that

$$
\begin{aligned}
(u_t, pu)|_0^T &= \int_0^T [(u_{tt}, pu) + (u_t, pu_t)]dt = \int_Q (fpu - pu\mathcal{A}^2 u + pu_t^2)dQ \\
&= \int_\Sigma [\mathcal{A}u(pu)_{\nu_A} - pu(\mathcal{A}u)_{\nu_A}]d\Sigma + \int_Q [pu_t^2 - \mathcal{A}(pu)\mathcal{A}u + fpu]dQ
\end{aligned}
$$

which gives the identity (3.20).

A similar computation gives the identity (3.21). $\qquad\square$

3.2   Escape Vector Fields for Plate

To have a real solution to the control problems for the systems with variable coefficients, it is necessary to obtain checkable geometric conditions. If the geometric conditions we have are uncheckable, we just turn one problem into another. Then the questions under consideration are still open. One of the main contributions of Riemann geometry is that it provides the control problems with the geometric conditions verifiable by the Riemann curvature theory. Now we introduce such a geometric condition for the control problems of the Euler-Bernoulli plate with variable coefficients.

Let coefficient matrix $A(x)$ be symmetric, positive for each $x \in \mathbb{R}^n$. Consider the Riemannian manifold $(\mathbb{R}^n, g)$ where the metric is given by

$$g = \langle A^{-1}(x)\cdot, \cdot \rangle \tag{3.23}$$

where $\langle \cdot, \cdot \rangle$ is the Euclidean metric of $\mathbb{R}^n$.

Definition 3.1 *Let g be the metric, given by (3.23), on $\mathbb{R}^n$ and let H be a vector field on Ω. Vector field H is said to be an escape vector field for plate if there is a function ϑ on $\overline{\Omega}$ such that*

$$DH(X, X) = \vartheta(x)|X|_g^2 \quad for \quad X \in \mathbb{R}_x^n, \quad x \in \Omega, \tag{3.24}$$

and

$$\vartheta_0 = \min_{x \in \Omega} \vartheta(x) > 0. \tag{3.25}$$

Remark 3.2 *Clearly, the conditions (3.24) and (3.25) imply that H is an escape vector field for the metric g on $\overline{\Omega}$. The converse is not true. In the case of constant coefficients where $a_{ij}(x) = \delta_{ij}$, for any $x_0 \in \mathbb{R}^n$ fixed, the radial field $H = x - x_0$ is an escape vector field for plate with $\vartheta = 1$.*

The condition (3.24) has the following useful properties which can help us simplify computation:

Theorem 3.4 *Let H be a vector field such that the condition (3.24) holds. Then the condition (3.24) is equivalent to*

$$DH + D^*H = 2\vartheta(x)g \quad for \quad x \in \Omega. \tag{3.26}$$

Moreover, if $T \in T^2(\Omega)$ is a symmetric tensor field of rank 2 in the metric g, then

$$\langle T, T(\cdot, D.H) \rangle_g = \vartheta(x)|T|_g^2 \quad for \quad x \in \Omega; \tag{3.27}$$

$$\mathrm{tr}_g T(\cdot, D.H) = \vartheta(x)\mathrm{tr}_g T \quad for \quad x \in \Omega \tag{3.28}$$

where tr_g is the trace in the metric g and "$\cdot$" denotes the position of variables.

Proof. Let $x \in \Omega$ be given. The condition (3.24) implies that

$$DH(X+Y, X+Y) - DH(X-Y, X-Y) = 4\vartheta(x)\langle X, Y \rangle_g$$

which yields the identity (3.26). The converse is clearly true.

Let $x \in \Omega$ be given. The symmetry of the tensor field T implies that there is an orthonormal basis $e_1, \cdots, e_n$ of the tangential space $(\mathbb{R}_x^n, g)$ such that

$$T(e_i, e_j) = 0 \quad \text{at} \quad x \quad \text{for} \quad i \neq j. \tag{3.29}$$

It follows from the formulas (3.29) and (3.24) that

$$T(e_i, D_{e_i}H) = T\left(e_i, \; DH(e_i, e_i)e_i\right) = \vartheta(x)T(e_i, e_i) \tag{3.30}$$

for $1 \leq i \leq n$. By the relations (3.30) and (3.29), we have

$$\langle T, T(\cdot, D.H)\rangle_g = \sum_{ij=1}^{n} T(e_i, e_j)T(e_i, D_{e_j}H)$$

$$= \sum_{i=1}^{n} T(e_i, e_i)T(e_i, D_{e_i}H) = \vartheta(x)|T|_g^2$$

and

$$\mathrm{tr}_g T(\cdot, D.H) = \sum_{i=1}^{n} T(e_i, D_{e_i}H) = \vartheta(x)\mathrm{tr}_g T.$$

$\square$

Remark 3.3 *If DH is symmetric, then $DH = \vartheta g$.*

Remark 3.4 *In the case of dimension 2, a real plate, the condition (3.24) or equivalently (3.26) is always true; see Theorem 4.8 in Chapter 4. Moreover, Theorem 4.8 will show that an escape vector field for plate in the case $n = 2$ always exists locally.*

In general, it is not easy to find a vector field verifying the conditions (3.24) and (3.25). The following theorem shows that they have a close relationship with the curvature.

For $\kappa \in \mathbb{R}$, we define

$$h(t) = \begin{cases} \sin \sqrt{\kappa}t & \text{for} \quad \kappa > 0; \\ t & \text{for} \quad \kappa = 0; \\ \sinh \sqrt{-\kappa}t & \text{for} \quad \kappa < 0, \end{cases} \quad \text{for} \quad t \in \mathbb{R}.$$

Theorem 3.5 *Let g be the metric given by (3.23). Let there be a function ω such that the Riemann manifold $(\mathbb{R}^n, e^{2\omega}g)$ has constant curvature κ. Given $x_0 \in \mathbb{R}^n$. Denote by $\hat{\rho}(x)$ the distance function from x_0 to x in the metric $\hat{g} = e^{2\omega}g$. Set*

$$H = h(\hat{\rho})\hat{D}\hat{\rho} \tag{3.31}$$

where $\hat{D}$ is the Levi-Civita connection of the Riemann manifold $(\mathbb{R}^n, \hat{g})$. Then

$$DH(X,X) = [h'(\hat{\rho}) - H(\omega)]|X|_g^2 \quad for \quad X \in \mathbb{R}_x^n, \ x \in \mathbb{R}^n. \tag{3.32}$$

Proof. Let $x \in \mathbb{R}^n$ be given and let $X \in \mathbb{R}_x^n$. We decompose X as a direct

$$X = \langle X, \hat{D}\hat{\rho} \rangle_{\hat{g}} \hat{D}\hat{\rho} + X^\perp$$

with $\langle \hat{D}\hat{\rho}, X^\perp \rangle_{\hat{g}} = 0$. Noting that $\hat{D}^2\hat{\rho}(\hat{D}\hat{\rho}, X^\perp) = 0$ and $(\mathbb{R}^n, \hat{g})$ has constant curvature κ, from Theorem 1.21, we have

$$\begin{aligned} \hat{D}H(X,X) &= [h'(\hat{\rho})\hat{D}\hat{\rho} \otimes \hat{D}\hat{\rho} + h(\hat{\rho})\hat{D}^2\hat{\rho}](X,X) \\ &= h'(\hat{\rho})|X|_{\hat{g}}^2. \end{aligned} \tag{3.33}$$

Denote by Γ_{ij}^k and $\hat{\Gamma}_{ij}^k$ the connection coefficients of the metrics g and $\hat{g}$, respectively. Set $\hat{g}_{ij} = \langle \frac{\partial}{\partial x_i}, \frac{\partial}{\partial x_j} \rangle_{\hat{g}}$. Then $g_{ij} = e^{-2\omega}\hat{g}_{ij}$. We have

$$\begin{aligned} \Gamma_{ij}^k &= \frac{1}{2}\sum_{l=1}^n g^{lk}\left(\frac{\partial g_{il}}{\partial x_j} + \frac{\partial g_{jl}}{\partial x_i} - \frac{\partial g_{ij}}{\partial x_l}\right) \\ &= \hat{\Gamma}_{ij}^k - \omega_{x_i}\delta_{kj} - \omega_{x_j}\delta_{ki} + \hat{g}_{ij}\sum_{l=1}^n \hat{g}^{kl}\omega_{x_l}. \end{aligned} \tag{3.34}$$

Denote $H = \sum_{i=1}^n h_i \frac{\partial}{\partial x_i}$. Let $X = \sum_{i=1}^n X_i \frac{\partial}{\partial x_i}$ be a vector field on $\mathbb{R}^n$. It follows from (3.34) that

$$\begin{aligned} D_X H &= \sum_{k=1}^n [X(h_k) + \sum_{ij=1}^n X_i h_j \Gamma_{ij}^k]\frac{\partial}{\partial x_k} \\ &= \hat{D}_X H - X(\omega)H - H(\omega)X + \langle X, H \rangle_{\hat{g}}\hat{D}\omega. \end{aligned} \tag{3.35}$$

Noting $e^{2\omega}g = \hat{g}$, from (3.35) and (3.33), we obtain

$$\begin{aligned} DH(X,X) &= e^{-2\omega}\langle D_X H, X \rangle_{\hat{g}} = e^{-2\omega}[\hat{D}H(X,X) - H(\omega)|X|_{\hat{g}}^2] \\ &= [h'(\hat{\rho}) - H(\omega)]|X|_g^2. \end{aligned}$$

$\square$

Remark 3.5 *The ideas in Theorem 3.5 are from Chapter 5 of the book [187].*

Let us see several examples. It follows from Theorem 3.5 with $\omega = 0$ that

Example 3.1 *Let the metric g have constant curvature κ. Given $x_0 \in \mathbb{R}^n$. Denote by ρ the distance function from x_0 to x in the metric g. If $\kappa \leq 0$, then $H = \sinh\sqrt{-\kappa}\rho D\rho$ is an escape vector field for plate for any $\Omega \subset \mathbb{R}^n$. Let $\kappa > 0$. If there is $x_0 \in \mathbb{R}^n$ such that $\overline{\Omega} \subset B(x_0, \pi/(2\sqrt{\kappa}))$, then $H = \sin\sqrt{\kappa}\rho D\rho$ is an escape vector field for plate on $\overline{\Omega}$ where $B(x_0, \pi/(2\sqrt{\kappa}))$ is the geodesic ball in $(\mathbb{R}^n, g)$ centered at x_0 with radius $\pi/(2\sqrt{\kappa})$.*

Example 3.2 *Consider an operator*

$$\mathcal{A}w = \sum_{i=1}^{n} (a(x)w_{x_i})_{x_i}$$

where $a(x)$ is a positive function on $\mathbb{R}^n$. Then $g = a^{-1}(x)I$. Let $\hat{g} = I$ be the Euclidean metric. Then $2\omega = \log a$. Let $x_0 \in \mathbb{R}^n$ be given. Then $\hat{\rho} = |x - x_0|$ and $H = \hat{\rho}\hat{D}\hat{\rho} = x - x_0$. It follows from Theorem 3.5 that

$$DH(X,X) = [1 - \frac{H(a)}{2a}]\|X\|_g^2.$$

If

$$\overline{\Omega} \subset \{\, x \in \mathbb{R}^n \mid \sum_{i=1}^{n} (x_i - x_{0i})a_{x_i}(x) < 2a(x)\,\},$$

then $H = x - x_0$ is an escape vector field for plate on $\overline{\Omega}$.

(a) Let $a(x) = 1 + |x|^2$ and $H = x$. Since $H(a) = 2|x|^2 < 2a(x)$ for all $x \in \mathbb{R}^n$, for any $\Omega \subset \mathbb{R}^n$, the vector field H is an escape vector field for plate.

(b) Let $a(x) = e^{-|x|^2}$ and $H = x$. Then $H(a) = -2|x|^2a < 2a$ for all $x \in \mathbb{R}^n$. Then H is an escape vector field for plate for any $\Omega \subset \mathbb{R}^n$.

(c) Let $a(x) = e^{|x|^2}$ and $H = x$. Then $H(a) = 2|x|^2a$. If $\overline{\Omega} \subset \{\, x \mid |x| < 1,\ x \in \mathbb{R}^n \,\}$, then H is an escape vector field for plate on $\overline{\Omega}$.

3.3 Exact Controllability from Boundary

Let Γ_0 be a portion of the boundary Γ which is relatively open. Set $\Gamma_1 = \Gamma \setminus \Gamma_0$.

Control in Fixed Boundary Condition Let $T > 0$ be given. We consider the control problem

$$\begin{cases} u_{tt} + \mathcal{A}^2 u = 0 & \text{in} \quad Q, \\ u(0) = u_0, \quad u_t(0) = u_1 & \text{on} \quad \Omega, \\ u = u_{\nu_A} = 0 & \text{on} \quad \Sigma_1 \\ u = \varphi, \quad u_{\nu_A} = \psi & \text{on} \quad \Sigma_0, \end{cases} \tag{3.36}$$

with controls φ and ψ, where $\mathcal{A}u = \operatorname{div} A(x)\nabla u$ and

$$Q = (0,T) \times \Omega, \quad \Sigma_1 = (0,T) \times \Gamma_1, \quad \Sigma_0 = (0,T) \times \Gamma_0.$$

The dual version for the above problem is in w

$$\begin{cases} w_{tt} + \mathcal{A}^2 w = 0 & \text{in} \quad Q, \\ w(0) = w_0, \quad w_t(0) = w_1 & \text{on} \quad \Omega, \\ w = w_{\nu_A} = 0 & \text{on} \quad \Sigma \end{cases} \tag{3.37}$$

where $\Sigma = (0,T) \times \Gamma$.

Remark 3.6 *In the case of constant coefficients where $\mathcal{A} = \Delta$, one control, $u_{\nu_A} = \psi$ is enough, see [109]. We here add another control function $u = \varphi$ in order to avoid the following uniqueness assumption: The problem*

$$\begin{cases} \mathcal{A}^2 w = \lambda w & in \quad \Omega, \\ w = w_{\nu_A} = \mathcal{A}w = 0 & on \quad \Gamma_0 \end{cases} \tag{3.38}$$

has the unique zero solution where λ is complex number, see [109]. Another uniqueness problem is considered by [119], that is, the conditions,

$$w = w_\nu = (\Delta w)_\nu = 0 \quad on \quad \Gamma_0,$$

imply that $w = 0$ if $\mathcal{A} = \Delta$. We do not know if the uniqueness problem (3.38) is true or not.

We introduce the energy of the system (3.37) by

$$2E(t) = \int_\Omega [w_t^2 + (\mathcal{A}w)^2] dx. \tag{3.39}$$

Consider the exact controllability of the system (3.36) on the space $L^2(\Omega) \times H^{-2}(\Omega)$.

Let w be the solution to the problem (3.37) for initial data $(w_0, w_1) \in H_0^2(\Omega) \times L^2(\Omega)$. For $T > 0$, we solve the following time-reverse problem

$$\begin{cases} u_{tt} + \mathcal{A}^2 u = 0 & in Q, \\ u(T) = u_t(T) = 0 & on \quad \Omega, \\ u = u_{\nu_A} = 0 & on \quad \Sigma_1, \\ u = -(\mathcal{A}w)_{\nu_A}, \quad u_{\nu_A} = \mathcal{A}w & on \quad \Sigma_0. \end{cases} \tag{3.40}$$

Then for the initial data $(u(0), u_t(0))$, we have found control actions $u = -(\mathcal{A}w)_{\nu_A}$ and $u_{\nu_A} = \mathcal{A}w$ on Σ_0 which steer the problem (3.36) to rest at a time T. Enlightened by [109], we define a map by

$$\Lambda(w_0, w_1) = \Big(u_t(0), -u(0) \Big) \tag{3.41}$$

Denote by $\hat{w}$ and by $\hat{u}$ the solutions to the problem (3.37) with initial data $(\hat{w}_0, \hat{w}_1)$ and to the problem (3.40) with boundary data $\hat{u} = -(\mathcal{A}\hat{w})_{\nu_A}$, $\hat{u}_{\nu_A} = \mathcal{A}\hat{w}$ on Σ_0, respectively.

Using the equations (3.37) and (3.40), we obtain

$$\Big(\Lambda(w_0, w_1), (\hat{w}_0, \hat{w}_1) \Big)_{L^2(\Omega) \times L^2(\Omega)} = (u_t(0), \hat{w}_0) - (u(0), \hat{w}_1)$$

$$= [(\hat{w}_t, u) - (u_t, \hat{w})]|_0^T = \int_Q (\hat{w}_{tt} u - u_{tt} \hat{w}) dQ$$

$$= \int_Q (\hat{w} \mathcal{A}^2 u - u \mathcal{A}^2 \hat{w}) dQ$$

$$= \int_\Sigma [\hat{w}(\mathcal{A}u)_{\nu_A} - \mathcal{A}u \hat{w}_{\nu_A} - u(\mathcal{A}\hat{w})_{\nu_A} + \mathcal{A}\hat{w} u_{\nu_A}] d\Sigma \tag{3.42}$$

$$= \int_{\Sigma_0} [\mathcal{A}w \mathcal{A}\hat{w} + (\mathcal{A}w)_{\nu_A} (\mathcal{A}\hat{w})_{\nu_A}] d\Sigma. \tag{3.43}$$

Let

$$H = H_0^2(\Omega) \times L^2(\Omega), \quad L = L^2(\Omega) \times L^2(\Omega),$$

$$\mathcal{R} = \{ \Lambda(w_0, w_1) \,|\, (\Lambda(w_0, w_1), (w_0, w_1))_{L^2(\Omega) \times L^2(\Omega)} < \infty \, \}.$$

Using Theorem 2.13, the formula (3.43), and Theorem 3.7 below, we obtain that $H^{-2}(\Omega) \times L^2(\Omega) = H^* \subset \mathcal{R}$. Since the system (3.36) is time-reverse, we have

Theorem 3.6 *Let $T > 0$ be given. Let H be an escape vector field on $\overline{\Omega}$ for plate. Then the control problem (3.36) is exactly controllable on the space $L^2(\Omega) \times H^{-2}(\Omega)$ with controls $(\varphi, \psi) \in L^2(\Sigma_0) \times L^2(\Sigma_0)$, where*

$$\Gamma_0 = \{ x \,|\, \langle H, \nu \rangle > 0, \ x \in \Gamma \}, \quad \Gamma_1 = \Gamma \setminus \Gamma_0.$$

We need

Lemma 3.2 *Let H be a vector field on Ω. Let $w \in H^2(\Omega)$ satisfy $w = w_{\nu_\mathcal{A}} = 0$ on $\hat{\Gamma}$ where $\hat{\Gamma}$ is a relatively open subset of Γ. Then*
(i) $Dw = 0$, in particular, $H(w) = 0$ on $\hat{\Gamma}$.
(ii) $(H(w))_{\nu_\mathcal{A}} = \langle H, \nu \rangle \mathcal{A} w$ on $\hat{\Gamma}$.

Proof. Let $e_1 = \nu_\mathcal{A}/|\nu_\mathcal{A}|_g$. Then e_1 is the unit normal along Γ in the metric g where g is given by (3.23). For $2 \le i \le n$, let e_i be the tangential vector fields along Γ such that $e_1, e_2, \cdots, e_n$ forms an orthonormal basis in $(\mathbb{R}^n, g)$. Then the conditions, $w_{\nu_\mathcal{A}}|_\Gamma = 0$ and $w|_\Gamma = 0$, imply that $e_i(w)|_\Gamma = 0$ for $1 \le i \le n$. Thus

$$Dw|_\Gamma = 0 \quad \text{and, therefore,} \quad D_{e_i} Dw|_\Gamma = 0 \quad \text{for} \ \ 2 \le i \le n. \tag{3.44}$$

It follows from the relations (3.44) and (3.5) that

$$(H(w))_{\nu_\mathcal{A}}/|\nu_\mathcal{A}|_g = e_1(H(w)) = \langle Dw, D_{e_1} H \rangle_g + \langle D_{e_1} Dw, H \rangle$$
$$= \langle H, e_1 \rangle_g D^2 w(e_1, e_1) = \langle H, \nu \rangle \Delta_g w/|\nu_\mathcal{A}|_g = \langle H, \nu \rangle \mathcal{A} w/|\nu_\mathcal{A}|_g,$$

that is, the formula (ii). $\qquad \square$

The following lemma is from [170], also see [188]. We omit the proof here.

Lemma 3.3 *Let $\hat{\Gamma}$ be a relatively open subset of Γ. If w solves the problem*

$$\begin{cases} \mathcal{A}^2 w = F(w, Dw, D^2 w, D^3 w) & \text{on} \ \ \Omega, \\ w = w_{\nu_\mathcal{A}} = \mathcal{A} w = (\mathcal{A} w)_{\nu_\mathcal{A}} = 0 & \text{on} \ \ \hat{\Gamma}, \end{cases}$$

then $w = 0$ on Ω.

Theorem 3.7 *Let $T > 0$ be given. Let H be an escape vector field on $\overline{\Omega}$ for plate. Then there is a constant $c_T > 0$ such that*

$$\int_{\Sigma_0} [(\mathcal{A} w)^2 + (\mathcal{A} w)_{\nu_\mathcal{A}}^2] d\Sigma \ge c_T E(0) \tag{3.45}$$

for all solutions w of the problem (3.37).

Proof. Let $x \in \Omega$ be given. Let $e_1, \cdots, e_n$ be an orthonormal basis of $(\mathbb{R}^n_x, g)$. Using the symmetry of D^2w as a tensor of rank 2 on $\mathbb{R}^n_x$, and the formulas (3.26) and (3.5), we obtain

$$\begin{aligned}
\langle DH, D^2w \rangle_g &= \sum_{ij=1}^{n} DH(e_i, e_j) D^2w(e_i, e_j) \\
&= \sum_{ij} [DH(e_i, e_j) + DH(e_j, e_i)] D^2w(e_i, e_j)/2 \\
&= \vartheta(x) \Delta_g w = \vartheta(x) \mathcal{A}w - \vartheta(x) D v_g(w).
\end{aligned}$$
(3.46)

Let w solve the problem (3.37). Then $E(t) = E(0)$ for all $t \geq 0$. Letting $p = \operatorname{div} H - 2\vartheta_0$, $f = 0$, and using the boundary conditions in the identity (3.20), we obtain

$$\int_Q [w_t^2 - (\mathcal{A}w)^2](\operatorname{div} H - 2\vartheta_0) \, dQ$$

$$= (w_t, pw)|_0^T + \int_Q \mathcal{A}w[2Dp(w) + w\mathcal{A}p]dQ$$

$$\leq \varepsilon[E(0) + E(T) + \int_0^T E(\tau)d\tau] + C_\varepsilon L(w)$$

$$= (2 + T)\varepsilon E(0) + C_\varepsilon L(w)$$
(3.47)

for $\varepsilon > 0$ given small, where ϑ_0 is given in (3.25) and

$$\begin{aligned}
L(w) &= \|w(0)\|^2 + \||Dw|_g(0)\|^2 + \|w(T)\|^2 + \||Dw|_g(T)\|^2 \\
&\quad + \|w\|_{L^2(Q)}^2 + \||Dw|_g\|_{L^2(Q)}^2 \quad \text{for} \quad t \geq 0.
\end{aligned}$$
(3.48)

Next, using the estimate (3.47), the formula (3.46), the boundary conditions of the problem (3.37), and the inequality (3.25) in the identity (3.9) yields

$$c \int_{\Sigma_0} (\mathcal{A}w)^2 d\Sigma \geq \int_{\Sigma_0} (\mathcal{A}w)^2 \langle H, \nu \rangle d\Sigma \geq \int_{\Sigma} (\mathcal{A}w)^2 \langle H, \nu \rangle d\Sigma$$

$$\geq 4\vartheta_0 \int_Q (\mathcal{A}w)^2 dQ + \int_Q [w_t^2 - (\mathcal{A}w)^2] \operatorname{div} H \, dQ$$

$$\quad - (2 + T)\varepsilon E(0) - C_\varepsilon L(w)$$

$$= 2\vartheta_0 \int_Q [w_t^2 + (\mathcal{A}w)^2]dQ + \int_Q [w_t^2 - (\mathcal{A}w)^2](\operatorname{div} H - 2\vartheta_0) \, dQ$$

$$\quad - (2 + T)\varepsilon E(0) - C_\varepsilon L(w)$$

$$\geq 2[2\vartheta_0 T - (2 + T)\varepsilon]E(0) - C_\varepsilon L(w)$$

which gives

$$c_T L(w) + c \int_{\Sigma_0} [(\mathcal{A}w)^2 + (\mathcal{A}w)_{\nu_A}^2] d\Sigma \geq 2\vartheta_0 T E(0)$$

where we have set $\varepsilon = \vartheta_0 T/(2+T)$. By Lemma 3.3, the lower terms in the right hand side of the above inequality can be absorbed as in the proof of Lemma 2.5. Then the inequality (3.45) holds true. $\qquad\square$

Control in Simply Supported Boundary Condition We consider the exact controllability problem

$$
\begin{cases}
u_{tt} + \mathcal{A}^2 u = 0 & \text{in} \quad (0,T) \times \Omega, \\
u(0) = u_0, \quad u_t(0) = u_1 & \text{on} \quad \Omega, \\
u = \mathcal{A}u = 0 & \text{on} \quad (0,T) \times \Gamma_1, \\
u = \varphi, \quad \mathcal{A}u = \psi & \text{on} \quad (0,T) \times \Gamma_0,
\end{cases}
\tag{3.49}
$$

with controls φ and ψ. The dual version for the above problem is in w

$$
\begin{cases}
w_{tt} + \mathcal{A}^2 w = 0 & \text{in} \quad (0,T) \times \Omega, \\
w(0) = w_0, \quad w_t(0) = w_1 & \text{on} \quad \Omega, \\
w = \mathcal{A}w = 0 & \text{on} \quad (0,T) \times \Gamma.
\end{cases}
\tag{3.50}
$$

This time, we define an energy of the system (3.50) by

$$
2E_{1/4}(t) = \int_\Omega (|\nabla_g w_t|_g^2 + |\nabla_g(\mathcal{A}w)|_g^2)dx.
\tag{3.51}
$$

Then $E_{1/4}(t) = E_{1/4}(0)$ for all $t \geq 0$.

We introduce a selfadjoint operator $\mathcal{A}$ on $L^2(\Omega)$ by

$$
\mathcal{A}w = \text{div}\, A(x)\nabla w, \quad D(\mathcal{A}) = H^2(\Omega) \cap H_0^1(\Omega).
\tag{3.52}
$$

The eigenvalues of $-\mathcal{A}$ are

$$
0 < \lambda_1 \leq \cdots \leq \lambda_k \leq \cdots, \quad \text{with} \quad \lim_{k\to\infty} \lambda_k = \infty.
$$

There is an orthonormal basis $\{\varphi_k\}$ of $L^2(\Omega)$ such that $-\mathcal{A}\varphi_k = \lambda_k \varphi_k$ for all $k \geq 1$. Moreover, By [87], Corollary 17.5.8, there are $c_1, c_2 > 0$ such that

$$
|N(\lambda) - c_1 \lambda^{n/2}| \leq c_2 \lambda^{(n-1)/2} \log \lambda
$$

for any $\lambda > 0$ where $N(\lambda)$ denotes the number of the eigenvalues $\leq \lambda$ of $-\mathcal{A}$. Letting $\lambda = \lambda_k$ in the above inequality yields a constant $c > 0$ such that, for k large enough,

$$
\lambda_k \geq ck^{2/n},
\tag{3.53}
$$

from which we obtain

$$
\sum_{k=1}^{\infty} \lambda_k^{-\beta} < \infty \quad \text{for} \quad \beta > n/2.
\tag{3.54}
$$

Remark 3.7 *A finer estimate than* (3.53) *was given by* [187] *for the eigenvalues of the Laplacian on Riemannian manifolds. For the constant coefficient problem with $A(x) = I$, a theorem due to H. Weyl (see* [2]) *states*

$$\lambda_k = (c + o(1))k^{4/n} \quad as \quad k \to \infty.$$

First, we have

Lemma 3.4 *Let w be a complex solution to the problem* (3.50). *Then w can be discomposed as*

$$w = w_+ + w_- \tag{3.55}$$

which satisfies

$$w_t = \mathrm{i}\,\mathcal{A}w_+ - \mathrm{i}\,\mathcal{A}w_-. \tag{3.56}$$

Proof. It is easy to check that a solution w to the problem (3.50) can be expressed by

$$w = \sum_{k=1}^{\infty} \alpha_{+,k} e^{-\mathrm{i}\,\lambda_k t}\varphi_k + \sum_{k=1}^{\infty} \alpha_{-,k} e^{\mathrm{i}\,\lambda_k t}\varphi_k \tag{3.57}$$

where

$$2\alpha_{\pm,k} = (w_0, \varphi_k) \pm \mathrm{i}\,\frac{1}{\lambda_k}(w_1, \varphi_k) \quad \text{for} \quad k \geq 1 \tag{3.58}$$

where $(\cdot, \cdot)$ is the complex product on $L^2(\Omega)$. The formula (3.56) is true if we let

$$w_\pm = \sum_{k=1}^{\infty} \alpha_{\pm,k} e^{\mp \mathrm{i}\,\lambda_k t}\varphi_k. \tag{3.59}$$

Lemma 3.5 *Let H be an escape vector field for the metric g where g is given by* (3.23). *Let $T > 0$ be given. Then there are constants $c_{1T} > 0$, $c_{2T} > 0$ such that*

$$c_{1T} E_{1/4}(0) \leq \int_{\Sigma_0} [w_{t\nu_\mathcal{A}}^2 + (\mathcal{A}w)_{\nu_\mathcal{A}}^2]d\Sigma \leq c_{2T} E_{1/4}(0) \tag{3.60}$$

for all solutions w of the problem (3.50), $\Gamma_0 = \{\, x \,|\, x \in \Gamma, \langle H, \nu \rangle > 0 \,\}$ *and $\Sigma_0 = (0, T) \times \Gamma_0$.*

Proof. Since H is escaping on $\overline{\Omega}$, there is $\varrho_0 > 0$ such that

$$DH(X, X) \geq \varrho_0 |X|_g^2 \quad \text{for} \quad X \in \mathbb{R}_x^n, \, x \in \overline{\Omega}. \tag{3.61}$$

By the boundary conditions, $w = \mathcal{A}w = 0$ on Γ, we have

$$|\nu_\mathcal{A}|_g^2 H(w_t) = \langle H, \nu \rangle w_{t\nu_\mathcal{A}}, \quad |\nu_\mathcal{A}|_g^2 \nabla_g w_t = w_{t\nu_\mathcal{A}}\nu_\mathcal{A}, \tag{3.62}$$

$$|\nu_\mathcal{A}|_g^2 H(\mathcal{A}w) = \langle H, \nu \rangle (\mathcal{A}w)_{\nu_\mathcal{A}}, \quad |\nu_\mathcal{A}|_g^2 \nabla_g(\mathcal{A}w) = (\mathcal{A}w)_{\nu_\mathcal{A}}\nu_\mathcal{A}. \tag{3.63}$$

Letting $p = \operatorname{div} H/2$ and $f = 0$, and using the boundary conditions of the problem (3.50) in the identity (3.21), we obtain

$$| \int_Q [|\nabla_g w_t|^2_g - |\nabla_g (\mathcal{A}w)|^2_g] \operatorname{div} H \, dQ| \le CL(w) \qquad (3.64)$$

where

$$L(w) = \|w_t(0)\|^2 + \|w_t(T)\|^2 + \||D^2 w|_g(0)\|^2 + \||D^2 w|_g(T)\|^2$$
$$+ \|w_t\|^2_{L^2(Q)} + \||D^2 w|_g\|^2_{L^2(Q)} \qquad (3.65)$$

are the lower order terms relatively to the energy $E_{1/4}(0)$.

Using the boundary conditions of the problem (3.50) and the relations (3.61)-(3.64) in the identity (3.15) gives

$$2c \int_{\Sigma_0} [w^2_{t\nu_A} + (\mathcal{A}w)^2_{\nu_A}] d\Sigma \ge 2 \int_{\Sigma} [w^2_{t\nu_A} + (\mathcal{A}w)^2_{\nu_A}] \frac{\langle H, \nu \rangle}{|\nu_A|^2_g} d\Sigma$$
$$\ge 2(\varrho_0 T - 2\varepsilon) E_{1/4}(0) - C_\varepsilon L(w).$$

Letting $\varepsilon = \varrho_0 T/4$ and using Lemma 3.3, as in Lemma 2.5, to absorb the lower order terms, we obtain the left hand side of the inequality (3.60).

Next, we choose a vector field $\hat{H}$ on $\overline{\Omega}$ such that

$$\hat{H} = A(x)\nu \quad \text{for} \quad x \in \Gamma.$$

Using the vector field $\hat{H}$ in the identity (3.15), and the relations (3.62) and (3.63), we obtain the right hand side of the inequality (3.60). $\qquad \square$

The following estimates are observability inequalities in the simply supported boundary conditions.

Theorem 3.8 *Let H be an escape vector field for the metric g. Let $T > 0$ be given. Then there is a constant $c_T > 0$ such that*

$$\int_{\Sigma_0} [w^2_{\nu_A} + (\mathcal{A}w)^2_{\nu_A}] d\Sigma \ge c_T E_{1/4}(0) \qquad (3.66)$$

for all solutions w of the problem (3.50) where Γ_0 is given in Lemma 3.5.

Proof. Let $T > \varepsilon > 0$ be given small. We consider complex solutions of the problem (3.50). Denote by $\mathbf{Z}_\pm$ all the solutions to the problem (3.50) which are given by the formulas (3.58) and (3.59), respectively. For $w_\pm \in \mathbf{Z}_\pm$, let

$$2E_{1/4}(w_\pm) = \int_\Omega (|\nabla_g(w_{\pm\,t}(t))|^2_g + |\nabla_g(\mathcal{A}w_\pm(t))|^2_g) dx.$$

Then $E_{1/4}(w_\pm)(t) = E_{1/4}(w_\pm(0))$. Since $w_{\pm\,t} = \pm i \mathcal{A} w_\pm$ for $x \in \Omega$, we apply Lemma 3.5 with the interval $[0, T]$ replaced by $[\varepsilon, T - \varepsilon]$ to obtain

$$c_{1T\varepsilon} E_{1/4}(w_\pm)(0) \le \int_\varepsilon^{T-\varepsilon} \int_{\Gamma_0} |(\mathcal{A}w_\pm)_{\nu_A}|^2 d\Sigma \le c_{2T\varepsilon} E_{1/4}(w_\pm)(0). \qquad (3.67)$$

In particular, letting $w_+ = e^{-i\lambda_k t}\varphi_k$ in (3.67) yields

$$c_1\lambda_k \leq \int_{\Gamma_0} |\varphi_{k\nu_A}|^2 d\Gamma \leq c_2\lambda_k \quad \text{for all} \quad k \geq 1. \tag{3.68}$$

Let $\phi \in C_0^\infty(\mathbb{R})$ be such that $0 \leq \phi \leq 1$ and

$$\phi(t) = 1 \quad \text{for} \quad t \in [\varepsilon, T - \varepsilon]; \quad \phi(t) = 0 \quad \text{for} \quad t \notin (0, T).$$

Let $\hat{\phi}$ denote the Fourier transform of ϕ. Since ϕ has a compact support, for $\beta > n/2$ given there is $c_\beta > 0$ such that

$$|\hat{\phi}(\zeta)| \leq c_\beta \zeta^{-\beta} \quad \text{for all} \quad \zeta \geq \lambda_1. \tag{3.69}$$

Let

$$\eta_k = \sum_{l=1}^\infty \frac{1}{(\lambda_k + \lambda_l)^\beta} \quad \text{for} \quad k \geq 1.$$

By (3.53) it is easy to prove that

$$\eta_k \to 0 \quad \text{as} \quad k \to \infty. \tag{3.70}$$

Let w and v be complex solutions to the problem (3.50). Set

$$p_\Gamma(w, v) = \int_{\Sigma_0} \phi(t)(\mathcal{A}w)_{\nu_A}\overline{(\mathcal{A}v)_{\nu_A}}\,d\Sigma,$$

$$p_1(w, v) = \int_Q \phi(t)\langle\nabla_g(\mathcal{A}w), \nabla_g(\overline{\mathcal{A}v})\rangle_g\,dQ,$$

$$p_2(w, v) = \int_Q \phi(t)\langle\nabla_g w_t, \nabla_g\overline{v_t}\rangle_g\,dQ.$$

By Lemma 3.4, $w = w_+ + w_-$ where $w_\pm \in \mathbf{Z}_\pm$. Then

$$p_2(w_\pm, w_\pm) = p_1(w_\pm, w_\pm), \quad p_2(w_+, w_-) = -p_1(w_+, w_-), \tag{3.71}$$

and the conservation of the energy implies

$$2p_1(w_\pm, w_\pm) = p_1(w_\pm, w_\pm) + p_2(w_\pm, w_\pm) = 2E_{1/4}(w_\pm)(0)\int_0^T \phi(\tau)d\tau, \tag{3.72}$$

$$p_1(w_+, w_+) + p_1(w_-, w_-) = E_{1/4}(w)(0)\int_0^T \phi(\tau)d\tau. \tag{3.73}$$

Then it follows from (3.67) and (3.72) that

$$c_1 p_i(w_\pm, w_\pm) \leq p_\Gamma(w_\pm, w_\pm) \leq c_2 p_i(w_\pm, w_\pm) \quad \text{for} \quad i = 1, 2. \tag{3.74}$$

Using the relations (3.59), (3.58), (3.68), (3.74), (3.69), and (3.70), we obtain

$$|p_\Gamma(w_+, w_-)| = |\sum_{kl=1}^{\infty} \int_{\Gamma_0} \lambda_k \lambda_l \alpha_{+,k} \overline{\alpha_{-,l}}(\varphi_k)_{\nu_A}(\varphi_l)_{\nu_A} d\Gamma \int_0^T \phi(t) e^{-i(\lambda_k+\lambda_l)\tau} d\tau|$$

$$= \sqrt{2\pi} |\sum_{kl=1}^{\infty} \int_{\Gamma_0} \lambda_k \lambda_l \alpha_{+,k} \overline{\alpha_{-,l}}(\varphi_k)_{\nu_A}(\varphi_l)_{\nu_A} d\Gamma \hat{\phi}(\lambda_k + \lambda_l)|$$

$$\leq C \sum_{kl=1}^{\infty} \frac{\lambda_k^{3/2} \lambda_l^{3/2}}{(\lambda_k + \lambda_l)^\beta} |\alpha_{+,k}||\alpha_{-,l}| \leq C \sum_{kl=1}^{\infty} \frac{1}{(\lambda_k + \lambda_l)^\beta} (\lambda_k^3 |\alpha_{+,k}|^2 + \lambda_l^3 |\alpha_{-,l}|^2)$$

$$= C \sum_{k=1}^{\infty} \eta_k \lambda_k^3 |\alpha_{+,k}|^2 + C \sum_{l=1}^{\infty} \eta_l \lambda_l^3 |\alpha_{-,l}|^2$$

$$\leq C_m(\|w_0\|^2 + \|w_1\|^2) + C\eta_m [p_1(w_+, w_+) + p_1(w_-, w_-)]$$

$$\leq C_m(\|w_0\|^2 + \|w_1\|^2) + C\eta_m E_{1/4}(w)(0) \text{(by (3.73))} \tag{3.75}$$

for m being a positive integer.

Finally, using the relations (3.74), (3.75), and (3.71), we obtain

$$\int_{\Sigma_0} |(\mathcal{A}w)_{\nu_A}|^2 d\Sigma \geq p_\Gamma(w, w)$$

$$\geq p_\Gamma(w_+, w_+) + p_\Gamma(w_-, w_-) - 2|p_\Gamma(w_+, w_-)|$$

$$\geq c_1 [p_1(w_+, w_+) + p_1(w_-, w_-)] - 2|p_\Gamma(w_+, w_-)|$$

$$\geq [c_1 - 2C\eta_m][p_1(w_+, w_+) + p_1(w_-, w_-)] - C_m(\|w_0\|^2 + \|w_1\|^2)$$

$$\geq \sigma_m E_{1/4}(0) - 2C_m(\|w_0\|^2 + \|w_1\|^2).$$

where $\sigma_m = c_1 - 2C\eta_m$, which gives

$$C_m(\|w_0\|^2 + \|w_1\|^2) + \int_{\Sigma_0} (|w_{\nu_A}|^2 + |(\mathcal{A}w)_{\nu_A}|^2) d\Sigma \geq \sigma_m E_{1/4}(0).$$

We fixed m large enough such that $\sigma_m > 0$ to have the inequality (3.66) by absorbing the lower order terms as in Lemma 2.5. □

Let $\mathcal{A}^{-1}$ denote the inverse of the selfadjoint operator in (3.52). Then there are constants $c_1 > 0$, $c_2 > 0$ such that

$$c_1(\|w_0\|^2_{H_0^1(\Omega)} + \|w_1\|^2_{H^{-1}(\Omega)})$$

$$\leq \||\nabla_g(\mathcal{A}(\mathcal{A}^{-1}w_0))|_g\|^2 + \||\nabla_g(\mathcal{A}^{-1}w_1)|_g\|^2$$

$$\leq c_2(\|w_0\|^2_{H_0^1(\Omega)} + \|w_1\|^2_{H^{-1}(\Omega)}) \tag{3.76}$$

for all $(w_0, w_1) \in H_0^1(\Omega) \times H^{-1}(\Omega)$. We define

$$\|(u, v)\|^2_{\mathcal{B}(\Omega)} = \||\nabla_g(\mathcal{A}u)|_g\|^2 + \||\nabla_g v|_g\|^2,$$

$$\mathcal{B}(\Omega) = \{ (u, v) \mid u, \, v \in L^2(\Omega), \, \|(u, v)\|_{\mathcal{B}(\Omega)} < \infty \}.$$

The inequality (3.76) shows that $\mathcal{A}^{-1} \colon H_0^1(\Omega) \times H^{-1}(\Omega) \to \mathcal{B}(\Omega)$ is an isomorphism.

Let $(v_0, v_1) \in H_0^1(\Omega) \times H^{-1}(\Omega)$ be given. Let w be the solution of the problem (3.50) with initial data $(w(0), w_t(0)) = (\mathcal{A}^{-1}v_0, \mathcal{A}^{-1}v_1) \in \mathcal{B}(\Omega)$. For $T > 0$, we solve the following time-reverse problem

$$\begin{cases} u_{tt} + \mathcal{A}^2 u = 0 & \text{in } Q, \\ u(T) = u_t(T) = 0 & \text{on } \Omega, \\ u = \mathcal{A}u = 0 & \text{on } \Sigma_1, \\ u = (\mathcal{A}w)_{\nu_\mathcal{A}}, \quad \mathcal{A}u = w_{\nu_\mathcal{A}} & \text{on } \Sigma_0. \end{cases} \tag{3.77}$$

Then for the initial data $(u(0), u_t(0))$, we have found control actions $u = (\mathcal{A}w)_{\nu_\mathcal{A}}$ and $\mathcal{A}u = w_{\nu_\mathcal{A}}$ on the portion Γ_0 of the boundary which steer the problem (3.36) to rest at a time T. We define a map by

$$\Lambda(v_0, v_1) = \left(- \mathcal{A}^{-1}u_t(0), \mathcal{A}^{-1}u(0) \right).$$

Then

$$\left(\Lambda(v_0, v_1), (v_0, v_1) \right)_{L^2(\Omega) \times L^2(\Omega)} = \int_{\Sigma_0} [w_{\nu_\mathcal{A}}^2 + (\mathcal{A}w)_{\nu_\mathcal{A}}^2] d\Sigma. \tag{3.78}$$

By Proposition 3.8 and the inequality (3.76), there is $c_T > 0$ such that

$$\left(\Lambda(v_0, v_1), (v_0, v_1) \right)_{L^2(\Omega) \times L^2(\Omega)} \geq c_T \|(v_0, v_1)\|_{H_0^1(\Omega) \times H^{-1}(\Omega)}^2$$

for $(v_0, v_1) \in H_0^1(\Omega) \times H^{-1}(\Omega)$. It follows from Theorem 2.13 that

Theorem 3.9 *Let H be an escape vector field for the metric g. For each $T > 0$, the problem (3.49) is exactly controllable on the space $H^{-1}(\Omega) \times \mathcal{A}H^{-1}(\Omega)$ with controls $(\varphi, \psi) \in [L^2(\Sigma_0)]^2$.*

Remark 3.8 *For the Euler-Bernoulli plate with the simply supported boundary conditions, the system behaves like a wave equation so that the existence of an escape vector field in the metric g can guarantee the exact controllability. However, for the fixed boundary conditions, an escape vector field for plate is needed.*

3.4   Controllability for Transmission of Plate

We consider exact boundary controllability of transmission of the plate with variable coefficients where the middle surface is assumed to be composed of two materials with connecting conditions between them.

Let Ω be a bounded domain in $\mathbb{R}^n$ with smooth boundary Γ. Let Ω be divided into two regions, Ω_1 and Ω_2, by a smooth hypersurface Γ_3 on Ω. Set $\Gamma_i = \partial\Omega \cap \partial\Omega_i$ for $i = 1$ and 2. Denote by ν_i the normal of $\partial\Omega_i$ in the Euclidean metric of $\mathbb{R}^n$ pointing outside Ω_i. Then

$$\nu_1 + \nu_2 = 0 \quad \text{for all} \quad x \in \Gamma_3.$$

We define

$$a(x) = \begin{cases} a_1 & \text{on} \quad \Omega_1, \\ a_2 & \text{on} \quad \Omega_2, \end{cases}$$

where $a_i > 0$ are given constants.

Definition 3.2 *A function w on Ω is said to satisfy connecting conditions on Γ_3 for plate if*

$$\begin{cases} w_1 = w_2, \quad w_{1\nu_{1A}} + w_{2\nu_{2A}} = 0 \quad on \quad \Gamma_3, \\ a_1 \mathcal{A}w_1 = a_2 \mathcal{A}w_2, \quad a_1(\mathcal{A}w_1)_{\nu_{1A}} + a_2(\mathcal{A}w_2)_{\nu_{2A}} = 0 \quad on \quad \Gamma_3, \end{cases} \tag{3.79}$$

where $w_i = w|_{\Omega_i}$ for $i = 1,\, 2$.

Let $T > 0$ be given and let $\Gamma_0 \subset \Gamma$ be a subset. We consider controllability of the problem

$$\begin{cases} u_{tt} + a(x)\mathcal{A}^2 u = 0 \quad \text{in} \quad (0, T) \times (\Omega/\Gamma_3), \\ u(0) = u^0(x), \quad u_t(0) = u^1(x) \quad \text{on} \quad \Omega, \\ u = u_{\nu_A} = 0 \quad \text{on} \quad \Sigma/\Sigma_0, \\ u \text{ satisfies the connecting conditions for plate on } \Sigma_3, \\ u = \varphi, \quad u_{\nu_A} = \psi \quad \text{on} \quad \Sigma_0, \end{cases} \tag{3.80}$$

where φ, ψ are boundary controls, $\Sigma_3 = (0, T) \times \Gamma_3$, and $\Sigma_0 = (0, T) \times \Gamma_0$. The duality problem of the system (3.80) is

$$\begin{cases} w_{tt} + a(x)\mathcal{A}^2 w = 0 \quad \text{in} \quad (0, T) \times (\Omega/\Gamma_3), \\ w(0) = w^0(x), \quad w_t(0) = w^1(x) \quad \text{on} \quad \Omega, \\ w = w_{\nu_A} = 0 \quad \text{on} \quad \Sigma, \\ w \text{ satisfies the connecting conditions for plate on } \Sigma_3. \end{cases} \tag{3.81}$$

First, we consider regularity of the system (3.81). Let

$$H^4(\Omega, \Gamma_3) = \{\, w \in H_0^2(\Omega),\, w_i \in H^4(\Omega_i),\, i = 1,\, 2,$$
$$w \text{ satisfies the connecting conditions for plate on } \Gamma_3 \,\}.$$

Define an operator

$$\mathcal{A}w = \text{div } A(x)\nabla w, \quad \mathcal{D}(\mathcal{A}) = H^2(\Omega) \cap H_0^2(\Omega).$$

Let $\mathbb{A}$ be given by

$$\mathbb{A} = \begin{pmatrix} 0 & I \\ -a\mathcal{A}^2 & 0 \end{pmatrix}, \quad \mathcal{D}(\mathbb{A}) = H^4(\Omega, \Gamma_3) \times H_0^1(\Omega). \tag{3.82}$$

It is readily shown that $\mathbb{A}$ is skew-adjoint on the space $H_0^2(\Omega) \times L^2(\Omega)$, i.e. $\mathbb{A}^* = -\mathbb{A}$, where the inner product on $H_0^2(\Omega)$ is given by

$$(w, v)_{H_0^2(\Omega)} = \int_\Omega a(x) \mathcal{A}w(x) \mathcal{A}v(x) \, dx. \qquad (3.83)$$

So $\mathbb{A}$ generates a C_0-group $e^{\mathbb{A}t}$ on $H_0^2(\Omega) \times L^2(\Omega)$. Then

Proposition 3.1 *For $(w^0, w^1) \in H_0^2(\Omega) \times L^2(\Omega)$, the system (3.81) admits a unique weak solution*

$$w \in C([0, T]; H_0^2(\Omega)) \cap C^1([0, T]; L^2(\Omega)).$$

Furthermore, if $(w^0, w^1) \in H^4(\Omega, \Gamma_3) \times H_0^2(\Omega)$, then

$$w \in C([0, T]; H^4(\Omega, \Gamma_3)) \cap C^1([0, T]; H_0^2(\Omega)).$$

We need

Lemma 3.6 *Let H be a vector field on Ω. Let a function w on Ω satisfy the connecting conditions for plate on Γ_3. Then*

$$(H(w_1))_{\nu_{1\mathcal{A}}} + (H(w_2))_{\nu_{2\mathcal{A}}} = \langle H, \nu_1 \rangle (\mathcal{A}w_1 - \mathcal{A}w_2) \quad \text{for} \quad x \in \Gamma_3 \qquad (3.84)$$

where $w_i = w|_{\Omega_i}$ for $i = 1. 2$.

Proof. Let $x \in \Gamma_3$ be given. Let $E_1, \cdots, E_n$ be a frame field normal at x where $E_1(x) = \nu_{1\mathcal{A}}(x)/|\mathcal{A}\nu|_g^2$. Then the connecting conditions imply that $Dw_1 = Dw_2$ for $x \in \Gamma_3$. Thus

$$\sum_{i=2}^n D^2 w_1(E_i, E_i) = \sum_{i=2}^n D^2 w_2(E_i, E_i) \quad \text{at} \quad x. \qquad (3.85)$$

In addition, by (3.5), we have

$$\mathcal{A}w_1 = \frac{1}{|\mathcal{A}\nu|_g^2} D^2 w_1(\nu_{1\mathcal{A}}, \nu_{1\mathcal{A}}) + \sum_{i=2}^n D^2 w_1(E_i, E_i) + Dw_1(v_g) \qquad (3.86)$$

at x. Using (3.85) and (3.86), we obtain

$$\begin{aligned}
(H(w_1))&_{\nu_{1\mathcal{A}}} + (H(w_2))_{\nu_{2\mathcal{A}}} \\
&= D^2 w_1(H, \nu_{1\mathcal{A}}) + D^2 w_2(H, \nu_{2\mathcal{A}}) + Dw_1(D_{\nu_{1\mathcal{A}}} H) + Dw_2(D_{\nu_{2\mathcal{A}}} H) \\
&= \langle H, \nu_1 \rangle (\mathcal{A}w_1 - \mathcal{A}w_2) \quad \text{at} \quad x.
\end{aligned}$$

Lemma 3.7 *Let H be a vector field on Ω. Let w solve the problem (3.81). Then*

$$\int_{\Sigma} a(x)(\mathcal{A}w)^2 \langle H, \nu \rangle d\Sigma + \int_{\Sigma_3} [a_1(\mathcal{A}w_1)^2 - a_2(\mathcal{A}w_2)^2] \langle H, \nu_1 \rangle d\Sigma$$

$$= 2(w_t, H(w))|_0^T + \int_Q [w_t^2 - a(x)(\mathcal{A}w)^2] \operatorname{div} H dQ$$

$$+2 \int_Q a(x)\mathcal{A}w[2\langle DH, D^2w \rangle_g + P(Dw)]dQ \qquad (3.87)$$

where

$$P(Dw) = \mathcal{A}H(w) + \operatorname{Ric}(H, Dw) - D^2 v_g(H, Dw),$$

$v_g = \log \det A(x)/2$, *and $\mathcal{A}H$ is given by (3.7).*

Proof. We multiply the equation in (3.81) by $2H(w_i)$ and integrate it over $Q_i = (0, T) \times \Omega_i$ for $i = 1, 2$. A similar computation as in the proof of Theorem 3.1 yields

$$\int_{\Sigma_i \cup \Sigma_3} \psi(w_i) d\Sigma = 2 \int_{\Omega_i} w_{it} H(w_i) dx|_0^T + \int_{Q_i} \{[w_{it}^2 - a_i(\mathcal{A}w_i)^2] \operatorname{div} H$$

$$+4a_i \mathcal{A}w_i \langle DH, D^2 w_i \rangle_g + 2a_i \mathcal{A}w_i P(Dw_i)\} dQ \qquad (3.88)$$

where $\Sigma_i = (0, T) \times \Gamma_i$ and

$$\psi(w_i) = 2a_i \mathcal{A}w_i (H(w_i))_{\nu_{i,\mathcal{A}}} - 2a_i H(w_i)(\mathcal{A}w_i)_{\nu_{i,\mathcal{A}}} + [w_{it}^2 - a_i(\mathcal{A}w_i)^2]\langle H, \nu_i \rangle,$$

for $i = 1, 2$.

By Lemma 3.2, the boundary conditions $w_i = w_{i\nu_{i,\mathcal{A}}} = 0$ on Γ_i imply that

$$\psi(w_i) = a_i(\mathcal{A}w_i)^2 \langle H, \nu_i \rangle \quad \text{for} \quad x \in \Gamma_i, \ i = 1, 2. \qquad (3.89)$$

Since $\nu_1 = -\nu_2$ for $x \in \Gamma_3$, the connecting conditions for plate yield $H(w_1) = H(w_2)$ for $x \in \Gamma_3$ and

$$\psi(w_1) + \psi(w_2)$$
$$= 2a_1 \mathcal{A}w_1 [(H(w_1))_{\nu_{1,\mathcal{A}}} + (H(w_2))_{\nu_{2,\mathcal{A}}}] + 2H(w_1)(a_1 \mathcal{A}w_1 - a_2 \mathcal{A}w_2)$$
$$-2H(w_1)[a_1(\mathcal{A}w_1)_{\nu_{1,\mathcal{A}}} + a_2(\mathcal{A}w_2)_{\nu_{2,\mathcal{A}}}] + [a_2(\mathcal{A}w_2)^2 - a_1(\mathcal{A}w_1)^2]\langle H, \nu_1 \rangle$$
$$= [a_1(\mathcal{A}w_1)^2 - a_2(\mathcal{A}w_2)^2]\langle H, \nu_1 \rangle. \qquad (3.90)$$

Summing up the identities (3.88) for $i = 1, 2$, and using the formulas (3.89) and (3.90), we obtain the identity (3.87). $\qquad \square$

Lemma 3.8 *Let p be a function. Let w solve the problem (3.81). Then*

$$\int_Q p[w_t^2 - a(x)(\mathcal{A}w)^2]dQ = (w_t, pw)|_0^T$$

$$+ \int_Q a(x)\mathcal{A}w[2Dp(w) + w\mathcal{A}p]dQ. \qquad (3.91)$$

Proof. A similar computation as in the proof of Theorem 3.3 gives

$$\int_{Q_i} p[w_{it}^2 - a_i(\mathcal{A}w_i)^2]dQ = \int_{Q_i} a_i \mathcal{A}w_i[2Dp(w_i) + w_i \mathcal{A}p]dQ$$

$$+ \int_{\Omega_i} pw_{it}w_i dx\big|_0^T + \int_{\Sigma_i \cup \Sigma_3} a_i[pw_i(\mathcal{A}w_i)_{\nu_{iA}} - \mathcal{A}w_i(pw_i)_{\nu_{iA}}]d\Sigma \quad (3.92)$$

for $i = 1$, 2. In addition, the boundary conditions and the connecting conditions imply

$$\sum_i a_i[pw_i(\mathcal{A}w_i)_{\nu_{iA}} - \mathcal{A}w_i(pw_i)_{\nu_{iA}}] = 0 \quad \text{for} \quad x \in \Gamma \cup \Gamma_3.$$

Then (3.91) follows by summing up (3.92) with $i = 1$, 2. $\square$

We define the energy of the problem (3.81) by

$$2E(t) = \int_\Omega [w_t^2 + a(x)(\mathcal{A}w)^2]dx.$$

Then it follows from (3.81) that

$$\frac{d}{dt}E(t) = \sum_{i=1}^2 \int_{\Omega_i}(w_{it}w_{itt} + a_i\mathcal{A}w_i\mathcal{A}w_{it})dx = \sum_{i=1}^2 a_i \int_{\Gamma_i \cup \Gamma_3} w_{it}(\mathcal{A}w_i)_{\nu_{iA}}d\Gamma$$

$$+ \sum_{i=1}^2 a_i \int_{\Omega_i}(\langle \mathcal{A}\nabla(\mathcal{A}w_i), \nabla w_{it}\rangle + \mathcal{A}w_i\mathcal{A}w_{it})dx$$

$$= \sum_{i=1}^2 a_i \int_{\Gamma_i \cup \Gamma_3} [(\mathcal{A}w_i)w_{it\nu_{iA}} - w_{it}(\mathcal{A}w_i)_{\nu_{iA}}]d\Gamma = 0,$$

that is,

$$E(t) = E(0) \quad \text{for} \quad t \ge 0.$$

The observability estimate is given by

Theorem 3.10 *Let H be an escape vector field for plate such that*

$$(a_1 - a_2)\langle H, \nu_1 \rangle \ge 0 \quad \text{for} \quad x \in \Gamma_3. \tag{3.93}$$

Let $T > 0$ be given. Then there is $c_T > 0$ satisfying

$$c_T \int_{\Sigma_0} \{[(\mathcal{A}w)^2 + [(\mathcal{A}w)_{\nu_A}]^2\}d\Sigma \ge E(0) \tag{3.94}$$

for all solutions w to the problem (3.81) where $\Sigma_0 = (0, T) \times \Gamma_0$ and

$$\Gamma_0 = \{x \mid x \in \Gamma, \ \langle H, \nu \rangle > 0\}.$$

Proof. Let $\vartheta_0 > 0$ be given by (3.25). By the formula (3.46), we have

$$\mathcal{A}w[2\langle DH, D^2w\rangle_g + P(Dw)] \geq \vartheta_1(\mathcal{A}w)^2 - C|Dw|_g^2 \qquad (3.95)$$

where $0 < \vartheta_1 < 2\vartheta_0$ is a constant and ϑ_0 is given in (3.25). Moreover, the connecting conditions and the assumption (3.93) on Σ_3 yield

$$\int_{\Sigma_3} [a_1(\mathcal{A}w_1)^2 - a_2(\mathcal{A}w_2)^2]\langle H, \nu_1\rangle d\Sigma$$
$$= \int_{\Sigma_3} \frac{a_2(\mathcal{A}w_2)^2}{a_1}(a_2 - a_1)\langle H, \nu_1\rangle d\Sigma \leq 0. \qquad (3.96)$$

Using the estimates (3.95) and (3.96) in the identity (3.87), we obtain

$$\int_{\Sigma_0} a(x)(\mathcal{A}w)^2\langle H, \nu\rangle d\Sigma$$
$$\geq \int_{\Sigma} a(x)(\mathcal{A}w)^2\langle H, \nu\rangle d\Sigma + \int_{\Sigma_3} [a_1(\mathcal{A}w_1)^2 - a_2(\mathcal{A}w_2)^2]\langle H, \nu_1\rangle d\Sigma$$
$$\geq 2\vartheta_1 \int_Q a(x)(\mathcal{A}w)^2 dQ + \int_Q [w_t^2 - a(x)(\mathcal{A}w)^2]\,\mathrm{div}\,H dQ$$
$$+2(w_t, H(w))|_0^T - C\int_Q |Dw|_g^2 dQ$$
$$\geq \vartheta_1 \int_Q [w_t^2 + a(x)(\mathcal{A}w)^2]dQ + \int_Q p[w_t^2 - a(x)(\mathcal{A}w)^2]dQ$$
$$-C_\varepsilon(\||Dw(T)|_g\|^2 + \||Dw(0)|_g\|^2 + \||Dw|_g\|_{L^2(Q)}^2)$$
$$-\varepsilon[E(T) + E(0)] \qquad (3.97)$$

where $p = \mathrm{div}\,H - \vartheta_1$ and $\varepsilon > 0$ can be chosen small. Using the identity (3.91) in the inequality (3.97) gives

$$c\int_{\Sigma_0} \{(\mathcal{A}w)^2 + [(\mathcal{A}w)_{\nu_A}]^2\}d\Sigma \geq (T\vartheta_1 - 4\varepsilon)E(0)$$
$$-C_\varepsilon[\||Dw(T)|_g\|^2 + \||Dw(0)|_g\|^2 + \||Dw|_g\|_{L^2(Q)}^2]$$
$$-C_\varepsilon[\|w(0)\|^2 + \|w(T)\|^2 + \|w\|_{L^2(Q)}^2],$$

which, by letting $0 < \varepsilon < T\vartheta_1/4$, yields

$$\int_{\Sigma_0} \{(\mathcal{A}w)^2 + [(\mathcal{A}w)_{\nu_A}]^2\}d\Sigma + L(w) \geq c_T E(0) \qquad (3.98)$$

for all solutions w to the problem (3.81), where

$$L(w) = \|w(0)\|^2 + \|w(T)\|^2 + \|w\|_{L^2(Q)}^2$$
$$+\||Dw(T)|_g\|^2 + \||Dw(0)|_g\|^2 + \||Dw|_g\|_{L^2(Q)}^2.$$

Finally, the lower order terms can be absorbed as in Lemma 2.5 and the inequality (3.94) follows.   □

It follows from Theorem 3.10 that

Theorem 3.11 *Let H be an escape vector field for plate such that the condition (3.93) holds true. Then for any $T > 0$, the transmission problem (3.80) is exactly $L^2(\Omega) \times H^{-2}(\Omega)$ controllable by $L^2(\Sigma_0) \times L^2(\Sigma_0)$ controls.*

3.5   Stabilization from Boundary for the Plate with a Curved Middle Surface

Model Throughout this section M is a surface of $\mathbb{R}^3$ with the induced metric g from the Euclidean metric of $\mathbb{R}^3$. We shall consider a curvy plate whose middle surface, Ω, is part of the surface M and where the extension effects along the tangential direction are neglected. Let boundary Γ of the middle surface Ω consist of two disjoint parts $\Gamma_0 \cup \Gamma_1 = \Gamma$. We assume that the material undergoing obeys Hooke's law. The potential energy is defined by

$$P(u) = \int_\Omega [(1-\mu)|D^2u|^2 + \mu(\operatorname{tr} D^2u)^2]dx$$

where u is the displacement of the plate along the normal, $0 < \mu < 1/2$ is the Poisson coefficient, D denotes the covariant differential of the metric g, and tr is the trace in the metric g.

If there is no external force, then the equations of motion for u are obtained by setting to zero the first variation of the Lagrangian:

$$\int_0^T [\|u_t\|^2 - P(u)]d\tau = 0$$

where $\|\cdot\|$ is the norm of $L^2(\Omega)$. The above variation is taken with respect to kinematically admissible displacements.

We obtain, as a result of calculation by the above variation, the following system:

$$\begin{cases} u_{tt} + \Delta^2 u + (1-\mu)\operatorname{div}(\kappa \nabla u) = 0 & \text{in} \quad (0, \infty) \times \Omega, \\ u = u_\nu = 0 & \text{on} \quad (0, \infty) \times \Gamma_1, \\ \Delta u + (1-\mu)B_1 u = 0 & \text{on} \quad (0, \infty) \times \Gamma_0, \\ (\Delta u)_\nu + (1-\mu)B_2 u = 0 & \text{on} \quad (0, \infty) \times \Gamma_0, \\ u(0) = u_0, \quad u_t(0) = u_1 & \text{on} \quad \Omega, \end{cases} \tag{3.99}$$

where Δ, div, ∇, κ, and ν are the Laplacian, the divergence, the gradient, the

Gauss curvature function of surface M, and the outside normal along Γ in the induced metric g, respectively. In the system (3.99), B_1 and B_2 are boundary operators, defined by

$$B_1 u = -D^2 u(\tau, \tau), \quad \text{and} \quad B_2 u = \frac{\partial}{\partial \tau}(D^2 u(\tau, \nu)) + \kappa u_\nu,$$

respectively, where τ is a unit tangential vector field on Γ such that ν, τ forms an orthonormal basis $(M_x, g(x))$ for each $x \in \Gamma$.

Remark 3.9 *The term* $\operatorname{div}(\kappa \nabla u)$ *in the system* (3.99) *comes from the curvedness of middle surface* Ω. *If* $M = \mathbb{R}^2$, *then* $\kappa = 0$ *and the system is the same as in* [109].

Lemma 3.9 *Let M be a surface of $\mathbb{R}^3$. Let* Ric *be the Ricci tensor of M. Then*

$$\operatorname{Ric} = \kappa g. \tag{3.100}$$

Proof. Let $x \in M$ be given. Let e_1, e_2 be an orthonormal basis of M_x. For any $X, Y \in \mathcal{X}(M)$, by Theorem 1.8, we have

$$\mathbf{R}(e_i, X, e_i, Y) = \langle X, e_j \rangle \langle Y, e_j \rangle \mathbf{R}(e_i, e_j, e_i, e_j) = \kappa \langle X, e_j \rangle \langle Y, e_j \rangle$$

for $i \neq j$, $1 \leq i, j \leq 2$, where $\mathbf{R}$ is the curvature tensor. Then

$$\operatorname{Ric}(X, Y) = \mathbf{R}(e_1, X, e_1, Y) + \mathbf{R}(e_2, X, e_2, Y) = \kappa \langle X, Y \rangle.$$

$\square$

We introduce

$$a(u, v) = (1 - \mu)\langle D^2 u, D^2 v \rangle_g + \mu \operatorname{tr} D^2 u \operatorname{tr} D^2 v, \tag{3.101}$$

$$Au = \Delta^2 u + (1 - \mu) \operatorname{div}(\kappa \nabla u), \tag{3.102}$$

$$\Gamma_1(u) = \Delta u + (1 - \mu) B_1 u, \quad \Gamma_2(u) = (\Delta u)_\nu + (1 - \mu) B_2 u. \tag{3.103}$$

The following formula plays an important role in our problems, which presents the relationship between the interior and the boundary.

Lemma 3.10 *Let the boundary Γ of the middle surface Ω be a closed curve. Then*

$$\int_\Omega a(u, v) dg = (Au, v) + \int_\Gamma [\Gamma_1(u) v_\nu - \Gamma_2(u) v] d\Gamma \tag{3.104}$$

for $u, v \in H^4(\Omega)$.

Proof. Using the formulas (1.155), (3.100), and (1.137), we have

$$\int_\Omega \langle D^2 u, D^2 v \rangle_g \, dg = (D^2 u, D^2 v)_{L^2(\Omega, T^2)}$$

$$= (\mathbf{\Delta} Du - \kappa Du, Dv)_{L^2(\Omega, \Lambda)} + \int_\Gamma \langle D_\nu Du, Dv \rangle d\Gamma$$

$$= (\delta \mathbf{\Delta} du - \delta(\kappa \nabla u), v) + \int_\Gamma [v \langle \mathbf{\Delta} Du - \kappa Du, \nu \rangle + \langle D_\nu Du, Dv \rangle] d\Gamma$$

$$= (\Delta^2 u + \operatorname{div}(\kappa \nabla u), v)$$

$$+ \int_\Gamma \{ D^2 u(\nu, \nu) v_\nu + D^2 u(\nu, \tau) v_\tau - v[(\Delta u)_\nu + \kappa u_\nu] \} d\Gamma. \tag{3.105}$$

Since Γ is a closed curve,

$$\int_\Gamma D^2 u(\nu, \tau) v_\tau d\Gamma = - \int_\Gamma [D^2 u(\nu, \tau)]_\tau v d\Gamma. \tag{3.106}$$

Moreover, the Green formula yields

$$\int_\Omega \operatorname{tr} D^2 u \, \operatorname{tr} D^2 v dg = (\Delta u, \Delta v) = (\Delta^2 u, v)$$

$$+ \int_\Gamma [\Delta u v_\nu - v(\Delta u)_\nu] d\Gamma. \tag{3.107}$$

The formula (3.111) follows from the relations (3.105)-(3.107). □

We define the energy of the system by

$$2E(t) = \int_\Omega [u_t^2 + a(u, u)] dx.$$

Let u satisfy the equation in (3.99) and the fixed boundary conditions, $u = u_\nu = 0$ on Γ_1. Using Lemma 3.10, we obtain

$$E'(t) = \int_\Gamma [\Gamma_1(u) u_{t\nu} - \Gamma_2(u) u_t] d\Gamma.$$

If we act on $\Sigma_0 = (0, \infty) \times \Gamma_0$ by

$$\Gamma_1(u) = -u_{t\nu}, \quad \Gamma_2(u) = u_t,$$

then

$$E(t) = E(s) - \int_s^t \int_{\Gamma_0} (u_t^2 + u_{t\nu}^2) d\Gamma d\tau \quad \text{for} \quad t \geq 0, \ s \geq 0. \tag{3.108}$$

We have a closed loop system

$$\begin{cases} u_{tt} + \Delta^2 u + (1 - \mu) \operatorname{div}(\kappa \nabla u) = 0 & \text{in} \quad (0, \infty) \times \Omega, \\ u = u_\nu = 0 & \text{on} \quad (0, \infty) \times \Gamma_1, \\ \Delta u + (1 - \mu) B_1 u = -u_{t\nu} & \text{on} \quad (0, \infty) \times \Gamma_0, \\ (\Delta u)_\nu + (1 - \mu) B_2 u = u_t & \text{on} \quad (0, \infty) \times \Gamma_0, \\ u(0) = u_0, \quad u_t(0) = u_1 & \text{on} \quad \Omega. \end{cases} \tag{3.109}$$

We shall establish the exponential decay of the energy for the above system.

Well Posedness of the System (3.109) We introduce bilinear forms

$$\mathcal{B}(u, v) = \int_\Omega a(u, v) \, dx, \quad \alpha(u, v) = \int_{\Gamma_0} (uv + u_\nu v_\nu) d\Gamma.$$

It follows from Lemma 3.10 that an appropriate variational formulation of the system (3.109) is as follows: Find $w \in C([0, \infty), H^2_{\Gamma_1}(\Omega)) \cap C^1([0, \infty), H^1_{\Gamma_1}(\Omega))$ such that

$$\begin{cases} [(w_t, v) + \alpha(w, v)]_t + \mathcal{B}(w, v) = 0 & \text{for all} \quad v \in H^2_{\Gamma_1}(\Omega), \\ w(0) = w_0 \in H^2_{\Gamma_1}(\Omega), \quad w_t(0) = w_1 \in L^2(\Omega). \end{cases} \tag{3.110}$$

We assume that

$$\Gamma_1 \neq \emptyset, \quad \overline{\Gamma_0} \cap \overline{\Gamma_1} = \emptyset. \tag{3.111}$$

Then the condition (3.111) implies that $\mathcal{B}(\cdot, \cdot)$ is an equivalent norm of $H^2_{\Gamma_1}(\Omega)$.

Let $H^{-2}_{\Gamma_1}(\Omega)$ denote *the dual space* of $H^2_{\Gamma_1}(\Omega)$ with respect to the L^2-topology. Let $\aleph$ be the canonical isomorphism from $H^2_{\Gamma_1}(\Omega)$ endowed with the inner product $\mathcal{B}(\cdot, \cdot)$ to $H^{-2}_{\Gamma_1}(\Omega)$. Thus

$$\mathcal{B}(u, v) = (\aleph u, v) \quad \text{for} \quad u, v \in H^2_{\Gamma_1}(\Omega). \tag{3.112}$$

Moreover, since $0 \leq \alpha(u, u) \leq c\|u\|^2_{H^2_{\Gamma_1}(\Omega)}$, there is a positive operator $\aleph_0$: $H^2_{\Gamma_1}(\Omega) \to H^{-2}_{\Gamma_1}(\Omega)$ such that

$$\alpha(u, v) = (\aleph_0 u, v) \quad \text{for} \quad u, v \in H^2_{\Gamma_1}(\Omega). \tag{3.113}$$

Then the variational problem (3.110) can be written as

$$(w_t + \aleph_0 w)_t + \aleph w = 0 \quad \text{in} \quad H^{-2}_{\Gamma_1}(\Omega). \tag{3.114}$$

Let us formally rewrite the system (3.114) as

$$\mathbb{C} Y_t + I\!N Y = 0 \tag{3.115}$$

where

$$\mathbb{C} = \begin{pmatrix} \aleph & 0 \\ 0 & I \end{pmatrix}, \quad I\!N = \begin{pmatrix} 0 & -\aleph \\ \aleph & \aleph_0 \end{pmatrix}, \quad Y = \begin{pmatrix} w \\ w_t \end{pmatrix}.$$

We shall solve the problem (3.115) in $H^2_{\Gamma_1}(\Omega) \times L^2(\Omega)$. Let

$$\mathcal{D}(I\!N) = \{\, (u, v) \,|\, u, v \in H^2_{\Gamma_1}(\Omega), \, \aleph u + \aleph_0 v \in L^2(\Omega) \,\}.$$

Then $I\!N: \mathcal{D}(I\!N) \to H^{-2}_{\Gamma_1}(\Omega) \times L^2(\Omega)$. Since $\mathbb{C}: H^2_{\Gamma_1}(\Omega) \times L^2(\Omega) \to H^{-2}_{\Gamma_1}(\Omega) \times L^2(\Omega)$ is the canonical isomorphism, the system (3.115) is equivalent to

$$Y_t + \mathbb{C}^{-1} I\!N Y = 0. \tag{3.116}$$

Theorem 3.12 *Let the condition (3.111) be true. Then* $-\mathbb{C}^{-1}I\!N$ *is the infinitesimal generator of a C_0-semigroup of contraction on* $H^2_{\Gamma_1}(\Omega) \times L^2(\Omega)$.

Proof. **Step 1** $\mathcal{D}(I\!N)$ *is dense in* $H^2_{\Gamma_1}(\Omega) \times L^2(\Omega)$.
By Lemma 3.10, we have

$$(\aleph u + \aleph_0 v, w) = \mathcal{B}(u, w) + \alpha(v, w) = (Au, w)$$
$$+ \int_{\Gamma_0} \{[\Gamma_1(u) + v_\nu]w_\nu + [v - \Gamma_2(u)]w\} \, d\Gamma$$

which implies

$$\mathcal{D}(I\!N) \supset \mathcal{D}_0$$

where

$$\mathcal{D}_0 = \{ \ (u,v) \in [H^4(\Omega) \cap H^2_{\Gamma_1}(\Omega)] \times H^2_{\Gamma_1}(\Omega), \ \Gamma_1(u) + v_\nu = 0,$$
$$v - \Gamma_2(u) = 0 \text{ on } \Gamma_0 \ \}.$$

$D(I\!N)$ is dense in $H^2_{\Gamma_1}(\Omega) \times L^2(\Omega)$ since $\mathcal{D}_0$ is.
    Step 2 $-\mathbb{C}^{-1}I\!N$ is dissipative and $\text{Range}\,(\lambda I + \mathbb{C}^{-1}I\!N) = H^2_{\Gamma_1}(\Omega) \times L^2(\Omega)$ for $\lambda > 0$.
    We need prove that for arbitrary $(\varphi, \ \psi) \in H^2_{\Gamma_1}(\Omega) \times L^2(\Omega)$, there is $(u, \ v) \in \mathcal{D}(I\!N)$, such that

$$(I + \mathbb{C}^{-1}I\!N)(u, \ v) = (\varphi, \ \psi).$$

It suffices to show that the map $I + \aleph + \aleph_0 : H^2_{\Gamma_1}(\Omega) \rightarrow H^{-2}_{\Gamma_1}(\Omega)$ is onto. Indeed, it is enough if there exists $v \in H^2_{\Gamma_1}$ satisfying

$$(I + \aleph + \aleph_0)v = \psi - \aleph\varphi.$$

Let the above relation be true. Choosing v in this way and setting $u = v + \varphi$, we will have $(u, \ v) \in \mathcal{D}(I\!N)$, and

$$(I + \mathbb{C}^{-1}I\!N)(u, \ v) = (u - v, \ \aleph u + \aleph_0 v + v) = (\varphi, \ \psi).$$

Now we prove the surjection of the map $I + \aleph + \aleph_0 : H^2_{\Gamma_1}(\Omega) \rightarrow H^{-2}_{\Gamma_1}(\Omega)$. For arbitrary $f \in H^{-2}_{\Gamma_1}(\Omega)$, consider the map $F : H^2_{\Gamma_1}(\Omega) \rightarrow \mathbb{R}$ defined by

$$F(u) = \frac{1}{2}\|u\|^2_{H^2_{\Gamma_1}(\Omega)} + \frac{1}{2}\|u\|^2_{L^2(\Omega)} + \frac{1}{2}\int_\Gamma (u^2 + u_\nu^2) \, d\Gamma - (f, u)_{(H^{-2}_{\Gamma_1}(\Omega), \, H^2_{\Gamma_1}(\Omega))}.$$

It can be verified that F is well-defined, continuously differentiable and that

$$F'(u)v = (u, \ v)_{H^2_{\Gamma_1}(\Omega)} + (u, \ v)_{L^2(\Omega)} + \int_\Gamma uv + u_\nu v_\nu \, d\Gamma - (f, \ v)_{(H^{-2}_{\Gamma_1}(\Omega), \, H^2_{\Gamma_1}(\Omega))}$$
$$= \mathcal{B}(u, \ v) + (u, \ v)_{L^2(\Omega)} + \alpha(u, \ v) - (f, \ v)_{(H^{-2}_{\Gamma_1}(\Omega), \, H^2_{\Gamma_1}(\Omega))}$$
$$= ((I + \aleph + \aleph_0)u - f, \ v)_{(H^{-2}_{\Gamma_1}(\Omega), \, H^2_{\Gamma_1}(\Omega))}.$$

The convexity of F is easily verified. Finally F is coercive: $F(u) \to \infty$ if $\|u\|_{H^2_{\Gamma_1}(\Omega)} \to \infty$. This follows from the inequality

$$F(u) \geq (\frac{1}{2}\|u\|_{H^2_{\Gamma_1}(\Omega)} - \|f\|_{H^{-2}_{\Gamma_1}(\Omega)})\|u\|_{H^2_{\Gamma_1}(\Omega)}.$$

It follows that the infimum of F is attained at some point $u \in H^2_{\Gamma_1}(\Omega)$. Then $F'(u) = 0$ i.e. $(I + \aleph + \aleph_0)u = f$. $\qquad\square$

Boundary Trace In order to save space, we introduce an important result of Theorem 2.3 in [123] without proofs, which allows us to get rid of some geometrical conditions on the shape of the boundary. It reads as follows.

Lemma 3.11 *Let u solve the problem*

$$\begin{cases} u_{tt} + \Delta^2 u + (1-\mu)\operatorname{div}(\kappa\nabla u) = 0 & in \quad (0,\infty) \times \Omega, \\ u = u_\nu = 0 & on \quad (0,\infty) \times \Gamma_1. \end{cases}$$

Let $0 < \alpha < T/2$ be given. Then there is $c_{T\alpha} > 0$ such that

$$\int_{(\alpha,T-\alpha)\times\Gamma_0} |D^2 u|^2 d\Gamma d\tau \leq c_{T\alpha}(\|\Gamma_1(u)\|^2_{L^2(\Sigma_0)} + \|\Gamma_2(u)\|^2_{L^2((0,T);H^{-1}(\Gamma_0))}$$

$$+\|u_t\|^2_{L^2(\Sigma_0)} + \|u_{t\nu}\|^2_{L^2((0,T);H^{-1}(\Gamma_0))} + L(u)) \tag{3.117}$$

where $L(u)$ are some lower order terms relative to the energy $\int_0^T E(t)dt$.

Stabilization of the Closed-Loop System (3.109) Let $u \in H^m(\Omega)$ be given. We say that $L(u)$ is a lower order term related to the norm of u in $H^m(\Omega)$ if for any $\varepsilon > 0$ there is $C_\varepsilon > 0$ such that

$$|L(u)| \leq \varepsilon\|u\|^2_{H^m(\Omega)} + C_\varepsilon\|u\|^2.$$

We need

Lemma 3.12 *Let H be a vector field satisfying the condition (3.24). Then*

$$2\int_\Omega a(u, H(u))dx = \int_\Gamma a(u,u)\langle H,\nu\rangle d\Gamma + 2\int_\Omega \vartheta a(u,u)dx + L(u) \tag{3.118}$$

where $L(u)$ denotes some lower order terms relative to u in the norm of $H^2(\Omega)$.

Proof. Let $x \in \Omega$ be given. Let E_1, E_2 be a frame field normal at x. We compute at x, via the identity (1.22),

$$D^2(H(u))(E_i, E_j) = E_j E_i(Du(H)) = E_j(D^2 u(H, E_i) + Du(D_{E_i}H))$$
$$= D^3 u(E_i, H, E_j) + D^2 u(D_{E_j}H, E_i) + D^2 u(D_{E_i}H, E_j) + Du(D_{E_j}D_{E_i}H)$$
$$= H(D^2 u(E_i, E_j)) + D^2 u(D_{E_j}H, E_i) + D^2 u(D_{E_i}H, E_j) + \text{lo}(u), \tag{3.119}$$

where lo (u) denotes some terms of order ≤ 1. Since $D^2 u(\cdot, \cdot)$ is symmetrical, it follows from the formulas (3.119), (3.27) and (3.28) that

$$2\langle D^2 u, D^2(H(u))\rangle = H(|D^2 u|^2) + 4\vartheta|D^2 u|^2 + \text{lo}\,(u) \tag{3.120}$$

and

$$2\,\text{tr}\,D^2 u\,\text{tr}\,D^2(H(u)) = H((\text{tr}\,D^2 u)^2) + 4\vartheta(\text{tr}\,D^2 u)^2 + \text{lo}\,(u). \tag{3.121}$$

Noting that the relation $H(a(u,u)) = \text{div}\,a(u,u)H - 2\vartheta a(u,u)$, we obtain the identity (3.118) from the formulas (3.120) and (3.121). $\qquad\square$

Lemma 3.13 *Let H be a vector field satisfying the condition* (3.24). *Let u solve the problem*

$$u_{tt} + Au = 0 \quad in \quad (0,\infty) \times \Omega \tag{3.122}$$

where Au is given by (3.102). *Then*

$$2\int_Q \vartheta[u_t^2 + a(u,u)]dQ = \int_\Sigma \text{SB}\,d\Sigma - 2(u_t, H(u))|_0^T + L(u) \tag{3.123}$$

where

$$\text{SB} = [u_t^2 - a(u,u)]\langle H, \nu\rangle + 2\Gamma_1(u)(H(u))_\nu - 2\Gamma_2(u)H(u) \tag{3.124}$$

and $L(u)$ denotes some lower order terms related to the energy $\int_0^T E(\tau)d\tau$.

Proof. Multiplying by $2H(u)$ the equation (3.122) and integrating over $Q = (0,T) \times \Omega$, we obtain

$$2\int_Q u_{tt}H(u)dQ = 2(u_t, H(u))|_0^T + 2\int_Q \vartheta u_t^2 dQ - \int_\Sigma u_t^2\langle H, \nu\rangle d\Sigma. \tag{3.125}$$

On the other hand, by using Lemmas 3.10 and 3.12, we have

$$2\int_Q AuH(u)dQ = \int_\Sigma a(u,u)\langle H, \nu\rangle d\Sigma + 2\int_Q \vartheta a(u,u)dQ$$

$$+2\int_\Sigma [\Gamma_2(u)H(u) - \Gamma_1(u)(H(u))_\nu]d\Sigma + L(u). \tag{3.126}$$

The identity (3.123) follows from the identities (3.125) and (3.126). $\qquad\square$

We have

Theorem 3.13 *Assume that the conditions* (3.111) *holds. Let H be an escape vector field on $\overline{\Omega}$ for plate such that*

$$\langle H, \nu\rangle \leq 0 \quad on \quad \Gamma_1. \tag{3.127}$$

Then there are constants $c_1, c_2 > 0$ such that

$$E(t) \leq c_1 e^{-c_2 t} E(0) \quad for \quad t \geq 0 \tag{3.128}$$

for all solutions u to the system (3.109).

Proof. To get the uniform stabilization, it will suffice to prove that there are a time T and a constant $c_T > 0$, which are independent of solution u such that

$$E(T) \leq c_T \int_{\Sigma_0} (|u_t|^2 + |u_{t\nu}|^2) d\Sigma. \tag{3.129}$$

Indeed, if the above inequality is true, then the identity (3.108) implies $E(T) \leq c_T(E(0) - E(T))$, that is,

$$E(T) \leq \frac{c_T}{1 + c_T} E(0).$$

We have the uniform stabilization.

In the following, we prove the inequality (3.129).

Step 1 On Γ_1, the boundary conditions $u = u_\nu = 0$ imply that

$$Du = 0, \quad D^2 u(\tau, \tau) = D^2 u(\tau, \nu) = 0. \tag{3.130}$$

It follows from (3.130) that

$$(H(u))_\nu = D^2 u(\nu, \nu)\langle H, \nu \rangle, \quad B_1 u = B_2 u = 0, \quad \text{for} \quad x \in \Gamma_1, \tag{3.131}$$

$$a(u, u) = [D^2 u(\nu, \nu)]^2 = (\Delta u)^2 \quad \text{for} \quad x \in \Gamma_1. \tag{3.132}$$

Let SB be given by (3.124). Then the condition (3.127) and the relations (3.130)-(3.132) yield

$$\text{SB} = (\Delta u)^2 \langle H, \nu \rangle \leq 0 \quad \text{on} \quad \Gamma_1. \tag{3.133}$$

Next, consider SB on Γ_0. Clearly, $|(H(u))_\nu| \leq c(|D^2 u| + |Du|)$ and $|H(u)| \leq c|Du|$ for $x \in \Gamma_0$. From the formula (3.124), we obtain

$$\text{SB} \leq c(u_t^2 + |D^2 u|^2 + |\Gamma_1(u)|^2 + |\Gamma_2(u)|^2) + L(u) \quad \text{on} \quad \Gamma_0. \tag{3.134}$$

Step 2 Let $0 < \alpha < T/2$ be given. Change the integral domain $(0, T)$ into $(\alpha, T - \alpha)$ in the identity (3.123) with respect to time variable and use the inequalities (3.133), (3.134), and (3.117) to give

$$2\vartheta_0 \int_\alpha^{T-\alpha} E(\tau) d\tau \leq c \int_\alpha^{T-\alpha} \int_{\Gamma_0} (u_t^2 + u_{t\nu}^2 + |D^2 u|^2) d\Sigma$$
$$+ c[E(\alpha) + E(T - \alpha)] + L(u)$$
$$\leq c_T \int_{\Sigma_0} (u_t^2 + u_{t\nu}^2) d\Sigma + c[E(\alpha) + E(T - \alpha)] + L(u) \tag{3.135}$$

where $\vartheta_0 = \min_{x \in \Omega} \vartheta$.

On the other hand, using the identity (3.108), we have

$$\int_\alpha^{T-\alpha} E(t) dt \geq (T - 2\alpha)E(T) - T \int_{\Sigma_0} (u_t^2 + u_{t\nu}^2) d\Sigma, \tag{3.136}$$

$$E(\alpha) \le \frac{1}{T - 2\alpha} \int_\alpha^{T-\alpha} E(t)dt + \int_{\Sigma_0} (u_t^2 + u_{t\nu}^2)d\Sigma. \tag{3.137}$$

Noting that $E(T - \alpha) \le E(\alpha)$, inserting (3.136) and (3.137) into (3.135) yields

$$E(T) \le c_T \int_{\Sigma_0} (u_t^2 + u_{t\nu}^2)d\Sigma + L(u) \tag{3.138}$$

when T is large enough.

The inequality (3.129) follows from the inequality (3.138) by the uniqueness argument as in Lemma 2.5, see Exercise 3.4.   □

Exercises

3.1 *Prove the formula* (3.78).

3.2 *Let the operator $\mathbb{A}$ be given by* (3.82). *Prove that $\mathbb{A}$ is skew-adjoint on the space $H_0^2(\Omega) \times L^2(\Omega)$ where the inner product on $H_0^2(\Omega)$ is given by* (3.83).

3.3 *Let conditions* (3.111) *hold true. Then*

$$\mathcal{B}^{1/2}(u, u) = \left(\int_\Omega a(u, u)dx \right)^{1/2}$$

defines an equivalent norm of $H_{\Gamma_1}^2(\Omega)$.

3.4 *Prove that the inequality* (3.138) *implies that the problem*

$$\begin{cases} u_{tt} + \Delta^2 u + (1 - \mu)\operatorname{div}(\kappa \nabla u) = 0 & in \quad (0, T) \times \Omega, \\ u = u_\nu = \Delta u = (\Delta u)_\nu = 0 & on \quad \Gamma_0 \end{cases}$$

has a finite dimensional space Ξ of solutions. Then it follows from Lemma 3.3 that $\Xi = \emptyset$.

3.6   Notes and References

Sections 3.1, 3.2, and 3.3 are from [209]; Section 3.4 is from [78]; Section 3.5 is from [77].

For the control problems of the Euler-Bernoulli plate with constant coefficients, we refer to [9], [10], [89], [99], [127], [120], [146], and many other authors.

[209] was the first paper to introduce the Riemannian geometrical approach to establish observability estimates for the Euler-Bernoulli plate with variable coefficients. The content of this chapter again shows the virtues of Riemannian geometry: The Bochner technique was used to simplify computation to obtain multiplier identities (Sections 3.1) and plate models (Section 3.5); the curvature theory provides us the global information on the existence of an escape vector field for plate which guarantees the exact controllability and the stabilization. An excellent survey in this direction is included in [75].

Recently, some further results in this direction have been obtained but not included in this chapter: [28] studied thermoelastic plates with variable coefficients; [81] studied the stabilization of elastic plates with dynamical boundary controls; [131] extended Carleman estimates to the Euler-Bernoulli plate with variable coefficients and lower order terms.

Behavior of the Euler-Bernoulli plate varies with its boundary conditions. For instance, an escape vector field for the metric is enough to guarantee controllability/stabilization for the simply supported boundary conditions. But the fixed (or free) boundary conditions need something more, that is, an escape vector field for plate. Normally those assumptions are locally true, that is, when the middle surface of the plate is small in some sense. However, as the author knows, only the sectional curvature theory of Riemannian geometry has provided a useful tool to yield the global information on those assumptions in the case of variable coefficients.

[179], based on the operator theory, established the exponential decay of the energy in the simply supported boundary conditions for the Euler-Bernoulli plate with variable coefficients. The stabilization here came down to verifying an assumption given by [207] for the wave equation. This assumption was not checkable until [64] proved that it is equivalent to an escape vector field for the metric. Thus its checkability is due to the curvature theory again.

Chapter 4

Linear Shallow Shells. Modeling, and Control

We view the middle surface of a shallow shell as a Riemannian manifold of dimension 2 to derive its mathematical models in the form of coordinates free with the help of the Bochner technique. Then analysis of controllability/stabilization is carried out.

4.1 Equations in Equilibrium. Green's Formulas

We keep all the notations as in Chapter 1. Let M be a surface in $\mathbb{R}^3$ with the normal field N. Suppose that g is the induced metric of the surface M from the standard metric of $\mathbb{R}^3$. We work on the standard Riemannian manifold (M, g) for the shell problem.

Let us assume that the middle surface of a shell occupies a bounded region Ω of the surface M. The shell, a body in $\mathbb{R}^3$, is defined by

$$S = \{ \, p \mid p = x + zN(x), \; x \in \Omega, \; -h/2 < z < h/2 \, \} \tag{4.1}$$

where h is the thickness of the shell, small.

Let the shell occupy a region $\mathcal{F}(S)$ in $\mathbb{R}^3$ after deformation, where

$$\mathcal{F}: \; S \to \mathbb{R}^3$$

is a differentiable mapping. After deformation, the middle surface becomes another surface, given by

$$\mathcal{F}(\Omega) = \Big\{ \mathcal{F}(x) \mid x \in \Omega \Big\}. \tag{4.2}$$

Suppose that the deformation of a shell obeys the basic kinematical assumption of the classical Kirchhoff-Love theory: *The normals to the undeformed middle surface move to the normals of the deformed middle surface without any change in length.*

Under the above Kirchhoff-Love assumption, the modeling of a shell is to find out the middle surface $\mathcal{F}(\Omega)$ of the deformed shell by physics laws where the middle surface Ω of the undeformed shell is given. To this end, by Theorem 1.15, it suffices to work out the first and the second fundamental forms of the surface $\mathcal{F}(\Omega)$. They are determined, respectively, by the strain tensor and the curvature tensor below.

The strain tensor of the middle surface is defined by

$$\Upsilon = \frac{1}{2}(\bar{g} - g), \tag{4.3}$$

where $\bar{g}$ is the induced metric of $F(\Omega)$ in $\mathbb{R}^3$. The change of curvature tensor of the middle surface is defined by

$$\rho = \overline{\Pi} - \Pi, \tag{4.4}$$

where Π and $\overline{\Pi}$ are the second fundamental forms of the surfaces M and $F(\Omega)$, respectively.

We recall that the classical Kirchhoff-Love assumption is contradicted by the statical assumption of approximate plane stress. For an isotropic material both assumptions agree in the statement that $\Upsilon(X, N) = 0$ for $x \in \Omega$ and $X \in \mathcal{X}(\Omega)$, but they disagree in their predictions of the direct normal strain component $\Upsilon(N, N)$. The assumption of plane stress implies that for a material obeying Hooke's law

$$\Upsilon(N, N) = -\frac{\mu}{1 - \mu} \operatorname{tr} \Upsilon \quad \text{for} \quad x \in \Omega$$

where μ is Poisson's coefficient, but the length of the normal keeping unchanged forces $\Upsilon(N, N) = 0$. However, we keep the classical Kirchhoff-Love assumption in this chapter for the shallow shell. We will use a complementary hypothesis of Koiter in Chapter 6 for the Koiter model.

Let $\zeta(x) = (\zeta_1, \zeta_2, \zeta_3)$ denote the displacement vector which carries a generic material point $x = (x_1, x_2, x_3)$ on the undeformed middle surface Ω to a new position in space in a deformed configuration $\mathcal{F}(\Omega)$ of the middle surface. The new position is thus given by

$$\mathcal{F}(x) = x + \zeta(x) \quad \forall\, x \in \Omega. \tag{4.5}$$

We decompose the displacement vector ζ into the sum

$$\zeta(x) = W(x) + w(x)N(x) \quad x \in \Omega,\ W(x) \in M_x, \tag{4.6}$$

where W and w are components of ζ on the tangent plane and on the normal of the undeformed middle surface Ω, respectively. Then $W \in \mathcal{X}(\Omega)$ is a vector field and w is a function on Ω, respectively.

Given $x \in \Omega$. Let E_1, E_2 be a frame field normal at x. Then E_1, E_2, N form a frame field in $\mathbb{R}^3$ at x. By (4.2), we have

$$\overline{g}(E_i, E_j) = \langle \mathcal{F}_* E_i, \mathcal{F}_* E_j \rangle \quad \text{for} \quad 1 \le i, \ j \le 2, \tag{4.7}$$

where $\mathcal{F}_*$ is the differential of the deformation mapping $\mathcal{F}$.

The second fundamental form Π of the surface M is defined by

$$\Pi(X,Y) = \langle \tilde{D}_X N, Y \rangle, \quad \forall \ X, \ Y \in \mathcal{X}(M), \tag{4.8}$$

where $\tilde{D}$ is the covariant differential of $\mathbb{R}^3$ in the Euclidean metric.

For any vector fields X and Y on M, since the manifold M is a submanifold of $\mathbb{R}^3$, we obtain from the formula (4.8)

$$\tilde{D}_X Y = D_X Y + \langle \tilde{D}_X Y, \ N \rangle N = D_X Y - \Pi(X,Y)N \tag{4.9}$$

where D is the covariant differential of the surface M in the induced metric g. It follows from (4.6) and (4.9) that

$$\begin{aligned} \tilde{D}_{E_i}\zeta &= \tilde{D}_{E_i}W + E_i(w)N + w\tilde{D}_{E_i}N \\ &= D_{E_i}W + \Big[E_i(w) - \Pi(E_i, W)\Big]N + w\tilde{D}_{E_i}N. \end{aligned} \tag{4.10}$$

We therefore obtain, by the formulas (4.5), (4.7), and (4.10),

$$\begin{aligned} \overline{g}(E_i, E_j) &= \langle E_i + \tilde{D}_{E_i}\zeta, \ E_j + \tilde{D}_{E_j}\zeta \rangle \\ &= \delta_{ij} + \langle E_i, \tilde{D}_{E_j}\zeta \rangle + \langle E_j, \tilde{D}_{E_i}\zeta \rangle + \langle \tilde{D}_{E_i}\zeta, \tilde{D}_{E_j}\zeta \rangle \tag{4.11} \\ &= g(E_i, E_j) + DW(E_i, E_j) + DW(E_j, E_i) \\ &\quad + 2w\Pi(E_i, E_j) + \langle \tilde{D}_{E_i}\zeta, \tilde{D}_{E_j}\zeta \rangle \tag{4.12} \end{aligned}$$

for $i = 1$, 2, since Π is symmetric.

The relations (4.3) and (4.12) then yield the following displacement-strain relation of the middle surface

$$\Upsilon(\zeta) = \frac{1}{2}(DW + D^*W) + w\Pi + T(\zeta), \tag{4.13}$$

where $T(\zeta)$ is a tensor field of rank 2 on Ω, defined by

$$T(\zeta)(X,Y) = \frac{1}{2}\langle \tilde{D}_X \zeta, \tilde{D}_Y \zeta \rangle \quad \text{for} \quad X, Y \in \mathcal{X}(\Omega). \tag{4.14}$$

After linearization with respect to the displacement, the formula (4.13) yields

$$\Upsilon(\zeta) = \frac{1}{2}(DW + D^*W) + w\Pi. \tag{4.15}$$

In addition, we define the change of curvature tensor of the middle surface Ω as

$$\rho(\zeta) = -D^2 w \tag{4.16}$$

in a form where $D^2 w$ is the Hessian of w. The choice of the tensor field (4.16) is justified for a *shallow shell*, see [94], p. 27, or [165], p. 355. Note that the formulas (4.15) and (4.16) are in a form coordinates free.

Let the material of the shell be isotropic and obey Hooke's law. The shell strain energy associated to a displacement vector field ζ of the middle surface Ω can be written as

$$\mathcal{B}_1(\zeta, \zeta) = hb \int_\Omega B(\zeta, \zeta)\, dx, \tag{4.17}$$

where

$$B(\zeta, \zeta) = a(\Upsilon(\zeta), \Upsilon(\zeta)) + \gamma a(\rho(\zeta), \rho(\zeta)), \tag{4.18}$$

$$b = E/(1 - \mu^2), \quad \gamma = h^2/12, \tag{4.19}$$

$$a(\Upsilon(\zeta), \Upsilon(\zeta)) = (1 - \mu)|\Upsilon(\zeta)|^2 + \mu(\operatorname{tr}\Upsilon(\zeta))^2 \tag{4.20}$$

for $x \in \Omega$, where E, μ respectively denote Young's modulus and Poisson's coefficient of the material. In the formula (4.20) the tensor product $\langle \Upsilon(\zeta), \Upsilon(\zeta) \rangle$ and the trace tr Υ are defined in Chapter 1 by (1.103) and (1.18), respectively, for each $x \in \Omega$.

The expression (4.17) is an *approximation* to the shell strain energy. Its derivation from the three-dimensional elasticity theory is carried out by integration on the thickness of the shell, following the methods of asymptotic expansions. For those methods, we refer to [95], [158], or [183].

Traditionally, the formula (4.20) is expressed in a coordinate form. Lemma 4.1 below shows that they are the same. Let $\mathbf{a}_\alpha \in \mathcal{X}(\Omega)$ be a vector field basis on Ω where $\alpha = 1, 2$. Then

$$\mathbf{a}_{\alpha\beta} = \mathbf{a}_\alpha^\tau \cdot \mathbf{a}_\beta = \langle \mathbf{a}_\alpha, \mathbf{a}_\beta \rangle = g(\alpha, \beta)$$

is the first fundamental form of the middle surface Ω and

$$\gamma_{\alpha\beta} = \Upsilon(\mathbf{a}_\alpha, \mathbf{a}_\beta)$$

is in the traditional symbols. Let

$$\left(\mathbf{a}^{\alpha\beta} \right) = \left(\mathbf{a}_{\alpha\beta} \right)^{-1}$$

and let

$$\gamma_\beta^\alpha = \sum_{\lambda=1}^{2} \mathbf{a}^{\alpha\lambda} \gamma_{\lambda\beta}.$$

We have

Lemma 4.1

$$\sum_{\alpha\beta=1}^{2} \gamma_\beta^\alpha \gamma_\alpha^\beta = \langle \Upsilon, \Upsilon \rangle \quad for \quad x \in \Omega;$$

$$\sum_{\alpha=1}^{2} \gamma_\alpha^\alpha = \operatorname{tr} \Upsilon \quad for \quad x \in \Omega.$$

Proof. Let $x \in \Omega$ be fixed. Let E_1, E_2 be an orthonormal basis of M_x. Let

$$C = \left(C_{ij} \right) \quad \text{where } C_{\alpha\beta} \text{ is such that} \quad \mathbf{a}_\alpha = \sum_i C_{i\alpha} E_i \quad \text{at} \quad x.$$

Then

$$\left(\mathbf{a}_{\alpha\beta} \right) = C^\tau C, \quad \left(\mathbf{a}^{\alpha\beta} \right) = C^{-1} C^{-\tau}.$$

We obtain

$$\begin{aligned}
\sum \gamma_\beta^\alpha \gamma_\alpha^\beta &= \sum (\mathbf{a}^{\alpha\lambda} C_{j_1\alpha})(\mathbf{a}^{\beta\lambda_1} C_{j\beta}) C_{i\lambda} C_{i_1\lambda_1} \Upsilon(E_i, E_j) \Upsilon(E_{i_1}, E_{j_1}) \\
&= \sum c^{\lambda j_1} c^{\lambda_1 j} C_{i\lambda} C_{i_1\lambda_1} \Upsilon(E_i, E_j) \Upsilon(E_{i_1}, E_{j_1}) \\
&= \sum \delta_{ij_1} \delta_{i_1 j} \Upsilon(E_i, E_j) \Upsilon(E_{i_1}, E_{j_1}) \\
&= \sum \left[\Upsilon(E_i, E_j) \right]^2 = \langle \Upsilon, \Upsilon \rangle \quad \text{at} \quad x
\end{aligned}$$

since the tensor field Υ is symmetric, where $\left(c^{\alpha\beta} \right) = \left(C_{\alpha\beta} \right)^{-1}$.

Similarly, it follows that

$$\sum_{\alpha=1}^{2} \gamma_\alpha^\alpha = \sum \mathbf{a}^{\alpha\lambda} C_{j\alpha} C_{i\lambda} \Upsilon(E_i, E_j) = \sum \Upsilon(E_i, E_i) = \operatorname{tr} \Upsilon$$

at x. □

Now, with the expression (4.17), we are able to associate the following symmetric bilinear form, directly defined on the middle surface Ω:

$$\mathcal{B}(\zeta, \eta) = \int_\Omega B(\zeta, \eta) \, dx, \tag{4.21}$$

where ζ is given by the formula (4.6) and

$$\eta = U + uN \quad \text{for} \quad U(x) \in M_x, \ x \in \Omega. \tag{4.22}$$

Next, we consider the expression of the work done by the external load. For simplification, we assume that the external load P is given by

$$P(p) = F(x) + f(x)N(x) \quad \text{for} \quad p = x + zN(x) \in \mathbf{S}, \quad x \in \Omega \tag{4.23}$$

where $F \in \mathcal{X}(\Omega)$ and f are a vector field and a function on Ω, respectively. Then, the work done by the external load associated with a displacement vector η of the middle surface can be written as

$$L(\eta) = \int_S [\langle F, U \rangle + fu] dv.$$

By the formula (4.8), for each $x \in \Omega$, there is a symmetric matrix $A(x)$: $M_x \to M_x$ such that

$$\Pi(X, Y) = \langle A(x)X, Y \rangle \quad \text{for all} \quad X, Y \in \Omega_x.$$

Let λ_1 and λ_2 be the two eigenvalues of the matrix $A(x)$. Then

$$\operatorname{tr} \Pi = \lambda_1 + \lambda_2 \quad \text{and} \quad \kappa = \lambda_1 \lambda_2 \tag{4.24}$$

are the mean curvature and the Gaussian curvature, respectively. The smallest principal radius of the curvature of the undeformed middle surface is given by

$$\mathbf{R} = \begin{cases} 1/\max\{|\lambda_1|, |\lambda_2|\} & \lambda_1^2 + \lambda_2^2 \neq 0, \\ \infty & \lambda_1 = \lambda_2 = 0, \end{cases} \quad \text{for} \quad x \in \Omega. \tag{4.25}$$

The following lemma is ready.

Lemma 4.2

$$|\kappa| \leq \frac{1}{\mathbf{R}^2}, \quad |\operatorname{tr} \Pi| \leq \frac{2}{\mathbf{R}}, \quad \text{and} \quad |\Pi| \leq \frac{\sqrt{2}}{\mathbf{R}} \tag{4.26}$$

for all $x \in \Omega$.

We have

$$L(\eta) = \int_{-h/2}^{h/2} \int_\Omega [\langle F, U \rangle + fu](1 + \operatorname{tr} \Pi z + \kappa z^2) \, dx dz. \tag{4.27}$$

For a thin shell, it is assumed that

$$h/\mathbf{R} \ll 1, \tag{4.28}$$

see [94], p.18. By the formula (4.27), $L(\eta)$ can further be simplified as

$$L(\eta) = h \int_\Omega [\langle F, U \rangle + fu] \, dx \tag{4.29}$$

since $|\operatorname{tr} \Pi z| \leq h/\mathbf{R}$ and $|\kappa z^2| \leq h^2/(4\mathbf{R}^2)$.

Let the boundary Γ of the middle surface Ω be nonempty. We assume that the shell is clamped along a portion Γ_0 of Γ and free on Γ_1, where $\Gamma_0 \cup \Gamma_1 = \Gamma$. Set

$$H^1_{\Gamma_0}(\Omega, \Lambda) = \{ W \mid W \in H^1(\Omega, \Lambda), \ W|_{\Gamma_0} = 0 \},$$

$$H_{\Gamma_0}^2(\Omega) = \{\, w \,|\, w \in H^2(\Omega),\ w|_{\Gamma_0} = \frac{\partial w}{\partial \nu}|_{\Gamma_0} = 0 \,\},$$

where $H^1(\Omega, \Lambda)$ is the Sobolev space defined in Section 1.4 of Chapter 1. Reset ζ, η, and P as

$$\zeta = (W, w), \quad \eta = (U, u), \quad \text{and } P = (F, f),$$

respectively. We then derive the following variational problem:

For $P \in L^2(\Omega, \Lambda) \times L^2(\Omega)$ find $\zeta \in H_{\Gamma_0}^1(\Omega, \Lambda) \times H_{\Gamma_0}^2(\Omega)$ such that

$$\mathcal{B}_1(\zeta, \eta) = L(\eta), \quad \forall\, \eta \in H_{\Gamma_0}^1(\Omega, \Lambda) \times H_{\Gamma_0}^2(\Omega). \tag{4.30}$$

We recall some notation in Chapter 1. Let $T^2(\Omega)$ be all tensor fields of rank 2 on Ω and let $\Lambda(\Omega)$ be all 1-forms on Ω. Let $\mathcal{X}(\Omega)$ be all vector fields on Ω. Let $X \in \mathcal{X}(\Omega)$. Then $X \in \Lambda(\Omega)$ in the following sense:

$$X(Y) = \langle X, Y \rangle \quad \text{for all} \quad Y \in \mathcal{X}(\Omega).$$

In the above sense $\mathcal{X}(\Omega) = \Lambda(\Omega)$. Let f be a function. Then $\nabla f = df$ where ∇f and df denote the gradient of f and the covariant differential of f (which is a 1-form), respectively.

We need the following:

Proposition 4.1 *We have*

$$\mathcal{Q}\Pi = d\operatorname{tr}\Pi, \tag{4.31}$$

where $\mathcal{Q}\colon T^2(\Omega) \to \mathcal{X}(\Omega)$ is an operator defined by the formula (1.144) of Chapter 1.

Proof. By Corollary 1.1, $D\Pi$ is symmetric, that is,

$$D\Pi(X, Y, Z) = D\Pi(Y, X, Z) = D\Pi(Y, Z, X) \tag{4.32}$$

for X, Y, $Z \in \mathcal{X}(\Omega)$. Given $x \in \Omega$. Let E_1, E_2 be a frame field normal at x. By the formulas (1.144) and (4.32), we have

$$\langle \mathcal{Q}\Pi, E_i \rangle = \operatorname{tr} l_{E_i} D\Pi = \sum_{j=1}^{2} D\Pi(E_i, E_j, E_j) = \sum_{j=1}^{2} D\Pi(E_j, E_j, E_i)$$

$$= \sum_{j=1}^{2} E_i(\Pi(E_j, E_j)) = E_i(\operatorname{tr}\Pi) \quad \text{at } x \tag{4.33}$$

since $D_{E_i} E_j = 0$ at x for $i = 1$, 2.

The formula (4.31) follows from (4.33). $\qquad\square$

The following formula is the Green formula which expresses the relationship between the boundary and the interior of a displacement vector field. This formula plays an important role in the boundary-valued problems.

Theorem 4.1 *Let the bilinear form* $\mathcal{B}(\cdot, \cdot)$ *be given in (4.21). For* $\zeta = (W, w)$, $\eta = (U, u) \in H^1(\Omega, \Lambda) \times H^2(\Omega)$, *we have*

$$\mathcal{B}(\zeta, \eta) = (\mathcal{A}\zeta, \eta)_{L^2(\Omega,\Lambda) \times L^2(\Omega)} + \int_\Gamma \partial(\mathcal{A}\zeta, \eta) \, d\Gamma \tag{4.34}$$

where

$$\begin{aligned}
\partial(\mathcal{A}\zeta, \eta) = \; & B_1(W, w)\langle U, \nu \rangle + B_2(W, w)\langle U, \tau \rangle \\
& + \gamma\Big(\Delta w + (1 - \mu)B_3 w\Big)\frac{\partial u}{\partial \nu} \\
& - \gamma\Big(\frac{\partial \Delta w}{\partial \nu} + (1 - \mu)B_4 w\Big)u,
\end{aligned} \tag{4.35}$$

ν, τ *are the normal and the tangential along curve* Γ, *respectively,*

$$\mathcal{A}\zeta = \begin{pmatrix} -\boldsymbol{\Delta}_\mu W - (1-\mu)\kappa W - \mathcal{H}(w) \\ \gamma[\Delta^2 w - (1-\mu)\delta(\kappa dw)] + (\mathrm{tr}^2 \Pi - 2(1-\mu)\kappa)w + \mathcal{G}(W) \end{pmatrix}, \tag{4.36}$$

$\boldsymbol{\Delta}_\mu$ *is of the Hodge-Laplacian type, applied to 1-forms (or equivalently vector fields), defined by*

$$\boldsymbol{\Delta}_\mu = -\left(\frac{1-\mu}{2}\delta d + d\delta\right), \tag{4.37}$$

d *the exterior differential,* δ *the formal adjoint of* d, Δ *the Laplacian on manifold* M,

$$\begin{cases} \mathcal{H}(w) = (1-\mu)\,\mathrm{i}\,(dw)\Pi + \mu \,\mathrm{tr}\,\Pi dw + w d\,\mathrm{tr}\,\Pi, \\ \mathcal{G}(W) = (1-\mu)\langle DW, \Pi \rangle - \mu \,\mathrm{tr}\,\Pi \delta W, \end{cases} \tag{4.38}$$

$$\begin{cases} B_1(W, w) = (1-\mu)\Upsilon(\zeta)(\nu, \nu) + \mu(wH - \delta W), \\ B_2(W, w) = (1-\mu)\Upsilon(\zeta)(\nu, \tau), \\ B_3 w = -D^2 w(\tau, \tau), \\ B_4 w = \dfrac{\partial}{\partial \tau}(D^2 w(\tau, \nu)) + \kappa(x)\dfrac{\partial w}{\partial \nu}, \end{cases} \tag{4.39}$$

$\mathrm{i}\,(dw)\Pi$ *is the interior product of* Π *by* dw, *and* Π *and* κ *are the second fundamental form and the Gaussian curvature of the surface* M, *respectively.*

Proof. Let $\mathcal{Q}$ be defined by (1.144). Let $x \in \Omega$ be given. Let E_1, E_2 be a frame field normal at x, that is,

$$D_{E_i} E_j = 0 \quad \text{at} \quad x \tag{4.40}$$

for $i = 1, 2$. By the formulas (1.144) and (4.40)

$$\begin{aligned}
\langle \mathcal{Q}(w\Pi), E_k \rangle &= D(w\Pi)(E_k, E_1, E_1) + D(w\Pi)(E_k, E_2, E_2) \\
&= E_1\Big(w\Pi(E_k, E_1)\Big) + E_2\Big(w\Pi(E_k, E_2)\Big) \\
&= \Pi(E_k, \nabla w) + w \,\mathrm{tr}\,\mathrm{i}\,(E_k)D\Pi \quad \text{at} \quad x
\end{aligned}$$

which yields, via Proposition 4.1,

$$\mathcal{Q}(w\Pi) = \mathrm{i}\,(dw)\Pi + wd\,\mathrm{tr}\,\Pi \tag{4.41}$$

since $Dw = dw$. In addition, by (4.24), we have

$$|\Pi|^2 = \overset{2}{\mathrm{tr}}\,\Pi - 2\kappa, \quad x \in M. \tag{4.42}$$

Using the formula (4.15), we obatin

$$\langle \Upsilon(\varsigma),\ \Upsilon(\eta)\rangle = \frac{1}{2}\langle DW,\ DU + D^*U\rangle + \langle w\Pi,\ DU\rangle$$
$$+((\langle \Pi, DW\rangle + w|\Pi|^2)u. \tag{4.43}$$

We integrate the identity (4.43) over Ω. It follows from the relations (1.155), (1.156), (3.100), (1.145), (4.41), and (4.42) that

$$\int_\Omega \langle \Upsilon(\varsigma),\ \Upsilon(\eta)\rangle\,dx$$

$$= \frac{1}{2}(DW,\ DU + D^*U)_{L^2(\Omega,T^2)} + (w\Pi,\ DU)_{L^2(\Omega,T^2)}$$

$$+((\langle \Pi, DW\rangle + w|\Pi|^2,\ u)$$

$$= \frac{1}{2}(\mathbf{\Delta}W + d\delta W - 2\kappa W,\ U)_{L^2(\Omega,\Lambda)} + \frac{1}{2}\int_\Gamma [\langle D_\nu W, U\rangle + \langle D_U W, \nu\rangle]\,d\Gamma$$

$$+(-\mathcal{Q}(w\Pi), U)_{L^2(\Omega,\Lambda)} + \int_\Gamma \langle \mathrm{i}\,(\nu)(w\Pi), U\rangle\,d\Gamma + ((\langle \Pi, DW\rangle + w|\Pi|^2,\ u)$$

$$= ((\frac{1}{2}\delta d + d\delta)W - \kappa W - \mathrm{i}\,(dw)\Pi - wd\,\mathrm{tr}\,\Pi,\ U)_{L^2(\Omega,\Lambda)}$$

$$+((\langle \Pi, DW\rangle + w(\overset{2}{\mathrm{tr}}\,\Pi - 2\kappa),\ u) + \int_\Gamma \Upsilon(\varsigma)(\nu, U)\,d\Gamma \tag{4.44}$$

where the following formula is used

$$\Upsilon(\nu,U) = \frac{1}{2}\Big[DW(\nu,U) + DW(U,\nu)\Big] + w\Pi(\nu,U)$$
$$= \frac{1}{2}\Big[\langle D_U W, \nu\rangle + \langle D_\nu W, U\rangle\Big] + \langle \mathrm{i}\,(\nu)(w\Pi), U\rangle.$$

On the other hand, the formulas (1.111) and (1.100) yield

$$\mathrm{tr}\,DW = -\delta W. \tag{4.45}$$

By the equations (4.15), (4.45), and (1.137), we obtain the following

$$\int_\Omega \mathrm{tr}\,\Upsilon(\varsigma)\ \mathrm{tr}\,\Upsilon(\eta)\,dx = \int_\Omega (-\delta W + w\,\mathrm{tr}\,\Pi)(-\delta U + u\,\mathrm{tr}\,\Pi)\,dx$$

$$= (\delta W - w\,\mathrm{tr}\,\Pi,\ \delta U) + (-\mathrm{tr}\,\Pi\delta W + w\overset{2}{\mathrm{tr}}\,\Pi,\ u)$$

$$= (d\delta W - d(w\,\mathrm{tr}\,\Pi),\ U)_{L^2(\Omega,\Lambda)} + (-\mathrm{tr}\,\Pi\delta W + w\overset{2}{\mathrm{tr}}\,\Pi,\ u)$$

$$+ \int_\Gamma \langle U, \nu\rangle\,\mathrm{tr}\,\Upsilon(\varsigma)\,d\Gamma. \tag{4.46}$$

Using the formulas (4.46), (4.44), (4.18), and (4.21), we have

$$
\int_\Omega a(\Upsilon(\zeta),\ \Upsilon(\eta))\,dx
$$
$$
= -(\mathbf{\Delta}_\mu W + (1-\mu)\kappa W + \mathcal{H}(w),\ U)_{L^2(\Omega,\Lambda)}
$$
$$
+(\mathcal{G}(W) + (\overset{2}{\operatorname{tr}}\Pi - 2(1-\mu)\kappa)w,\ u)
$$
$$
+ \int_\Gamma [B_1(W,w)\langle U,\nu\rangle + B_2(W,w)\langle U,\tau\rangle]\,d\Gamma \tag{4.47}
$$

where the following formula is used

$$
\Upsilon(\nu,U) = \Upsilon(\nu,\nu)\langle U,\nu\rangle + \Upsilon(\nu,\tau)\langle U,\tau\rangle \quad \text{for} \quad x \in \Gamma.
$$

Next, we compute the integration $\int_\Omega a(\rho(\zeta),\ \rho(\eta))\,dx$.
Since $\delta w = 0$, the relations $\mathbf{\Delta} = d\delta + \delta d$ and $d^2 = 0$ imply

$$
\delta\mathbf{\Delta}dw = \delta d\delta dw = \Delta^2 w. \tag{4.48}
$$

Noting that $dw = Dw$, we obtain from (4.16), (1.155), (1.137), and (4.48) that

$$
\int_\Omega \langle \rho(\zeta),\ \rho(\eta)\rangle\,dx = (Ddw, Ddu)_{L^2(\Omega,T^2)}
$$
$$
= (\mathbf{\Delta}dw - \kappa dw,\ du)_{L^2(\Omega,\Lambda)} + \int_\Gamma \langle D_\nu dw,\ du\rangle\,d\Gamma
$$
$$
= (\Delta^2 w - \delta(\kappa dw),\ u) + \int_\Gamma u[\langle \nu,\ \mathbf{\Delta}dw\rangle - \kappa\frac{\partial w}{\partial\nu}]\,d\Gamma + \int_\Gamma D^2w(\nu, du)\,d\Gamma
$$
$$
= (\Delta^2 w - \delta(\kappa dw),\ u) + \int_\Gamma [D^2w(\nu,\nu)\frac{\partial u}{\partial\nu} + D^2w(\nu,\tau)\frac{\partial u}{\partial\tau}]\,d\Gamma
$$
$$
- \int_\Gamma u[\frac{\partial\Delta w}{\partial\nu} + \kappa\frac{\partial w}{\partial\nu}]\,d\Gamma, \tag{4.49}
$$

since $\mathbf{\Delta} = -\Delta$ if it is applied to functions and $\mathbf{\Delta}dw = -d\Delta w$. In addition, since $\operatorname{tr}\rho(\zeta) = -\Delta w$, by Green's formula, we have

$$
\int_\Omega \operatorname{tr}\rho(\zeta)\operatorname{tr}\rho(\eta)\,dx = (\Delta w, \Delta u) = (\Delta^2 w, u)
$$
$$
+ \int_\Gamma [\Delta w\frac{\partial u}{\partial\nu} - u\frac{\partial\Delta w}{\partial\nu}]\,d\Gamma. \tag{4.50}
$$

It is generally assumed that Γ is a closed curve so that

$$
\int_\Gamma D^2w(\nu,\tau)\frac{\partial u}{\partial\tau}\,d\Gamma = -\int_\Gamma u\frac{\partial}{\partial\tau}(D^2w(\nu,\tau))\,d\Gamma. \tag{4.51}
$$

Furthermore, we have

$$
\Delta w = D^2w(\nu,\nu) + D^2w(\tau,\tau), \quad \text{on } \Gamma. \tag{4.52}
$$

By the formulas (4.49)-(4.52), and (4.20), we obtain

$$\int_\Omega a(\rho(\zeta), \rho(\eta)) \, dx = (\Delta^2 w - (1 - \mu)\delta(\kappa dw), \ u)$$

$$+ \int_\Gamma [(\Delta w + (1 - \mu)B_3 w) \frac{\partial u}{\partial \nu} - (\frac{\partial \Delta w}{\partial \nu} + (1 - \mu)B_4 w)u] \, d\Gamma. \qquad (4.53)$$

Finally, the identity (4.34) follows from the formulas (4.21), (4.18), (4.47), and (4.53). □

By Theorem 4.1 the following proposition is immediate.

Proposition 4.2 *The variational problem* (4.30) *is equivalent to solving the following boundary value problem in unknown* $\zeta = (W, w)$

$$b\mathcal{A}\zeta = (F, f) \qquad (4.54)$$

subject to the boundary conditions

$$\begin{cases} W|_{\Gamma_0} = 0, \\ w|_{\Gamma_0} = \frac{\partial w}{\partial \nu}|_{\Gamma_0} = 0, \end{cases} \qquad (4.55)$$

$$\begin{cases} B_1(W, w) = B_2(W, w) = 0 \quad on \quad \Gamma_1, \\ \Delta w + (1 - \mu)B_3 w = 0 \quad on \quad \Gamma_1, \\ \frac{\partial \Delta w}{\partial \nu} + (1 - \mu)B_4 w = 0 \quad on \quad \Gamma_1, \end{cases} \qquad (4.56)$$

where $\mathcal{A}$ *and* B_i $(1 \le i \le 4)$ *are defined in* (4.36) *and in* (4.39), *respectively.*

4.2 Ellipticity of the Strain Energy of Shallow Shells

The ellipticity of the strain energy of shells is an important issue necessary to all the control problems. In this section we derive the ellipticity results for shallow shells.

Let $\Gamma_0 \subset \Gamma$ be a portion of the boundary of the middle surface Ω with a positive length. Consider the Sobolev spaces

$$H^1_{\Gamma_0}(\Omega, \Lambda) = \{ W \mid W \in H^1(\Omega, \Lambda), \ W|_{\Gamma_0} = 0 \} \qquad (4.57)$$

and

$$H^2_{\Gamma_0}(\Omega) = \{ w \mid w \in H^2(\Omega), \ w|_{\Gamma_0} = \frac{\partial w}{\partial \nu}|_{\Gamma_0} = 0 \} \qquad (4.58)$$

with the norms of $H^1(\Omega, \Lambda)$ and $H^2(\Omega)$, respectively. In the following, we will show that the bilinear form $\mathcal{B}(\cdot, \cdot)$, given in (4.21), induces an equivalent norm on the space $H^1_{\Gamma_0}(\Omega, \Lambda) \times H^2_{\Gamma_0}(\Omega)$, i.e., the energy norm for a shallow shell.

We need several lemmas first.

Lemma 4.3 *Let W be a vector field (or equivalently a 1-form). Set*

$$U = DW + D^*W. \tag{4.59}$$

Then

$$
\begin{aligned}
D^2W(X,Y,Z) &= \frac{1}{2}[DU(X,Y,Z) + DU(Z,X,Y) - DU(Y,Z,X)] \\
&\quad + \mathbf{R}(X,Y,Z,W), \quad \forall\, X,\,Y,\,Z \in \mathcal{X}(M).
\end{aligned} \tag{4.60}
$$

Proof. Given $x \in M$. Let E_1, E_2 be a frame field normal at x. By (4.59), we have

$$
\begin{aligned}
DU(E_i, E_j, E_k) &= E_k\,(U(E_i, E_j)) \\
&= D^2W(E_i, E_j, E_k) + D^2W(E_j, E_i, E_k)
\end{aligned} \tag{4.61}
$$

at x since $(D_{E_i}E_j)(x) = 0$. Changing i, j, and k into k, i, and j, respectively, in the formula (4.61), we obtain from the formula (1.22)

$$
\begin{aligned}
DU(E_k, E_i, E_j) &= D^2W(E_k, E_i, E_j) + D^2W(E_i, E_k, E_j) \\
&= D^2W(E_k, E_i, E_j) + D^2W(E_i, E_j, E_k) + \mathbf{R}(E_k, E_j, W, E_i)
\end{aligned} \tag{4.62}
$$

at x where $\mathbf{R}(\cdot,\cdot,\cdot,\cdot)$ is the curvature tensor. A similar argument gives

$$
\begin{aligned}
DU(E_j, E_k, E_i) &= D^2W(E_j, E_i, E_k) + D^2W(E_k, E_i, E_j) \\
&\quad + \mathbf{R}(E_k, E_i, W, E_j) + \mathbf{R}(E_j, E_i, W, E_k)
\end{aligned} \tag{4.63}
$$

at x. By the equations (4.61), (4.62), and (4.63), we obtain

$$
\begin{aligned}
&2D^2W(E_i, E_j, E_k) \\
&= [DU(E_i, E_j, E_k) + DU(E_k, E_i, E_j) - DU(E_j, E_k, E_i)] \\
&\quad - \mathbf{R}(E_k, E_j, W, E_i) + \mathbf{R}(E_k, E_i, W, E_j) + \mathbf{R}(E_j, E_i, W, E_k)
\end{aligned} \tag{4.64}
$$

at x. By the first identity of Bianchi, (2) in Theorem 1.8,

$$\mathbf{R}(X,Y,Z,W) + \mathbf{R}(Y,Z,X,W) + \mathbf{R}(Z,X,Y,W) = 0$$

and the properties of the curvature tensor

$$\mathbf{R}(X,Y,Z,W) = -\mathbf{R}(Y,X,Z,W) = \mathbf{R}(Y,X,W,Z) = \mathbf{R}(W,Z,Y,X),$$

we obtain

$$
\begin{aligned}
&-\mathbf{R}(E_k, E_j, W, E_i) + \mathbf{R}(E_k, E_i, W, E_j) + \mathbf{R}(E_j, E_i, W, E_k) \\
&= -\mathbf{R}(E_j, E_k, E_i, W) - \mathbf{R}(E_k, E_i, E_j, W) + \mathbf{R}(E_i, E_j, E_k, W) \\
&= 2\mathbf{R}(E_i, E_j, E_k, W) \quad \text{at } x.
\end{aligned} \tag{4.65}
$$

Inserting the formula (4.65) into the formula (4.64), we obtain the identity (4.60). $\qquad\square$

Lemma 4.4 *Let $W \in \mathcal{X}(\Omega)$ such that*

$$DW + D^*W = 0 \quad on \quad \Omega. \tag{4.66}$$

Then

$$\mathbf{\Delta}W = 2\kappa W \quad on \quad \Omega \tag{4.67}$$

where $\mathbf{\Delta}$ is the Hodge-Laplace operator and κ is the Gauss curvature.

Proof. Let $x \in \Omega$ be given. Let E_1, E_2 be a frame field normal at x such that $D_{E_i} E_j(x) = 0$ for $1 \leq i, j \leq 2$. It follows from (4.60), (4.66), and (3.100) that

$$
\sum_{i=1}^{2} D^2 W(E_j, E_i, E_i) = \sum_{i=1}^{2} R(E_j, E_i, E_i, W)
$$
$$
= - \operatorname{Ric}(E_j, W) = -\kappa \langle W, E_j \rangle. \tag{4.68}
$$

By (1.114) and (4.68), we obtain at x

$$
\mathbf{\Delta}W = -\sum_{i=1}^{2} D^2_{E_i E_i} W + \sum_{ij=1}^{2} \mathbf{R}(E_i, E_j, W, E_j)E_i
$$
$$
= -\sum_{ij=1}^{2} D^2 W(E_j, E_i, E_i)E_j + \kappa W = 2\kappa W.
$$

$\square$

Remark 4.1 *In the flat case where $\kappa = 0$, if $W = (w_1, w_2)$, then*

$$
DW + D^*W = \left(\frac{\partial w_i}{\partial x_j} + \frac{\partial w_j}{\partial x_i} \right).
$$

Therefore the identity (4.67) is equivalent to

$$
\sum_{i=1}^{2} \frac{\partial^2 w_1}{\partial x_i^2} = \sum_{i=1}^{2} \frac{\partial^2 w_2}{\partial x_i^2} = 0.
$$

Lemma 4.5 *There is constant $c > 0$ such that*

$$\|DW + D^*W\|_{L^2(\Omega, T^2)} \geq c \|W\|_{H^1(\Omega, \Lambda)} \quad for \quad W \in H^1_{\Gamma_0}(\Omega, \Lambda). \tag{4.69}$$

Proof. Set

$$\mathcal{X} = \{ W \mid W \in L^2(\Omega, \Lambda), \ DW + D^*W \in L^2(\Omega, T^2) \}. \tag{4.70}$$

Given $W \in \mathcal{X}$. By the formula (4.60), $D^2W \in H^{-1}(\Omega, T^3)$ since $DW + D^*W \in L^2(\Omega, T^2)$, where

$$H^{-1}(\Omega, T^3) = \left(H^1_\Gamma(\Omega, T^3) \right)^*. \tag{4.71}$$

By a lemma of J. L. Lions (see [18]) in a local coordinate, we obtain $W \in H^1(\Omega, \Lambda)$. Then

$$\mathcal{X} = H^1(\Omega, \Lambda). \tag{4.72}$$

In order to get the inequality (4.69), it suffices by the open mapping theorem to prove the following uniqueness:

$$DW + D^*W = 0 \quad \text{implies} \quad W = 0 \quad \text{for } W \in H^1_{\Gamma_0}(\Omega, \Lambda). \tag{4.73}$$

Suppose that

$$DW + D^*W = 0 \quad \text{on} \quad \Omega. \tag{4.74}$$

Then W satisfies the equation (4.67) by Lemma 4.4.

Let ν and τ be the normal and the tangential along Γ, respectively. The condition $W|_{\Gamma_0} = 0$ implies that $D_\tau W|_{\Gamma_0} = 0$, which gives, by combining the formula (4.74),

$$DW|_{\Gamma_0} = W|_{\Gamma_0} = 0. \tag{4.75}$$

Applying [5], or [170] to the system of the equation (4.67) and the boundary condition (4.75), we obtain that $W = 0$ on Ω, since, in the standard coordinate system $x = (x_1, x_2)$, $\mathbf{\Delta}W = (-\Delta w_1 + \text{first order terms}, -\Delta w_2 + \text{first order terms})$ if $W = (w_1, w_2)$. □

We now assume that the shell is clamped on a portion Γ_0 of boundary Γ with length $\Gamma_0 > 0$. We have the following:

Theorem 4.2 *There is a constant $c > 0$ such that*

$$\mathcal{B}(\zeta, \zeta) \geq c\|\zeta\|^2_{H^1_{\Gamma_0}(\Omega, \Lambda) \times H^2_{\Gamma_0}(\Omega)} \tag{4.76}$$

for all $\zeta = (W, w) \in H^1_{\Gamma_0}(\Omega, \Lambda) \times H^2_{\Gamma_0}(\Omega)$. Hence $\mathcal{B}(\cdot, \cdot)$ induces an equivalent norm over space $H^1_{\Gamma_0}(\Omega, \Lambda) \times H^2_{\Gamma_0}(\Omega)$.

Proof. By means of Schwartz's inequality, it follows from the formula (4.15) that

$$\|\Upsilon(\zeta)\|^2_{L^2(\Omega, T^2)} = \frac{1}{4}\|DW + D^*W\|^2_{L^2(\Omega, T^2)}$$
$$+ (w\Pi, DW + D^*W)_{L^2(\Omega, T^2)} + \|w|\Pi|\|^2$$
$$\geq \frac{1}{8}\|DW + D^*W\|^2_{L^2(\Omega, T^2)} - 2\max_{x \in \Omega}|\Pi|^2\|w\|^2. \tag{4.77}$$

By the formulas (4.20) and (4.77), we obtain

$$\int_\Omega a(\Upsilon(\zeta), \Upsilon(\zeta))\, dx$$
$$\geq \frac{1-\mu}{8}[\|DW + D^*W\|^2_{L^2(\Omega, T^2)} - 16\max_{x \in \Omega}|\Pi|^2\|w\|^2]. \tag{4.78}$$

In addition, by the relations (4.16) and (4.20), we obtain

$$\int_\Omega a(\rho(\zeta), \rho(\zeta)) \, dx \geq (1 - \mu) \|D^2 w\|^2_{L^2(\Omega, T^2)}. \tag{4.79}$$

From the inequalities (4.78) and (4.79) and Lemma 4.5, it is obvious that there are C_1, $C_2 > 0$ such that

$$\mathcal{B}(\zeta, \zeta) + C_1 \|w\|^2 \geq C_2 \|\zeta\|^2_{H^1_{\Gamma_0}(\Omega, \Lambda) \times H^2_{\Gamma_0}(\Omega)}. \tag{4.80}$$

From the inequality (4.80), to get the inequality (4.76), it suffices by the compactness-uniqueness argument to prove the following uniqueness:

$$\mathcal{B}(\zeta, \zeta) = 0 \quad \text{implies} \quad \zeta = 0.$$

However, this is almost immediate. If $\mathcal{B}(\zeta, \zeta) = 0$, the inequality (4.79) yields $w = 0$ so that we obtain from the inequality (4.78) that $DW + D^*W = 0$. Finally, $W = 0$ follows from the inequality (4.69), i.e., $\zeta = 0$. $\qquad \square$

By Theorem 4.2 the following result is ready.

Theorem 4.3 *For $(F, f) \in L^2(\Omega, \Lambda) \times L^2(\Omega)$, the boundary value problem (4.54)-(4.56) has a unique solution $\zeta \in H^1_{\Gamma_0}(\Omega, \Lambda) \times H^2_{\Gamma_0}(\Omega)$.*

4.3 Equations of Motion

Let us compute the kinetic energy for a shallow shell.

Suppose that $\mathcal{F}(\Omega)$ is the middle surface of the deformed shell which is given by (4.2). Given $x \in \Omega$. Denote by $\overline{N}(\mathcal{F}(x))$ the normal field of the surface $\mathcal{F}(\Omega)$ at $\mathcal{F}(x)$. Let E_1, E_2 be a frame field normal at x, positively orientated, i.e., E_1, E_2, N is a frame field of $\mathbb{R}^3$ at x such that

$$N = E_1 \times E_2. \tag{4.81}$$

Then we have

$$\overline{N}(\mathcal{F}(x)) = \frac{\mathcal{F}_* E_1 \times \mathcal{F}_* E_2}{|\mathcal{F}_* E_1 \times \mathcal{F}_* E_2|}. \tag{4.82}$$

Set

$$\zeta_{ij} = \langle \tilde{D}_{E_i} \zeta, E_j \rangle, \quad 1 \leq i, j \leq 3, \tag{4.83}$$

where ζ is the displacement vector of the undeformed middle surface Ω, given in (4.6), and $E_3 = N$. Using the formula (4.10), we have

$$\zeta_{13} E_1 + \zeta_{23} E_2 = dw - i(W)\Pi \tag{4.84}$$

where $i(W)\Pi$ is *the interior product of tensor field Π by vector field W*. It follows from the formulas (4.5), (4.81), (4.82), and (4.84) that

$$
\begin{aligned}
\mathcal{F}_* E_1 \times \mathcal{F}_* E_2 &= (E_1 + \tilde{D}_{E_1}\zeta) \times (E_2 + \tilde{D}_{E_2}\zeta) \\
&= N + \tilde{D}_{E_1}\zeta \times E_2 + E_1 \times \tilde{D}_{E_2}\zeta + \tilde{D}_{E_1}\zeta \times \tilde{D}_{E_2}\zeta \\
&= (1 + \zeta_{11} + \zeta_{22})N - dw + i(W)\Pi + \mathbf{N}(\zeta) \qquad (4.85)
\end{aligned}
$$

where $\mathbf{N}(\zeta)$ denote all the nonlinear terms with respect to displacement field ζ. The equations (4.11) and (4.83) give

$$
\begin{aligned}
|\mathcal{F}_* E_1 \times \mathcal{F}_* E_2|^2 &= |\mathcal{F}_* E_1|^2 |\mathcal{F}_* E_2|^2 - \langle \mathcal{F}_* E_1, \mathcal{F}_* E_2 \rangle^2 \\
&= (1 + 2\zeta_{11} + |\tilde{D}_{E_1}\zeta|^2)(1 + 2\zeta_{22} + |\tilde{D}_{E_2}\zeta|^2) \\
&\quad - (\zeta_{12} + \zeta_{21} + \langle \tilde{D}_{E_1}\zeta, \tilde{D}_{E_2}\zeta \rangle)^2 \\
&= 1 + 2\zeta_{11} + 2\zeta_{22} + \mathbf{N}(\zeta),
\end{aligned}
$$

that is,

$$
|\mathcal{F}_* E_1 \times \mathcal{F}_* E_2| = 1 + \zeta_{11} + \zeta_{22} + \mathbf{N}(\zeta). \qquad (4.86)
$$

Let us substitute the expressions (4.85) and (4.86) into the formula (4.82). After linearization, we obtain

$$
\overline{N}(\mathcal{F}(x)) = -dw + i(W)\Pi + N. \qquad (4.87)
$$

Let S be given in the formula (4.1). Given $p = x + zN(x) \in$ S where $x \in \Omega$ and $-h/2 < z < h/2$. It follows from the classical Love-Kirchhoff assumptions that

$$
\mathcal{F}(p) = \mathcal{F}(x) + z\overline{N}(\mathcal{F}(x)) \quad \text{for} \quad p = x + zN(x). \qquad (4.88)
$$

Denote $\overline{\zeta}(p)$ the displacement vector of the point p. Then, the relations (4.87) and (4.88) yield

$$
\overline{\zeta}(p) = \zeta(x) - z(dw - i(W)\Pi) \quad \text{for} \quad p = x + zN(x),\ x \in \Omega. \qquad (4.89)
$$

Now, we assume that displacement vector $\overline{\zeta}$ depends on time t and, for simplicity, the mass density per unit of volume is unit. Then the kinetic energy per unit area of the undeformed middle surface is obtained by integration with respect to z

$$
\mathcal{H}(t) = \int_{-h/2}^{h/2} |\overline{\zeta}_t|^2 (1 + \operatorname{tr}\Pi z + \kappa z^2)\, dz. \qquad (4.90)
$$

Set

$$
Y = dw - i(W)\Pi. \qquad (4.91)
$$

By the formula (4.89), we obtain

$$
\begin{aligned}
|\overline{\zeta}_t|^2 (1 + \operatorname{tr}\Pi z + \kappa z^2) &= |\zeta_t|^2 - (-\operatorname{tr}\Pi|\zeta_t|^2 + 2\langle \zeta_t, Y_t \rangle)z \\
&\quad + (\kappa|\zeta_t|^2 - 2\operatorname{tr}\Pi\langle \zeta_t, Y_t \rangle + |Y_t|^2)z^2 \\
&\quad - (2\kappa\langle \zeta_t, Y_t \rangle - \operatorname{tr}\Pi|Y_t|^2)z^3 + \kappa|Y_t|^2 z^4. \qquad (4.92)
\end{aligned}
$$

Integrating both the sides of the equation (4.92) over $(-h/2, h/2)$ with respect to z yields

$$
\begin{aligned}
\mathcal{H}(t) &= h|\zeta_t|^2 + \frac{h^3}{12}(\kappa|\zeta_t|^2 - 2\operatorname{tr}\Pi\langle\zeta_t, Y_t\rangle + |Y_t|^2) + \frac{h^5}{80}\kappa|Y_t|^2 \\
&= h|W_t|^2 + hw_t^2 + \frac{h^3}{12}|Dw_t|^2 + \frac{h^3}{12}\kappa|W_t|^2\frac{h^3}{12}\kappa w_t^2 \\
&\quad + \frac{h^3}{12}|\operatorname{i}(W_t)\Pi|^2 + \frac{h^3}{6}\Pi(W_t, W_t)\operatorname{tr}\Pi - \frac{h^3}{6}\operatorname{tr}\Pi\langle W_t, Dw_t\rangle - \frac{h^3}{6}\Pi(W_t, Dw_t) \\
&\quad + \frac{h^5}{80}\kappa|Dw_t|^2 - \frac{h^5}{40}\kappa\Pi(W_t, Dw_t) + \frac{h^5}{80}\kappa|\operatorname{i}(W_t)\Pi|^2
\end{aligned}
\tag{4.93}
$$

since $dw = Dw$.

By using some ideas in [94], however, we can further simplify $\mathcal{H}(t)$ as

$$
\mathcal{H}(t) = h|W_t|^2 + hw_t^2 + \frac{h^3}{12}|Dw_t|^2.
\tag{4.94}
$$

In order to justify the approximation (4.94) it will be shown that the order of magnitude of the terms, which have been omitted in the formula (4.93), is indeed negligible compared with the expression (4.94). All the neglected terms are listed below:

$$
\frac{h^3}{12}\kappa|W_t|^2;
\tag{4.95}
$$

$$
\frac{h^3}{12}\kappa w_t^2;
\tag{4.96}
$$

$$
\frac{h^3}{12}|\operatorname{i}(W_t)\Pi|^2;
\tag{4.97}
$$

$$
\frac{h^3}{6}\Pi(W_t, W_t)\operatorname{tr}\Pi;
\tag{4.98}
$$

$$
-\frac{h^3}{6}\operatorname{tr}\Pi\langle W_t, Dw_t\rangle;
\tag{4.99}
$$

$$
-\frac{h^3}{6}\Pi(W_t, Dw_t);
\tag{4.100}
$$

$$
\frac{h^5}{80}\kappa|Dw_t|^2;
\tag{4.101}
$$

$$
-\frac{h^5}{40}\kappa\Pi(W_t, Dw_t);
\tag{4.102}
$$

$$
\frac{h^5}{80}\kappa|\operatorname{i}(W_t)\Pi|^2.
\tag{4.103}
$$

It is almost obvious that the neglected terms (4.95), (4.96), and (4.101) are indeed negligible in a thin shell. In fact, if $\mathbf{R}$ is the smallest principal radius of curvature of the undeformed middle surface given by (4.25), we have the estimates

$$|(4.95)| \leq \frac{h^2}{12\mathbf{R}^2}(h|W_t|^2); \quad |(4.96)| \leq \frac{h^2}{12\mathbf{R}^2}(hw_t^2);$$

$$|(4.101)| \leq \frac{3}{20}\left(\frac{h}{\mathbf{R}}\right)^2\left(\frac{h^3}{12}|Dw_t|^2\right). \tag{4.104}$$

In addition, by Lemma 4.2, we have

$$|(4.97)| \leq \frac{1}{6}\left(\frac{h}{\mathbf{R}}\right)^2(h|W_t|^2); \quad |(4.98)| \leq \frac{\sqrt{2}}{3}\left(\frac{h}{\mathbf{R}}\right)^2(h|W_t|^2);$$

$$|(4.103)| \leq \frac{1}{5}\left(\frac{h}{\mathbf{R}}\right)^4(h|W_t|^2). \tag{4.105}$$

Finally, by means of Schwartz's inequality and Lemma 4.2, we obtain estimates

$$|(4.99)| \leq \frac{\sqrt{3}}{3}\left(\frac{h}{\mathbf{R}}\right)(h|W_t|^2 + \frac{h^3}{12}|Dw_t|^2); \tag{4.106}$$

$$|(4.100)| \leq \frac{\sqrt{6}}{3}\left(\frac{h}{\mathbf{R}}\right)(h|W_t|^2 + \frac{h^3}{12}|Dw_t|^2); \tag{4.107}$$

$$|(4.102)| \leq \frac{\sqrt{6}}{20}\left(\frac{h}{\mathbf{R}}\right)^3(h|W_t|^2 + \frac{h^3}{12}|Dw_t|^2). \tag{4.108}$$

By the assumption (4.28) for a thin shell and by the estimates (4.104)-(4.108), the approximation (4.94) is justified so that we obtain the kinetic energy of the shell

$$\mathcal{P} = \int_\Omega \mathcal{H}(t)\,dx = \int_\Omega (h|W_t|^2 + hw_t^2 + \frac{h^3}{12}|Dw_t|^2)\,dx. \tag{4.109}$$

The equations of motion for ζ are obtained by setting to zero the first variation of the Lagrangian

$$\int_0^T [\mathcal{P}(t) + L(\zeta) - \mathcal{B}_1(\zeta,\zeta)]\,dt \tag{4.110}$$

(the "Principle of Virtual Work"), where $L(\cdot)$ and $\mathcal{B}_1(\cdot,\cdot)$ are given by the formulas (4.29) and (4.17), respectively. We assume that there is no external loading on the shell. Then the variation of (4.110) is taken with respect to kinematically admissible displacement fields.

We obtain, as a result of the calculation by the formulas (4.34) and (4.110), the following

Theorem 4.4 *We assume that there are no external loads on the shell and that the shell is clamped along a portion* Γ_0 *of* Γ *and free on* Γ_1, *where* $\Gamma_0 \cup \Gamma_1 = \Gamma$. *Then the displacement vector* $\zeta = (W, w)$ *satisfies the following boundary value problem:*

$$
\begin{cases}
W_{tt} - b[\boldsymbol{\Delta}_\mu W + (1 - \mu)\kappa W + F(w)] = 0, \\
w_{tt} - \gamma \Delta w_{tt} + b[\gamma \left(\Delta^2 w - (1 - \mu)\delta(\kappa dw)\right) \quad \text{in} \quad Q^\infty, \\
+(\text{tr}^2 \Pi - 2(1 - \mu)\kappa)w + \mathcal{G}(W)] = 0,
\end{cases}
\tag{4.111}
$$

$$
\begin{cases}
W = 0, \\
w = \dfrac{\partial w}{\partial \nu} = 0,
\end{cases}
\quad \text{on} \quad \Sigma_0^\infty
\tag{4.112}
$$

$$
\begin{cases}
B_1(W, w) = B_2(W, w) = 0, \\
\Delta w + (1 - \mu)B_3 w = 0, \\
b\gamma[\dfrac{\partial \Delta w}{\partial \nu} + (1 - \mu)B_4 w] - \gamma \dfrac{\partial w_{tt}}{\partial \nu} = 0,
\end{cases}
\quad \text{on} \quad \Sigma_1^\infty,
\tag{4.113}
$$

where

$$
Q^\infty = (0, \infty) \times \Omega, \quad \Sigma_0^\infty = (0, \infty) \times \Gamma_0, \quad \Sigma_1^\infty = (0, \infty) \times \Gamma_1.
\tag{4.114}
$$

If the shell is flat, we have

Corollary 4.1 *Let the shell be flat, a plate. The above equations in* (4.111) *are uncoupled. In this case* $M = \mathbb{R}^2$ *with the Euclidean metric in* $\mathbb{R}^2$. *In the natural coordinate, setting* $W = (w_1, w_2)$, *we then have*

$$
\begin{cases}
W_{tt} - b\boldsymbol{\Delta}_\mu W = 0 \quad \text{in} \quad Q_\infty, \\
W = 0, \quad \text{on } \Sigma_0^\infty, \\
(1 - \mu)(\dfrac{\partial w_1}{\partial \nu}\nu_1 + \dfrac{\partial w_2}{\partial \nu}\nu_2) + \mu(\dfrac{\partial w_1}{\partial x} + \dfrac{\partial w_2}{\partial y}) = 0 \quad \text{on} \quad \Sigma_1^\infty, \\
(\dfrac{\partial w_2}{\partial \tau} + \dfrac{\partial w_1}{\partial \nu})\nu_2 + (\dfrac{\partial w_1}{\partial \tau} - \dfrac{\partial w_2}{\partial \nu})\nu_1 \quad \text{on} \quad \Sigma_1^\infty,
\end{cases}
\tag{4.115}
$$

where $\nu = (\nu_1, \nu_2)$, $\tau = (\nu_2, -\nu_1)$,

$$
\boldsymbol{\Delta}_\mu W = \begin{pmatrix}
\dfrac{\partial^2 w_1}{\partial x^2} + \dfrac{1 - \mu}{2}\dfrac{\partial^2 w_1}{\partial y^2} + \dfrac{1 + \mu}{2}\dfrac{\partial^2 w_2}{\partial x \partial y} \\
\dfrac{1 - \mu}{2}\dfrac{\partial^2 w_2}{\partial x^2} + \dfrac{\partial^2 w_2}{\partial y^2} + \dfrac{1 + \mu}{2}\dfrac{\partial^2 w_1}{\partial x \partial y}
\end{pmatrix},
\tag{4.116}
$$

and

$$
\begin{cases}
w_{tt} - \gamma \Delta w_{tt} + b\gamma \Delta^2 w = 0, \quad \text{in } Q_\infty, \\
w = \dfrac{\partial w}{\partial \nu} = 0 \quad \text{on} \quad \Sigma_0^\infty, \\
\Delta w + (1 - \mu)B_3 w = 0 \quad \text{on} \quad \Sigma_1^\infty \\
b\gamma[\dfrac{\partial \Delta w}{\partial \nu} + (1 - \mu)B_4 w] - \gamma \dfrac{\partial w_{tt}}{\partial \nu} = 0 \quad \text{on} \quad \Sigma_1^\infty,
\end{cases}
\tag{4.117}
$$

where

$$\Delta w = \frac{\partial^2 w}{\partial x^2} + \frac{\partial^2 w}{\partial y^2}. \tag{4.118}$$

The equations (4.117) are the same as in [105], pp. 15-16, a Kirchhoff plate.

   Proof. Since $M = \mathbb{R}^2$, we have

$$\Pi = 0 \quad \text{and} \quad \text{tr}\,\Pi = \kappa = 0. \tag{4.119}$$

It follows from the formulas (4.119) and (4.43) that

$$\mathcal{H}(w) = 0 \quad \text{and} \quad \mathcal{G}(W) = 0. \tag{4.120}$$

It is easily checked that

$$-d\delta W = \left(\frac{\partial^2 w_1}{\partial x^2} + \frac{\partial^2 w_2}{\partial x \partial y}\right)\frac{\partial}{\partial x} + \left(\frac{\partial^2 w_1}{\partial x \partial y} + \frac{\partial^2 w_2}{\partial y^2}\right)\frac{\partial}{\partial y}; \tag{4.121}$$

$$-\delta dW = \left(\frac{\partial^2 w_1}{\partial y^2} - \frac{\partial^2 w_2}{\partial x \partial y}\right)\frac{\partial}{\partial x} + \left(\frac{\partial^2 w_2}{\partial x^2} - \frac{\partial^2 w_1}{\partial x \partial y}\right)\frac{\partial}{\partial y}, \tag{4.122}$$

since $W = w_1 \dfrac{\partial}{\partial x} + w_2 \dfrac{\partial}{\partial y}$.

   Inserting the formulas (4.121) and (4.122) into the formula (4.37) yields the formula (4.116).

   By the formulas (4.39) and (4.119), we obtain

$$\begin{aligned}
B_1(W, w) &= (1 - \mu)DW(n, n) - \mu W \\
&= (1 - \mu)\left(\frac{\partial w_1}{\partial \nu}\nu_1 + \frac{\partial w_2}{\partial \nu}\nu_2\right)\left(\frac{\partial w_1}{\partial x} + \frac{\partial w_2}{\partial y}\right);
\end{aligned} \tag{4.123}$$

$$\begin{aligned}
B_2(W, w) &= \frac{1 - \mu}{2}\left(\langle D_\nu W, \tau \rangle + \langle D_\tau W, \nu \rangle\right) \\
&= \frac{1 - \mu}{2}\left[\left(\frac{\partial w_1}{\partial \tau} - \frac{\partial w_2}{\partial \nu}\right)\nu_1 + \left(\frac{\partial w_1}{\partial \nu} + \frac{\partial w_2}{\partial \tau}\right)\nu_2\right].
\end{aligned} \tag{4.124}$$

   By the formulas (4.120), (4.123), and (4.124), we obtain the equations (4.115) and (4.117) by Theorem 4.4.

   We check B_3 and B_4 on the boundary Γ in order to show that the equations (4.117) are the same as in [105], pp. 15-16. In fact, we have

$$\begin{aligned}
B_3 w &= -D^2 w(\tau, \tau) = -(\nu_2, -\nu_1)\begin{pmatrix} \dfrac{\partial^2 w}{\partial x^2} & \dfrac{\partial^2 w}{\partial x \partial y} \\ \dfrac{\partial^2 w}{\partial x \partial y} & \dfrac{\partial^2 w}{\partial y^2} \end{pmatrix}\begin{pmatrix} \nu_2 \\ -\nu_1 \end{pmatrix} \\
&= 2\nu_1\nu_2\frac{\partial^2 w}{\partial x \partial y} - \nu_1\frac{\partial^2 w}{\partial y^2} - \nu_2\frac{\partial^2 w}{\partial x^2}
\end{aligned}$$

and

$$B_4 w = \frac{\partial}{\partial \tau}D^2 w(\tau, n) = \frac{\partial}{\partial \tau}\left[(\nu_2^2 - \nu_1^2)\frac{\partial^2 w}{\partial x \partial y} + \nu_1\nu_2\left(\frac{\partial^2 w}{\partial x^2} - \frac{\partial^2 w}{\partial y^2}\right)\right].$$

4.4 Multiplier Identities

We will establish some multiplier identities by the Bochnner technique which are useful to control problems. Let M be a surface in $\mathbb{R}^3$ with the induced metric g and let $\Omega \subset M$ be the middle surface of a shell. Let $\zeta = (W, w)$ be a displacement vector field and let V be a vector field on the middle surface Ω. The main multipliers we will use are

$$m(\zeta) = \Big(D_V W, V(w) \Big) \quad \text{and} \quad q\zeta$$

where q is a function on Ω. For $\zeta = (W, w)$, let

$$e(\zeta) = |W_t|^2 + |w_t|^2 + \gamma |Dw_t|^2 + B(\zeta, \zeta) \tag{4.125}$$

for each $x \in \Omega$ where $\gamma = h^2/12$ and $B(\zeta, \zeta)$ is given by (4.18). If $\zeta = (W, w)$ solves the shallow shell problem (4.111), the energy of the displacement vector field ζ is given by

$$E(t) = \frac{1}{2} \int_\Omega e(\zeta) \, dx. \tag{4.126}$$

In the sequel, lo (ζ) denotes all the *lower order terms with respect to the energy* to which for any $\varepsilon > 0$ there exists $C_\varepsilon > 0$ such that

$$| \operatorname{lo}(\zeta)|^2 \le \varepsilon e(\zeta) + C_\varepsilon l(\zeta) \quad \text{for all} \quad x \in \Omega \tag{4.127}$$

where

$$l(\zeta) = |W|^2 + |w|^2 + |Dw|^2 + \gamma w_t^2. \tag{4.128}$$

So lo (ζ) may be different from line to line and page to page.

We need several lemmas for our computation.

Lemma 4.6 *Let $V, W \in \mathcal{X}(\Omega)$ be vector fields on the middle surface Ω. Then*

$$D\Big(D_V W\Big) = D_V(DW) + \mathbf{R}(V, \cdot, W, \cdot) + DW(\cdot, D.V) \tag{4.129}$$

where $\mathbf{R}(\cdot, \cdot, \cdot, \cdot)$ is the curvature tensor of the middle surface Ω and "$\cdot$" denotes the position of variables.

Proof. Given $x \in \Omega$. Let E_1, E_2 be a frame field normal at x. Using the relations $D_{E_i} E_j(x) = 0$ and (1.22), we have

$$
\begin{aligned}
D(D_V W)(E_i, E_j) &= E_j \Big(D_V W(E_i) \Big) = E_j \Big(DW(E_i, V) \Big) \\
&= D^2 W(E_i, V, E_j) + DW(E_i, D_{E_j} V) \\
&= D^2 W(E_i, E_j, V) + \mathbf{R}_{V E_j} W(E_i) + DW(E_i, D_{E_j} V) \\
&= D_V(DW)(E_i, E_j) + \mathbf{R}(V, E_j, W, E_i) \\
&\quad + DW(E_i, D_{E_j} V) \quad \text{at} \quad x
\end{aligned}
\tag{4.130}
$$

for $1 \leq i, j \leq 2$. The above identity (4.130) means that

$$D(D_V W)(X, Y) = D_V(DW)(X, Y) + \mathbf{R}(V, X, W, Y) + DW(X, D_Y V)$$

for all $X, Y \in \mathcal{X}(\Omega)$ since $R(V, \cdot, W, \cdot)$ is symmetric.   □

Let $V \in \mathcal{X}(\Omega)$ be a vector field and let $T \in T^2(\Omega)$ be a tensor field of rank 2. We define a tensor field of rank 2 on Ω by

$$G(V, T) = \frac{1}{2} \Big[T(\cdot, D.V) + \big(T(\cdot, D.V) \big)^* \Big] \tag{4.131}$$

Let where $\big(T(\cdot, D.V) \big)^*$ denotes the transpose of the tensor field $T(\cdot, D.V)$.

Lemma 4.7 *Let $V \in \mathcal{X}(\Omega)$ and $\zeta = (W, w)$ be given. Let $m(\zeta) = \big(D_V W, V(w) \big)$. Then*

$$\Upsilon \big(m(\zeta) \big) = D_V \Upsilon(\zeta) + G(V, DW) + \mathrm{lo}\,(\zeta) \tag{4.132}$$

where Υ is the strain tensor of the middle surface, given by (4.15). Moreover,

$$\rho \big(m(\zeta) \big) = D_V \rho(\zeta) + 2G(V, \rho(\zeta)) + \mathrm{lo}\,(\zeta) \tag{4.133}$$

where ρ is the change of curvature tensor of the middle surface, given by (4.16).

Proof. We make the covariant differential on the formula (4.15) with respect to the vector field V, use the formula (4.129), and have

$$D_V \Upsilon(\zeta) = \frac{1}{2} \Big[D_V(DW) + D_V(D^* W) \Big] + V(w)\Pi + w D_V \Pi. \tag{4.134}$$

On the other hand, we make a transpose on the formula (4.129) to obtain

$$D^* \big(D_V W \big) = D_V(D^* W) + \mathbf{R}(V, \cdot, W, \cdot) + DW(D.V, \cdot). \tag{4.135}$$

Inserting the formulas (4.129) and (4.135) into the formula (4.134), we obtain the identity (4.132) where

$$\mathrm{lo}\,(\zeta) = \mathbf{R}(V, \cdot, W, \cdot) - w D_V \Pi$$

since $\mathbf{R}(V, \cdot, W, \cdot)$ is symmetric.

Let $x \in \Omega$ be given. Let E_1, E_2 be a frame field normal at x. Using the relations $D_{E_i} E_j(x) = 0$ and the relation (1.22) , we have

$$
\begin{aligned}
D^2 \big(V(w) \big)(E_i, E_j) &= E_j E_i \big(Dw(V) \big) = E_j \big(D^2 w(E_i, V) + Dw(D_{E_i} V) \big) \\
&= D^3 w(E_i, V, E_j) + D^2 w(E_i, D_{E_j} V) \\
&\quad + D^2 w(D_{E_i} V, E_j) + Dw(D_{E_j} D_{E_i} V) \\
&= D^2 w(E_i, E_j, V) + \mathbf{R}(V, E_j, Dw, E_i) \\
&\quad + D^2 w(E_i, D_{E_j} V) + D^2 w(D_{E_i} V, E_j) \\
&\quad + Dw(D_{E_j} D_{E_i} V).
\end{aligned}
\tag{4.136}
$$

The last term in the right hand side of (4.136) can be computed as follows:

$$
\begin{aligned}
Dw(D_{E_j}D_{E_i}V) &= \langle D_{E_j}D_{E_i}V, Dw\rangle \\
&= E_j\Big(DV(Dw, E_i)\Big) - \langle D_{E_i}V, D_{E_j}Dw\rangle \\
&= D^2V(Dw, E_i, E_j) + DV(D_{E_j}Dw, E_i) \\
&\quad -\langle D_{E_i}V, D_{E_j}Dw\rangle \\
&= i\,(Dw)D^2V(E_i, E_j). \tag{4.137}
\end{aligned}
$$

Inserting the formula (4.137) into the formula (4.136) yields the identity

$$
\begin{aligned}
D^2\Big(V(w)\Big) &= D_V\Big(D^2w\Big) + D^2w(\cdot, D.V) + \Big(D^2w(\cdot, D.V)\Big)^{*} \\
&\quad + i\,(Dw)\Pi + \mathbf{R}(V, \cdot, W, \cdot). \tag{4.138}
\end{aligned}
$$

Noting that $\rho(\zeta) = -D^2w$ and $\rho\big(m(\zeta)\big) = -D^2\Big(V(w)\Big)$, we obtain the identity (4.133) from the identity (4.138) where

$$
\mathrm{lo}\,(\zeta) = -i\,(Dw)\Pi - \mathbf{R}(V, \cdot, W, \cdot).
$$

$$\square$$

Lemma 4.8 *Let $V \in \mathcal{X}(\Omega)$ be a vector field and let $\zeta = (W, w)$ be given. Let $m(\zeta) = \Big(D_V W, V(w)\Big)$. Suppose that the bilinear forms $B(\cdot, \cdot)$ and $a(\cdot, \cdot)$ are given by (4.18) and (4.20), respectively. Then*

$$
\begin{aligned}
2B\Big(\zeta, m(\zeta)\Big) &= \mathrm{div}\,\Big[B(\zeta, \zeta)V\Big] - B(\zeta, \zeta)\,\mathrm{div}\,V \\
&\quad + 2a\Big(\Upsilon(\zeta), G(V, DW)\Big) + 4\gamma a\Big(\rho(\zeta), G(V, \rho(\zeta))\Big) + \mathrm{lo}\,(\zeta) \tag{4.139}
\end{aligned}
$$

where $G(V, DW)$ and $G(V, \rho(\zeta))$ are defined by the formula (4.131).

Proof. By the expression (4.132), we have

$$
\begin{aligned}
2\langle \Upsilon(\zeta), \Upsilon\big(m(\zeta)\big)\rangle &= V\Big(|\Upsilon(\zeta)|^2\Big) + 2\langle \Upsilon(\zeta), G(V, DW)\rangle + \mathrm{lo}\,(\zeta) \\
&= \mathrm{div}\,(|\Upsilon(\zeta)|^2 V) - |\Upsilon(\zeta)|^2\,\mathrm{div}\,V \\
&\quad + 2\langle \Upsilon(\zeta), G(V, DW)\rangle + \mathrm{lo}\,(\zeta). \tag{4.140}
\end{aligned}
$$

Let $x \in \Omega$ be given and let E_1, E_2 be a frame field normal at x. Then

$$
V\Big(\mathrm{tr}\,\Upsilon(\zeta)\Big) = V[\sum_{i=1}^{2}\Upsilon(\zeta)(E_i, E_i)] = \sum_{i=1}^{2}D_V\Upsilon(E_i, E_i) = \mathrm{tr}\,D_V\Upsilon \tag{4.141}
$$

at x. From the formulas (4.141) and (4.132), we obtain

$$
\begin{aligned}
2\operatorname{tr}\Upsilon(\zeta)\operatorname{tr}\Upsilon\big(m(\zeta)\big) &= 2\operatorname{tr}\Upsilon(\zeta)V\big(\operatorname{tr}\Upsilon(\zeta)\big) + 2\operatorname{tr}\Upsilon(\zeta)\operatorname{tr}G(V,DW) \\
&\quad + \operatorname{lo}(\zeta) \\
&= \operatorname{div}\Big[\big(\operatorname{tr}\Upsilon(\zeta)\big)^2 V\Big] - \big(\operatorname{tr}\Upsilon(\zeta)\big)^2 \operatorname{div}V \\
&\quad + 2\operatorname{tr}\Upsilon(\zeta)\operatorname{tr}G(V,DW) + \operatorname{lo}(\zeta).
\end{aligned} \tag{4.142}
$$

Using the formulas (4.140) and (4.142), we have

$$
\begin{aligned}
2a\big(\Upsilon(\zeta),\Upsilon\big(m(\zeta)\big)\big) &= \operatorname{div}\Big[a\big(\Upsilon(\zeta),\Upsilon(\zeta)\big)V\Big] - a\big(\Upsilon(\zeta),\Upsilon(\zeta)\big)\operatorname{div}V \\
&\quad + 2a\big(\Upsilon(\zeta),G(V,DW)\big) + \operatorname{lo}(\zeta).
\end{aligned} \tag{4.143}
$$

Next, we compute $a\big(\rho(\zeta),\rho\big(m(\zeta)\big)\big)$. It follows from the expression (4.133) that

$$
\begin{aligned}
2\langle\rho(\zeta),\rho\big(m(\zeta)\big)\rangle &= \operatorname{div}\big(|\rho(\zeta)|^2 V\big) - |\rho(\zeta)|^2 \operatorname{div}V \\
&\quad + 4\langle\rho(\zeta),\,G(V,\rho(\zeta))\rangle^2 + \operatorname{lo}(\zeta);
\end{aligned} \tag{4.144}
$$

$$
\begin{aligned}
2\operatorname{tr}\rho(\zeta)\operatorname{tr}\rho\big(m(\zeta)\big) &= \operatorname{div}\Big[(\operatorname{tr}\rho(\zeta))^2 V\Big] - (\operatorname{tr}\rho(\zeta))^2\operatorname{div}V \\
&\quad + 4\operatorname{tr}\rho(\zeta)\operatorname{tr}G(V,\rho(\zeta)) + \operatorname{lo}(\zeta).
\end{aligned} \tag{4.145}
$$

The formulas (4.144) and (4.145) yield

$$
\begin{aligned}
2a\big(\rho(\zeta),\rho(m(\zeta))\big) &= \operatorname{div}\Big[a\big(\rho(\zeta),\rho(\zeta)\big)V\Big] - a\big(\rho(\zeta),\rho(\zeta)\big)\operatorname{div}V \\
&\quad + 4a\big(\rho(\zeta),G(V,\rho(\zeta))\big) + \operatorname{lo}(\zeta).
\end{aligned} \tag{4.146}
$$

Finally, the formula (4.139) follows from the identities (4.143) and (4.146).
□

Now we consider multiplier identities for a displacement vector field to the equations of motion for the shallow shell. Let $\zeta = (W, w)$ be a solution to the following problem

$$
\zeta_{tt} - \gamma(0, \Delta w_{tt}) + \mathcal{A}\zeta = \varphi \quad \text{in} \quad \Omega \tag{4.147}
$$

where $\mathcal{A}$ is the shallow shell operator, given by (4.36), and γ is given in (4.19), respectively and $\varphi = (F, f)$ denotes an external force acting on the middle surface of the shallow shell. In the case of zero external load, the equation (4.147) is different from the equations (4.111) by a coefficient b of the operator $\mathcal{A}$ but by a change of variable they are the same. For simplicity, we study the problem (4.147).

Let V be a vector field and let u be a function on Ω. By (2.6),

$$2\langle Du, D(V(u))\rangle = 2DV(Du, Du) + \operatorname{div}(|Du|^2 V)$$
$$-|Du|^2 \operatorname{div} V. \tag{4.148}$$

In order to simplify computation in obtaining multiplier identities we have to choose a special vector field V for the multiplier $m(\zeta)$. We make the following: Vector field V on Ω satisfies

$$DV(X, X) = \vartheta(x)|X|^2 \quad \text{for} \quad X \in M_x, \quad x \in \Omega \tag{4.149}$$

where ϑ is a function on Ω.

In the following $L(\zeta)$ denotes the lower order term with respect to the energy level, that is, for any $\varepsilon > 0$ there is $C_\varepsilon > 0$ such that

$$0 \le L(\zeta) \le \varepsilon \int_0^T E(t)dt + C_\varepsilon[\|l(\zeta)(0)\|^2 + \|l(\zeta)(T)\|^2 + \|l(\zeta)\|_{L^2(Q)}^2] \tag{4.150}$$

where $l(\zeta)$ is given by (4.128).

Let

$$Q = (0, T) \times \Omega, \quad \Sigma = (0, T) \times \Gamma.$$

Theorem 4.5 *Let $\zeta = (W, w)$ solve the problem* (4.147) *and let a vector field V satisfy the assumption* (4.149). *Then*

$$\int_\Sigma [|\zeta_t|^2 + \gamma|Dw_t|^2 - B(\zeta, \zeta)]\langle V, \nu\rangle \, d\Sigma + 2\int_\Sigma [\partial(A\zeta, m(\zeta)) + \gamma V(w)\frac{\partial w_{tt}}{\partial \nu}] \, d\Sigma$$
$$= 2Z(t)|_0^T + 2\int_Q [a(\Upsilon(\zeta), G(V, DW)) + 2\gamma a(\rho(\zeta), G(V, \rho(\zeta)))] \, dQ$$
$$+ \int_Q \{[|\zeta_t|^2 + \gamma|Dw_t|^2 - B(\zeta, \zeta)] \operatorname{div} V - 2\gamma DV(Dw_t, Dw_t)\} \, dQ$$
$$- \int_Q \langle\varphi, m(\zeta)\rangle \, dQ + L(\zeta) \tag{4.151}$$

where

$$Z(t) = (\zeta_t, m(\zeta))_{L^2(\Omega, \Lambda) \times L^2(\Omega)} + \gamma(Dw_t, D(V(w))) \tag{4.152}$$

and the tensor field $G(V, DW)$ of rank 2 is defined by (4.131).

Proof. We multiply the equation (4.147) by $2m(\zeta)$ and integrate over Q by parts.

Letting $u = w_t$ in the formula (4.148) yields

$$2\langle\nabla w_t, \nabla V(w_t)\rangle = 2DV(Dw_t, Dw_t) - |Dw_t|^2 \operatorname{div} V$$
$$+ \operatorname{div}(|Dw_t|^2 V). \tag{4.153}$$

Let $P = \zeta - \gamma(0, \Delta w)$. By the formula (4.153), we have

$$
\begin{aligned}
2\langle P_{tt}, m(\zeta)\rangle &= 2\langle \zeta_{tt}, m(\zeta)\rangle - 2\gamma \Delta w_{tt} V(w) \\
&= -2\gamma \operatorname{div}\left[V(w)\nabla w_{tt}\right] + 2\langle \zeta_{tt}, m(\zeta)\rangle + 2\gamma\langle \nabla w_{tt}, \nabla V(w)\rangle \\
&= -2\gamma \operatorname{div}\left[V(w)\nabla w_{tt}\right] + 2[\langle \zeta_t, m(\zeta)\rangle + \gamma\langle \nabla w_t, \nabla V(w)\rangle]_t \\
&\quad -2[\langle \zeta_t, m(\zeta_t)\rangle + \gamma\langle \nabla w_t, \nabla V(w_t)\rangle] \\
&= -\operatorname{div}\left[(|\zeta_t|^2 + \gamma|Dw_t|^2)V + 2\gamma V(w)Dw_{tt}\right] \\
&\quad + (|\zeta_t|^2 + \gamma|Dw_t|^2)\operatorname{div} V - 2\gamma DV(Dw_t, Dw_t) \\
&\quad + 2[\langle \zeta_t, m(\zeta)\rangle + \gamma\langle \nabla w_t, \nabla V(w)\rangle]_t.
\end{aligned}
\tag{4.154}
$$

Integrating the identity (4.154) over $Q = (0, T) \times \Omega$ gives

$$
2\int_Q \langle P_{tt}, m(\zeta)\rangle \, dQ = 2Z(t)|_0^T
$$

$$
+ \int_Q [(|\zeta_t|^2 + \gamma|Dw_t|^2)\operatorname{div} V - 2\gamma DV(Dw_t, Dw_t)]\, dQ
$$

$$
- \int_\Sigma [2\gamma V(w)\langle Dw_{tt}, \nu\rangle + (|\zeta_t|^2 + \gamma|Dw_t|^2)\langle V, \nu\rangle]\, d\Sigma. \tag{4.155}
$$

On the other hand, by the Green formula (4.34) and the formula (4.139)

$$
2\int_Q \langle A\zeta, m(\zeta)\rangle \, dQ = 2\int_0^T (A\zeta, m(\zeta))_{L^2(\Omega,\Lambda) \times L^2(\Omega)}\, dt
$$

$$
= 2\int_Q B(\zeta, m(\zeta))\, dQ - 2\int_\Sigma \partial(A\zeta, m(\zeta))\, d\Sigma
$$

$$
= \int_\Sigma [B(\zeta, \zeta)\langle V, \nu\rangle - 2\partial(A\zeta, m(\zeta))]\, d\Sigma + 2\int_Q a(\Upsilon(\zeta), G(V, DW))\, dQ
$$

$$
+ \int_Q [4\gamma a(\rho(\zeta), G(V, \rho(\zeta))) - B(\zeta, \zeta)\operatorname{div} V]\, dQ + L(\zeta). \tag{4.156}
$$

Finally, using the formulas (4.147), (4.155), and (4.156), we obtain the identity (4.151). □

Theorem 4.6 *Let $\zeta = (W, w)$ solve the problem (4.147) and let q be a function on Ω. Then*

$$
\int_Q q[|W_t|^2 - a(\Upsilon(\zeta), \Upsilon(\zeta))]\, dQ + \int_Q \langle \varphi, q(W, 0)\rangle \, dQ
$$

$$
= -\int_\Sigma \partial(A\zeta, q(W, 0))\, d\Sigma + L(\zeta) \tag{4.157}
$$

and

$$
\int_Q q[w_t^2 + \gamma|Dw_t|^2 - \gamma a(\rho(\zeta), \rho(\zeta))]\, dQ + \int_Q \langle \varphi, q(0, w)\rangle \, dQ
$$

$$
= -\int_\Sigma [\partial(A\zeta, q(0, w)) + \gamma q w \frac{\partial w_{tt}}{\partial \nu}]\, d\Sigma + L(\zeta). \tag{4.158}
$$

Proof. Set $\eta_1 = (W, 0)$. We multiply the equation (4.147) by $q\eta_1$ and integrate it over Q.

Let $P = \zeta - \gamma(0, \Delta w)$. Then

$$\langle P_{tt}, q\eta_1 \rangle = q\langle P_t, \eta_1 \rangle_t - q\langle P_t, \eta_{1t} \rangle = q\langle W_t, W \rangle_t - q|W_t|^2. \qquad (4.159)$$

On the other hand, it is easy to check that

$$D(qW) = qDW + W \otimes Dq. \qquad (4.160)$$

Then

$$\Upsilon(q\eta_1) = q\Upsilon(\eta_1) + \frac{1}{2}(W \otimes Dq + Dq \otimes W) = q\Upsilon(\zeta) + \mathrm{lo}\,(\zeta). \qquad (4.161)$$

Thus

$$B(\zeta, q\eta_1) = a(\Upsilon(\zeta), \Upsilon(q\eta_1)) = qa(\Upsilon(\zeta), \Upsilon(\zeta)) + \mathrm{lo}\,(\zeta). \qquad (4.162)$$

Using the formula (4.162) and the Green formula (4.34), we obtain

$$\int_Q \langle \mathcal{A}\zeta, q\eta_1 \rangle \, dQ = \int_Q qa(\Upsilon(\zeta), \Upsilon(\zeta)) \, dQ$$
$$- \int_\Sigma \partial(\mathcal{A}\zeta, q(W, 0)) \, d\Sigma + L(\zeta). \qquad (4.163)$$

Then the identity (4.157) follows from the relations (4.163), (4.159), and (4.147).

Next, we use a multiplier $q\eta_2$ where $\eta_2 = (0, w)$. First, we have

$$\langle P_{tt}, q\eta_2 \rangle = -\gamma \,\mathrm{div}\,(qw\nabla w_{tt}) + \gamma\langle \nabla w_{tt}, \nabla(qw) \rangle + qw_{tt}w$$
$$= -\gamma \,\mathrm{div}\,(qw\nabla w_{tt}) + \gamma\langle \nabla w_t, \nabla(qw) \rangle_t + q(w_t w)_t$$
$$- qw_t^2 + \mathrm{lo}\,(\zeta). \qquad (4.164)$$

It is easy to check that

$$D^2(qw) = qD^2 w + Dw \otimes Dq + Dq \otimes Dw + wD^2 q. \qquad (4.165)$$

Then

$$\rho(q\eta_2) = q\rho(\zeta) + \mathrm{lo}\,(\zeta). \qquad (4.166)$$

Since $\Upsilon(q\eta_2) = \mathrm{lo}\,(\zeta)$, the formula (4.166) yields

$$B(\zeta, q\eta_2) = q\gamma a(\rho(\zeta), \rho(\zeta)) + \mathrm{lo}\,(\zeta) \qquad (4.167)$$

where the bilinear form $B(\cdot, \cdot)$ is given by (4.18). By the Green formula (4.34) and the relation (4.167), we obtain

$$\int_Q \langle \mathcal{A}\zeta, q\zeta \rangle \, dQ = \int_Q q\gamma a(\rho(\zeta), \rho(\zeta)) \, dQ$$
$$- \int_\Sigma (\mathcal{A}\zeta, q(0, w)) \, d\Sigma + L(\zeta). \qquad (4.168)$$

Finally, the identity (4.158) follows from the formulas (4.164) and (4.168).

□

Theorem 4.7 *Let* $\zeta = (W, w)$ *solve the problem* (4.147) *and let a vector field* V *satisfy the assumption* (4.149). *Then*

$$\int_\Sigma [|\zeta_t|^2 + \gamma|Dw_t|^2 - B(\zeta, \zeta)]\langle V, \nu \rangle \, d\Sigma$$

$$+ 2 \int_\Sigma [\partial(\mathcal{A}\zeta, m(\zeta)) + \gamma V(w)\frac{\partial w_{tt}}{\partial \nu}] \, d\Sigma$$

$$= 2Z(t)|_0^T + 2 \int_Q a(\Upsilon(\zeta), G(V, DW)) \, dQ$$

$$+ 2 \int_Q \vartheta(x)[|\zeta_t|^2 + \gamma a(\rho(\zeta), \rho(\zeta)) - a(\Upsilon(\zeta), \Upsilon(\zeta))] \, dQ$$

$$- \int_Q \langle \varphi, m(\zeta) \rangle \, dQ + \text{lo}\,(\zeta) \tag{4.169}$$

where $Z(t)$ *is given by* (4.152) *and the tensor field* $G(V, DW)$ *is defined by* (4.131).

Proof. Since $\rho(\zeta)$ is symmetric, we have from (3.27) and (3.28) that

$$\langle \rho(\zeta), G(V, \rho(\zeta)) \rangle = \langle \rho(\zeta), \rho(\zeta)(\cdot, D.V) \rangle = \vartheta(x)|\rho(\zeta)|^2;$$

$$\operatorname{tr} \rho(\zeta) \operatorname{tr} G(V, \rho(\zeta)) = \operatorname{tr} \rho(\zeta) \operatorname{tr} \rho(\zeta)(\cdot, D.V) = \vartheta(\operatorname{tr} \rho(\zeta))^2,$$

which yields

$$a(\rho(\zeta), G(V, \rho(\zeta))) = \vartheta a(\rho(\zeta), \rho(\zeta)). \tag{4.170}$$

Moreover, the assumption (4.149) yields

$$DV(Dw_t, Dw_t) = \vartheta|Dw_t|^2 \quad \text{and} \quad \operatorname{div} V = 2\vartheta. \tag{4.171}$$

Using the relations (4.170) and (4.171) in the identity (4.151), we obtain the identity (4.169). $\quad\square$

4.5 Escape Vector Field and Escape Region for the Shallow Shell

Some geometric conditions on the middle surface are necessary to obtain continuous observability estimates for a shallow shell. In this section we will introduce such conditions, study their existence, and present some examples.

We assume that the ellipticity of the strain energy of the shallow shell holds, that is, there is a $\lambda_0 > 0$ such that

$$\lambda_0 B(\zeta, \zeta) \geq \|DW\|_{L^2(\Omega, T^2)}^2 + \gamma\|D^2 w\|_{L^2(\Omega, T^2)}^2 \tag{4.172}$$

for all $\zeta = (W, w) \in H^1(\Omega, \Lambda) \times H^2(\Omega)$ where the bilinear form $\mathcal{B}(\cdot, \cdot)$ is defined by (4.21). The above inequality is established by Theorem 4.2 in Section 4.2 if displacement vector fields of the shallow shell are fixed along a portion Γ_0 of the boundary with a positive length.

Geometric Condition for Boundary Control

Definition 4.1 *A vector field V on Ω is said to be an escape vector field for the shallow shell if the following assumptions hold:*
(i) *There is a function ϑ on Ω such that*

$$DV(X, X) = \vartheta(x)|X|^2 \quad \text{for all} \quad X \in M_x, \quad x \in \Omega. \tag{4.173}$$

(ii) *Let*
$$\iota(x) = \langle DV, \mathcal{E} \rangle / 2 \quad \text{for} \quad x \in \Omega \tag{4.174}$$

where $\mathcal{E}$ denotes the volume element of the surface M. Suppose that the functions $\vartheta(x)$ and $\iota(x)$ satisfy the inequality

$$2 \min_{x \in \Omega} \vartheta(x) > \lambda_0(1 + \mu) \max_{x \in \Omega} |\iota(x)| \tag{4.175}$$

where λ_0 is given in (4.172) and μ is Poisson's coefficient of the material.

Remark 4.2 *Clearly, if a vector field is escaping for the shallow shell, then it is also escaping for plate, see Section 3.2.*

Remark 4.3 *In Section 4.6 we shall show that existence of an escape vector field for the shallow shell can guarantee controllability/stabilization.*

Remark 4.4 *We shall prove that on the 2-dimensional middle surface Ω there always exists a vector field V satisfying the assumption (4.173), see Theorem 4.8 later. Then the key to being an escape vector field for the shallow shell is the assumption (4.175). By Theorem 2.3, a necessary condition to the assumption (4.175) is that there is no closed geodesics inside the middle surface Ω.*

Remark 4.5 *The function $\iota(x)$, given by (4.174), depicts the symmetry of the covariant differential DV of the vector field V in some sense. Clearly, if DV is symmetric, then $\iota(x) = 0$ for all $x \in \Omega$.*

If the shell is flat, a plate, then $M = \mathbb{R}^2$. For any $x^0 \in \mathbb{R}^2$, set $V(x) = x - x^0$. It is easy to check that

$$DV = g \quad \text{and} \quad \iota(x) = 0.$$

Then $V = x - x^0$ is an escape vector field for the flat shell where $\vartheta(x) = 1$ and $\iota(x) = 0$.

We shall show that if M is a surface of $\mathbb{R}^3$ that is of constant curvature or of revolution, then there exists a vector field V on the whole M with $\iota(x) = 0$

for all $x \in M$, see Theorems 4.9 and 4.10 below. We will also present in such cases curvature conditions for existence of an escape vector field for the shallow shell, see Corollaries 4.3 and 4.4 below.

First, we have

Lemma 4.9 *Let $V \in \mathcal{X}(\Omega)$ satisfy the assumption (4.173). Then the two-order tensor field DV can be decomposed as*

$$DV = \vartheta(x)g + \iota(x)\mathcal{E} \quad for \quad x \in \Omega. \tag{4.176}$$

Proof. We decompose DV into a sum of the symmetric part and the skew part as

$$DV = \frac{1}{2}(DV + D^*V) + \frac{1}{2}(DV - D^*V). \tag{4.177}$$

Since the space $\Lambda^2(M)$ of 2-forms on the surface M is one-dimensional there is a function $q(x)$ on Ω such that

$$(DV - D^*V)/2 = q(x)\mathcal{E} \quad for \quad x \in \Omega. \tag{4.178}$$

Using the formula $\langle D^*V, \mathcal{E} \rangle = \langle D, \mathcal{E}^* \rangle = -\langle DV, \mathcal{E} \rangle$ and the formulas (4.174) and (4.178), we obtain

$$\begin{aligned} q(x) &= q(x)\langle \mathcal{E}, \mathcal{E} \rangle/2 = (\langle DV, \mathcal{E} \rangle - \langle D^*V, \mathcal{E} \rangle)/2 \\ &= \langle DV, \mathcal{E} \rangle/2 = \iota(x) \end{aligned} \tag{4.179}$$

since $\langle \mathcal{E}, \mathcal{E} \rangle = 2$.

Finally the formula (4.176) follows from (3.26), (4.179), (4.178) and (4.177).   □

Consider a set Θ which consists of all the C^∞ functions q such that there is a region $\aleph \subset M$ satisfying that $\overline{\Omega} \subset \aleph$ and

$$\Delta q = \kappa(x) \quad for all \quad x \in \aleph \tag{4.180}$$

where κ is the Gaussian curvature function of the surface M. It is easily checked that Θ is nonempty by an elliptic boundary value problem for the Laplacian since $\Omega \subset M$ is a bounded region with a nonempty boundary Γ. For a given $q \in \Theta$, consider a metric, given by

$$\hat{g} = e^{2q}g \tag{4.181}$$

where g is the induced metric of M in $\mathbb{R}^3$. Denote by $(\aleph, \hat{g})$ the Riemannian manifold with the metric (4.181) and by $\hat{\rho}(x)$ the distance function on $(\aleph, \hat{g})$ from $x_0 \in \aleph$ to $x \in \aleph$, respectively. Let $\hat{D}$ be the covariant differential of $(\aleph, \hat{g})$ in the metric $\hat{g}$.

The following theorem shows that there always exists a vector field satisfying the assumption (4.173) on the whole middle surface Ω and this vector field is locally escaping for the shallow shell.

Theorem 4.8 *Let $q \in \Theta$ and $x_0 \in \aleph$ be given. Let*

$$V(x) = \hat{\rho}(x)\hat{D}\rho(x) \quad \text{for} \quad x \in \aleph. \tag{4.182}$$

Then the vector field V satisfies the assumption (4.173) where $\vartheta = 1 - V(q)$ and $\iota(x) = \langle Dq \otimes V, \mathcal{E} \rangle$ such that

$$\lim_{x \to x_0} \vartheta(x) = 1 \quad \text{and} \quad \lim_{x \to x_0} \iota(x) = 0. \tag{4.183}$$

Locally, V is an escape vector field for the shallow shell.

Proof. We compute in a coordinate. Let (x_1, x_2) be a coordinate on $\aleph$. Set

$$g_{ij} = \langle \frac{\partial}{\partial x_i}, \frac{\partial}{\partial x_j} \rangle \quad \text{and} \quad \hat{g}_{ij} = \hat{g}(\frac{\partial}{\partial x_i}, \frac{\partial}{\partial x_j}). \tag{4.184}$$

for $1 \leq i, j \leq 2$. It follows from (4.181) that $\hat{g}_{ij} = e^{2q} g_{ij}$.

Denote by Γ_{ij}^k and by $\hat{\Gamma}_{ij}^k$ the coefficients of connections D and $\hat{D}$, respectively. As a computation in (3.34) in Chapter 3, we have the following formulas

$$\Gamma_{ij}^k = \hat{\Gamma}_{ij}^k - \delta_{ik} q_{x_j} - \delta_{jk} q_{x_i} + g_{ij} \sum_{l=1}^{2} g^{kl} q_{x_l}, \tag{4.185}$$

$$D_X V = \hat{D}_X V - V(q)X - X(q)V + \langle X, V \rangle Dq. \tag{4.186}$$

From the formulas (4.185), we obtain

$$\Delta q - \kappa + \hat{\kappa} e^{2q} = 0 \quad \text{in} \quad \aleph \tag{4.187}$$

where $\hat{\kappa}$ is the Gaussian curvature function of $(\aleph, \hat{g})$ in the metric $\hat{g}$ and $\left(g^{ij} \right) = \left(g_{ij} \right)^{-1}$. The relations (4.181) and (4.187) imply that

$$\hat{\kappa} = 0 \quad \text{for} \quad x \in \aleph, \tag{4.188}$$

that is, the Riemannian manifold $(\aleph, \hat{g})$ has zero curvature. It follows from Theorem 3.5 that

$$DV(X, X) = \langle D_X V, X \rangle = [1 - V(q)]|X|^2. \tag{4.189}$$

The assumption (4.173) is true for the vector field V where $\vartheta(x) = 1 - V(q)$.

Let $x \in \aleph$ be given. Let e_1, e_2 be an orthonormal basis of M_x with the positive orientation. From the formulas (4.186) and (4.181), we have

$$\begin{aligned} DV(e_1, e_2) &= \langle D_{e_2} V, e_1 \rangle = \langle \hat{D}_{e_2} V, e_1 \rangle + e_1(q)\langle V, e_2 \rangle - e_2(q)\langle V, e_1 \rangle \\ &= e^{-2q} \hat{g}(e_1, e_2) + e_1(q)\langle V, e_2 \rangle - e_2(q)\langle V, e_1 \rangle \\ &= e_1(q)\langle V, e_2 \rangle - e_2(q)\langle V, e_1 \rangle; \end{aligned} \tag{4.190}$$

$$DV(e_2, e_1) = e_2(q)\langle V, e_1\rangle - e_1(q)\langle V, e_2\rangle. \qquad (4.191)$$

It follows from the formulas (4.190) and (4.191) that

$$\iota(x) = \langle DV, \mathcal{E}\rangle/2 = \Big(DV(e_1, e_2) - DV(e_2, e_1)\Big)/2 = \langle Dq \otimes V, \mathcal{E}\rangle.$$

Finally, from the relations (4.182) and (4.181), we have

$$|V| = e^{-q}|V|_{\hat{g}} = \hat{\rho}(x)e^{-q}$$

so that the limits in (4.183) follow.   □

Remark 4.6 *For a detailed computation for the formula (4.187), we refer to Chapter 5 of the book [187].*

Theorem 4.9 *Let a surface M of $\mathbb{R}^3$ be of constant curvature κ. For $x_0 \in M$ given, let ρ be the distance function from $x \in M$ to $x_0 \in M$ on (M, g), i.e., $\rho(x) = d_g(x_0, x)$. Set $V = h(\rho)D\rho$ where h is defined by*

$$h(\rho) = \begin{cases} \sin\sqrt{\kappa}\rho, & \kappa > 0, \\ \rho, & \kappa = 0, \\ \sinh\sqrt{-\kappa}\rho, & \kappa < 0, \end{cases} \qquad for \quad x \in \exp_{x_0}(\Sigma(x_0)) \qquad (4.192)$$

where $\exp_{x_0}(\Sigma(x_0))$ is the interior of the cut locus of the point x_0. Then

$$DV = \vartheta(x)g \quad and \quad \iota(x) = 0 \quad for \quad x \in \exp_{x_0}(E(x_0)) \qquad (4.193)$$

where

$$\vartheta(x) = \begin{cases} \sqrt{\kappa}\cos\sqrt{\kappa}\rho(x), & \kappa > 0, \\ 1, & \kappa = 0, \\ \sqrt{-\kappa}\cosh\sqrt{-\kappa}\rho(x), & \kappa < 0. \end{cases} \qquad (4.194)$$

Proof. It follows from Theorem 1.21 that

$$D^2\rho(\tau, \tau) = \begin{cases} \sqrt{\kappa}\cot\sqrt{\kappa}\rho(x), & \kappa > 0, \\ \frac{1}{\rho(x)}, & \kappa = 0, \\ \sqrt{-\kappa}\coth\sqrt{-\kappa}\rho(x), & \kappa < 0, \end{cases} \qquad (4.195)$$

where $\tau \in M_x$ is such that τ, $D\rho$ forms an orthonormal basis of M_x and $D^2\rho$ is the Hessian of the distance ρ. Since $D^2\rho(D\rho, X) = 0$ for any $X \in M_x$, it follows from the formula (4.195) that

$$\begin{aligned} DV(X, Y) &= h'(\rho)X(\rho)Y(\rho) + h(\rho)D^2\rho(X, Y) \\ &= h'(\rho)\langle X, D\rho\rangle\langle Y, D\rho\rangle + h(\rho)D^2\rho(\tau, \tau)\langle X, \tau\rangle\langle Y, \tau\rangle \\ &= \vartheta(x)\langle X, Y\rangle. \end{aligned}$$

Moreover, the symmetry of DV implies $\iota(x) = 0$.   □

Remark 4.7 *It should be noted that, even if the surface M has zero Gaussian curvature, the vector field V, defined in Theorem 4.9, is only escaping on the open set $\exp_{x_0} \Sigma(x_0)$. It may not be escaping on the whole surface M. For example, let M be a cylindrical surface. Then M has zero Gaussian curvature. Since a big circle on the cylinder is a closed geodesic for the cylinder, a necessary condition, Theorem 2.3, implies that there is no escape vector field on the middle surface Ω if Ω contains a big circle. In fact it is easy to show that for any $x_0 \in M$, $\exp_{x_0} \Sigma(x_0) \neq M$.*

The following corollaries are immediate from Theorem 4.9.

Corollary 4.2 *Let M be a cylinder and let $\Omega \subset M$ be a middle surface. If there is a generator of the cylinder which does not intersect Ω, then there exists an escape vector field for the shallow shell.*

Corollary 4.3 *Let M be a surface in $\mathbb{R}^3$ with constant curvature κ which is simply connected.*

If $\kappa < 0$, then for any middle surface $\Omega \subset M$ there exists an escape vector field for the shallow shell.

Let $\kappa > 0$. If there is a geodesic ball $B(x_0, \delta) \subset M$ centered at some point $x_0 \in M$ with radius $0 < \delta < \pi/(2\sqrt{\kappa})$ such that the middle surface $\Omega \subset B(x_0, \delta)$, then an escape vector field for the shallow shell exists.

Theorem 4.10 *Let f be a function on $\mathbb{R}$. Consider a surface of revolution, given by*

$$M = \{ (x,y,z) \,|\, z = f(r),\, r = \sqrt{x^2 + y^2},\, (x,y) \in \mathbb{R}^2 \}. \tag{4.196}$$

The Gaussian curvature is

$$\kappa(p) = \frac{f'(r)f''(r)}{r(1 + f'^2)^2} \quad for \quad p = (x,y,z) \in M. \tag{4.197}$$

Then there exist a vector field V and a function ϑ such that

$$DV = \vartheta(p)g \quad and \quad \iota(p) = 0 \quad for \quad p \in M \tag{4.198}$$

where g is the induced metric of M in $\mathbb{R}^3$. Furthermore, if

$$\int_{\kappa(t)>0} t\kappa(t)\,dt \leq 1, \tag{4.199}$$

then

$$\vartheta(p) > 0 \quad for \ all \quad p \in M, \tag{4.200}$$

where

$$\kappa(t) = \frac{f'(r(t))f''(r(t))}{r(t)(1 + f'^2(r(t)))^2} \quad for \quad t > 0, \tag{4.201}$$

and the function $r(t)$ is given by the equation

$$t = \int_0^{r(t)} \sqrt{1 + f'^2(s)}\,ds \quad for \quad t \geq 0. \tag{4.202}$$

Proof. Denote $p_0 = (0, 0, f(0)) \in M$ and denote by $\rho(p) = d_g(p, p_0)$ the distance function in the induced metric g from $p \in M$ to p_0.

Let $p = (x, y, f(r)) \in M$ where $r^2 = x^2 + y^2$. Since M is of revolution, it is easily checked that a curve : $[0, \rho(p)] \to M$, given by

$$\sigma(t) = \left(r(t)\frac{x}{r}, r(t)\frac{y}{r}, f(r(t)) \right) \qquad (4.203)$$

is a unique minimizing geodesic, parameterized by arc length, which joins p_0 to p.

Let $\tau_0 \in M_0$ be such that $\dot{\sigma}(0), \tau_0$ forms an orthonormal basis of M_0. Let $\tau = \tau(t) \in M_{\sigma(t)}$ be the parallel transport of τ_0 along the geodesic $\sigma(t)$ for $t > 0$. By Lemma 1.5, there is a normal Jacobi field $J(t) = h(t)\tau(t)$ along σ such that

$$\begin{cases} h''(t) + \kappa(t)h(t) = 0 & t > 0, \\ h(0) = 0, \quad h'(0) = 1, \end{cases} \qquad (4.204)$$

and

$$D^2\rho(\tau, \tau) = \frac{h'(\rho)}{h(\rho)} \quad \text{for} \quad \rho > 0. \qquad (4.205)$$

Set $V = h(\rho)D\rho$. We shall prove that $DV = h'(\rho)g$ for all $p \in M$. By (4.205), we obtain

$$\begin{aligned} DV(X, Y) &= h'(\rho)X(\rho)Y(\rho) + h(\rho)D^2\rho(X, Y) \\ &= h'(\rho)\langle X, D\rho\rangle\langle Y, D\rho\rangle + h(\rho)D^2\rho(\tau, \tau)\langle X, \tau\rangle\langle Y, \tau\rangle \\ &= h'(\rho)\langle X, Y\rangle \end{aligned}$$

where the formula $D^2\rho(D\rho, X) = 0$ for $X \in M_p$ is used.

Let the assumption (4.199) hold. We will prove that $h'(t) > 0$ for all $t \geq 0$. Set $\tilde{\kappa}(t) = \max(\kappa(t), 0)$. Then the assumption (4.199) means that

$$\int_0^\infty t\tilde{\kappa}(t)\, dt \leq 1. \qquad (4.206)$$

Let ϕ be the solution to the problem

$$\begin{cases} \phi''(t) + \tilde{\kappa}(t)\phi(t) = 0 & t > 0, \\ \phi(0) = 0, \quad \phi'(0) = 1. \end{cases} \qquad (4.207)$$

It follows from the above equations that

$$\phi(t) > 0 \quad \text{and} \quad \phi'(t) > 0 \quad \text{on} \quad (0, \infty). \qquad (4.208)$$

Then Sturm's theorem implies that

$$h(t) \geq \phi(t) > 0 \quad \text{on} \quad (0, \infty) \qquad (4.209)$$

since $\tilde{\kappa}(t) \geq \kappa(t)$ for $t > 0$.

Set $u(t) = h'(t)\phi(t) - h(t)\phi'(t)$. By the formulas (4.204), (4.207), and (4.209), we obtain

$$u'(t) = (\tilde{\kappa} - \kappa)\phi h \geq 0 \quad \text{on} \quad (0, \infty)$$

which yields $u(t) \geq u(0) = 0$, that is, by (4.208) and (4.209),

$$h'(t) \geq h(t)\frac{\phi'(t)}{\phi(t)} > 0 \quad \text{on} \quad (0, \infty).$$

□

From Theorem 4.10, we have

Corollary 4.4 *Let a surface M of revolution be given by (4.196). If the curvature condition (4.199) holds true, then for any middle surface $\Omega \subset M$, there exists an escape vector field for the shallow shell.*

Geometric Condition for Internal Control Next, we consider an open region of the middle surface where a control action will take place. This control action may be for exact controllability in the open-loop case, or for stabilization by an interior feedback. We seek geometric conditions on such regions G. Clearly, one of the choices is $G = \Omega$ but this is trivial. We are particularly interested in the case where G is not the whole middle surface and is as small as possible in some sense. The goal of the following treatment is to try to choose a control region small.

Definition 4.2 *A region $G \subset \Omega$ is said to be an escape region for the shallow shell if the following assumptions hold: There are finite subregions $\Omega_i \subset \Omega$ with boundaries Γ_i for $1 \leq i \leq J$ where J is some natural number such that*
(i)
$$\Omega_i \cap \Omega_j = \emptyset \quad \text{for all} \quad 1 \leq i < j \leq J; \tag{4.210}$$

(ii) *for each Ω_i there is an escape vector field V_i such that*

$$DV_i(X, X) = \vartheta_i(x)|X|^2 \quad \text{on} \quad \Omega_i, \tag{4.211}$$

$$2 \min_{x \in \Omega_i} \vartheta_i(x) > \lambda_0(1 + \mu) \max_{x \in \Omega_i} \iota_i(x), \tag{4.212}$$

where $\iota_i(x) = \langle DV_i, \mathcal{E}\rangle/2$ for all $1 \leq i \leq J$;
(iii)
$$G \supset \overline{\Omega} \cap \mathcal{N}_\varepsilon\left[\cup_{i=1}^J \Gamma_{i0} \cup \left(\Omega/\cup_{i=1}^J \Omega_j\right)\right] \tag{4.213}$$

where $\varepsilon > 0$ small and

$$\mathcal{N}_\varepsilon(S) = \cup_{x \in S}\{y \in \Omega \,|\, d_g(y, x) < \varepsilon\} \quad \text{for} \quad S \subset M, \tag{4.214}$$

$$\Gamma_{i0} = \{x \in \Gamma_i \,\langle V_i(x), \nu_i(x)\rangle > 0\},$$

ν_i *are normals of Ω_i pointing outside of Ω_i,*

for all $1 \leq i \leq J$

Remark 4.8 *If there exists an escape vector field V on the whole middle surface Ω for the shallow shell, then the relations (4.211) and (4.212) hold true for $V_1 = V$. In this case, we can take $J = 1$ and $\Omega_1 = \Omega$. An escape region G needs only to be supported in a neighborhood of Γ_0, where*

$$\Gamma_0 = \{\, x \in \partial\Omega \,|\, \langle V(x), n(x) \rangle > 0 \,\}. \tag{4.215}$$

This roughly means that we can dig out almost the whole Ω from the domain $\overline{\Omega}$ and that the escape region is supported only in a neighborhood of a subset of the boundary.

The following theorem is clearly true.

Theorem 4.11 *If there exists an escape vector field V for the shallow shell, then an escape region can be supported in a neighborhood of Γ_0 where Γ_0 is given by (4.215).*

In general an escape vector field for the shallow shell does not exist on the whole middle surface Ω. However, by Theorem 4.8, on a geodesic ball with a small radius an escape vector field exists. Then we can dig out from Ω as many as possible geodesic balls with radii small and let an escape region be supported in a neighborhood of the remaining. Therefore the following results are immediate.

Theorem 4.12 *For $\varepsilon > 0$ given, we can choose an escape region $G \subset \overline{\Omega}$ with*

$$\mathrm{meas}\,(G) < \varepsilon$$

where $\mathrm{meas}\,(G)$ *is the 2-dimensional Lebesgue measure of G.*

Here we give some examples verifying existence of an escape vector field for the shallow shell and showing the structure of an escape region.

Example 4.1 *Let a surface M be a cylinder given by*

$$M = \{\, (x,y,z) \,|\, x^2 + y^2 = 1,\, z \in \mathbb{R} \,\}.$$

Let

$$\Phi(x,y) = \{\, (x,y,z) \in M \,|\, z \in \mathbb{R} \,\}$$

be a generatrix of M for (x,y) fixed with $x^2 + y^2 = 1$.

(a) Let a middle surface $\Omega \subset M$ be such that there is $(x_0, y_0) \in \mathbb{R}^2$ with $x_0^2 + y_0^2 = 1$ satisfying $\Omega \subset M/\Phi(x_0, y_0)$. Let $p_0 = (-x_0, -y_0, 0) \in M$ and let $\rho(p) = d_g(p, p_0)$ be the distance function from $p \in M$ to p_0. Clearly, for any $p \in M/\Phi(x_0, y_0)$ there is a unique minimizing geodesic on $M/\Phi(x_0, y_0)$ connecting p and p_0. Then $M/\Phi(x_0, y_0) \subset \exp_{p_0} \Sigma(p_0)$. Let $V = \rho D\rho$. By Theorem 4.9, the relations (4.173) and (4.175) hold true. This means that there exists an escape vector field for the shallow shell.

(b) *Let a middle surface* Ω *be a segment of the cylinder, given by*

$$\Omega = \{\,(x, y, z)\,|\,x^2 + y^2 = 1,\ z \in (-1, 1)\,\}$$

with boundary

$$\Gamma = \{\,(x, y, -1)\,|\,x^2 + y^2 = 1\,\} \cup \{\,(x, y, 1)\,|\,x^2 + y^2 = 1\,\}$$

Since big circles $\omega(z_0) = \{\,(x, y, z_0)\,|\,x^2 + y^2 = 1\,\} \subset \Omega$ *are closed geodesics for all* $z_0 \in (-1, 1)$, *an escape vector field on the whole middle surface* Ω *does not exit, by Theorem 2.3.*

We consider the structure of an escape region for this middle surface. Let $p_0 = (0, -1, 0) \in \Omega$. *Let* $\Omega_1 = \Omega/\Phi(0, 1)$ *and* $V_1 = \rho D \rho$. *By Theorem 4.9, the relations (4.211) and (4.212) hold true with* $J = 1$. *Thus an escape region of the middle surface can be supported in a neighborhood of* $\Gamma \cup \Phi(0, 1)$, *see Figure 4.1.*

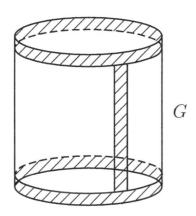

FIGURE 4.1: An escape region for a cylinder.

Example 4.2 *Let M be the unit sphere, given by*

$$M = \{\,(x, y, z)\,|\,x^2 + y^2 + z^2 = 1\,\}.$$

For $p \in M$ *and* $\delta > 0$, *let*

$$B(p, \delta) = \{\,q\,|\,q \in M,\ d_g(p, q) < \delta\,\}$$

be the geodesic ball centered at p with radius δ.

(a) *Let a middle surface* $\Omega \subset M$ *be a spherical cap such that there is* $p \in M$ *satisfying*

$$\overline{\Omega} \subset B(p, \pi/2).$$

By Theorem 4.9, there exists an escape vector field for the shallow shell.

(b)   *Let* $\Omega = M$ *be a middle surface without boundary. Then by Theorem 2.3 there is no escape vector field on the whole* Ω. *We consider the structure of an escape region of* Ω. *Let*

$$\omega = \{\, (x, y, 0) \,|\, x^2 + y^2 = 1 \,\}.$$

Then ω *is a big circle on* Ω. *For* $\varepsilon > 0$ *given small, let*

$$\Omega_1 = B\Big((0,0,1), \pi/2 - \varepsilon\Big) \quad and \quad \Omega_2 = B\Big((0,0,-1), \pi/2 - \varepsilon\Big).$$

By Theorem 4.9, there are escape vector fields V_i *satisfying the relations (4.211) and (4.212) with* $i = 1, 2$. *Thus an escape region of* Ω *can be supported in a neighborhood of the big circle* ω, *see Figure 4.2.*

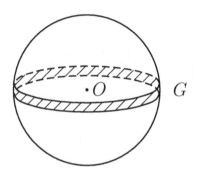

FIGURE 4.2: An escape region for a closed spherical shell.

Example 4.3 *Let a middle surface* Ω *be a portion of a surface* M *of revolution, given by*

$$M = \{\, (x, y, z) \,|\, z = \log(1 + x^2 + y^2), \; (x, y) \in \mathbb{R}^2 \,\}. \tag{4.216}$$

Set $f(r) = \log(1 + r^2)$. *It is easy to check that the Gaussian curvature is*

$$\kappa(p) = -\frac{2(1 + r^2)}{[1 + (1 + r^2)^2]^2} < 0 \tag{4.217}$$

where $p = (x, y, f(r)) \in M$.

Then the curvature condition (4.199) holds for the surface M *because the set* $\{\, t \,|\, \kappa(t) > 0 \,\}$ *is empty. By Corollary 4.4, an escape vector field for the shallow shell exists for any middle surface* $\Omega \subset M$.

Example 4.4 *Consider a helicoid, defined by*

$$M = \{\, \alpha(t, s) \,|\, (t, s) \in \mathbb{R}^2, \; t > 0 \,\} \tag{4.218}$$

where

$$\alpha(t, s) = (t \cos s, t \sin s, c_0 s), \quad c_0 > 0. \tag{4.219}$$

The Gaussian curvature is $-c_0^2/(t^2 + c_0^2)^2$.
   We set

$$E_1 = \alpha_t = (\cos s, \sin s, 0);$$

$$E_2 = \alpha_s / \sqrt{t^2 + c_0^2} = (-t \sin s, \cos s, c_0) / \sqrt{t^2 + c_0^2}.$$

Then E_1, E_2 *forms a frame field on the whole surface M. We obtain*

$$D_{E_1} E_1 = 0, \quad D_{E_2} E_1 = t E_2 / (t^2 + c_0^2),$$

$$D_{E_1} E_2 = 0, \quad D_{E_2} E_2 = -t E_1 / (t^2 + c_0^2),$$

where D is the Levi-Civita connection of the surface M.
   For any $c > 0$, *set*

$$V_c = f_c E_1 + h E_2$$

where

$$f_c = \sqrt{t^2 + c_0^2} \Big(\int_0^t (t^2 + c_0^2)^{-1/2} \, dt + c \Big), \quad h = \sqrt{t^2 + c_0^2} \, s.$$

A computation yields

$$DV_c = \vartheta_c(x) g + \iota(x) \mathcal{E} \quad \text{for} \quad c > 0$$

where

$$\vartheta_c = 1 + t f_c / (t^2 + c_0^2) \quad \text{and} \quad \iota(x) = -st / \sqrt{t^2 + c_0^2}.$$

Clearly, for any bounded middle surface $\Omega \subset M$ *with* $\overline{\Omega} \subset M$ *and for any* $c_1 > 0$, *there is* $c > 0$ *large enough such that*

$$\min_{x \in \Omega} \vartheta_c(x) \geq c_1 \max_{x \in \Omega} \iota(x).$$

Then an escape vector field for the shallow shell exists.

4.6 Observability Inequalities. Exact Controllability

   In this section we shall derive some observability estimates of the shallow shell from boundary and also from interior. Those estimates are necessary both for exact controllability and stabilization of the shallow shell.
   Suppose that the ellipticity (4.172) of the strain energy of the shallow shell

holds.

Observability Estimate from Boundary Let V be an escape vector field for the shallow shell with functions ϑ and ι such that the assumptions (4.173) and (4.175) hold. Set

$$\sigma_0 = \max_{x \in \Omega} |V|, \quad \sigma_1 = \min_{x \in \Omega} \vartheta(x) - \lambda_0(1 + \mu) \max_{x \in \Omega} |\iota(x)|/2. \tag{4.220}$$

For $T > 0$ given, set

$$Q = \Omega \times (0, T), \quad \Sigma = \Gamma \times (0, T),$$

$$\Sigma_0 = \Gamma_0 \times (0, T), \quad \Sigma_1 = \Gamma_1 \times (0, T).$$

Let $\zeta = (W, \zeta) \in H^1(\Omega, \Lambda) \times L^2(\Omega)$ be a solution to the following problem

$$\zeta_{tt} - \gamma(0, \Delta w_{tt}) + \mathcal{A}\zeta = 0 \quad \text{in} \quad \Omega \tag{4.221}$$

where $\mathcal{A}$ is the shallow shell operator, given by (4.36), and γ is given in (4.19), respectively.

We define, for $(t, x) \in \Sigma$,

$$\begin{aligned}
\text{SB} &= [|\zeta_t|^2 + \gamma|Dw_t|^2 - B(\zeta, \zeta)]\langle V, \nu \rangle \\
&+ \partial\left(\mathcal{A}\zeta, 2m(\zeta) + (\varrho W, -\vartheta w)\right) + \gamma\left(2V(w) - \vartheta w\right)\frac{\partial w_{tt}}{\partial \nu} \tag{4.222}
\end{aligned}$$

where $m(\zeta) = (D_V W, V(w))$ and $\varrho = 2\vartheta - \sigma_1$. By (4.220)

$$\varrho \geq \lambda_0(1 + \mu) \max_{x \in \Omega} |\iota(x)|/2. \tag{4.223}$$

We need the following two lemmas in order to obtain observability estimates.

Lemma 4.10 *Let $Z(t)$ be given by (4.152). Then*

$$|Z(t)| \leq \sigma_0 \lambda_0 E(t) + L(\zeta) \tag{4.224}$$

where $L(\zeta)$ are the lower terms satisfying the estimate (4.150).

Proof. It is easy to check that

$$|D(V(w))| \leq |D^2 w||V| + c|Dw| \leq \sigma_0|D^2 w| + c|Dw|.$$

We have

$$\begin{aligned}
2\langle \zeta_t, m(\zeta) \rangle &+ 2\gamma\langle Dw_t, D(V(w)) \rangle \\
&\leq 2\sigma_0|W_t||DW| + 2\sigma_0|w_t||Dw| + 2\gamma|Dw_t||D(V(w))| \\
&\leq \sigma_0 \, e(\zeta) + \text{lo}\,(\zeta) \tag{4.225}
\end{aligned}$$

for $\varepsilon > 0$ small. The inequality (4.224) follows from the ellipticity (4.172) and the formula (4.225). $\qquad \square$

In the following lemma the assumption (4.175) plays a key role.

Lemma 4.11 *Let V be an escape vector field for the shallow shell. Then*

$$\sigma_1 \int_\Omega a\Big(\Upsilon(\zeta), \Upsilon(\zeta)\Big) \, dx \le \int_\Omega a\Big(\Upsilon(\zeta), G(V, DW)\Big) \, dx + CL(\zeta) \qquad (4.226)$$

where $\Upsilon(\zeta)$ is the strain tensor of the middle surface, given by (4.15), $a(\cdot, \cdot)$ is a bilinear form, given by (4.20), the tensor field $G(V, DW)$ of rank 2 is defined by (4.131) and $\sigma_1 > 0$ is given in (4.220).

Proof. Given $x \in \Omega$. We compute at x.

Since $DW + D^*W$ is symmetric, there is an orthnonrmal basis e_1, e_2 of M_x with the positive orientation such that

$$DW(e_1, e_2) + DW(e_2, e_1) = 0 \quad \text{at} \quad x \qquad (4.227)$$

which yields

$$\Upsilon(\zeta)(e_1, e_2) = \text{lo}\,(\zeta) \quad \text{at} \quad x. \qquad (4.228)$$

On the other hand, the formula (4.176) in Lemma 4.9 implies that

$$\begin{aligned}
D_{e_1} V &= \langle D_{e_1} V, e_1 \rangle e_1 + \langle D_{e_1} V, e_2 \rangle e_2 \\
&= DV(e_1, e_1)e_1 + DV(e_2, e_1)e_2 \\
&= \vartheta(x)e_1 - \iota(x)e_2 \quad \text{at} \quad x
\end{aligned} \qquad (4.229)$$

since $\mathcal{E}(e_2, e_1) = -1$. Similarly, we have

$$D_{e_2} V = \iota(x)e_1 + \vartheta(x)e_2 \quad \text{at} \quad x. \qquad (4.230)$$

Set

$$W_{ij} = DW(e_i, e_j) \quad \text{at} \quad x$$

for $1 \le i, j \le 2$. Using the formulas (4.131), (4.229) and (4.230), we obtain

$$G(V, DW)(e_1, e_1) = DW(e_1, D_{e_1} V) = \vartheta(x)W_{11} - \iota(x)W_{12}; \qquad (4.231)$$

$$G(V, DW)(e_2, e_2) = \iota(x)W_{21} + \vartheta(x)W_{22} \quad \text{at} \quad x. \qquad (4.232)$$

It follows from the formulas (4.228), (4.231), (4.232) and (4.227) that

$$\begin{aligned}
&\langle \Upsilon(\zeta), G(V, DW) \rangle \\
&= \Upsilon(e_1, e_1)G(V, DW)(e_1, e_1) + \Upsilon(e_2, e_2)G(V, DW)(e_2, e_2) + \text{lo}\,(\zeta) \\
&= \vartheta(x)(W_{11}^2 + W_{22}^2) + \iota(x)(W_{11} + W_{22})W_{21} + \text{lo}\,(\zeta) \\
&= \vartheta(x)|\Upsilon(\zeta)|^2 + \iota(x)(W_{11} + W_{22})W_{21} + \text{lo}\,(\zeta)
\end{aligned} \qquad (4.233)$$

at x. Similarly, we obtain

$$\begin{aligned}
\text{tr}\,\Upsilon(\zeta)\,\text{tr}\,G(V, DW) &= \vartheta(x)\Big(\text{tr}\,\Upsilon(\zeta)\Big)^2 \\
&\quad + 2\iota(x)(W_{11} + W_{22})W_{21} + \text{lo}\,(\zeta) \quad \text{at} \quad x.
\end{aligned} \qquad (4.234)$$

From the relations (4.233), (4.234) and (4.20), we have

$$
\begin{aligned}
a\Big(\Upsilon(\zeta), G(V, DW)\Big) &= \vartheta(x)a\Big(\Upsilon(\zeta), \Upsilon(\zeta)\Big) + \mathrm{lo}\,(\zeta) \\
&\quad +(1+\mu)\iota(x)(W_{11}+W_{22})W_{21} \\
&\geq \vartheta(x)a\Big(\Upsilon(\zeta), \Upsilon(\zeta)\Big) - (1+\mu)|\iota(x)||DW|^2/2 + \mathrm{lo}\,(\zeta) \\
&\geq \min_{x\in\Omega} |\vartheta(x)| a\Big(\Upsilon(\zeta), \Upsilon(\zeta)\Big) \\
&\quad -\frac{1+\mu}{2}\max_{x\in\Omega} |\iota(x)||DW|^2 + \mathrm{lo}\,(\zeta).
\end{aligned}
\tag{4.235}
$$

Furthermore, letting $\zeta = (W, 0)$ in the ellipticity assumption (4.172) yields

$$
\int_\Omega |DW|\, dx \leq \lambda_0 \int_\Omega a\Big(\Upsilon(\zeta), \Upsilon(\zeta)\Big)\, dx
\tag{4.236}
$$

Finally, using the escaping assumption (4.175) and the inequality (4.236), we integrate the inequality (4.235) over Ω to have the estimate (4.226). $\quad\square$

First, we have the estimate from boundary

Theorem 4.13 *Let V be an escape vector field for the shallow shell. Let $\zeta = (W, w) \in H^1(\Omega, \Lambda) \times L^2(\Omega)$ solve the problem (4.221) such that $\mathrm{SB}\,|_\Sigma$ is finite where $\mathrm{SB}\,|_\Sigma$ is defined by (4.222). Let $T > 0$ be given. Then*

$$
2\sigma_1 \int_0^T E(t)\, dt \leq \int_\Sigma \mathrm{SB}\, d\Sigma + \sigma_0 \lambda_0 [E(0) + E(T)] + C_T L(\zeta)
\tag{4.237}
$$

where $\lambda_0 > 0$ is the ellipticity number given by (4.172), $\sigma_0 > 0$ and $\sigma_1 > 0$ are given in (4.220), and SB is given by (4.222).

Proof. Set

$$
\Psi = 2[\sigma_1 - \vartheta(x)]a(\Upsilon(\zeta), \Upsilon(\zeta)) + 2\vartheta(x)[|\zeta_t|^2 + \gamma a(\rho(\zeta), \rho(\zeta))]
$$

where σ_1 is given in (4.220). Then

$$
\begin{aligned}
\Psi &= \sigma_1[|\zeta_t|^2 + \gamma|Dw_t|^2 + B(\zeta, \zeta)] \\
&\quad +(2\vartheta - \sigma_1)[|\zeta_t|^2 + \gamma a(\rho(\zeta), \rho(\zeta))] \\
&\quad +(\sigma_1 - 2\vartheta)a(\Upsilon(\zeta), \Upsilon(\zeta)) - \sigma_1 \gamma |Dw_t|^2 \\
&= \sigma_1 e(\zeta) + (2\vartheta - \sigma_1)[|W_t|^2 - a(\Upsilon(\zeta), \Upsilon(\zeta))] \\
&\quad +(\vartheta - \sigma_1)[w_t^2 + \gamma|Dw_t|^2 + \gamma a(\rho(\zeta), \rho(\zeta))] \\
&\quad +\vartheta(x)[\gamma a(\rho(\zeta), \rho(\zeta)) - \gamma|Dw_t|^2 - w_t^2] + 2\vartheta w_t^2
\end{aligned}
\tag{4.238}
$$

where $e(\zeta)$ is defined by (4.125).

We use the identity (4.169) where φ is set zero and obtain, via the formula (4.226) and the relation (4.238),

$$\int_\Sigma [|\zeta_t|^2 + \gamma|Dw_t|^2 - B(\zeta,\zeta)]\langle V,\nu\rangle \, d\Sigma + 2\int_\Sigma [\partial(\mathcal{A}\zeta, m(\zeta)) + \gamma V(w)\frac{\partial w_{tt}}{\partial\nu}] \, d\Sigma$$

$$= 2Z(t)|_0^T + 2\int_Q a(\Upsilon(\zeta), G(V, DW)) \, dQ$$

$$+2\int_Q \vartheta(x)[|\zeta_t|^2 + \gamma a(\rho(\zeta), \rho(\zeta)) - a(\Upsilon(\zeta), \Upsilon(\zeta))] \, dQ + \mathrm{lo}\,(\zeta)$$

$$\geq 2Z(t)|_0^T + \int_Q \Psi \, dQ + \mathrm{lo}\,(\zeta)$$

$$\geq 2Z(t)|_0^T + 2\sigma_1 \int_0^T E(t) \, dt + \int_Q (2\vartheta - \sigma_1)[|W_t|^2 - a(\Upsilon(\zeta), \Upsilon(\zeta))] \, dQ$$

$$+\int_Q \vartheta(x)[\gamma a(\rho(\zeta), \rho(\zeta)) - \gamma|Dw_t|^2 - w_t^2] \, dQ + \mathrm{lo}\,(\zeta) \qquad (4.239)$$

since $\min_{x\in\Omega} \vartheta(x) > \sigma_1$.

Next, we let $q = 2\vartheta(x) - \sigma_1$ in the identity (4.157) and let $q = \vartheta(x)$ in the identity (4.158), respectively, and have

$$\int_Q (2\vartheta - \sigma_1)[|W_t|^2 - a(\Upsilon(\zeta), \Upsilon(\zeta))] \, dQ$$

$$= -\int_\Sigma \partial(\mathcal{A}\zeta, \varrho(W, 0)) \, d\Sigma + \mathrm{lo}\,(\zeta) \qquad (4.240)$$

and

$$\int_Q \vartheta[w_t^2 + \gamma|Dw_t|^2 - \gamma a(\rho(\zeta), \rho(\zeta))] \, dQ$$

$$= -\int_\Sigma [\partial(\mathcal{A}\zeta, \vartheta(0, w)) + \gamma\vartheta w\frac{\partial w_{tt}}{\partial\nu}] \, d\Sigma + \mathrm{lo}\,(\zeta). \qquad (4.241)$$

Finally, we insert the identities (4.240) and (4.241) into the inequality (4.239), use the estimate (4.224), and obtain the inequality (4.237). $\qquad\square$

In order to absorb the lower order terms in the estimate (4.237) through a compactness-uniqueness argument as in Lemma 2.5, we need the following uniqueness result on a the shallow shell.

Proposition 4.3 *Let λ be a complex number and let $\hat\Gamma \subset \Gamma$ be relatively open. Suppose that $\zeta = (W, w)$ solves the problem*

$$\lambda^2\zeta - \lambda^2\gamma(0, \Delta w) + \mathcal{A}\zeta = 0 \quad on \quad \Omega \qquad (4.242)$$

subject to boundary conditions

$$\begin{cases} W = DW = 0 \quad on \quad \hat\Gamma, \\ w = \dfrac{\partial w}{\partial\nu} = \Delta w = \dfrac{\partial\Delta w}{\partial\nu} = 0 \quad on \quad \hat\Gamma. \end{cases} \qquad (4.243)$$

Then

$$W = w = 0 \quad on \quad \Omega. \tag{4.244}$$

Proof. It will suffice to show that the equation (4.242) is equivalent to a system of three equations of the fourth order differential operators with the same principal part Δ^2 where Δ is the Laplacian on the surface M in the induced metric g and the boundary conditions (4.243) yield the zero Cauchy data on $\hat{\Gamma}$ for this system. Therefore, this proposition is covered by [188].

Denote by $\mathrm{LF}\,(\zeta)$ all the terms for which there is a constant $C > 0$ such that

$$|\mathrm{LF}\,(\zeta)| \leq C \sum_{i=1}^{3} (|D^i W| + |D^i w|) \quad on \quad \Omega. \tag{4.245}$$

From the formulas (4.36) and (4.37) the first equation of the system (4.242) is

$$\left(\frac{1-\mu}{2}\delta d + d\delta\right) W = (\kappa - \mu\kappa + \lambda^2)W + \mathcal{H}(w) \quad on \quad \Omega \tag{4.246}$$

where $\mathcal{H}(w)$ is given in (4.38). Applying the operator $d\delta$ to both the sides of the equation (4.246), we have, via Theorem 1.25,

$$d\delta d\delta W = \mathrm{LF}\,(\zeta) \quad on \quad \Omega. \tag{4.247}$$

Similarly,

$$\delta d\delta d W = \mathrm{LF}\,(\zeta) \quad on \quad \Omega. \tag{4.248}$$

Then the relations (4.247) and (4.248) yield

$$\Delta^2 W = \mathrm{LF}\,(\zeta) \quad on \quad \Omega \tag{4.249}$$

where $\Delta = d\delta + \delta d$ is the Hodge-Laplacian.

Let (x_1, x_2) be a coordinate on M and let $W = (w_1, w_2)$ in this coordinate. It is easy to check that

$$\Delta^2 W = (\Delta^2 w_1, \Delta^2 w_2) + \mathrm{LF}\,(w_1, w_2).$$

Then in this coordinate the system (4.242) is equivalent to a problem

$$\begin{cases} \Delta^2 w_1 = \mathrm{LF}\,(w_1, w_2, w), \\ \Delta^2 w_2 = \mathrm{LF}\,(w_1, w_2, w), \\ \Delta^2 w = \mathrm{LF}\,(w_1, w_2, w). \end{cases} \tag{4.250}$$

To complete the proof, we have to show that the boundary conditions in (4.243) mean that the system (4.250) has the zero Cauchy data on $\hat{\Gamma}$. From the boundary conditions in (4.243) it will suffice to prove

$$D^2 W|_{\hat{\Gamma}} = D^3 W|_{\hat{\Gamma}} = 0. \tag{4.251}$$

Let $x \in \hat{\Gamma}$ be given. Let E_1, E_2 be a frame field normal at x such that

$E_1(x) = \nu(x)$ and $E_2(x) = \tau(x)$ where $\nu(x)$ and $\tau(x)$ are respectively the normal and the tangential of the boundary point x. Then

$$D_{\nu(x)}E_i = D_{\tau(x)}E_i = 0 \quad \text{at} \quad x. \qquad (4.252)$$

Since the vector τ is in a tangential direction along boundary, the boundary conditions (4.243) imply

$$\tau\Big(DW(E_i, E_j)\Big) = 0 \quad \text{at} \quad x. \qquad (4.253)$$

By (4.253) and (4.252), we obtain at x

$$D^2W(E_i, E_j, \tau) = \tau\Big(DW(E_i, E_j)\Big) - DW(D_\tau E_i, E_j)$$
$$-DW(E_i, D_\tau E_j) = 0. \qquad (4.254)$$

Next, we shall prove

$$D^2W(E_i, E_j, \nu) = 0 \quad \text{at} \quad x. \qquad (4.255)$$

If $E_j = \tau$, then the formula (4.254) and Theorem 1.10 yield

$$D^2W(E_i, \tau, \nu) = D^2W(E_i, \nu, \tau) + \mathbf{R}_{\tau\nu}W = 0 \quad \text{at} \quad x. \qquad (4.256)$$

Therefore, to get (4.255) true, what remains is to prove

$$D^2W(E_i, \nu, \nu) = 0 \quad \text{at} \quad x. \qquad (4.257)$$

Using the formulas (1.117), (1.119), (4.254), and (4.256), we have at x

$$\begin{aligned}
\delta dW &= -\sum_i D_{E_i}D_{E_i}W + \sum_{ij}\langle D_{E_j}D_{E_i}W, E_j\rangle E_i \\
&= -\sum_{ij} D^2W(E_j, E_i, E_i)E_j + \sum_{ij} D^2W(E_j, E_i, E_j)E_i \\
&= -\sum_j D^2W(E_j, \nu, \nu)E_j + \sum_i D^2W(\nu, E_i, \nu)E_i \\
&= -D^2W(\tau, \nu, \nu)\tau \qquad (4.258)
\end{aligned}$$

and

$$\begin{aligned}
d\delta W &= -\sum_{ij} D^2W(E_j, E_j, E_i)E_i = -\sum_j D^2W(E_j, E_j, \nu)\nu \\
&= -D^2W(\nu, \nu, \nu)\nu. \qquad (4.259)
\end{aligned}$$

Using the relations (4.258), (4.259), and (4.246), one has

$$-\frac{1-\mu}{2}D^2W(\tau, \nu, \nu)\tau - D^2W(\nu, \nu, \nu)\nu = \Big(\frac{1-\mu}{2}\delta d + d\delta\Big)W = 0$$

at x, that is, the formula (4.257) holds true.

A similar computation produces $D^3 W = 0$ on $\hat{\Gamma}$.   □

We now go back to boundary control.

Control in Fixed Boundary Condition Let $\Gamma = \partial\Omega \neq \emptyset$ with $\Gamma = \Gamma_0 \cup \Gamma_1$ and $\Gamma_0 \cap \Gamma_1 = \emptyset$. Consider the Dirichlet control problem in unknown $\varpi = (\Phi, \phi)$

$$
\begin{cases}
\varpi_{tt} - \gamma(0, \Delta\phi_{tt}) + \mathcal{A}\varpi = 0 & \text{on} \quad Q, \\
\varpi(0) = \varpi_0, \quad \varpi_t(0) = \varpi_1 & \text{on} \quad \Omega, \\
\Phi|_{\Gamma_1} = 0, \quad \Phi|_{\Gamma_0} = U \quad 0 < t < T, \\
\phi|_{\Gamma_1} = \dfrac{\partial\phi}{\partial\nu}\Big|_{\Gamma_1} = 0 \quad 0 < t < T, \\
\phi|_{\Gamma_0} = u, \quad \dfrac{\partial\phi}{\partial\nu}\Big|_{\Gamma_0} = v \quad 0 < t < T,
\end{cases}
\tag{4.260}
$$

with a control action (U, u, v) on the portion Γ_0 of the boundary. The question is to find some constant $T_0 > 0$ such that for $T > T_0$, the following steering property of (4.260) holds true: for all initial data in an appropriate Hilbert space which will be specified later, there exists a suitable control function (U, u, v) such that the corresponding solution of (4.260) satisfies

$$
\varpi(x, T) \equiv 0, \quad \varpi_t(x, T) \equiv 0.
$$

Exact controllability of (4.260) will be carried out by the dual method. The dual version of (4.260) in unknown $\zeta = (W, w)$ follows as

$$
\begin{cases}
\zeta_{tt} - \gamma(0, \Delta w_{tt}) + \mathcal{A}\zeta = 0 & \text{on} \quad Q \\
\zeta(0) = \zeta_0, \quad \zeta_t(0) = \zeta_1 & \text{on} \quad \Omega, \\
W = 0 & \text{on} \quad \Sigma, \\
w = \dfrac{\partial w}{\partial\nu} = 0 & \text{on} \quad \Sigma.
\end{cases}
\tag{4.261}
$$

Remark 4.9 *In the flat case, for the normal component, one control function* $\frac{\partial\phi}{\partial\nu} = v$ *is enough; see* [109]. *We here add another control function* $\phi = u$ *for the problem (4.260) in order to avoid the following uniqueness assumption: The problem*

$$
\begin{cases}
\lambda^2\zeta - \lambda^2\gamma(0, \Delta w) + \mathcal{A}\zeta = 0 & \text{on} \quad \Omega, \\
W = D_\nu W = 0 & \text{on} \quad \Gamma, \\
w = \dfrac{\partial w}{\partial\nu} = \Delta w = 0 & \text{on} \quad \Gamma
\end{cases}
\tag{4.262}
$$

admits the unique zero solution. The above uniqueness result does not fall into a class of systems to which the Holmgren theorem may be applied even if the coefficients are analytic since it is not the Cauchy problem for component w *(it does for component* W*). For the flat case, it has been proved in* [109].

Lemma 4.12 *Let* $\zeta = (W, w)$ *be a solution to the Dirichlet problem (4.261). Then*

$$
E(t) = E(0) \quad \text{for} \quad t \geq 0
\tag{4.263}
$$

where $E(t)$ is defined by (4.126);

$$\int_Q (|\zeta_t|^2 + \gamma |Dw_t|^2)\, dQ = \int_Q B(\zeta, \zeta)\, dQ + \mathrm{lo}\,(\zeta); \qquad (4.264)$$

$$TE(0) = 2 \int_Q B(\zeta, \zeta)\, dxdt + \mathrm{lo}\,(\zeta); \qquad (4.265)$$

where $B(\cdot, \cdot)$ is the bilinear from, given in (4.18).
  Moreover,

$$\begin{cases} B_1(W, w) = DW(\nu, \nu), \\ B_2(W, w) = (1 - \mu)DW(\tau, \nu)/2, \\ B_3 w = 0, \\ B_4 w = 0 \end{cases} \quad on \quad \Gamma \qquad (4.266)$$

where B_i are given by (4.39) and

$$B(\zeta, \zeta) = (DW(\nu, \nu))^2 + (1 - \mu)(DW(\tau, \nu))^2/2 + \gamma(\Delta w)^2, \qquad (4.267)$$

$$\mathrm{SB} = B(\zeta, \zeta)\langle V, \nu \rangle \qquad (4.268)$$

for $x \in \Gamma$ where SB is defined by (4.222).

Proof. It is easy to check that $E'(t) = 0$, that is, $E(t) = E(0)$. We let $q = 1$ and $\varphi = 0$ in the identities (4.157) and (4.158), respectively, and then add them up to obtain the formula (4.264), via the boundary conditions (4.261). The formula (4.265) follows from the formulas (4.263) and (4.264).

Moreover, the conditions $W|_\Gamma = 0$ and $w|_\Gamma = \dfrac{\partial w}{\partial \nu}|_\Gamma = 0$ imply that on Γ, $D_\tau W = 0$, and

$$\delta W = -DW(\nu, \nu), \quad \Upsilon(\zeta)(\nu, \tau) = DW(\nu, \nu),$$

$$\Upsilon(\zeta)(\nu, \tau) = DW(\tau, \nu)/2, \quad Dw = 0, \quad D^2 w(\tau, \tau) = 0,$$

$$D^2 w(\nu, \tau) = 0, \quad \Delta w = D^2 w(\nu, \nu),$$

$$\text{and} \quad \frac{\partial V(w)}{\partial \nu} = D^2 w(V, \nu) = \langle V, \nu \rangle \Delta w,$$

which yield the formula (4.266) from the formula (4.39). Using the above formulas, we obtain by (4.20) and (4.35), respectively, the formula (4.267) and

$$\partial(\mathcal{A}\zeta, 2m(\zeta)) = 2B(\zeta, \zeta)\langle V, \nu \rangle \quad \text{on} \quad \Gamma. \qquad (4.269)$$

Finally, using the formulas (4.269) and (4.267) in the formula (4.222), we obtain the formula (4.268). □

For convenience, set

$$\mathrm{ECSD} = [H_0^1(\Omega, \Lambda) \times H_0^2(\Omega)] \times [L^2(\Omega, \Lambda) \times H_0^1(\Omega)]. \qquad (4.270)$$

Let $\zeta = (W, w)$ be the solution to the problem (4.261) for the initial data $(\zeta_0, \zeta_1) \in \text{ECSD}$. For $T > 0$ we solve the following time-reverse problem in unknown $\eta = (\Psi, \psi)$

$$
\begin{cases}
\eta_{tt} - \gamma(0, \Delta\psi_{tt}) + \mathcal{A}\eta = 0 & \text{on} \quad Q, \\
\eta(T) = \eta_t(T) = 0 & \text{on} \quad \Omega, \\
\Psi|_{\Gamma_1} = 0, \quad \Psi|_{\Gamma_0} = -D_\nu W & 0 < t < T, \\
\psi|_{\Gamma_1} = \dfrac{\partial\psi}{\partial\nu}\Big|_{\Gamma_1} = 0 & 0 < t < T, \\
\psi|_{\Gamma_0} = \dfrac{\partial\Delta w}{\partial\nu}, \quad \dfrac{\partial\psi}{\partial\nu}\Big|_{\Gamma_0} = -\Delta w & 0 < t < T.
\end{cases}
\tag{4.271}
$$

Then for the initial data $(\eta(0), \eta_t(0))$, we have found a control action $(-D_\nu W, \dfrac{\partial\Delta w}{\partial\nu}, -\Delta w)$ on the portion Γ_0 of the boundary which steers the problem (4.260) to rest at a time T. Enlightened by [109], we define a map by

$$
\Lambda(\zeta_0, \zeta_1) = \Big(\eta_t(0) - (0, \gamma\Delta\psi_t(0)), -\eta(0) + (0, \gamma\Delta\psi(0))\Big).
\tag{4.272}
$$

We multiply the equation (4.271) by ζ, multiply the equation (4.261) by η, integrate over $Q = \Omega \times (0, T)$, and obtain, via the Green formula (4.34) and Lemma 4.12,

$$
\begin{aligned}
&\Big(\Lambda(\zeta_0, \zeta_1), (\zeta_0, \zeta_1)\Big)_{[L^2(\Omega,\Lambda)\times L^2(\Omega)]^2} \\
&= \Big(\eta_t(0), \zeta_0\Big)_{L^2(\Omega,\Lambda)\times L^2(\Omega)} - \Big(\eta(0), \zeta_1\Big)_{L^2(\Omega,\Lambda)\times L^2(\Omega)} \\
&\quad + \gamma(\Delta\psi(0), w_1)_{L^2(\Omega)} - \gamma(\Delta\psi_t(0), w_0)_{L^2(\Omega)} \\
&= \int_0^T \Big[(\mathcal{A}\eta, \zeta)_{L^2(\Omega,\Lambda)\times L^2(\Omega)} - (\mathcal{A}\zeta, \eta)_{L^2(\Omega,\Lambda)\times L^2(\Omega)}\Big] dt \\
&= \int_0^T \Big[\partial(\mathcal{A}\eta, \zeta) - \partial(\mathcal{A}\zeta, \eta)\Big] dt \\
&= \int_{\Sigma_0} [(DW(n, n))^2 + (1 - \mu)(DW(\tau, n))^2/2] \, d\Sigma \\
&\quad + \gamma\int_{\Sigma_0} [(\Delta w)^2 + (\dfrac{\partial\Delta w}{\partial\nu})^2] \, d\Sigma \\
&= \int_{\Sigma_0} \Big[B(\zeta, \zeta) + \gamma(\dfrac{\partial\Delta w}{\partial\nu})^2\Big] \, d\Sigma
\end{aligned}
$$

where $\Sigma_0 = \Gamma_0 \times (0, T)$.

Since $-\Delta: H_0^1(\Omega) \to H^{-1}(\Omega)$ and $-\Delta: L^2(\Omega) \to H^{-2}(\Omega)$ are isomorphisms, respectively, it follows from Theorem 2.13 that

Theorem 4.14 *The problem* (4.260) *is exactly*

$$
[L^2(\Omega, \Lambda) \times H_0^1(\Omega)] \times [H^{-1}(\Omega, \Lambda) \times L^2(\Omega)]
$$

controllable by $L^2(\Sigma_0, \Lambda) \times L^2(\Sigma_0) \times L^2(\Sigma_0)$ controls if the following observability inequality is true: There is a $T_0 > 0$ such that, for any $T > T_0$, there exists $c > 0$ satisfying

$$\int_{\Sigma_0} \left[B(\zeta,\zeta) + \gamma(\frac{\partial \Delta w}{\partial \nu})^2 \right] d\Sigma \geq cE(0) \tag{4.273}$$

where $\zeta = (W, w)$ are solutions to the problem (4.261) with initial data $(\zeta_0, \zeta_1) \in$ ECSD where the space ECSD is given by (4.270).

We now establish *continuous observability inequality in the fixed boundary conditions.*

Theorem 4.15 *Suppose that there is an escape vector field V for the shallow shell. Then for any $T > T_0$ there is $c > 0$ such that the observability estimate* (4.273) *holds where*

$$T_0 = \lambda_0 \sigma_0 / \sigma_1, \tag{4.274}$$

$$\Gamma_0 = \{ x \in \Gamma \mid \langle V, \nu(x) \rangle > 0 \}, \tag{4.275}$$

and constants λ_0, σ_0 and σ_1 are given in (4.220).

Proof. Let $T > T_0$ be given. Let $\varepsilon > 0$ be given such that $2\varepsilon < \sigma_1(T - T_0)$. Using Lemma 4.12 in Theorem 4.13, we obtain

$$\max_{x \in \Gamma_0} \langle V, \nu \rangle \int_{\Sigma_0} B(\zeta,\zeta) \, d\Sigma + CL(\zeta)$$

$$\geq [2\sigma_1(T - T_0) - 2\varepsilon]E(0). \tag{4.276}$$

Next, we shall use Proposition 4.3 to absorb the lower order terms in (4.276) to obtain (4.273) by a compactness-uniqueness argument as in Lemma 2.5. □

Control in Free Boundary Condition We let $\Gamma_1 \neq \emptyset$, $\overline{\Gamma}_1 \cap \overline{\Gamma}_0 = \emptyset$, and consider the control problem in unknown $\varpi = (\Phi, \phi)$

$$\begin{cases} \varpi_{tt} - \gamma(0, \Delta\phi_{tt}) + \mathcal{A}\varpi = 0 & \text{on} \quad Q, \\ \varpi(0) = \varpi_0, \quad \varpi_t(0) = \varpi_1 & \text{on} \quad \Omega. \end{cases} \tag{4.277}$$

The unknown ϖ is subject to boundary condition on $\Sigma_1 = \Gamma_1 \times (0, T)$

$$\Phi = \phi = \frac{\partial \phi}{\partial \nu} = 0 \quad \text{on} \quad \Sigma_1. \tag{4.278}$$

We act on $\Sigma_0 = \Gamma_0 \times (0, T)$ by

$$\begin{cases} B_1(\Phi, \phi) = u_1, \quad B_2(\Phi, \phi) = u_2, \\ \Delta\phi + (1 - \mu)B_3\phi = v_1, \\ \dfrac{\partial\Delta\phi}{\partial\nu} + (1 - \mu)B_4\phi - \dfrac{\partial\phi_{tt}}{\partial\nu} = v_2 \end{cases} \quad \text{on} \quad \Sigma_0. \tag{4.279}$$

The dual problem for the above is the following in $\zeta = (W, w)$.

$$\begin{cases} \zeta_{tt} - \gamma(0, \Delta w_{tt}) + \mathcal{A}\zeta = 0 & \text{on} \quad Q, \\ \zeta(0) = \zeta_0, \quad \zeta_t(0) = \zeta_1 & \text{on} \quad \Omega, \end{cases} \tag{4.280}$$

subject to boundary condition on $\Sigma_1 = \Gamma_1 \times (0, T)$

$$W = w = \frac{\partial w}{\partial \nu} = 0 \quad \text{on} \quad \Sigma_1, \tag{4.281}$$

and on $\Sigma_0 = \Gamma_0 \times (0, T)$

$$\begin{cases} B_1(W, w) = B_2(W, w) = 0, \\ \Delta w + (1 - \mu) B_3 w = 0, \\ \dfrac{\partial \Delta w}{\partial \nu} + (1 - \mu) B_4 w - \dfrac{\partial w_{tt}}{\partial \nu} = 0 \end{cases} \quad \text{on} \quad \Sigma_0. \tag{4.282}$$

We use the boundary conditions (4.281) and (4.282) in the formula (4.222) and obtain

Lemma 4.13 *Let V be an escape vector field for the shallow shell. Let ζ solve the problem (4.280)-(4.282). Then*

$$\int_{\Sigma} \mathrm{SB} \, d\Sigma = \int_{\Sigma_0} [|\zeta_t|^2 + \gamma|Dw_t|^2 - B(\zeta, \zeta)] \, d\Sigma + \int_{\Sigma_1} B(\zeta, \zeta)\langle V, \nu\rangle \, d\Sigma. \tag{4.283}$$

By an argument as in the fixed boundary conditions the exact controllability of the problem (4.277)-(4.279) leads to the following *continuous observability inequality in the free boundary conditions*: Seek $T_0 > 0$ such that for any $T > T_0$, there is $c > 0$ satisfying

$$\int_{\Sigma_0} [|\zeta_t|^2 + \gamma|Dw_t|^2] \, d\Sigma \geq cE(0) \tag{4.284}$$

for all initial data $(\zeta_0, \zeta_1) \in \mathrm{ECSN}_{\Gamma_1}$ for which the left hand side of (4.284) is finite where

$$\mathrm{ECSN}_{\Gamma_1} = [H^1_{\Gamma_1}(\Omega, \Lambda) \times H^2_{\Gamma_1}(\Omega)] \times [L^2(\Omega, \Lambda) \times L^2(\Omega)]. \tag{4.285}$$

Using an argument as in the proof of Theorem 4.15, we have

Theorem 4.16 *Let V be an escape vector field for the shallow shell. Then for any $T > T_0$, there is $c > 0$ such that the inequality (4.284) holds where T_0 and Γ_0 are defined by (4.274) and (4.275), respectively.*

Remark 4.10 *Let the shell be flat, that is, $M = \mathbb{R}^2$. Then the control system (4.260) or (4.277)-(4.279) becomes two systems, where one is a wave equation on the component Φ and the other is a plate equation on the component ϕ. We take $\lambda_0 = 1$. If we set $V = x - x^0$, x^0 a fixed point in $\mathbb{R}^2$, then inequalities (4.273) and (4.284) on the component w are exactly the same as in [109]. In this case, $\sigma_1 = 1$. It follows that $T_0 = 2\mathrm{diameter}(\Omega)$, which is the best for wave component W; see [99]. In this sense, T_0, given by (4.274), is the best.*

Finally, we study **the internal exact controllability**.

Consider a control problem in unknown $\varpi = (\Phi, \phi)$

$$\begin{cases} \varpi_{tt} - \gamma(0, \Delta\phi_{tt} + \mathcal{A}\varpi = b(x)(F, f) & \text{on } \Omega \times (0, T), \\ \varpi(0) = \varpi_0, \quad \varpi_t(0) = \varpi_1 & \text{on } \Omega, \\ \Phi = \phi = \dfrac{\partial\phi}{\partial\nu} = 0 & \text{on } \Gamma \times (0, T), \end{cases} \quad (4.286)$$

where $b \geq 0$ and $b \in C^2(\Omega)$. In this case, the boundary can be empty. If $\Gamma = \emptyset$, there will be no boundary term in the system (4.286).

The question is to find some constant $T_0 > 0$ such that for $T > T_0$, the following steering property of (4.286) holds true: for all initial data $\varpi_0 \in L^2(\Omega, \Lambda) \times H_0^1(\Omega)$, $\varpi_1 \in H^{-1}(\Omega, \Lambda) \times L^2(\Omega)$, there exists a suitable control function $(F, f) \in L^2(0, T; L^2(\Omega, \Lambda) \times H^{-1}(\Omega))$ such that the corresponding solution of (4.286) satisfies

$$\varpi(x, T) \equiv 0, \quad \varpi_t(x, T) \equiv 0.$$

In this case, we say that the dynamics (4.286) are *exactly controllable* in the interval $[0, T]$ on $(L^2(\Omega, \Lambda) \times H_0^1(\Omega)) \times (H^{-1}(\Omega, \Lambda) \times L^2(\Omega))$ by means of the internal control $b(F, f)$.

Let

$$G = \{\, x \in \overline{\Omega} \,|\, b(x) > 0 \,\}$$

be the region where the control acts. In the design of the controller, an important issue is how to choose G to be small in some sense out of all the open subsets of the middle surface Ω for which the exact controllability of (4.286) can be achieved. In fact, the structure of such a G has been studied in Section 4.5. In this section, we always assume that G is an escape region for the shallow shell with the structure (4.210)-(4.213). Set

$$\sigma_{i0} = \max_{x \in \Omega_i} |V_i|, \quad \sigma_{i1} = \min_{x \in \Omega_i} \vartheta_i(x) - \lambda_0(1 + \mu)\max_{x \in \Omega_i}|\iota_i(x)|/2, \quad (4.287)$$

$$T_0 = 2\lambda_0 \frac{\max_i \sigma_{i0}}{\min_i \sigma_{i1}}. \quad (4.288)$$

We will obtain the internal exact controllability by the dual method through the following observability inequality from the interior.

Theorem 4.17 *Let the middle surface Ω have a boundary or no boundary. Let G be an escape region for the shallow shell with the structure (4.210)-(4.213). Then for any $T > T_0$ there is $c > 0$ such that*

$$\int_0^T \int_G [B(\zeta, \zeta) + |\zeta|^2]\, dx dt \geq cE(0) \quad (4.289)$$

for all solutions ζ to the problem (4.261) where T_0 is given by (4.288).

Proof. For $\varepsilon > \varepsilon_0 > \varepsilon_1 > \varepsilon_2 > 0$, set

$$\Theta_i = \mathcal{N}_{\varepsilon_i}\left[\cup_{i=1}^J \Gamma_{i0} \cup \left(\Omega/\cup_{i=1}^J \Omega_j\right)\right] \tag{4.290}$$

for $i = 0, 1, 2$, where a set $\mathcal{N}_\varepsilon$ is given by (4.214). Clearly, we have

$$\Theta_2 \subset \Theta_1 \subset \overline{\Theta}_0 \subset G. \tag{4.291}$$

For each $1 \leq i \leq J$, let $h_i \in C^\infty(M)$ be such that

$$\begin{cases} 0 \leq h_i \leq 1, \\ h_i = 1 \quad \text{in} \quad \overline{\Omega}_i/\Theta_1; \\ h_i = 0 \quad \text{in} \quad \Theta_2. \end{cases} \tag{4.292}$$

For each i, we apply the identity (4.151) with $\Omega := \Omega_i$, $V := h_i V_i$, and $\varphi = 0$, to obtain

$$\int_0^T \int_{\partial\Omega_i} [Q_1(\zeta)\langle h_i V_i, \nu_i\rangle + 2\Sigma(\zeta, h_i V_i)]\, d\Gamma dt = 2Z_i(t)|_0^T$$
$$+ \int_0^T \int_{\Omega_i} [Q_1(\zeta)\, \mathrm{div}\,(h_i V_i) + Q_2(\zeta, h_i V_i)]\, dx dt \tag{4.293}$$

where

$$Q_1(\zeta) = |\zeta_t|^2 + \gamma |Dw_t|^2 - B(\zeta, \zeta); \tag{4.294}$$

$$\Sigma(\zeta, h_i V_i) = \partial(A\zeta, m_i(\zeta)) + \gamma h_i V_i(w)\frac{\partial w_{tt}}{\partial\nu}; \tag{4.295}$$

$$Z_i(t) = (\zeta_t, m_i(\zeta))_{L^2(\Omega_i,\Lambda) \times L^2(\Omega_i)} + \gamma(Dw_t, D(h_i V_i(w)))_{L^2(\Omega_i)}; \tag{4.296}$$

$$m_i(\zeta) = h_i(D_{V_i}W, V_i(w)); \tag{4.297}$$

$$Q_2(\zeta, h_i V_i) = 2[a(\Upsilon(\zeta), G(h_i V_i, DW)) + 2\gamma a(\rho(\zeta), G(h_i V_i, \rho(\zeta)))]$$
$$- 2\gamma D(h_i V_i)(Dw_t, Dw_t) + \mathrm{lo}\,(\zeta) \tag{4.298}$$

and the tensor field $G(V, DW)$ is defined by (4.131).

Set

$$\partial\Omega_i = I_1 \cup I_2$$

where $I_1 = \Gamma_{i0} \cup (\partial\Omega_i/\Gamma)$ and $I_2 = (\partial\Omega_i/\Gamma_{i0}) \cap \Gamma$. Since $\partial\Omega_i/\Gamma \subset \Omega/\cup_{j=1}^J \Omega_j \subset \Theta_2$ and $\Gamma_{i0} \subset \Theta_2$, we have $h_i = 0$ for $x \in I_1$ and then obtain

$$\int_0^T \int_{I_1} [Q_1(\zeta)\langle h_i V_i, \nu_i\rangle + 2\Sigma(\zeta, h_i V_i)]\, d\partial\Omega_i dt = 0. \tag{4.299}$$

In addition, since $I_2 \subset \Gamma \cap \{x \in \partial\Omega_i \,|\, \langle V_i, \nu_i\rangle \leq 0\}$ by the definition of Γ_{i0}, the fixed boundary conditions in (4.261) and the formula (4.269) imply that

$$Q_1(\zeta)\langle h_i V_i, \nu_i\rangle + 2\Sigma(\zeta, h_i V_i) = h_i B(\zeta, \zeta)\langle V_i, \nu_i\rangle \leq 0 \tag{4.300}$$

for $x \in I_2$. It follows from the relations (4.299) and (4.300) that

$$\int_0^T \int_{\partial\Omega_i} [Q_1(\zeta)\langle h_i V_i, \nu_i\rangle + 2\Sigma(\zeta, h_i V_i)] \, d\partial\Omega_i dt \le 0 \qquad (4.301)$$

for $1 \le i \le J$. We obtain, from the identity (4.293) and the inequality (4.301),

$$2Z_i(t)|_0^T + \int_0^T \int_{\Omega_i} Q_1(\zeta) \, \mathrm{div}\,(h_i V_i) \, dxdt$$

$$+ \int_0^T \int_{\Omega_i} Q_2(\zeta, h_i V_i) \, dxdt \le 0. \qquad (4.302)$$

Now, we deal with the first integral in the left hand side of (4.302). We take $\Omega := \Omega_i$, $q = \mathrm{div}\,(h_i V_i)$, and $\varphi = 0$ in the identities (4.157) and (4.158), respectively, and then add them up to have

$$\int_0^T \int_{\Omega_i} Q_1(\zeta) \, \mathrm{div}\,(h_i V_i) \, dxdt + \mathrm{lo}\,(\zeta)$$

$$= -\int_0^T \int_{\partial\Omega_i} [\partial(\mathcal{A}\zeta, \mathrm{div}\,(h_i V_i)\zeta) + \gamma w \frac{\partial w_{tt}}{\partial\nu} \mathrm{div}\,(h_i V_i)] \, d\partial\Omega_i dt. \qquad (4.303)$$

Let $\partial\Omega_i = (\partial\Omega_i/\Omega) \cup (\partial\Omega_i \cap \Omega)$. Since $\partial\Omega_i/\Omega \subset \Gamma$, the integrand in the right hand side of (4.303) is zero by the Dirichlet boundary conditions of ζ on Γ. In addition, $\partial\Omega_i \cap \Omega \subset \Theta_2$ implies that the integrand is also zero by $h_i = 0$ on Θ_2. Therefore

$$\int_0^T \int_{\Omega_i} Q_1(\zeta) \, \mathrm{div}\,(h_i V_i) \, dxdt = L(\zeta). \qquad (4.304)$$

Next, we deal with the second integral in the left hand side of (4.302). On Ω_i/Θ_1, since $h_i = 1$ and the vector field V_i is escaping with the structure (4.210)-(4.213), using the formulas (4.170) and (4.171) in the relation (4.298) we obtain

$$Q_2(\zeta, h_i V_i) = 2a(\Upsilon(\zeta), G(V_i, DW)) + 4\gamma\vartheta_i a(\rho(\zeta), \rho(\zeta))$$
$$- 2\gamma\vartheta_i |Dw_t|^2 + \mathrm{lo}\,(\zeta) \qquad (4.305)$$

for $x \in \Omega_i/\Theta_1$. We integrate (4.305) over Ω_i/Θ_1 and, by Lemma 4.11, have

$$2\sigma_{i1} \int_0^T \int_{\Omega_i/\Theta_1} B(\zeta, \zeta) \, dxdt + 2\gamma \int_0^T \int_{\Omega_i/\Theta_1} \vartheta_i [a(\rho(\zeta), \rho(\zeta)) - |Dw_t|^2] \, dxdt$$

$$\le \int_0^T \int_{\Omega_i/\Theta_1} Q_2(\zeta, h_i V_i) \, dxdt + L(\zeta) \qquad (4.306)$$

where σ_{i1} is given in (4.287).

Using the relations (4.306) and (4.304) in the inequality (4.302), we obtain

$$2\sigma_{i1} \int_0^T \int_{\Omega_i/\Theta_1} B(\zeta,\zeta)\,dxdt \leq 2(|Z_i(0)| + |Z_i(T)|)$$

$$+2\gamma \int_0^T \int_{\Omega_i/\Theta_1} \vartheta_i[|Dw_t|^2 - a(\rho(\zeta),\rho(\zeta))]\,dxdt$$

$$- \int_{\Omega_i \cap \Theta_1} Q_2(\zeta, h_i V_i)\,dxdt + L(\zeta). \tag{4.307}$$

We estimate the first integral in the right hand side of the inequality (4.307) as follows. Applying the identity (4.158) with $\Omega := \Omega_i$, $q = \operatorname{div}(h_i V_i)$, and $\varphi = 0$ yields

$$\gamma \int_0^T \int_{\Omega_i} q[|Dw_t|^2 - a(\rho(\zeta),\rho(\zeta))]\,dxdt$$

$$= -\int_0^T \int_{\partial\Omega_i} [\partial(A\zeta, q(0,w)) + \gamma qw \frac{\partial w_{tt}}{\partial\nu}]\,d\partial\Omega_i dt + L(\zeta) \tag{4.308}$$

where the term qw_t^2 in (4.158) is eliminated as a lower order term. The same argument as that for (4.303) shows that the boundary integral in the right hand side of (4.308) is zero. Therefore

$$2\gamma \int_0^T \int_{\Omega_i/\Theta_1} \vartheta_i[|Dw_t|^2 - a(\rho(\zeta),\rho(\zeta))]\,dxdt$$

$$= \gamma \int_0^T \int_{\Omega_i \cap \Theta_1} q[|Dw_t|^2 - a(\rho(\zeta),\rho(\zeta))]\,dxdt + L(\zeta)$$

$$\leq c\gamma \int_0^T \int_{\Omega_i \cap \Theta_1} [|Dw_t|^2 + a(\rho(\zeta),\rho(\zeta))]\,dxdt + L(\zeta)$$

$$\leq c \int_0^T \int_{\Omega_i \cap \Theta_1} [\gamma|Dw_t|^2 + B(\zeta,\zeta)]\,dxdt + L(\zeta). \tag{4.309}$$

We consider the second integral in the right hand side of (4.307). Clearly, it follows from (4.298) that

$$|Q_2(\zeta, h_i V_i)| \leq C_\varepsilon[\gamma|Dw_t|^2 + B(\zeta,\zeta)] + \varepsilon(|DW|^2 + |D^2 w|^2) \tag{4.310}$$

for $x \in \Omega_i$.

We apply Lemma 4.10 with $\Omega := \Omega_i$ and $V := h_i V_i$ to obtain

$$2|Z_i(t)| \leq (\sigma_{i0}\lambda_0 + \varepsilon) \int_{\Omega_i} e(\zeta)\,dx + \operatorname{lo}(\zeta). \tag{4.311}$$

Using the formulas (4.310), (4.309), (4.311), and (4.172) in the inequality

(4.307), we obtain

$$2\sigma_{i1} \int_0^T \int_{\Omega_i/\Theta_1} B(\zeta,\zeta)\,dxdt \leq 2\sigma_{i0}\lambda_0 \int_{\Omega_i} e(\zeta)\,dxdt + \varepsilon TE(0)$$

$$+c \int_0^T \int_{\Omega_i \cap \Theta_1} [\gamma|Dw_t|^2 + B(\zeta,\zeta)]\,dxdt + L(\zeta). \qquad (4.312)$$

Since $\Omega = (\cup_{i=1}^J \Omega_i/\Theta_1) \cup \Theta_1$, we add (4.312) from $i=1$ to $i=J$ to have

$$2\min_i \sigma_{i1} \int_0^T \int_\Omega B(\zeta,\zeta)\,dxdt \leq 2\lambda_0 \max_i \sigma_{i0} E(0) + \varepsilon TE(0)$$

$$+c \int_0^T \int_{\Theta_1} B(\zeta,\zeta)\,dxdt + c\gamma \int_0^T \int_{\Theta_1} |Dw_t|^2\,dxdt + L(\zeta). \qquad (4.313)$$

To finish the proof it remains to estimate the last integral in the right hand side of (4.313). Let $q \in C^\infty(\Omega)$ be $0 \leq q \leq 1$ such that

$$q = 1 \quad \text{in} \quad \Theta_1; \quad q = 0 \quad \text{in} \quad G/\Theta_1.$$

We add the identity (4.157) and the identity (4.158) up with the above q and $\varphi = 0$, and then obtain, by the Dirichlet boundary conditions of ζ,

$$\gamma \int_0^T \int_{\Theta_1} |Dw_t|^2\,dx \leq \int_0^T \int_\Omega q(|\zeta_t|^2 + \gamma|Dw_t|^2)\,dxdt$$

$$= \int_0^T \int_\Omega qB(\zeta,\zeta)\,dxdt + L(\zeta)$$

$$\leq c \int_0^T \int_G B(\zeta,\zeta)\,dxdt + L(\zeta). \qquad (4.314)$$

Finally, using the relations (4.314), (4.263), (4.265), and Lemma 4.14 below in the inequality (4.313), we obtain the inequality (4.289). □

By a compactness-uniqueness argument as in Lemma 2.5, we have

Lemma 4.14 *Let $T > 0$ and $C_1 > 0$ be such that*

$$C_1 E(0) \leq \int_0^T \int_G [B(\zeta,\zeta) + |\zeta|^2]\,dxdt + L(\zeta) \qquad (4.315)$$

for all solutions ζ to the problem (4.261). Then there is $c > 0$ such that the inequality (4.289) holds.

Internal Exact Controllability Let $\zeta = (W, w)$ be a solution to the Dirichlet problem (4.261). We solve the following problem for $\eta = (\Psi, \psi)$

$$\begin{cases} \eta_{tt} - \gamma(0, \Delta\psi_{tt}) + \mathcal{A}\eta = -b(x)(\mathcal{A}\zeta + \zeta) & \text{on} \quad \Omega \times (0,T), \\ \eta(T) = \eta_t(T) = 0 & \text{on} \quad \Omega, \\ \Psi = \psi = \dfrac{\partial\psi}{\partial\nu} = 0 & \text{on} \quad \Gamma \times (0,T). \end{cases} \qquad (4.316)$$

Therefore, we define a map Λ: ECSD $\to$ (ECSD)$'$ by

$$\Lambda(\zeta_0, \zeta_1) = \Big(\eta_t(0) - \gamma(0, \Delta\psi_t(0)), \ -\eta(0) + \gamma(0, \Delta\psi(0))\Big) \tag{4.317}$$

where the space ECSD is given by (4.270). We have

Proposition 4.4 *Let* $(\zeta_0, \zeta_1) \in$ ECSD. *Then*

$$\Big(\Lambda(\zeta_0, \zeta_1), \ (\zeta_0, \zeta_1)\Big)_{[L^2(\Omega, \Lambda) \times L^2(\Omega)]^2}$$
$$= \int_0^T [B(\zeta, b(x)\zeta) + b(x)|\zeta|^2] \, dxdt. \tag{4.318}$$

Proof. By the problems (4.316), (4.261), and the Green formula (4.34), the left hand side of (4.318) equals

$$\Big(\eta_t(0), \zeta_0\Big)_{L^2(\Omega,\Lambda) \times L^2(\Omega)} - \Big(\eta(0), \zeta_1\Big)_{L^2(\Omega,\Lambda) \times L^2(\Omega)}$$
$$+ \gamma(\Delta\psi(0), w_1)_{L^2(\Omega)} - \gamma(\Delta\psi_t(0), w_0)_{L^2(\Omega)}$$
$$= \int_0^T (b\mathcal{A}\zeta + b\zeta, \zeta)_{L^2(\Omega,\Lambda) \times L^2(\Omega)} \, dt$$
$$= \int_0^T \int_\Omega [B(\zeta, b\zeta) + b|\zeta|^2] \, dxdt. \tag{4.319}$$

$\square$

Clearly, we have

$$\int_0^T \int_G [B(\zeta, \zeta) + |\zeta|^2] \, dxdt \le C_T E(0)$$

for all solutions ζ to the Dirichlet problem (4.261). It follows from Theorem 4.17 that

Theorem 4.18 *Let the middle surface* Ω *have a boundary or no boundary. Let* G *be an escape region for the shallow shell with the structure (4.210)-(4.213). Let* $T_0 > 0$ *be given by (4.288). Then for any* $T > T_0$ *the internal control problem (4.286) is exactly* $\{[H_0^1(\Omega, \Lambda) \times H_0^2(\Omega)] \times [L^2(\Omega, \Lambda) \times H_0^1(\Omega)]\}^*$ *controllable.*

4.7   Exact Controllability for Transmission

Let a middle surface Ω be divided into two regions, Ω_1 and Ω_2, by a smooth curve Γ_3 on Ω. Set $\Gamma_i = \partial\Omega \cap \partial\Omega_i$ for $i = 1$ and 2. Denote by ν_i the normal of

$\partial\Omega_i$ (pointing outside Ω_i) and by τ_i the tangential on $\partial\Omega_i$, respectively. Then

$$\nu_i = \nu, \quad \tau_i = \tau \quad \text{for} \quad x \in \Gamma_1 \cup \Gamma_2 = \partial\Omega,$$

$$\nu_1(x) = -\nu_2(x), \quad \tau_1(x) = -\tau_2(x) \quad \text{for} \quad x \in \Gamma_3. \tag{4.320}$$

Suppose that the type of material in Ω_1 is different from that in Ω_2, where Young's modulus and Poisson's coefficient have a jump across the separatrix Γ_3.

First, let us clarify transmission conditions for the shallow shell. We assume that the notation here is the same as in Theorem 4.1. Let E_i and μ_i, respectively, denote Young's modulus and Poisson's coefficient of the material corresponding to the region Ω_i, $i = 1, 2$. Let $\mathcal{A}_i$ denote the shallow shell operator, given by the formula (4.36) in which E and μ have been replaced with E_i and μ_i, respectively, for $i = 1, 2$. Let $\zeta_i = (W_i, w_i)$ be a displacement vector on Ω_i for $i = 1$ and 2, respectively. Set on Γ_3

$$\begin{cases} TC_{i1}(\zeta_i) = (1 - \mu_i)\Upsilon(\zeta_i)(\nu_i, \nu_i) + \mu_i(w_i H - \delta W_i), \\ TC_{i2}(\zeta_i) = (1 - \mu_i)\Upsilon(\zeta_i)(\nu_i, \tau_i) \\ TC_{i3}(\zeta_i) = \gamma[\Delta w_i - (1 - \mu_i)D^2 w_i(\tau_i, \tau_i)] \\ TC_{i4}(\zeta_i) = -(-1)^i \gamma\{\dfrac{\partial \Delta w_i}{\partial \nu_i} \\ \qquad\qquad + (1 - \mu_i)[\frac{\partial}{\partial \tau_i}D^2 w_i(\tau_i, \nu_i) + \kappa\dfrac{\partial w_i}{\partial \nu_i}]\}. \end{cases} \tag{4.321}$$

Vector fields $\zeta_i = (W_i, w_i)$ on Ω_i are said to satisfy *the transmission conditions for the shallow shell on* Γ_3 if the following relations hold:

$$\begin{cases} W_1 = W_2, \\ w_1 = w_2, \quad \dfrac{\partial w_1}{\partial \nu_1} + \dfrac{\partial w_2}{\partial \nu_2} = 0 \qquad\qquad on \quad \Gamma_3. \\ b_1 TC_{1j}(\zeta_1) = b_2 TC_{2j}(\zeta_2) \quad \text{for} \quad j = 1, 2, 3, 4, \end{cases} \tag{4.322}$$

It is easy to check that

Lemma 4.15 *Let* $\zeta_i = (W_i, w_i)$ *satisfy the transmission* (4.322) *and let* $\eta_i = (\Psi_i, \psi_i)$ *satisfy the transmission conditions* (4.322) *on* Γ_3, *respectively. Let* $\partial(\mathcal{A}_i\cdot, \cdot)$ *be given by* (4.35) *in which* μ *is replaced by* μ_i, *for* $i = 1, 2$. *Then*

$$b_1 \partial(\mathcal{A}_1\eta_1, \zeta_1) + b_2 \partial(\mathcal{A}_2\eta_2, \zeta_2) = 0 \quad on \quad \Gamma_3. \tag{4.323}$$

Let $\varpi_i = (\Phi_i, \phi_i)$ denote a displacement vector field on Ω_i, for $i = 1, 2$. Let ω be a region such that $\Gamma_3 \subset \omega \subset \Omega$. Let $\omega_i = \Omega_i \cap \omega$. We consider the control problem of transmission

$$\begin{cases} \varpi_{itt} - \gamma(0, \Delta\phi_{itt}) + b_i\mathcal{A}_i\varpi_i = \chi_{\omega_i}(F_i, f_i) \quad \text{on} \quad \Omega_i \times (0, T), \\ \varpi_i(0) = \varpi_{i0}, \quad \varpi_{it}(0) = \varpi_{i1} \quad \text{on} \quad \Omega, \\ \Phi_1 = \phi_1 = \dfrac{\partial\phi_1}{\partial\nu} = 0 \quad \text{on} \quad \Gamma_2 \times (0, T), \\ \varpi_i \text{ satisfy the transmission conditions (4.322) on } \Gamma_3, \\ \Phi_1 = U, \quad \phi_1 = u, \quad \dfrac{\partial\phi_1}{\partial\nu} = v \quad \text{on} \quad \Gamma_1 \times (0, T), \end{cases} \tag{4.324}$$

where χ_{ω_i} are the characteristic functions of ω_i and $b_i = E_i/(1-\mu_i^2)$ for $i = 1$ and 2. In the system (4.324) an internal control acts on a neighborhood of the transmission part of the two portions of the middle surface and another control (U, u, v) acts on the boundary Γ_1.

The dual version of the system (4.324) in unknown $\zeta = (W, w)$ is given by

$$
\begin{cases}
\zeta_{itt} - \gamma(0, \Delta w_{itt}) + b_i \mathcal{A}_i \zeta_i = 0 & \text{on} \quad \Omega_i \times (0, T), \\
\zeta_i(0) = \zeta_{i0}, \quad \zeta_{it}(0) = \zeta_{i1} & \text{on} \quad \Omega_i, \\
W = w = \dfrac{\partial w}{\partial \nu} = 0 & \text{on} \quad \Gamma_2 \times (0, T), \\
\zeta_i \text{ satisfy the transmission conditions (4.322) on } \Gamma_3, \\
W = w = \dfrac{\partial w}{\partial \nu} = 0 & \text{on} \quad \Gamma_1 \times (0, T).
\end{cases}
\tag{4.325}
$$

Let ζ be a solution to the transmission problem (4.325) such that $\zeta_i = \zeta|_{\Omega_i}$. The total energy of the system (4.325) is defined by

$$
E(t) = \sum_{i=1}^{2} \int_{\Omega_i} [|\zeta_{it}|^2 + \gamma|Dw_{it}|^2 + b_i B_i(\zeta_i, \zeta_i)] \, dx
\tag{4.326}
$$

where $B_i(\cdot, \cdot)$ are bilinear forms, given by (4.17) in which μ is replaced by μ_i for $i = 1, 2$.

Let ω be a region such that $\Gamma_3 \subset \omega$. Let $\zeta = (W, w)$ be a solution to the problem (4.325) with an initial data (ζ_0, ζ_1) satisfying $E(0) < \infty$. Let $\kappa(x) \in C^\infty(\Omega)$ be given such that $\operatorname{supp} \kappa \subset \omega$. We then solve the problem in $\eta = (\Psi, \psi)$

$$
\begin{cases}
\eta_{itt} - \gamma(0, \Delta \psi_{itt}) + b_i \mathcal{A} \eta_i = -\kappa(x) \mathcal{A}_i \zeta_i & \text{on} \quad \Omega_i \times (0, T), \\
\eta_i(T) = \zeta_{it}(T) = 0 & \text{on} \quad \Omega_i, \\
\Psi = \psi = \dfrac{\partial \psi}{\partial \nu} = 0 & \text{on} \quad \Gamma_2 \times (0, T), \\
\eta_i \text{ satisfy the transmission conditions (4.322) on } \Gamma_3, \\
\Psi = -D_\nu W, \quad \psi = \dfrac{\partial \Delta w}{\partial \nu}, \quad \dfrac{\partial \psi}{\partial \nu} = -\Delta w & \text{on} \quad \Gamma_1 \times (0, T).
\end{cases}
\tag{4.327}
$$

Therefore, we define a map Λ by

$$
\Lambda(\zeta_0, \zeta_1) = \Big(\eta_t(0) - \gamma(0, \Delta \psi_t(0)), \; -\eta(0) + \gamma(0, \Delta \psi(0)) \Big).
\tag{4.328}
$$

Let

$$
b(x) = b_i = E_i/(1 - \mu_i^2), \quad B(\zeta, \zeta) = B_i(\zeta_i, \zeta_i),
$$
$$
\text{and} \quad \mathcal{A}\zeta = \mathcal{A}_i \zeta_i \quad \text{for} \quad x \in \Omega_i, \quad i = 1, 2.
$$

Then

$$
E(t) = \frac{1}{2} \int_\Omega [|\zeta_t|^2 + \gamma|Dw_t|^2 + b(x) B(\zeta, \zeta)] \, dx.
$$

We have

Proposition 4.5 *Let ζ be a solution to the problem (4.325). Then*

$$\Big(\Lambda(\zeta_0, \zeta_1), \ (\zeta_0, \zeta_1) \Big)_{[L^2(\Omega, \Lambda) \times L^2(\Omega)]^2} = \int_0^T \int_\Omega b(x) B(\zeta, \kappa\zeta) \, dx dt$$

$$+ \int_0^T \int_{\Gamma_1} [b(x) B(\zeta, \zeta) + \gamma (\frac{\partial \Delta w}{\partial \nu})^2] \, d\Gamma dt, \tag{4.329}$$

where $B(\cdot, \cdot)|_{\Omega_i} = B_i(\cdot, \cdot)$ for $i = 1$ and 2.

Proof. Using the relations (4.325), (4.327) and the Green formula (4.34), we obtain

$$\Big(\Lambda(\zeta_0, \zeta_1), \ (\zeta_0, \zeta_1) \Big)_{[L^2(\Omega, \Lambda) \times L^2(\Omega)]^2}$$

$$= \sum_{i=1}^2 \int_0^T [(\eta_i, \zeta_{itt})_{L^2(\Omega_i, \Lambda) \times L^2(\Omega_i)} - (\eta_{itt}, \zeta_i)_{L^2(\Omega_i, \Lambda) \times L^2(\Omega_i)}] \, dt$$

$$+ \gamma \sum_{i=1}^2 [(\Delta\psi_i(0), w_{it}(0))_{L^2(\Omega_i)} - (\Delta\psi_{it}(0), w_i(0))_{L^2(\Omega_i)}]$$

$$= \sum_{i=1}^2 b_i \int_0^T \int_{\partial\Omega_i} [\partial(\mathcal{A}_i \eta_i, \zeta_i) - \partial(\mathcal{A}_i \zeta_i, \eta_i)] \, d\partial\Omega_i dt$$

$$+ \sum_{i=1}^2 \int_0^T (\mathcal{A}_i \zeta_i, \kappa\zeta_i)_{L^2(\Omega_i, \Lambda) \times L^2(\Omega_i)} dt. \tag{4.330}$$

Let us compute the integrals in the right hand side of (4.330) on Ω_1 and Ω_2, separately. It follows from the formulas (4.266) and (4.267) with $\mu = \mu_1$ that

$$\int_{\partial\Omega_1} [\partial(\mathcal{A}_1\eta_1, \zeta_1) - \partial(\mathcal{A}_1\zeta_1, \eta_1)] \, d\partial\Omega_1$$

$$= \int_{\Gamma_3} [\partial(\mathcal{A}_1\eta_1, \zeta_1) - \partial(\mathcal{A}_1\zeta_1, \eta_1)] \, d\Gamma_3$$

$$+ \int_{\Gamma_1} [B_1(\zeta_1, \zeta_1) + \gamma(\frac{\partial\Delta w}{\partial\nu})^2] \, d\Gamma \tag{4.331}$$

and

$$b_2 \int_{\partial\Omega_2} [\partial(\mathcal{A}_2\eta_2, \zeta_2) - \partial(\mathcal{A}_2\zeta_2, \eta_2)] \, d\partial\Omega_2$$

$$= b_2 \int_{\Gamma_3} [\partial(\mathcal{A}_2\eta_2, \zeta_2) - \partial(\mathcal{A}_2\zeta_2, \eta_2)] \, d\Gamma_3$$

$$= -b_1 \int_{\Gamma_3} [\partial(\mathcal{A}_1\eta_1, \zeta_1) - \partial(\mathcal{A}_1\zeta_1, \eta_1)] \, d\Gamma. \tag{4.332}$$

In the above computation, we have used the relation (4.320) and Lemma 4.15 on the common boundary Γ_3.

Using the formulas (4.332) and (4.331) in the formula (4.330), we have the formula (4.329).   □

We have proven

Theorem 4.19 *If there is* $\lambda_0 \geq 1$ *such that*

$$\lambda_0 \sum_{i=1}^{2} b_i \int_{\Omega_i} B_i(\zeta, \zeta) \, dx \geq \|DW\|_{L^2(\Omega, T^2)}^2 + \gamma \|D^2 w\|_{L^2(\Omega, T^2)} \qquad (4.333)$$

for all $\zeta = (W, w) \in H^1(\Omega, \Lambda) \times H^2(\Omega)$, *then the problem* (4.327) *is exactly* ECSD* *controllable if there is* $T_0 > 0$ *such that for any* $T > T_0$, *there exists* $c > 0$ *satisfying*

$$\int_Q b(x) B(\zeta, \kappa \zeta) \, dQ + \int_{\Sigma_1} [b(x) B(\zeta, \zeta) + \gamma (\frac{\partial \Delta w}{\partial \nu})^2] \, d\Gamma dt \geq cE(0) \quad (4.334)$$

for all solutions ζ *to the problem* (4.325), *where the space* ECSD *is given by* (4.270).

Next, we present appropriate geometric conditions on the material of the shallow shell for the observability estimate (4.334).

Definition 4.3 *A vector field* V *on* $\overline{\Omega}$ *is said to be an escape vector field for transmission of the shallow shell if the relation* (4.173) *is true such that*

$$2 \min_{x \in \Omega} \vartheta(x) > \lambda_0 b_0 (1 + \mu_0) \max_{x \in \Omega} \iota(x) \qquad (4.335)$$

where $\mu_0 = \max(\mu_1, \mu_2)$ *and* $b_0 = \max(b_1, b_2)$.

Existence of such vector fields has been studied and a number of examples are given in Section 4.5.

Assumption (H1) There is $\lambda_0 \geq 1$ such that the ellipticity condition (4.333) holds true.

Assumption (H2) Let V be an escape vector field for transmission of the shallow shell such that

$$\Gamma_1 = \{ x \mid x \in \Gamma, \langle V, \nu \rangle \geq 0 \}. \qquad (4.336)$$

Set

$$\sigma_0 = \max_{x \in \Omega} |V|, \quad \sigma_1 = \min_{x \in \Omega} \vartheta(x) - \lambda_0 b_0 (1 + \mu_0) \max_{x \in \Omega} \iota(x)/2. \qquad (4.337)$$

To prove the inequality (4.334), we need to clarify some multiplier identities for the transmission problem as follows.

Lemma 4.16 *Let the assumptions* (**H1**) *and* (**H2**) *hold. Let* $\zeta = (W, w)$ *be a solution to the problem*

$$\zeta_{tt} - \gamma(0, \Delta w_{tt}) + b\mathcal{A}\zeta = 0 \quad in \quad (\Omega/\Gamma_3) \times (0, T) \qquad (4.338)$$

and $\zeta_i = \zeta|_{\Omega_i}$ *satisfy the transmission conditions* (4.322) *on* $\Gamma_3 \times (0, T)$. *Let* $T > 0$ *be given. Then*

$$2\sigma_1 \int_0^T E(t), dt \leq \sigma_0 \lambda_0 [E(0) + E(T)] + \int_\Sigma \Sigma(\zeta, b) d\Sigma$$

$$+ \int_{\Sigma_3} \Sigma_3(\zeta, b) \, d\Sigma + CL(\zeta) \qquad (4.339)$$

where

$$\Sigma(\zeta, b) = \Big[|\zeta_t|^2 + \gamma |Dw_t|^2 - b(x)B(\zeta, \zeta) \Big] \langle V, \nu \rangle$$

$$+ \partial \Big(b(x)\mathcal{A}\zeta, 2m(\zeta) + (\varrho W, -\vartheta w) \Big) + \gamma \Big(2V(w) - \vartheta w \Big) \frac{\partial w_{tt}}{\partial \nu}, \quad (4.340)$$

$$\Sigma_3(\zeta, b) = 2b_1 \partial(\mathcal{A}_1\zeta_1, m(\zeta_1)) + 2b_2 \partial(\mathcal{A}_2\zeta_2, m(\zeta_2))$$

$$+ \Big[b_2 B_2(\zeta_2, \zeta_2) - b_1 B_1(\zeta_1, \zeta_1) \Big] \langle V, \nu_1 \rangle, \qquad (4.341)$$

$$m(\zeta) = (D_V W, V(w)), \quad \varrho = 2\vartheta - \sigma_1,$$

$$\Sigma = (0, T) \times \Gamma, \quad \Sigma_3 = (0, T) \times \Gamma_3.$$

Proof. Multiply (4.335) by $2m(\zeta)$ and integrate by parts.

Let $P = \zeta - \gamma(0, \Delta w)$. Using the identity (4.153), the relations (4.320) and (4.322), we obtain

$$\int_Q \langle P_{tt}, 2m(\zeta) \rangle dQ = \sum_{i=1}^2 \int_{Q_i} \langle P_{tt}, 2m(\zeta) \rangle dQ$$

$$= 2Z|_0^T + 2 \int_Q \vartheta |\zeta_t|^2 \, dQ$$

$$- \sum_i \int_{\Sigma_i} [(|\zeta_{it}|^2 + \gamma |Dw_{it}|^2)\langle V, \nu_i \rangle + 2\gamma V(w_i)\langle Dw_{itt}, \nu_i \rangle] d\Sigma$$

$$= 2Z|_0^T + 2 \int_Q \vartheta |\zeta_t|^2 \, dQ$$

$$- \int_\Sigma [(|\zeta_t|^2 + \gamma |Dw_t|^2)\langle V, \nu_i \rangle + 2\gamma V(w)\langle Dw_{tt}, \nu \rangle] d\Sigma, \qquad (4.342)$$

where $\Sigma_i = \partial \Omega_i \times (0, T)$ and

$$Z = \int_\Omega [\langle \zeta_t, m(\zeta) \rangle + \gamma \langle \nabla w_t, \nabla V(w) \rangle] \, dx.$$

In addition it follows, similar to Lemma 4.10, that

$$|Z| \leq \sigma_0 \lambda_0 E(t) + CL(\zeta). \tag{4.343}$$

On the other hand, using the relation (4.170) in the identity (4.156) in which Ω and $\mathcal{A}$ are replaced by Ω_i and $\mathcal{A}_i$, respectively, we have

$$2 \int_{Q_i} \langle \mathcal{A}_i \zeta_i, m(\zeta_i) \rangle \, dQ$$

$$= \int_{\Sigma_i} [B_i(\zeta_i, \zeta_i)\langle V, \nu_i \rangle - 2\partial(\mathcal{A}_i \zeta_i, m(\zeta_i))] \, d\Sigma + 2 \int_{Q_i} a(\Upsilon(\zeta_i), G(V, DW_i)) \, dQ$$

$$+ 2 \int_{Q_i} \vartheta[\gamma a_i(\rho(\zeta_i), \rho(\zeta_i)) - a_i(\Upsilon(\zeta_i), \Upsilon(\zeta_i))] \, dQ + L(\zeta), \tag{4.344}$$

where $Q_i = \Omega_i \times (0, T)$. Moreover, a similar argument as in Lemma 4.11 yields

$$\sigma_1 \int_{\Omega} b(x) a(\Upsilon(\zeta), \Upsilon(\zeta)) \, dx \leq \int_{\Omega} b(x) a(\Upsilon(\zeta), G(V, DW)) \, dx + CL(\zeta), \tag{4.345}$$

where $a(\Upsilon(\zeta), \Upsilon(\zeta))|_{\Omega_i} = a_i(\Upsilon(\zeta_i), \Upsilon(\zeta_i))$ and $a(\Upsilon(\zeta), G(V, DW))|_{\Omega_i} = a_i(\Upsilon(\zeta_i), G(V, DW_i))$. It follows from (4.344) and (4.345) that

$$\int_Q \langle b(x)\mathcal{A}\zeta, 2m(\zeta) \rangle \, dQ$$

$$\geq \int_{\Sigma} [b(x)B(\zeta, \zeta) - \partial(b(x)\mathcal{A}\zeta, 2m(\zeta))] \, d\Sigma - \int_{\Sigma_3} \Sigma_3(\zeta, b) \, d\Sigma$$

$$+ 2 \int_Q [(\sigma_1 - \vartheta)b(x)a(\Upsilon(\zeta), \Upsilon(\zeta)) + \vartheta b(x)\gamma a(\rho(\zeta), \rho(\zeta))] dQ. \tag{4.346}$$

From the relations (4.338), (4.342), and (4.346), we obtain

$$\int_{\Sigma} [|\zeta_t|^2 + \gamma|Dw_t|^2 - b(x)B(\zeta, \zeta)]\langle V, \nu \rangle \, d\Sigma$$

$$+ 2 \int_{\Sigma} [\partial(b(x)\mathcal{A}\zeta, m(\zeta)) + \gamma V(w)\langle Dw_{tt}, \nu \rangle] d\Sigma + \int_{\Sigma_3} \Sigma_3(\zeta, b) \, d\Sigma$$

$$\geq Z|_0^T + \int_Q \Psi_b \, dQ + L(\zeta) \tag{4.347}$$

where

$$\Psi_b = 2[\sigma_1 - \vartheta(x)]b(x)a(\Upsilon(\zeta), \Upsilon(\zeta)) + 2\vartheta(x)[|\zeta_t|^2 + \gamma b(x)a(\rho(\zeta), \rho(\zeta))].$$

A similar argument as in (4.238) yields

$$\int_Q \Psi_b \, dQ \geq \sigma_1 \int_0^T E(t) \, dt + \int_Q \varrho[|W_t|^2 - b(x)a(\Upsilon(\zeta), \Upsilon(\zeta))] \, dQ$$

$$+ \int_Q \vartheta[\gamma b(x)a(\rho(\zeta), \rho(\zeta)) - \gamma|Dw_t|^2 - w_t^2] \, dQ. \tag{4.348}$$

Next, we multiply the equation (4.338) by $\varrho(W, 0)$ and $\vartheta(0, w)$, respectively, and use the transmission conditions on Γ_3 to obtain

$$\int_Q \varrho[|W_t|^2 - b(x)a(\Upsilon(\zeta), \Upsilon(\zeta))] \, dQ$$
$$= -\int_\Sigma \partial(b(x)\mathcal{A}\zeta, \varrho(W, 0))d\Sigma + L(\zeta), \tag{4.349}$$

$$\int_Q \vartheta[\gamma b(x)a(\rho(\zeta), \rho(\zeta)) - \gamma|Dw_t|^2 - w_t^2] \, dQ$$
$$= \int_\Sigma [\partial(b(x)\mathcal{A}\zeta, \vartheta(0, w)) + \gamma \vartheta w \frac{\partial w_{tt}}{\partial \nu}] \, d\Sigma + L(\zeta). \tag{4.350}$$

Finally, we use the relations (4.343), (4.348), (4.349), and (4.350) in the equation (4.347) to have the estimate (4.339). □

Let ω_i be regions such that

$$\Gamma_3 \subset \omega_1 \subset \overline{\omega}_1 \subset \omega_2 \subset \overline{\omega}_2 \subset \omega.$$

Let $\theta \in C_0^\infty(\omega_1)$ be such that

$$0 \le \theta(x) \le 1; \quad \theta(x) = 1 \quad \text{for} \quad x \in \Gamma_3. \tag{4.351}$$

Lemma 4.17 *Let the assumptions* (**H1**) *and* (**H2**) *hold and let the transmission conditions* (4.322) *be true. Let* ζ *solve the problem* (4.325). *Then*

$$\int_{\Sigma_3} \Sigma_3(\zeta, b) \, d\Sigma + \int_\Sigma \theta b(x) B(\zeta, \zeta) \langle V, \nu \rangle \, d\Sigma \le \sigma_0 \lambda_0 [E(0) + E(T)]$$
$$+ c \int_0^T \int_\omega b(x) B(\zeta, \zeta) \, dx dt + L(\zeta). \tag{4.352}$$

Proof. Using the multiplier $\theta(x)m(\zeta)$ to the equation (4.338) and by integration by parts, we obtain, via the boundary conditions and the transmission conditions (4.322),

$$\int_{\Sigma_3} \Sigma_3(\zeta, b) \, d\Sigma + \int_\Sigma \theta[\partial(b(x)\mathcal{A}\zeta, 2m(\zeta)) - b(x)B(\zeta, \zeta)] \langle V, \nu \rangle \, d\Sigma$$
$$= 2Z_1|_0^T + \int_Q [|\zeta_t|^2 + \gamma|Dw_t|^2 - b(x)B(\zeta, \zeta)] \, \text{div} \, (\theta V) \, dQ$$
$$+ 2 \int_Q [b(x)a(\Upsilon(\zeta), G(\theta V, DW)) + \gamma b(x)a(\rho(\zeta), G(\theta V, \rho(\zeta)))] \, dQ$$
$$- \int_Q \gamma D(\theta V)(Dw_t, Dw_t) \, dQ + L(\zeta) \tag{4.353}$$

where

$$Z_1 = \int_\Omega \theta[\langle\zeta_t, m(\zeta)\rangle + \gamma\langle Dw_t, D(V(w))\rangle]\,dx.$$

The condition $0 \le \theta \le 1$ implies that

$$2|Z_1| \le \sigma_0\lambda_0 E(t) + L(\zeta). \qquad (4.354)$$

In addition, a similar computation as in (4.269) yields

$$\partial(\mathcal{A}_i\zeta_i, m(\zeta_i)) = B_i(\zeta_i, \zeta_i)\langle V, \nu\rangle \quad \text{for} \quad x \in \Gamma_i \quad i = 1,\, 2. \qquad (4.355)$$

On the other hand, using the multiplier $(W, w)\operatorname{div}(\theta V)$ to the problem (4.325), we obtain that the second integral in the right hand side of (4.353) is a lower order term. Thus the relation $\operatorname{supp}\theta \subset \omega_1$ implies that

the right hand side of (4.353)

$$\le Z_1|_0^T + C\int_0^T\int_{\Omega\cap\omega_1} [b(x)B(\zeta,\zeta) + \gamma|Dw_t|^2]\,dxdt + L(\zeta). \qquad (4.356)$$

Next, we estimate the term $\gamma\int_0^T\int_{\omega_1}|Dw_t|^2dxdt$. Let $q \in C^\infty(M)$ be such that

$$0 \le q \le 1; \quad q = 1 \quad \text{on} \quad \omega_1; \quad q = 0 \quad \text{on} \quad \omega_2.$$

Using the multiplier $q(0, w)$ to the problem (4.325), we have

$$\gamma\int_0^T\int_{\Omega\cap\omega_1}|Dw_t|^2dxdt \le \int_Q q[w_t^2 + \gamma|Dw_t|^2 - \gamma b(x)a(\rho(\zeta), \rho(\zeta))]\,dQ$$

$$+\gamma\int_Q qb(x)B(\zeta,\zeta)dQ \le c\int_0^T\int_\omega b(x)B(\zeta,\zeta)\,dxdt + L(\zeta). \qquad (4.357)$$

Finally, the inequality (4.352) follows after inserting (4.357), (4.356), (4.355), and (4.354) into (4.353). □

We shall prove

Theorem 4.20 *Let the assumptions* **(H1)** *and* **(H2)** *hold. Let V be an escape vector field for transmission of the shallow shell. Let $\omega \subset M$ be an open set such that $\Gamma_3 \subset \omega$. Let $\theta \in C_0^\infty(\omega)$ be such that the condition (4.351) holds. Then for any $T > 2T_0$ there is $c > 0$ such that the observability estimate (4.334) holds where*

$$T_0 = 2\lambda_0\sigma_0/\sigma_1. \qquad (4.358)$$

Proof. Let ζ be a solution to the problem (4.325). Then it follows from the Dirichlet boundary conditions in the problem (4.325) and the formulas (4.340) and (4.355) that

$$\int_\Sigma \Sigma(\zeta, b)\,d\Sigma = \int_\Sigma b(x)B(\zeta, \zeta)\,d\Sigma. \qquad (4.359)$$

Let $0 < \varepsilon_0 < 1$ be given. Set $\omega_1 = \{\, x \mid x \in \omega,\ \theta(x) > \varepsilon_0 \,\}$. Clearly, ω_1 is an open set such that $\Gamma_3 \subset \omega$. Let $\hat{\theta} \in C_0^\infty(\omega_1)$ be such that $0 \leq \hat{\theta} \leq 1$ and $\hat{\theta} = 1$ for $x \in \Gamma_3$. We apply Lemma 4.17 with $\theta := \hat{\theta}$ and $\omega := \omega_1$, and obtain

$$\int_{\Sigma_3} \Sigma_3(\zeta, b)\, d\Sigma + \int_\Sigma \hat{\theta} b(x) B(\zeta, \zeta)\langle V, \nu\rangle\, d\Sigma \leq \sigma_0 \lambda_0 [E(0) + E(T)]$$
$$+ \frac{c}{\varepsilon_0} \int_0^T \int_\Omega b(x) B(\zeta, \theta\zeta)\, dx dt + L(\zeta). \tag{4.360}$$

Using the relations (4.360), (4.359), and (4.336) in the inequality (4.339), we obtain the observability (4.334) by a compactness-uniqueness argument as in Proposition 4.3. $\qquad\square$

If the materials have a special property and the division curve Γ_3 has some special shape as follows, then boundary control may be enough for the exact controllability of the transmission problem. We make the following

Assumption (H3) $b_2(1 - \mu_2) < b_1(1 - \mu_1)$, $b_2(1 + \mu_2) < b_1(1 + \mu_1)$, and $\langle V, \nu_1\rangle \geq 0$ for all $x \in \Gamma_3$.

Remark 4.11 *The assumption* (**H3**) *is a condition on the Young's modulus and Poisson's coefficient of the material of the shell. It may be necessary to establish the observability inequality if only the boundary controls are allowed. In the case of wave equations, [152] gave an example to show that the similar condition as the assumption* (**H3**) *is necessary for the stabilization and controllability by the boundary control.*

Consider the boundary control problem

$$\begin{cases} \eta_{tt} - \gamma(0, \Delta\psi_{tt}) + b(x)\mathcal{A}\eta = 0 & \text{on} \quad Q, \\ \eta(T) = \eta_t(T) = 0 & \text{on} \quad \Omega, \\ \Psi = \psi = \dfrac{\partial\psi}{\partial\nu} = 0 & \text{on} \quad \Sigma_2, \\ \eta_i \text{ satisfy the transmission conditions on } \Gamma_3, \\ \Psi = -D_\nu W, \quad \psi = \dfrac{\partial\Delta w}{\partial\nu}, \quad \dfrac{\partial\psi}{\partial\nu} = -\Delta w & \text{on} \quad \Sigma_1. \end{cases} \tag{4.361}$$

We have

Theorem 4.21 *Let the assumptions* (**H1**), (**H2**) *and* (**H3**) *hold. Then for any $T > T_0$, the problem (4.22) is exactly* ECSD* *controllable by $L^2(\Sigma, \Lambda) \times L^2(\Sigma) \times L^2(\Sigma)$ where T_0 and* ECSD *are given by (4.358) and (4.270), respectively.*

The proof of the above theorem follows from the following observability estimate

Theorem 4.22 *Let the assumptions* (**H1**), (**H2**) *and* (**H3**) *hold. Then for any $T > T_0$ there is $c > 0$ such that*

$$\int_0^T \int_{\Gamma_1} [b(x)B(\zeta, \zeta) + \gamma(\frac{\partial \Delta w}{\partial \nu})^2] \, d\Gamma dt \geq cE(0) \tag{4.362}$$

for all solutions ζ to the problem (4.325).

Proof. By Theorem 4.16 it will suffice to prove that

$$\Sigma_3(\zeta, b) \leq \text{lo}(\zeta) \tag{4.363}$$

where $\Sigma_3(\zeta, b)$ is given by (4.341) and $\text{lo}(\zeta)$ is the lower order terms given by (4.127).

The transmission conditions (4.322) imply that on Γ_3

$$\begin{cases} DW_1(\tau_1, \tau_1) = DW_2(\tau_2, \tau_2), \\ DW_1(\nu_1, \tau_1) = DW_2(\nu_2, \tau_2); \\ Dw_1 = Dw_2, \quad V(w_1) = V(w_2), \\ D^2w_1(\tau_1, \tau_1) = D^2w_2(\tau_2, \tau_2), \\ D^2w(\nu_1, \tau_1) = D^2w_2(\nu_2, \tau_2), \end{cases} \tag{4.364}$$

and

$$\begin{cases} b_1 DW_1(\nu_1, \nu_1) + b_1\mu_1 DW_1(\tau_1, \tau_1) + \text{lo}(\zeta_1) \\ = b_2 DW_2(\nu_2, \nu_2) + b_2\mu_2 DW_2(\tau_2, \tau_2) + \text{lo}(\zeta_2), \\ b_1(1 - \mu_1)\Upsilon(\zeta_1)(\nu_1, \tau_1) = b_2(1 - \mu_2)\Upsilon(\zeta_2)(\nu_2, \tau_2), \\ b_1 D^2w_1(\nu_1, \nu_1) + b_1\mu_1 D^2w_1(\tau_1, \tau_1) \\ = b_2 D^2w_2(\nu_2, \nu_2) + b_2\mu_2 D^2w_2(\tau_2, \tau_2). \end{cases} \tag{4.365}$$

Furthermore the relations (4.365) and (4.364) reach on Γ_3

$$\begin{cases} b_2 DW_2(\nu_2, \nu_2) = b_1 DW_1(\nu_1, \nu_1) \\ \quad + (b_1\mu_1 - b_2\mu_2)DW_1(\tau_1, \tau_1) + \text{lo}(\zeta), \\ b_2(1 - \mu_2)DW_2(\tau_2, \nu_2) = b_1(1 - \mu_1)DW_1(\tau_1, \nu_1) \\ \quad + [b_1(1 - \mu_1) - b_2(1 - \mu_2)]DW_1(\nu_1, \tau_1) + \text{lo}(\zeta), \\ b_2 D^2w_2(\nu_2, \nu_2) = b_1 D^2w_1(\nu_1, \nu_1) \\ \quad + (b_1\mu_1 - b_2\mu_2)D^2w_1(\tau_1, \tau_1). \end{cases} \tag{4.366}$$

Using the relations (4.320), (4.322) and (4.364) in the formula (4.35), we obtain

$$
b_1\partial(\mathcal{A}_1\zeta_1, m(\zeta_1)) + b_2\partial(\mathcal{A}_2\zeta_2, m(\zeta_2))
$$
$$
= [b_1\, TC_{11}(\zeta_1)DW_1(\nu_1,\nu_1) - b_2\, TC_{21}(\zeta_2)DW_2(\nu_2,\nu_2)]\langle V,\nu_1\rangle
$$
$$
+ [b_1\, TC_{12}(\zeta_1)DW_1(\tau_1,\nu_1) - b_2\, TC_{22}(\zeta_2)DW_2(\tau_2,\nu_2)]\langle V,\nu_1\rangle
$$
$$
+ [b_1\, TC_{13}(\zeta_1)D^2w_1(\nu_1,\nu_1) - b_2\, TC_{23}(\zeta_2)D^2w_2(\nu_2,\nu_2)]\langle V,\nu_1\rangle
$$
$$
= [b_1(DW_1(\nu_1,\nu_2))^2 - b_2(DW_2(\nu_2,\nu_2))^2]\langle V,\nu_1\rangle
$$
$$
+ [b_1\mu_1 DW_1(\nu_1,\nu_1) - b_2\mu_2 DW_2(\nu_2,\nu_2)]DW_1(\tau_1,\tau_1)\langle V,\nu_1\rangle
$$
$$
+ 2[b_1(1-\mu_1)(\Upsilon(\zeta_1)(\nu_1,\tau_1))^2 - b_2(1-\mu_2)(\Upsilon(\zeta_2)(\nu_2,\tau_2))^2]\langle V,\nu_1\rangle
$$
$$
+ \gamma[b_1(D^2w_1(\nu_1,\nu_1))^2 - b_2(D^2w_2(\nu_2,\nu_2))^2]\langle V,\nu_1\rangle
$$
$$
+ \gamma[b_1\mu_1 D^2w_1(\nu_1,\nu_1) - b_2\mu_2 D^2w_2(\nu_2,\nu_2)]D^2w_1(\tau_1,\tau_1)\langle V,\nu_1\rangle. \qquad (4.367)
$$

Moreover, by a similar computation, we have

$$
b_2 B_2(\zeta_2,\zeta_2) - b_1 B_1(\zeta_1,\zeta_1)
$$
$$
= b_2(DW_2(\nu_2,\nu_2))^2 - b_1(DW_1(\nu_1,\nu_1))^2 + (b_2 - b_1)(DW_1(\tau_1,\tau_1))^2
$$
$$
+ 2b_2(1-\mu_2)(\Upsilon(\zeta_2)(\nu_2,\tau_2))^2 - 2b_1(1-\mu_1)(\Upsilon(\zeta_1)(\nu_1,\tau_1))^2
$$
$$
+ 2b_2\mu_2 DW_2(\nu_2,\nu_2)DW_2(\tau_2,\tau_2) - 2b_1\mu_1 DW_1(\nu_1,\nu_1)DW_1(\tau_1,\tau_1)
$$
$$
+ \gamma[b_2(D^2w_2(\nu_2,\nu_2))^2 - b_1(D^2w_1(\nu_1,\nu_1))^2 + (b_2 - b_1)(D^2w_1(\tau_1,\tau_1))^2]
$$
$$
+ 2\gamma[b_2\mu_2 D^2w_2(\nu_2,\nu_2)D^2w_2(\tau_2,\tau_2) - b_1\mu_1 D^2w_1(\nu_1,\nu_1)D^2w_1(\tau_1,\tau_1)]
$$
$$
+ 2\gamma[b_2(1-\mu_2) - b_1(1-\mu_1)](D^2w_1(\nu_1,\tau_1))^2. \qquad (4.368)
$$

It follows from the relations (4.368), (4.367), and (4.366) that

$$
2b_1\partial(\mathcal{A}_1\zeta_1, m(\zeta_1)) + 2b_2\partial(\mathcal{A}_2\zeta_2, m(\zeta_2))
$$
$$
+ [b_2 B_2(\zeta_2,\zeta_2) - b_1 B_1(\zeta_1,\zeta_1)]\langle V,\nu_1\rangle
$$
$$
= [b_1(DW_1(\nu_1,\nu_2))^2 - b_2(DW_2(\nu_2,\nu_2))^2 + (b_2 - b_1)(DW_1(\tau_1,\tau_1))^2]\langle V,\nu_1\rangle
$$
$$
+ 2[b_1(1-\mu_1)(\Upsilon(\zeta_1)(\nu_1,\tau_1))^2 - b_2(1-\mu_2)(\Upsilon(\zeta_2)(\nu_2,\tau_2))^2]\langle V,\nu_1\rangle
$$
$$
+ \gamma[b_1(D^2w_1(\nu_1,\nu_1))^2 - b_2(D^2w_2(\nu_2,\nu_2))^2]\langle V,\nu_1\rangle
$$
$$
+ 2\gamma[b_2(1-\mu_2) - b_1(1-\mu_1)](D^2w_1(\nu_1,\tau_1))^2]\langle V,\nu_1\rangle
$$
$$
+ \gamma(b_2 - b_1)(D^2w_1(\tau_1,\tau_1))^2]\langle V,\nu_1\rangle + \mathrm{lo}\,(\zeta)
$$

$$
\begin{aligned}
&= \Big\{ \frac{1}{b_2}[b_1(b_2 - b_1)(DW_1(\nu_1, \nu_1))^2 \\
&\quad -2b_1(b_1\mu_1 - b_2\mu_2)DW_1(\nu_1, \nu_1)DW_1(\tau_1, \tau_1) \\
&\quad +(b_2(b_2 - b_1) - (b_1\mu_1 - b_2\mu_2)^2)(DW_1(\tau_1, \tau_1))^2] \\
&\quad +2\frac{b_1(1 - \mu_1)}{b_2(1 - \mu_2)}[b_2(1 - \mu_2) - b_1(1 - \mu_1)](\Upsilon(\zeta_1)(\nu_1, \tau_1))^2 \\
&\quad +\frac{2\gamma}{b_2}[b_1(b_2 - b_1)(D^2 w_1(\nu_1, \nu_1))^2 \\
&\quad -2b_1(b_1\mu_1 - b_2\mu_2)D^2 w_1(\nu_1, \nu_1)D^2 w_1(\tau_1, \tau_1) \\
&\quad +(b_2(b_2 - b_1) - (b_1\mu_1 - b_2\mu_2)^2)(D^2 w_1(\tau_1, \tau_1))^2] \\
&\quad +2\gamma[b_2(1 - \mu_2) - b_1(1 - \mu_1)](D^2 w_1(\nu_1, \tau_1))^2 \Big\}\langle V, \nu_1 \rangle \\
&\quad + \mathrm{lo}\,(\zeta).
\end{aligned}
\tag{4.369}
$$

Let us estimate the terms in (4.369) separately. The assumption (**H3**) implies that

$$
b_2 - b_1 < b_2\mu_2 - b_1\mu_1 \quad \text{and} \quad b_2 - b_1 < -(b_2\mu_2 - b_1\mu_1),
$$

that is,

$$
b_2 - b_1 < 0.
$$

Thus

$$
\begin{aligned}
&b_1(b_2 - b_1)(DW_1(\nu_1, \nu_1))^2 \\
&-2b_1(b_1\mu_1 - b_2\mu_2)DW_1(\nu_1, \nu_1)DW_1(\tau_1, \tau_1) \\
&+(b_2(b_2 - b_1) - (b_1\mu_1 - b_2\mu_2)^2)(DW_1(\tau_1, \tau_1))^2 \\
&< -\varepsilon_0[(DW_1(\nu_1, \nu_1))^2 + (DW_1(\tau_1, \tau_1))^2]
\end{aligned}
\tag{4.370}
$$

for some $\varepsilon_0 > 0$;

$$
[b_2(1 - \mu_2) - b_1(1 - \mu_1)](\Upsilon(\zeta_1)(\nu_1, \tau_1))^2 < -\varepsilon_0(\Upsilon(\zeta_1)(\nu_1, \tau_1))^2;
\tag{4.371}
$$

$$
\begin{aligned}
&b_1(b_2 - b_1)(D^2 w_1(\nu_1, \nu_1))^2 \\
&-2b_1(b_1\mu_1 - b_2\mu_2)D^2 w_1(\nu_1, \nu_1)D^2 w_1(\tau_1, \tau_1) \\
&+(b_2(b_2 - b_1) - (b_1\mu_1 - b_2\mu_2)^2)(D^2 w_1(\tau_1, \tau_1))^2 \\
&< -\varepsilon_0[(D^2 w_1(\nu_1, \nu_1))^2 + (D^2 w_1(\tau_1, \tau_1))^2];
\end{aligned}
\tag{4.372}
$$

$$
[b_2(1 - \mu_2) - b_1(1 - \mu_1)](D^2 w_1(\nu_1, \tau_1))^2 < -\varepsilon_0(D^2 w_1(\nu_1, \tau_1))^2.
\tag{4.373}
$$

Finally, the estimate (4.363) follows from (4.369)-(4.373).

4.8 Stabilization by Linear Boundary Feedback

The goal of this section is to introduce some uniform stabilization results for the shallow shell model with suitable, natural, linear or nonlinear dissipative boundary feedback in the form of moments and shears applied to an edge of the shell. More explicitly, what this means is the following. First, with homogeneous boundary conditions, the shell model is conservative (energy preserving). Next, we impose suitable linear dissipative terms (tractions/shears/moments) in physical boundary conditions exercised only on a portion Γ_0 of the boundary Γ of the shell and then seek to force the energy of the new corresponding closed loop, well-posed (Theorem 4.23) dissipative problem to decay to zero at a certain rate. The rate depends explicitly on pre-assigned growth properties of the dissipative terms. This is the content of Theorem 4.26.

Consider the shallow shell in unknown $\zeta = (W, w)$

$$\zeta_{tt} - \gamma(0, \Delta w_{tt}) + \mathcal{A}\zeta = 0 \quad \text{in} \quad Q^\infty = \Omega \times (0, \infty) \tag{4.374}$$

and define the total energy of the shell by

$$2E(t) = \|W_t\|^2_{L^2(\Omega,\Lambda)} + \|w_t\|^2_{L^2(\Omega)} + \gamma\|Dw_t\|^2_{L^2(\Omega,\Lambda)} + \mathcal{B}(\zeta, \zeta) \tag{4.375}$$

for $t \geq 0$ where $\mathcal{A}$ and $\mathcal{B}(\cdot, \cdot)$ are given by (4.36) and (4.21), respectively. By the Green formula (4.34) and the equation (4.374), we obtain

$$
\begin{aligned}
\frac{d}{dt}E(t) &= 2(W_{tt}, W_t)_{L^2(\Omega,\Lambda)} + 2(w_{tt}, w_t)_{L^2(\Omega)} \\
&\quad + 2\gamma(Dw_{tt}, Dw_t)_{L^2(\Omega,\Lambda)} + 2\mathcal{B}(\zeta, \zeta_t) \\
&= 2\int_\Gamma \left[w_t \frac{\partial w_{tt}}{\partial \nu} + \partial(\mathcal{A}\zeta, \zeta_t) \right] d\Gamma \\
&= 2\int_\Gamma \Big[v_1(\zeta)\langle W_t, \nu \rangle + v_2(\zeta)\langle W_t, \tau \rangle \\
&\quad + v_3(\zeta)\frac{\partial w_t}{\partial \nu} + (v_4(\zeta) + \frac{\partial w_{tt}}{\partial \nu})w_t \Big] d\Gamma
\end{aligned}
\tag{4.376}
$$

where

$$
\begin{cases}
v_1(\zeta) = (1 - \mu)\Upsilon(\zeta)(\nu, \nu) + \mu(wH - \delta W), \\
v_2(\zeta) = (1 - \mu)\Upsilon(\zeta)(\nu, \tau), \\
v_3(\zeta) = \gamma[\Delta w - (1 - \mu)D^2 w(\tau, \tau)], \\
v_4(\zeta) = -\gamma\{\frac{\partial \Delta w}{\partial \nu} + (1 - \mu)[\frac{\partial}{\partial \tau}D^2 w(\tau, \nu) + \theta\frac{\partial w}{\partial \nu}]\}.
\end{cases}
\tag{4.377}
$$

We assume that $\Gamma = \Gamma_0 \cup \Gamma_1$ such that

$$\overline{\Gamma_0} \cap \overline{\Gamma_1} = \emptyset. \tag{4.378}$$

Then the condition (4.378) implies that Γ_i are closed curves for $i = 0$ and 1.

We assume that a solution to (4.374) is subject to the boundary conditions

$$W = 0, \quad w = \frac{\partial w}{\partial \nu} = 0 \quad \text{on} \quad \Sigma_1^\infty = \Gamma_1 \times (0, \infty). \tag{4.379}$$

We will act on $\Sigma_0^\infty = \Gamma_0 \times (0, \infty)$ by a feedback as follows. For $\zeta = (W, w)$, we set

$$\hat{\zeta} = (\langle W, \nu \rangle, \langle W, \tau \rangle, \langle Dw, \nu \rangle, \langle Dw, \tau \rangle, w). \tag{4.380}$$

We consider feedback laws to be defined by

$$\begin{cases} v_i(\zeta) = \mathcal{J}_i(\zeta_t) & i = 1, 2, 3, \\ v_4(\zeta) + \gamma \langle Dw_{tt}, \nu \rangle = \mathcal{J}_4(\zeta_t) \end{cases} \quad \text{on} \quad \Sigma_0^\infty = \Gamma_0 \times (0, \infty) \tag{4.381}$$

where the feedback operators are given by

$$\begin{cases} \mathcal{J}_i(\zeta) = -\langle \hat{\zeta}, F_i \rangle_{\mathbb{R}^5} & i = 1, 2, 3, \\ \mathcal{J}_4(\zeta) = -\langle \hat{\zeta}, F_5 \rangle_{\mathbb{R}^5} + \frac{\partial}{\partial \tau} \langle \hat{\zeta}, F_4 \rangle_{\mathbb{R}^5} \end{cases} \tag{4.382}$$

where $F_i = F_i(x) \in \mathbb{R}^5$ for $x \in \Gamma_0$. If the 5×5 matrix $F := (F_1, F_2, F_3, F_4, F_5)$ satisfies that

$$F \text{ is symmetric and positive semidefinite on } \Gamma_0, \tag{4.383}$$

then using the relations (4.377)-(4.382) in the formula (4.376), we have

$$\frac{d}{dt} E(t) = -2 \int_{\Gamma_0} \langle F\hat{\zeta}_t, \hat{\zeta}_t \rangle_{\mathbb{R}^5} \, d\Gamma \leq 0. \tag{4.384}$$

Therefore, the following closed-loop system under the feedback laws of (4.381) and (4.382) is dissipative in the sense that $E(t)$ is nonincreasing if the condition (4.383) is true:

$$\begin{cases} \zeta_{tt} - \gamma(0, \Delta w_{tt}) + \mathcal{A}\zeta = 0 & \text{in} \quad Q^\infty, \\ W = w = \frac{\partial w}{\partial \nu} = 0 & \text{on} \quad \Sigma_1^\infty, \\ v_i(\zeta) + \langle \hat{\zeta}_t, F_i \rangle = 0 & i = 1, 2, 3 \quad \text{on} \quad \Sigma_0^\infty, \\ v_4(\zeta) + \gamma \langle Dw_{tt}, \nu \rangle + \langle \hat{\zeta}_t, F_4 \rangle = \frac{\partial}{\partial \tau} \langle \hat{\zeta}_t, F_5 \rangle & \text{on} \quad \Sigma_0^\infty, \\ \zeta(0) = \zeta_0, \quad \zeta_t(0) = \zeta_1 & \text{on} \quad \Omega. \end{cases} \tag{4.385}$$

Remark 4.12 *When the tangent component W of the shell is zero, the feedback laws of the system (4.385) are what the paper [105] presented for the uniform stabilization of the Kirchhoff plate.*

We have proven

Lemma 4.18 *Let ζ be a solution to the problem (4.385) with finite energy. Then, for any $s \leq t$, the following identity holds true for the energy $E(t)$ defined by (4.375):*

$$E(t) + \int_s^t \int_{\Gamma_0} \langle F\hat{\zeta}_t, \hat{\zeta}_t \rangle_{\mathbb{R}^5} \, d\Gamma dt = E(s). \tag{4.386}$$

The above lemma roughly tells us that a solution to the closed-loop system (4.385) can be obtained by a C_0 semigroup of contraction on the space $[H_{\Gamma_1}(\Omega, \Lambda) \times H^2_{\Gamma_1}(\Omega)] \times [L^2(\Omega, \Lambda) \times H^1_{\Gamma_1}(\Omega)]$ if the condition (4.383) holds true. Following some ideas in [105] for the Kirchhoff plate, we shall deal with this point in detail. Then the regularity of solutions we need for the stabilization is worked out by [3].

Variational formulation Set

$$W = H^1_{\Gamma_1}(\Omega, \Lambda) \times H^2_{\Gamma_1}(\Omega), \quad V = L^2(\Omega, \Lambda) \times H^1_{\Gamma_1}(\Omega),$$

$$\mathcal{L} = L^2(\Omega, \Lambda) \times L^2(\Omega).$$

Introduce bilinear forms

$$\alpha(\zeta, \eta) = \int_{\Omega} [\langle \zeta, \eta \rangle + \gamma \langle Dw, Du \rangle] \, dx \tag{4.387}$$

and

$$\alpha_0(\zeta, \eta) = \int_{\Gamma_0} \langle F\hat{\zeta}, \hat{\eta} \rangle_{\mathbb{R}^5} \, d\Gamma \tag{4.388}$$

for $\zeta = (W, w)$ and $\eta = (U, u)$. It follows from Green's formula (4.34) that an appropriate variational formulation of the system (4.385) is as follows: Find a vector field $\zeta \in C([0, \infty); W) \cap C^1([0, \infty); V)$ such that

$$\begin{cases} [\alpha(\zeta_t, \eta) + \alpha_0(\zeta, \eta)]_t + \mathcal{B}(\zeta, \eta) = 0 \quad \text{for all} \quad \eta \in W, \\ \zeta(0) = \zeta_0 \in W, \quad \zeta_t(0) = \zeta_1 \in V. \end{cases} \tag{4.389}$$

Well-posedness of the system (4.385) Let the ellipticity (4.172) of the shallow shell hold. Suppose that the condition (4.383) holds true. Then $\mathcal{B}(\cdot, \cdot)$ and $\alpha(\cdot, \cdot)$ are equivalent inner products on W and V, respectively. We identify $\mathcal{L}$ with its dual $\mathcal{L}^*$ so that we have the dense and continuous embeddings

$$W \subset V \subset \mathcal{L} \subset V^* \subset W^*. \tag{4.390}$$

Let A (respectively, P) denote *the canonical isomorphism* of V (respectively, W) endowed with the inner product $\alpha(\cdot, \cdot)$ (respectively, $\mathcal{B}(\cdot, \cdot)$) onto V^* (respectively, W^*). Thus

$$\alpha(\zeta, \eta) = (A\zeta, \eta)_{\mathcal{L}} \quad \text{for all} \quad \zeta, \eta \in V, \tag{4.391}$$

$$\mathcal{B}(\zeta, \eta) = (P\zeta, \eta)_{\mathcal{L}} \quad \text{for all} \quad \zeta, \eta \in W, \tag{4.392}$$

where $(\cdot, \cdot)_{\mathcal{L}}$ denotes the inner product of $\mathcal{L}$. Moreover, clearly $0 \le \alpha_0(\zeta, \zeta)| \le c\|\zeta\|^2_W$. Then there is a nonnegative continuous operator $A_0: W \to W^*$ such that

$$\alpha_0(\zeta, \eta) = (A_0\zeta, \eta)_{\mathcal{L}} \quad \text{for all} \quad \zeta, \eta \in W. \tag{4.393}$$

Then the variational problem (4.389) can be written as

$$(A\zeta_t + A_0\zeta)_t + P\zeta = 0 \quad \text{in} \quad W^*. \tag{4.394}$$

Let us formally rewrite (4.394) as the system

$$
\begin{pmatrix} P & 0 \\ 0 & A \end{pmatrix} \begin{pmatrix} \varsigma \\ \varsigma_t \end{pmatrix}_t + \begin{pmatrix} 0 & -P \\ P & A_0 \end{pmatrix} \begin{pmatrix} \varsigma \\ \varsigma_t \end{pmatrix} = 0
$$

or

$$
\mathbb{C} Y_t + I\!N Y = 0 \tag{4.395}
$$

where

$$
\mathbb{C} = \begin{pmatrix} P & 0 \\ 0 & A \end{pmatrix}, \quad I\!N = \begin{pmatrix} 0 & -P \\ P & A_0 \end{pmatrix}, \quad \text{and} \quad Y = \begin{pmatrix} \varsigma \\ \varsigma_t \end{pmatrix}.
$$

We wish to solve the problem (4.395) in the space $\mathcal{W} \times \mathcal{V}$. In order to let the problem (4.395) make sense in that space it is natural to introduce

$$
D(I\!N) = \{ (\varsigma, \eta) \mid \varsigma, \, \eta \in \mathcal{W}, \, P\varsigma + A_0 \eta \in \mathcal{V}^* \}. \tag{4.396}
$$

Then $I\!N \colon D(I\!N) \to \mathcal{W}^* \times \mathcal{V}^*$. Since $\mathbb{C}$ is the canonical isomorphism of $\mathcal{W} \times \mathcal{V}$ onto $\mathcal{W}^* \times \mathcal{V}^*$, we rewrite the system (4.395) in the form

$$
Y_t + \mathbb{C}^{-1} I\!N Y = 0 \quad \text{in} \quad \mathcal{W} \times \mathcal{V}. \tag{4.397}
$$

Solutions of the system (4.385) are therefore defined via (4.397).

Theorem 4.23 *Let the ellipticity (4.172) of the shallow shell hold. Suppose that the condition (4.383) holds true. Then $-\mathbb{C}^{-1} I\!N$ is the infinitesimal generator of a C_0-semigroup of contraction on $\mathcal{W} \times \mathcal{V}$.*

Proof. (i) We claim that $D(I\!N)$ is dense in $\mathcal{W} \times \mathcal{V}$.

By the definitions of P and A_0, for $\varsigma = (U, u) \in \mathcal{W}$, we obtain by the Green formula (4.34) and the relation (4.382)

$$
(P\varsigma + A_0\eta, \, s)_{\mathcal{L}} = \mathcal{B}(\varsigma, s) + a_0(\eta, s)
$$

$$
= \int_\Omega \langle A\varsigma, s \rangle \, dx + \int_{\Gamma_0} [(v_1(\varsigma) - \mathcal{J}_1(\eta))\langle U, \nu \rangle + (v_2(\varsigma) - \mathcal{J}_2(\eta))\langle U, \tau \rangle
$$

$$
+ (v_3(\varsigma) - \mathcal{J}_3(\eta))\langle Du, \nu \rangle + (v_4(\varsigma) - \mathcal{J}_4(\eta))u] \, d\Gamma \tag{4.398}
$$

where $\mathcal{J}_i(\cdot)$ are given by the formula (4.382) for $1 \le i \le 4$.

The expression on the right-hand side of the formula (4.398) implies the relation

$$
D(I\!N) \supset D_0 \tag{4.399}
$$

where

$$
D_0 = \{ (\varsigma, \eta) \mid \varsigma \in \mathcal{W} \cap (H^2(\Omega, \Lambda) \times H^4(\Omega)), \, \eta \in \mathcal{W},
$$
$$
v_i(\varsigma) = \mathcal{J}_i(\eta), \, i = 1, \, 2, \, 3 \}.
$$

Indeed, if $(\varsigma, \eta) \in D_0$, it follows from the relations (4.398) and (4.399) that

$$
|(P\varsigma + A_0\eta, \, s)_{\mathcal{L}}| \le \|A\varsigma\|_{\mathcal{L}} \|s\|_{\mathcal{L}} + c\|v_4(\varsigma) - \mathcal{J}_4(\eta)\|_{H^{-1/2}(\Gamma_0)} \|u\|_{H^{1/2}(\Gamma_0)}
$$
$$
\le c\|\varsigma\|_{H^2(\Omega, \Lambda) \times H^4(\Omega)} \|s\|_{\mathcal{L}} + c(\|w\|_{H^4(\Omega)} + \|\eta\|_{\mathcal{W}}) \|u\|_{H^1(\Omega)}
$$
$$
\le c(\|\varsigma\|_{H^2(\Omega, \Lambda) \times H^4(\Omega)} + \|\eta\|_{\mathcal{W}}) \|s\|_{\mathcal{V}} \quad \text{for all} \quad \varsigma \in \mathcal{V}, \tag{4.400}
$$

that is, $P\zeta + A_0\eta \in \mathcal{V}^*$.

Then $D(I\!N)$ is dense in $\mathcal{W} \times \mathcal{V}$ since D_0 is dense in $\mathcal{W} \times \mathcal{V}$.

(ii) $-\mathbb{C}^{-1}I\!N$ is dissipative. Indeed, using the relations (4.391)-(4.393), we obtain

$$
\begin{aligned}
\left(\mathbb{C}^{-1}I\!N(\zeta,\eta),\ (\zeta,\eta)\right)_{\mathcal{W}\times\mathcal{V}} &= \left((-\eta,\ A^{-1}(P\zeta+A_0\eta)),\ (\zeta,\eta)\right)_{\mathcal{W}\times\mathcal{V}} \\
&= -\mathcal{B}(\eta,\zeta) + \alpha(A^{-1}(P\zeta+A_0\eta),\eta) \\
&= -\mathcal{B}(\eta,\zeta) + (P\zeta+A_0\eta,\ \eta)_{\mathcal{L}} \\
&= \alpha_0(\eta,\eta) \geq 0
\end{aligned}
$$

for $(\zeta,\eta) \in D(I\!N)$ where the inner products of $\mathcal{W}$ and $\mathcal{V}$ are $\mathcal{B}(\cdot,\cdot)$ and $\alpha(\cdot,\cdot)$, respectively.

(iii) We also have $\text{Range}\,(\lambda I + \mathbb{C}^{-1}I\!N) = \mathcal{W} \times \mathcal{V}$ for $\lambda > 0$. In fact, this is equivalent to

$$
\text{Range}\,(\lambda^2 A + \lambda A_0 + P) = \mathcal{V}^*.
$$

But, by the Lax-Milgram theorem, it is actually true. $\square$

As a consequence of Theorem 4.23, we have the following result.

Theorem 4.24 *Let the ellipticity (4.172) of the shallow shell hold. Suppose that the condition (4.383) holds true. If initial data $\zeta_0 \in \mathcal{W}$ and $\zeta_1 \in \mathcal{W}$ are such that*

$$
P\zeta_0 + A_0\zeta_1 \in \mathcal{V}^*, \tag{4.401}
$$

then the problem (4.385) admits a unique solution satisfying

$$
\zeta \in C^1([0,\infty); \mathcal{W}) \cap C^2([0,\infty); \mathcal{V}), \quad \zeta_{tt} \in C([0,\infty); \mathcal{V}),
$$

$$
A\zeta_{tt} + A_0\zeta_t + P\zeta = 0, \quad t \geq 0, \tag{4.402}
$$

$$
\zeta(0) = \zeta_0, \quad \zeta_t(0) = \zeta_1.
$$

Trace Estimate on Boundary In order to remove some unnecessary restrictions on the control portion Γ_0, some trace estimates of the displacement vector field $\zeta = (W, w)$ of the shell have to be introduced. This will be done for the W-component and the w-component, separately.

Consider the problem

$$
\begin{cases}
W_{tt} - \boldsymbol{\Delta}_\mu W = F & \text{in} \quad Q, \\
W|_{\Gamma_1} = 0 & \text{on} \quad \Sigma_1
\end{cases} \tag{4.403}
$$

where $\boldsymbol{\Delta}_\mu$ is given by (4.37) and

$$
Q = (0,T) \times \Omega, \quad \Sigma_1 = (0,T) \times \Gamma_1.
$$

Set

$$
\begin{cases}
v_1(W,0) = \dfrac{1-\mu}{2}(DW + D^*W)(\nu,\nu) - \mu\delta W, \\
v_2(W,0) = \dfrac{1-\mu}{2}(DW + D^*W)(\nu,\tau).
\end{cases} \tag{4.404}
$$

We localize the problem (4.403) and use Proposition 3.2.2 in [126] to obtain

Proposition 4.6 *Let* $0 < \alpha < T/2$. *Let* W *be a solution to the problem* (4.403). *Then there is* $C_{\alpha T} > 0$ *such that*

$$\int_\alpha^{T-\alpha} \int_{\Gamma_0} |D_\tau W|^2 \, d\Gamma dt \le C_{\alpha T} \int_0^T \|F\|^2_{H^{-1/2}(\Omega,\Lambda)} \, dt$$

$$+ C_{\alpha T} \int_{\Sigma_0} (|W_t|^2 + |v_1(W,0)|^2 + |v_2(W,0)|^2) \, d\Sigma + L(W) \qquad (4.405)$$

where $\Sigma_0 = (0,T) \times \Gamma_0$ *and* $L(W)$ *denotes some lower order terms related to the energy level* $\|W\|^2_{H^1(Q)}$.

Remark 4.13 *The proof of Proposition 4.6 in* [126] *was given by the microlocal arguments. To avoid introducing the microlocal analysis theory we refer the readers to* [126] *for the proof of Proposition 4.6 in detail. For the same reason we refer the readers to* [123] *for a detailed proof of Proposition 4.7 below.*

Consider the problem

$$\begin{cases} w_{tt} - \gamma \Delta w_{tt} + \Delta^2 w = f & \text{in} \quad Q, \\ w = \dfrac{\partial w}{\partial \nu} = 0 & \text{on} \quad \Sigma_1. \end{cases} \qquad (4.406)$$

Let on Σ_0

$$\begin{cases} v_3(0,w) = \gamma[\Delta w - (1-\mu)D^2 w(\tau,\tau)], \\ v_4(0,w) = -\gamma\{\dfrac{\partial \Delta w}{\partial \nu} + (1-\mu)[\frac{\partial}{\partial \tau} D^2 w(\tau,\nu) + \theta \dfrac{\partial w}{\partial \nu}]\}. \end{cases} \qquad (4.407)$$

Applying Theorem 2.1 in [123] in the case of dimension 2 to the problem (4.406), we obtain

Proposition 4.7 *Let* $0 < \alpha < T/2$. *Let* $0 < \varepsilon < 1/2$ *and* $0 < s_0 < 1/2$. *Then the following inequality holds true for solutions* w *of the problem* (4.406):

$$\int_\alpha^{T-\alpha} \int_{\Gamma_0} |D^2 w|^2 \, d\Gamma dt \le C_{\alpha,\varepsilon,s_0,T}[\|f\|^2_{H^{-s_0}(Q)} + \|v_3(0,w)\|^2_{L^2(\Sigma_0)}$$

$$+ \|v_4(0,w) + \gamma\langle Dw_{tt}, \nu\rangle\|^2_{H^{-1}(\Sigma_0)} + \|Dw_t\|^2_{L^2(\Sigma_0,\Lambda)}$$

$$+ \|w_t\|^2_{L^2(\Sigma_0)} + \|w\|^2_{L^2(0,T;H^{3/2+\varepsilon}(\Omega))}]. \qquad (4.408)$$

Now we are ready to prove the trace estimates:

Theorem 4.25 *Let* $T/2 > \alpha > 0$ *be given. Then*

$$\int_\alpha^{T-\alpha} \int_{\Gamma_0} (|DW|^2 + |D^2 w|^2) \, d\Gamma dt$$

$$\le C_{T,\alpha} \int_{\Sigma_0} (|\zeta_t|^2 + |Dw_t|^2) \, d\Sigma + L(\zeta) \qquad (4.409)$$

for all solutions to the problem (4.385).

Proof. By the formulas (4.36) and (4.38), the W-component of ζ solves a problem (4.403) with F satisfying the estimate

$$|F| \le c(|W| + |Dw| + |w|) \quad \text{in} \quad Q,$$

which implies that

$$\int_0^T \|F\|^2_{H^{-1/2}(\Omega,\Lambda)} \, dt = L(\zeta).$$

Noting that $v_i(W, 0) = v_i(\zeta) + \text{lo}(\zeta)$, applying Proposition 4.6 to the component W yields, via the feedback laws in (4.385),

$$\int_\alpha^{T-\alpha} \int_{\Gamma_0} |D_\tau W|^2 \, d\Gamma dt \le C_{\alpha T} \int_{\Sigma_0} (|W_t|^2 + |v_1(\zeta)|^2 + |v_2(\zeta)|^2) \, d\Sigma$$

$$+ \text{lo}(W) \le C_{\alpha T} \int_{\Sigma_0} (|\zeta_t|^2 + |Dw_t|^2) \, d\Sigma + L(\zeta). \tag{4.410}$$

Moreover, by the formula (4.377) and the feedback laws (4.381), we obtain

$$\begin{cases} DW(\nu,\nu) = \mathcal{F}_1(\zeta_t) - [(1 - \mu)\Pi(\nu,\nu)/2 + \mu H]w - \mu DW(\tau,\tau) \\ DW(\tau,\nu) = \dfrac{2}{1-\mu} \mathcal{F}_2(\zeta_t) - DW(\nu,\tau) \end{cases}$$

which implies

$$\int_\alpha^{T-\alpha} \int_{\Gamma_0} |D\nu_W|^2 \, d\Gamma dt \le C \int_{\Sigma_0} (|\zeta_t|^2 + |D_\tau W|^2 + |w|^2) \, d\Sigma. \tag{4.411}$$

Then the estimates (4.410) and (4.411) together give

$$\int_\alpha^{T-\alpha} \int_{\Gamma_0} |DW|^2 \, d\Gamma dt \le C_{\alpha T} \int_{\Sigma_0} (|\zeta_t|^2 + |Dw_t|^2) \, d\Sigma + L(\zeta). \tag{4.412}$$

Next, we estimate the component w of ζ. By the formulas (4.36) and (4.38), the w-component of ζ solves a problem (4.406) with f satisfying the estimate

$$|f| \le C(|DW| + |Dw| + |w|) \quad \text{in} \quad Q$$

which implies that

$$\|f\|^2_{H^{-s_0}(Q^T)} \le C\|DW\|^2_{H^{-s_0}(\Omega,T^2)}. \tag{4.413}$$

Let us prove

$$\|DW\|^2_{H^{-s_0}(\Omega,T^2)} \le C \int_0^T \|W\|^2_{H^{1-s_0}(\Omega,\Lambda)} \, dt. \tag{4.414}$$

For simplicity, we assume that Ω is a coordinate patch U with the coordinate system $x = (x_1, x_2)$. Then $Q = (0, T) \times U$ and $W = (w_1, w_2)$. Denote the Fourier transform variable of (t, x) by (s, y). By definition

$$H^{-s_0}(Q) = \left(H_0^{s_0}(Q)\right)^*.$$

For $u \in H^{-s_0}(Q)$ given, we have

$$
\begin{aligned}
\|u\|_{H^{-s_0}(Q)}^2 &= \|(1 + s^2 + |y|^2)^{-s_0/2}\hat{u}\|_{L^2(\mathbb{R}^3)}^2 \\
&\leq \|(1 + |y|^2)^{-s_0/2}\hat{u}\|_{L^2(\mathbb{R}^3)}^2 \\
&= \int_{\mathbb{R}^2} (1 + |y|^2)^{-s_0} \int_{\mathbb{R}} |\hat{u}|^2 \, ds dy \\
&= \int_{\mathbb{R}^2} (1 + |y|^2)^{-s_0} \int_0^T |\hat{u}^x|^2 \, dt dy \\
&= \int_0^T \|u\|_{H^{-s_0}(\Omega)}^2 dt
\end{aligned}
\tag{4.415}
$$

where $\hat{u}^x$ denotes the Fourier transform on the variable x. It follows from (4.415) that

$$
\begin{aligned}
\|DW\|_{H^{-s_0}(Q,\Lambda)}^2 &\leq C \sum_{i=1}^2 (\|w_{1x_i}\|_{H^{-s_0}(Q)}^2 + \|w_{2x_i}\|_{H^{-s_0}(Q)}^2) \\
&\leq C \sum_{i=1}^2 \int_0^T (\|w_{1x_i}\|_{H^{-s_0}(\Omega)}^2 + \|w_{2x_i}\|_{H^{-s_0}(\Omega)}^2) \, dt \\
&\leq C \int_0^T (\|w_1\|_{H^{1-s_0}(\Omega)}^2 + \|w_2\|_{H^{1-s_0}(\Omega)}^2) \, dt \\
&\leq C \int_0^T \|W\|_{H^{1-s_0}(\Omega,\Lambda)}^2 dt.
\end{aligned}
$$

By the inequalities (4.414) and (4.413), the term $\|f\|_{H^{-s_0}(Q)}^2$ is a lower order term.

Noting that $v_i(0, w) = v_i(\zeta)$ for $i = 1, 2$, we apply Proposition 4.7 to the component w and obtain, via the feedback laws in (4.385),

$$
\begin{aligned}
\int_\alpha^{T-\alpha} &\int_{\Gamma_0} |D^2 w|^2 \, d\Gamma dt \\
&\leq C_{\alpha,T}[\|\langle \hat{\zeta}_t, F_3 \rangle\|_{L^2(\Sigma_0)}^2 + \|\frac{\partial}{\partial \tau}\langle \hat{\zeta}_t, F_5 \rangle - \langle \hat{\zeta}_t, F_4 \rangle\|_{H^{-1}(\Sigma_0)}^2 \\
&\quad + \|Dw_t\|_{L^2(\Sigma_0,\Lambda)}^2 + \|w_t\|_{L^2(\Sigma_0)}^2 + \|w\|_{L^2(0,T;H^{3/2+\epsilon}(\Omega))}^2] + L(\zeta) \\
&\leq C_{\alpha,T} \int_{\Sigma_0} (|\zeta_t|^2 + |Dw_t|^2) \, d\Sigma + L(\zeta).
\end{aligned}
\tag{4.416}
$$

The inequality (4.409) follows from (4.412) and (4.416).   □

Stabilization of the Closed-Loop System (4.385) For stabilization some geometric conditions are necessary. We make the following assumptions:

Assumption H1 The ellipticity (4.172) holds;

Assumption H2 There exists an escape vector field V for the shallow shell such that the conditions (4.173) and (4.175) hold;

Assumption H3 Γ_0 and Γ_1 satisfy the following conditions

$$\Gamma_1 \neq \emptyset, \quad \overline{\Gamma}_1 \cap \overline{\Gamma}_0 = \emptyset, \quad \text{and} \quad \langle V, \nu \rangle \leq 0 \quad \text{for} \quad x \in \Gamma_1; \tag{4.417}$$

Assumption H4 F, given in (4.382), is symmetric and positive on $\overline{\Gamma}_0$.

Remark 4.14 *The assumptions* **H1**-**H3** *are geometric conditions on the middle surface of the shell, while the assumption* **H4** *is on the feedback. For a plate the assumptions* **H1**-**H2** *are automatically satisfied, if we set* $V = x - x_0$. *For the general case, the assumptions* **H1**-**H2** *can be verified by the geometric method; see Section 4.5. Thanks to [123] and [126], the geometric assumption* **H3** *is, generally, considered to be much weaker than the following:*

$$\langle V, \nu \rangle \leq 0 \quad \text{in} \quad \Gamma_1 \quad \text{and} \quad \langle V, \nu \rangle > 0 \quad \text{in} \quad \Gamma_0 \tag{4.418}$$

which is used to avoid the complex trace estimates.

We have

Theorem 4.26 *Let the assumptions* **H1**-**H4** *hold. Then the closed-loop system (4.385) is exponentially stable: There are constants* $C_i > 0$ *such that*

$$E(t) \leq C_1 e^{-C_2 t} E(0) \quad t \geq 0 \tag{4.419}$$

for all initial data $\zeta_0 \in H^1_{\Gamma_1}(\Omega, \Lambda) \times H^2_{\Gamma_1}(\Omega)$ *and* $\zeta_1 \in L^2(\Omega, \Lambda) \times H^1_{\Gamma_1}(\Omega)$.

Proof. By Theorem 4.13, we have

$$2\sigma_1 \int_0^T E(t)dt \leq \int_\Sigma \text{SB} \, d\Sigma + \sigma_0 \lambda_0 [E(0) + E(T)] + L(\zeta) \tag{4.420}$$

where SB is given by (4.222) and σ_0, σ_1 are given in (4.172).

On Σ_1 the boundary conditions in (4.385) and the assumption **H3** imply that

$$\int_{\Sigma_1} \text{SB} \, d\Sigma = \int_{\Sigma_1} B(\zeta, \zeta) \langle V, \nu \rangle d\Sigma \leq 0 \tag{4.421}$$

(see the proof of Lemma 4.12).

Let

$$\eta = (U, u) := (2D_V W + \varrho W, 2V(w) - \vartheta w) \tag{4.422}$$

where ϱ is given in (4.222). Then by the formulas (4.377), (4.35) and the feedback laws in (4.385) on Σ_0

$$\partial(\mathcal{A}\zeta, 2m(\zeta) + (\varrho W, -\vartheta w)) + \gamma(2V(w) - \vartheta w)\frac{\partial w_{tt}}{\partial \nu}$$

$$= \partial(\mathcal{A}\zeta, \eta) + \gamma\langle Dw_{tt}, \nu\rangle u$$

$$= -\langle\hat{\zeta}_t, F_1\rangle\langle U, \nu\rangle - \langle\hat{\zeta}_t, F_2\rangle\langle U, \tau\rangle - \langle\hat{\zeta}_t, F_3\rangle\frac{\partial u}{\partial \nu}$$

$$+(\frac{\partial}{\partial \tau}\langle\hat{\zeta}_t, F_5\rangle - \langle\hat{\zeta}_t, F_4\rangle)u$$

which yields, via (4.388),

$$\int_{\Gamma_0} [\partial(\mathcal{A}\zeta, 2m(\zeta) + (\varrho W, -\vartheta w)) + \gamma(2V(w) - \vartheta w)\frac{\partial w_{tt}}{\partial \nu}]d\Gamma$$

$$= -\int_{\Gamma_0} \langle\hat{\zeta}_t, F\hat{\eta}\rangle d\Gamma_0 = -\alpha_0(\zeta_t, \eta). \tag{4.423}$$

Using the formulas (4.423) and (4.222), we obtain

$$\int_{\Sigma_0} \mathrm{SB}\, d\Sigma = \int_{\Sigma_0} [|\zeta_t|^2 + \gamma|Dw_t|^2 - B(\zeta, \zeta)]\langle V, \nu\rangle\, d\Sigma$$

$$- \int_0 \alpha_0(\zeta_t, \eta)dt. \tag{4.424}$$

We estimate SB on Σ_0. Let $s_i > 0$ be given such that

$$s_1|X|^2 \leq \langle FX, X\rangle \leq s_2|X|^2 \quad \text{for} \quad X \in \mathbb{R}^5,\ x \in \Gamma_0. \tag{4.425}$$

Then the definition of (4.388) yields

$$s_1\int_{\Gamma_0} (|\zeta|^2 + |Dw|^2)\, d\Gamma \leq \alpha_0(\zeta, \zeta) \leq s_2\int_{\Gamma_0} (|\zeta|^2 + |Dw|^2)\, d\Gamma. \tag{4.426}$$

It follows from the inequality (4.426) that

$$|\alpha_0(\zeta_t, \eta)| \leq [\alpha_0(\zeta_t, \zeta_t)]^{1/2}[\alpha_0(\eta, \eta)]^{1/2}$$

$$\leq c\int_{\Gamma_0} (|\zeta_t|^2 + |Dw_t|^2 + |DW|^2 + |D^2w|^2)d\Gamma + L(\zeta). \tag{4.427}$$

Using (4.427) in (4.424), we have

$$\int_{\Sigma_0} \mathrm{SB}\, d\Sigma \leq \int_{\Sigma_0} (|\zeta_t|^2 + |Dw_t|^2 + |DW|^2 + |D^2w|^2)d\Gamma + L(\zeta). \tag{4.428}$$

Next, change the integral domain $(0, T)$ into $(\alpha, T - \alpha)$ in both sides of the inequality (4.420) and use Theorem 4.25 to give

$$\sigma_1\int_\alpha^{T-\alpha} E(t)dt \leq C_T[E(\alpha) + E(T - \alpha)]$$

$$+ \int_{\Sigma_0} (|\zeta_t|^2 + |Dw_t|^2)d\Sigma + L(\zeta). \tag{4.429}$$

Note the relation $\frac{d}{dt}E(t) = -\alpha_0(\zeta_t, \zeta_t)$ and the inequality (4.426) and we find, for any $T > t > 0$,

$$
\begin{aligned}
E(t) &= E(T) + \int_t^T \alpha_0(\zeta_t, \zeta_t)dt \\
&\leq E(T) + s_2 \int_{\Sigma_0} (|\zeta_t|^2 + |Dw_t|^2)d\Sigma; \qquad\qquad (4.430) \\
&\geq E(T) + s_1 \int_{\Sigma_0} (|\zeta_t|^2 + |Dw_t|^2)d\Sigma. \qquad\qquad (4.431)
\end{aligned}
$$

Using the inequalities (4.430) and (4.431) in the inequality (4.429), we obtain for $T > 0$ large enough

$$
E(T) \leq C_T \int_{\Sigma_0} (|\zeta_t|^2 + |Dw_t|^2)d\Sigma + L(\zeta).
$$

By the compactness and uniqueness (Proposition 4.3) approach, we now have

$$
E(T) \leq C_T \int_{\Sigma_0} (|\zeta_t|^2 + |Dw_t|^2)d\Sigma. \qquad\qquad (4.432)
$$

Finally, using the inequality (4.432) and the left-hand side of the inequality (4.426)

$$
E(T) \leq C_T \int_0 \alpha_0(\zeta_t, \zeta_t)dt = C_T(E(0) - E(T)),
$$

that is,

$$
E(T) \leq \frac{C_T}{1 + C_T}E(0). \qquad\qquad (4.433)
$$

The estimate (4.419) follows from the inequality (4.433). $\qquad\qquad\square$

4.9 Stabilization by Nonlinear Boundary Feedback

We consider the stabilization problem of nonlinear feedbacks

$$
\begin{cases}
\zeta_{tt} - \gamma(0, \Delta w_{tt}) + \mathcal{A}\zeta = 0 & \text{in} \quad Q^\infty, \\
W = w = \dfrac{\partial w}{\partial \nu} = 0 & \text{on} \quad \Sigma_1^\infty, \\
v_1(\zeta) = -g_1(\langle W_t, \nu\rangle), \quad v_2(\zeta) = -g_2(\langle W_t, \tau\rangle) & \text{on} \quad \Sigma_0^\infty, \\
v_3(\zeta) = -g_3(\frac{\partial w_t}{\partial \nu}), & \text{on} \quad \Sigma_0^\infty, \\
v_4(\zeta) + \gamma\langle Dw_{tt}, \nu\rangle = \frac{\partial}{\partial \tau}g_4(\frac{\partial w_t}{\partial \tau}) & \text{on} \quad \Sigma_0^\infty, \\
\zeta(0) = \zeta_0, \quad \zeta_t(0) = \zeta_1 & \text{on} \quad \Omega
\end{cases} \qquad (4.434)
$$

where $\mathcal{A}$ is given by the formula (4.36) and $v_i(\zeta)$ are given in (4.377) for $1 \leq i \leq 4$. In the above feedback laws g_i are nonlinear functions for $1 \leq i \leq 4$.

A starting point is, as usual, an equality which states how the energy $E(t)$ in (4.375) of the entire system (4.434) is affected by feedback laws h_i. This fact will tell us how to make assumptions on nonlinear functions g_i such that the energy $E(t)$ is decreasing.

Using the boundary conditions of (4.434) in the formula (4.376), we obtain

Lemma 4.19 *Let ζ be a finite energy solution of system (4.434). Then, for any $0 \leq s \leq t$, the following identity holds true for the energy $E(t)$ defined by (4.375):*

$$E(t) + \int_s^t \int_{\Gamma_0} \beth(\zeta_t, \zeta_t) \, d\Gamma dt = E(s) \tag{4.435}$$

where

$$\beth(\zeta_t, \zeta_t) = \sum_{i=1}^4 g_i(\hat{\zeta}_{it}) \hat{\zeta}_{it}, \tag{4.436}$$

$$\hat{\zeta}_1 = \langle W, \nu \rangle, \quad \hat{\zeta}_2 = \langle W, \tau \rangle, \quad \hat{\zeta}_3 = \langle Dw, \nu \rangle, \quad \hat{\zeta}_4 = \langle Dw, \tau \rangle. \tag{4.437}$$

We make the following assumption on nonlinear function g_i:
Assumption $\tilde{\mathbf{H}}4$ Let $g_i \in C^1(\mathbb{R})$ be such that $g_i(0) = 0$ and

$$g_i(s)s > 0 \quad \text{for} \quad s \in \mathbb{R}, \, s \neq 0 \tag{4.438}$$

for $1 \leq i \leq 4$. Moreover, there exist positive constants $m_i > 0$ for $1 \leq i \leq 3$, such that for all $s \in \mathbb{R}$ with $|s| \geq m_3$, we have

$$m_1|s|^2 \leq g_i(s)s \leq m_2|s|^2 \quad \text{for} \quad 1 \leq i \leq 4. \tag{4.439}$$

The well-posedness of the system (4.434) can be established by the nonlinear semigroup theory (see [6]). We only introduce the regularity results on this issue in Proposition 4.8 below. For the proofs, we refer the readers to [122] and [115].

Proposition 4.8 *Let the ellipticity (4.172) and the assumption $\tilde{\mathbf{H}}4$ hold. Then there exists a unique, global solution of finite energy to the problem (4.434). This is to say that for any initial data $\zeta_0 \in H^1_{\Gamma_1}(\Omega, \Lambda) \times H^2_{\Gamma_1}(\Omega)$ and $\zeta_1 \in L^2(\Omega, \Lambda) \times H^1_{\Gamma_1}(\Omega)$, there exists a unique solution*

$$\zeta \in C([0, T]; H^1_{\Gamma_1}(\Omega, \Lambda) \times H^2_{\Gamma_1}(\Omega)),$$

$$\zeta_t \in C([0, T]; L^2(\Omega, \Lambda) \times H^1_{\Gamma_1}(\Omega)),$$

where $T > 0$ is arbitrary.

In addition, as in the case of the linear feedbacks, to avoid some unnecessary geometric assumptions on the uncontrolled part of the boundary, the trace estimates in [126] below are crucial, which can be proven by the similar arguments as in the proof of Theorem 4.25.

Theorem 4.27 *Let $0 < \alpha < T/2$ be given. Then the inequality (4.409) holds true for all solutions $\zeta = (W, w)$ to the problem (4.434).*

Uniform Stabilization Let the assumption $\tilde{H}4$ hold. Following [115] we construct a function h below. Let a function h be concave and strictly increasing, and vanishing at the origin: $h(0) = 0$ such that the following inequalities are satisfied:

$$h(g_i(s)s) \geq |s|^2 + |g_i(s)|^2 \quad \text{for} \quad |s| \leq m_3, \quad 1 \leq i \leq 4 \tag{4.440}$$

where $m_3 > 0$ is given in (4.439). We define first the function $h_0(\cdot)$ by

$$h_0(\cdot) = C_1 h(\frac{\cdot}{8 \, \text{meas} \, \Sigma_0}) \tag{4.441}$$

for some $C_1 > 0$ which will be specified later. Since h_0 is monotone increasing, for $C > 0$, $CI + h_0$ is invertible. We next define the function

$$p(s) = (C_2 I + h_0)^{-1}(s) \tag{4.442}$$

for some $C_2 > 0$ which will be specified later. Then p is a positive, continuous, strictly increasing function with $p(0) = 0$. Since p is positive, increasing, $q(x)$ is. Finally, let

$$q = I - (I + p)^{-1}. \tag{4.443}$$

We need

Lemma 4.20 *Consider a sequence $\{ s_k \, | \, k = 0, 1, \cdots \}$ of positive numbers which satisfies*

$$s_{k+1} + p(s_{k+1}) \leq s_k \quad \text{for} \quad k \geq 0 \tag{4.444}$$

where the function p is given by (4.442). Then

$$s_k \leq S(k) \quad \text{for all} \quad k \geq 0 \tag{4.445}$$

where $S(t)$ is the solution to the problem

$$\frac{d}{dt}S(t) + q(S(t)) = 0, \quad S(0) = s_0. \tag{4.446}$$

Moreover, $\lim_{t \to \infty} S(t) = 0$.

Proof. Use induction. Assume that $s_k \leq S(k)$ and prove that $s_{k+1} \leq S(k+1)$.

The inequality (4.444) is equivalent to

$$s_{k+1} \leq (I + p)^{-1}(s_k). \tag{4.447}$$

On the other hand, the equation (4.446) gives

$$S(t) \leq S(t) + \int_s q(S(t))dt = S(s) \quad \text{for} \quad 0 \leq s \leq t, \tag{4.448}$$

since q is positive. Since $(I+p)^{-1}$ is monotone increasing, from (4.446)-(4.448), we have

$$
\begin{aligned}
S(k+1) &= S(k) - \int_k^{k+1} q(S(t))\, dt \geq S(k) - q(S(k)) \\
&= (I - q)(S(k)) = (I + p)^{-1}(S(k)) \geq (I + p)^{-1}(s_k) \\
&\geq s_{k+1}.
\end{aligned}
$$

Finally, it is obvious that the limit $c := \lim_{t\to\infty} S(t)$ exists. Then the fact that the integral $\int_0^\infty q(S(t))dt$ converges implies $0 = \lim_{t\to\infty} q(S(t)) = q(c)$, that is, $c = 0$. $\qquad\square$

We are ready to prove

Theorem 4.28 *Assume that the assumptions* **H1-H3** *in Section* 4.8 *hold. Let the assumption* **H̃4** *in this section hold. Then there exists a constant* $T_0 > 0$ *such that the following estimate holds true:*

$$
E(t) \leq S(\frac{t}{T_0} - 1) \quad \text{for} \quad t > T_0 \tag{4.449}
$$

where $S(t)$ *is the solution of the ordinary differential equation*

$$
\frac{d}{dt} S(t) + q(S(t)) = 0 \quad S(0) = E(0), \tag{4.450}
$$

where the function $q(x)$ *is given by* (4.443).

Proof. As in the proof of Theorem 4.26, we will use the observability inequality (4.237) in Theorem 4.13 as a base.

Step 1 Let $T/2 > \alpha > 0$ be given. Applying the observability inequality (4.237) in Theorem 4.13 with Q replaced by $\Omega \times [\alpha, T - \alpha]$ yields

$$
2\sigma_1 \int_\alpha^{T-\alpha} E(t)\, dt \leq \int_{(\alpha, T-\alpha)\times\Gamma} SB\, d\Gamma dt + \sigma_0\lambda_0[E(\alpha) + E(T-\alpha)]
$$
$$
+L(\zeta) \tag{4.451}
$$

where SB is given by (4.222).

The key is to deal with the boundary integral in the right hand side of (4.451). Clearly, the inequality (4.421) still holds true for solution ζ of the problem (4.434) in which Σ_1 is replaced by $\Gamma_1 \times (\alpha, T - \alpha)$, that is,

$$
\int_{(\alpha, T-\alpha)\times\Gamma_1} SB\, d\Gamma dt \leq 0. \tag{4.452}
$$

Therefore we need to estimate $SB|_{\Gamma_0\times(\alpha, T-\alpha)}$ where the feedback laws in (4.434) will enter the formula (4.222). Let $\eta = (U, u)$ be given by (4.422):

$$
U = 2D_V W + \beta W, \quad u = 2V(w) - \vartheta w.
$$

Let

$$\hat{\eta}_1 = \langle U, \nu \rangle, \quad \hat{\eta}_2 = \langle U, \tau \rangle, \quad \hat{\eta}_3 = \langle Du, \nu \rangle, \quad \hat{\eta}_4 = \langle Du, \tau \rangle. \tag{4.453}$$

It follows that

$$\sum_{i=1}^{4} |\hat{\eta}_i|^2 \le c(|DW|^2 + |D^2 w|^2) + \mathrm{lo}\,(\zeta) \quad \text{for} \quad (x, t) \in \Sigma_0. \tag{4.454}$$

A similar computation as in (4.424) yields

$$\int_{\Gamma_0 \times (\alpha, T-\alpha)} \mathrm{SB}\, d\Sigma = \int_{\Gamma_0 \times (\alpha, T-\alpha)} [|\zeta_t|^2 + \gamma |Dw_t|^2 - B(\zeta, \zeta)] \langle V, \nu \rangle\, d\Sigma$$

$$- \int_{\Gamma_0 \times (\alpha, T-\alpha)} \beth(\zeta_t, \eta) d\Sigma \tag{4.455}$$

where $\beth(\cdot, \cdot)$ is given in (4.436) and $\hat{\zeta}_i$, $\hat{\eta}_i$ are given in (4.437) and (4.453), respectively.

Using the assumption $\tilde{\mathbf{H}}\mathbf{4}$ for g_i and the formulas (4.436) and (4.454), we have

$$|\beth(\zeta_t, \eta)|^2 \le (\sum_{i=1}^{4} |g_i(\hat{\zeta}_{it})|^2 \sum_{i=1}^{4} |\hat{\eta}_i|^2)^{1/2} \le \sum_{i=1}^{4} (|g_i(\hat{\zeta}_{it})|^2 + |\hat{\eta}_i|^2)$$

$$\le c(\sum_{i=1}^{4} |g_i(\hat{\zeta}_{it})|^2 + |DW|^2 + |D^2 w|^2) + \mathrm{lo}\,(\zeta). \tag{4.456}$$

It follows from the relations (4.456), (4.455), and the trace estimates (4.409) that

$$\int_{\Gamma_0 \times (\alpha, T-\alpha)} \mathrm{SB}\, d\Sigma$$

$$\le C_{\alpha, T} \int_{\Gamma_0 \times (\alpha, T-\alpha)} (|\zeta_t|^2 + |Dw_t|^2 + \sum_{i=1}^{4} |g_i(\hat{\zeta}_{it})|^2) d\Sigma + L(\zeta)$$

$$= C_{\alpha, T} \sum_{i=1}^{4} \int_{\Sigma_0} (|g_i(\hat{\zeta}_{it})|^2 + |\hat{\zeta}_{it}|^2)\, d\Sigma + L(\zeta). \tag{4.457}$$

On the other hand, we use the identity (4.435) where $t = T$ and $s = T - \alpha$ to obtain

$$E(T - \alpha) = E(T) + \int_{T-\alpha} \int_{\Gamma_0} \beth(\zeta_t, \zeta_t) g d\Gamma dt$$

$$\le E(T) + c \sum_{i=1}^{4} \int_{\Sigma_0} (|g_i(\hat{\zeta}_{it})|^2 + |\hat{\zeta}_{it}|^2)\, d\Sigma. \tag{4.458}$$

Similarly,

$$E(\alpha) \leq E(T) + c \sum_{i=1}^{4} \int_{\Sigma_0} (|g_i(\hat{\zeta}_{it})|^2 + |\hat{\zeta}_{it}|^2) \, d\Sigma. \tag{4.459}$$

Using the inequalities (4.459), (4.458), (4.457), and (4.452) in the inequality (4.451), we obtain

$$2\sigma_1 \int_{\alpha}^{T-\alpha} E(t) \, dt \leq 2\sigma_0 \lambda_0 E(T) + L(\zeta)$$

$$+ C_T \sum_{i=1}^{4} \int_{\Sigma_0} (|g_i(\hat{\zeta}_{it})|^2 + |\hat{\zeta}_{it}|^2) \, d\Sigma. \tag{4.460}$$

Moreover, we let $t = T$ in the identity (4.435), integrate it with respect to the variable s over $(\alpha, T - \alpha)$, and have, via (4.460)

$$(T - 2\alpha)E(T) = \int_{\alpha}^{T-\alpha} E(s)ds - \int_{\alpha}^{T-\alpha} \int_{s} \int_{\Gamma_0} \beth(\zeta_t, \zeta_t) d\Gamma dt ds$$

$$\leq \int_{\alpha}^{T-\alpha} E(s)ds + C_T \sum_{i=1}^{4} \int_{\Sigma_0} (|g_i(\hat{\zeta}_{it})|^2 + |\hat{\zeta}_{it}|^2) \, d\Sigma$$

$$\leq \frac{2\sigma_0\lambda_0}{\sigma_1} E(T) + C_T \sum_{i=1}^{4} \int_{\Sigma_0} (|g_i(\hat{\zeta}_{it})|^2 + |\hat{\zeta}_{it}|^2) \, d\Sigma + \text{lo}\,(\zeta). \tag{4.461}$$

Combining the inequality (4.461) and the compactness/uniqueness argument by Proposition 4.3 yields, for T large enough,

$$E(T) \leq C_T \sum_{i=1}^{4} \int_{\Sigma_0} (|g_i(\hat{\zeta}_{it})|^2 + |\hat{\zeta}_{it}|^2) \, d\Sigma. \tag{4.462}$$

Step 2   Set

$$\Sigma_{iA} = \{\, \hat{\zeta}_{it} \in L^2(\Sigma_0) \,|\, |\hat{\zeta}_{it}| \geq m_3 \,\}, \quad \Sigma_{iB} = \Sigma_0 - \Sigma_{iA},$$

where $m_3 > 0$ is given in (4.439). By the assumption (4.439), we have

$$\int_{\Sigma_{iA}} (|g_i(\hat{\zeta}_{it})|^2 + |\hat{\zeta}_{it}|^2) \, d\Sigma \leq \frac{m_2^2 + 1}{m_1} \int_{\Sigma_0} g_i(\hat{\zeta}_{it})\hat{\zeta}_{it} d\Sigma \tag{4.463}$$

for $1 \leq i \leq 4$.

On the other hand, it follows from the assumption (4.440) that

$$\int_{\Sigma_{iB}} (|g_i(\hat{\zeta}_{it})|^2 + |\hat{\zeta}_{it}|^2) \, d\Sigma \leq \int_{\Sigma_0} h(g_i(\hat{\zeta}_{it})\hat{\zeta}_{it}) d\Sigma \tag{4.464}$$

for $1 \leq i \leq 4$. Since the function h is concave, the Jensen inequality states that

$$\sum_{i=1}^{4} \int_{\Sigma_0} h(g_i(\hat{\zeta}_{it})\hat{\zeta}_{it})d\Sigma \leq \sum_{i=1}^{4}(\operatorname{meas}\Sigma_0)h\Big(\frac{1}{\operatorname{meas}\Sigma_0}\int_{\Sigma_0} g_i(\hat{\zeta}_{it})\hat{\zeta}_{it}d\Sigma\Big)$$

$$\leq h_0\Big(2\sum_{i=1}^{4}\int_{\Sigma_0} g_i(\hat{\zeta}_{it})\hat{\zeta}_{it}d\Sigma\Big)/C_T \tag{4.465}$$

where h_0 is given in (4.441) with $C_1 = 4C_T\operatorname{meas}\Sigma_0$ where C_T is given in (4.462).

Using the inequalities (4.465), (4.464), and (4.463) in the inequality (4.462), we obtain

$$E(T) \leq \Big(C_T\frac{m_2^2+1}{2m_1}I + h_0\Big)\Big(2\sum_{i=1}^{4}\int_{\Sigma_0} g_i(\hat{\zeta}_{it})\hat{\zeta}_{it}d\Sigma\Big),$$

that is, via (4.442) and (4.435),

$$p(E(T)) \leq 2\sum_{i=1}^{4}\int_{\Sigma_0} g_i(\hat{\zeta}_{it})\hat{\zeta}_{it}d\Sigma = E(0) - E(T),$$

or

$$E(T) + p(E(T)) \leq E(0), \tag{4.466}$$

where $C_2 = C_T(m_2^2+1)/(2m_1)$ in (4.442).

Step 3 Applying the inequality (4.466) yields

$$E((k+1)T) + p(E((k+1)T)) \leq E(kT) \quad \text{for} \quad k \geq 0. \tag{4.467}$$

We now take

$$s_k = E(kT) \quad \text{for} \quad k \geq 0.$$

Thus Lemma 4.20 gives

$$E(kT) \leq S(k) \quad \text{for} \quad k \geq 0. \tag{4.468}$$

Setting $t = kT + \tau$ where $0 \leq \tau < T$, the decreasing of $E(t)$ and $S(t)$ yields

$$E(t) \leq E(kT) \leq S(k) = S(\frac{t-\tau}{T}) \leq S(\frac{t}{T}-1)$$

for $t \geq T$. $\qquad\square$

If we, in addition, assume that the functions g_i are of a polynomial growth at the origin, then the following explicit decay rates are ready:

Corollary 4.5 *Assume that the assumptions* **H1**-**H3** *in Section 4.8 hold. Let the assumption* **H̃4** *in this section hold. Moreover, suppose that there are* $\sigma_1 > 0$ *and* $\sigma_2 > 0$ *such that*

$$g_i(s)s \leq \sigma_1 s^2 \quad \text{for all} \quad s \in \mathbb{R},$$

$$g_i(s)s \geq \sigma_2 |s|^{p+1} \quad \text{for} \quad |s| \leq 1, \quad 1 \leq i \leq 4, \quad \text{some } p \geq 1.$$

Then

$$E(t) \leq Ce^{-\sigma_3 t} \quad \text{if } p = 1; \quad E(t) \leq Ct^{\frac{2}{1-p}} \quad \text{if } p > 1$$

where both constants $C > 0$ *and* $\sigma_3 > 0$ *depend in general on* $E(0)$.

Proof. Let us construct a function h to meet the assumption (4.440). Set

$$h(s) = \frac{1 + \sigma_1^2}{\sigma_2^{2/(p+1)}} s^{2/(p+1)}.$$

Then the inequalities (4.440) hold true with $m_3 = 1$. Thus $p(s) = (C_2 I + h_0)^{-1}(s)$ and $C_2 p(s) + C_3 p^m(s) = s$ for $m = 2/(1+p)$ and a suitable constant $C_3 > 0$. Near $s = 0$, we have a asymptotic behavior

$$p(s) \sim Cs^{1/m} \quad \text{and} \quad q(x) \sim Cs^{1/m}.$$

Then $S(t)$ behaves like

$$S(t) \sim \begin{cases} Ct^{\frac{2}{1-p}} & p > 1, \\ Ce^{-\sigma_3 t} & p = 1 \end{cases} \quad \text{near} \quad \infty.$$

$\square$

4.10   Notes and References

Sections 4.1, 4.2, and 4.3 are from [210]; Sections 4.4, 4.5, and 4.6 are from [211] and [67]; Section 4.7 is from [34]; Section 4.8 is from [32]; Section 4.9 is from [126].

Control problems of shallow shells were studied by several papers where the middles have special shapes. For circular cylindrical shells, the exponential decay of the energy by boundary feedbacks was obtained by [42]. For spherical shells uniform stabilization by boundary dissipation was given by [129] (for the linear case) and by [113] (for the nonlinear case).

Classically, the topics of thin shells were covered by many books, for example, see [59], [94], and [95]. In addition, [19], [49]-[54], by constructing the geometric properties of the middle surface through the oriented boundary distance function in $\mathbb{R}^n$ and by using tangential differential operators on a neighborhood of a submanifold of $\mathbb{R}^n$, presented the modeling and analysis of thin shells.

In classical shell theory the middle surface of a shell is described by one coordinate patch: A map from a domain of $\mathbb{R}^2$ to $\mathbb{R}^3$. Therefore shell models end up with highly complicated resultant equations where the strain tensor and the change of curvature tensor of the middle surface are as unknowns. In these formulas, the explicit presence of the Christoffel symbols makes them unsuitable for energy method computations of the type needed for the control problems, such as continuous observability/stabilization estimates. For example, in [95], p. 33, Koiter commented on this point, "Expressing the strain measures in terms of the displacement components and their derivatives, we obtain three equations in terms of the three displacement components as unknowns. This approach is not very attractive because the resulting so-called displacement equations of equilibrium are undoubtedly extremely complicated. As far as we are aware, such general displacement equations have never appeared in print, not even in the linear theory of shells." However, the above points of Koiter were just right in the context of classical geometry, that is, those were just the limits of classical geometry.

We view the middle surface of a shell as a Riemannian manifold of dimension 2 to derive its mathematical models in the form of coordinates free. Then the Bochner technique helps us overcome the complicated computation which is necessary in the modeling and analysis of controllability/stabilization. More importantly, it seems clear that, without this differential geometric tool, many of these more sophisticated theorems would probably not have been discovered or, at least, their discoveries would otherwise have been much delayed.

The shell model in the present form of free coordinates (Theorems 4.1, 4.4) was worked out by the author more than 10 years ago but those equations were just published in [210] recently. Based on those equations and with the help of the Bochner technique a series of results on control of shallow shells have been obtained.

The ellipticity of the shallow shell was first obtained by [14] under the "shallowness" assumptions:

$$\Pi \quad \text{and} \quad D\Pi$$

are small enough. Then the ellipticity was given by [210] which was based on some curvature assumptions. The results of Theorem 4.2 in Section 4.2 are new, first published here where we have removed the "shallowness" assumptions in [14] and the curvature assumptions in [210] due to Lemma 4.5.

Observability estimates from boundary (Theorems 4.15, 4.16) were given in [211] where the assumptions (4.173) and (4.175) of *an escape vector field for the shallow shell* were introduced and a number of examples were given.

The internal exact controllability of Theorem 4.17 was obtained by [67] where the geometric conditions (4.210)-(4.214) of an escape region for the shallow shell were introduced.

As we can see in Section 4.7, the excellent work [34] made the important contribution: Established new geometric conditions **(H3)** which reflect material properties and shapes of the division curve if we have exact controllability only from boundary for the transmission problem.

Uniform stabilization by linear boundary feedbacks was given by [32] which closely followed [105] in seeking stabilization results. We use the trace estimates (Proposition 4.6) in [126] directly to clarify some unsatisfactory reasoning in the proofs of those in [32] which are crucial for avoiding geometric assumptions on uncontrolled portion.

Uniform stabilization by nonlinear boundary feedbacks came from an excellent work [126] in which the nonlinear feedback laws are taken from [115]. One of the main contributions was a sharp trace theory of the elastic W-component (Proposition 4.6 or Section 3.2 of [126]) by extending the microlocal argument which was originated in [122] for the wave equation.

Chapter 5

Naghdi's Shells. Modeling and Control

The strain tensors for Naghdi's shells are from [157] and [158]. By viewing the middle surface of the shell as a Riemannian manifold with the induced metric in $\mathbb{R}^3$, Naghdi's shells will be formulated mathematically as in the case of shallow shells. Then the multiplier scheme is carried out with help of the Bochner technique. Boundary exact controllability and boundary stabilization are derived. Stabilization of transmission is also presented.

5.1 Equations of Equilibrium. Green's Formulas. Ellipticity of the Strain Energy. Equations of Motion

Green's Formulas and Equations of Equilibrium We keep all notations as before. Let us assume that the middle surface of the shell occupies a bounded region Ω of a surface M in $\mathbb{R}^3$ with a normal field N. The shell, a body in $\mathbb{R}^3$, is defined by

$$S = \{\, p \mid p = x + zN(x),\ x \in \Omega,\ -h/2 < z < h/2 \,\} \tag{5.1}$$

where h is the thickness of the shell, small.

The second fundamental form Π of M is given by

$$\Pi(X,Y) = \langle \tilde{D}_X N, Y \rangle \quad \forall\, X, Y \in \mathcal{X}(M)$$

where $\tilde{D}$ is the covariant differential of the standard metric of $\mathbb{R}^3$. For each $x \in \Omega$, Π is a symmetric, bilinear form over $M_x \times M_x$ so that there is a symmetric, linear operator $\mathcal{S}\colon M_x \to M_x$ such that

$$\Pi(X,Y) = \langle \mathcal{S}X, Y \rangle \quad \forall\, X, Y \in M_x,\ x \in \Omega. \tag{5.2}$$

The third fundamental form of the surface M is defined by

$$c(X,Y) = \langle \tilde{D}_X N, \tilde{D}_Y N \rangle \quad \forall X, Y \in \mathscr{X}(M). \tag{5.3}$$

In the Naghdi model, the displacement vector $\zeta(p)$ of point $p = x + zN(x) \in$ S of the three-dimensional shell can be approximated as follows:

$$\zeta(p) = \zeta_1(x) + z \sqcap (x) \quad x \in \Omega \tag{5.4}$$

(see (7.67) in [157]) where $\zeta_1(x) \in \mathbb{R}^3$ is the displacement vector field of the middle surface and $\sqcap(x) \in \mathbb{R}^3$ represents deformation of the normal $N(x)$ for each $x \in \Omega$. In the literature, $\sqcap(x)$ are called the *director displacement vector* which is capable of rotation and stretches independently of the deformation of material points.

We decompose vector fields ζ_1 and $\sqcap$ into sums

$$\zeta_1(x) = W_1(x) + w_1(x)N(x) \quad \text{and} \quad \sqcap(x) = V(x) + w_2(x)N(x), \tag{5.5}$$

respectively, where $W_1, V \in \mathscr{X}(\Omega)$. [157] ([7.59] and [7.55] in [157]) yields the following tensor fields, directly defined on the middle surface,

$$\Upsilon(\zeta) = \frac{1}{2}(DW_1 + D^*W_1) + w_1 \Pi, \tag{5.6}$$

$$\mathcal{X}_0(\zeta) = \frac{1}{2}[DV + D^*V + \Pi(\cdot, D.W_1) + \Pi(D.W_1, \cdot)] + w_2 \Pi + w_1 c, \tag{5.7}$$

and

$$\varphi_0(\zeta) = \frac{1}{2}[Dw_1 + V - i(W_1)\Pi], \tag{5.8}$$

where $\Upsilon(\zeta)$, $\mathcal{X}_0(\zeta)$, and $\varphi_0(\zeta)$ denote respectively *the strain tensor of the middle surface, the change of curvature tensor of the middle surface, and the rotation of the normal N.* In the above expressions we have made the use of the *canonical isomorphism: $M_x = M_x^*$* as M_x are inner product spaces of dimension 2 for $x \in M$. $i(W_1)\Pi$ is *the interior product of the tensor field Π by the vector field W_1.* The formulas $(5.6) - (5.8)$ are also called the strain tensors of Naghdi.

We make a change of variables by

$$W_2 = V + i(W_1)\Pi \quad \text{for} \quad x \in \Omega. \tag{5.9}$$

Let $x \in \Omega$ be given and let E_1, E_2 be a frame field normal at x. Using the relations $D_{E_i} E_j(x) = 0$, we obtain

$$
\begin{aligned}
DW_2(E_i, E_j) &= E_j \langle W_2, E_i \rangle = E_j \langle V, E_i \rangle + E_j(\Pi(W_1, E_i)) \\
&= DV(E_i, E_j) + \Pi(E_i, D_{E_j} W_1) + D\Pi(W_1, E_i, E_j),
\end{aligned}
$$

that is,

$$DW_2 = DV + \Pi(\cdot, D.W_1) + i(W_1)D\Pi \quad \text{for} \quad x \in \Omega. \tag{5.10}$$

Inserting (5.10) into (5.7) and (5.9) into (5.8), respectively, we have

$$\mathcal{X}_0(\zeta) = \frac{1}{2}(DW_2 + D^*W_2) + \mathcal{X}_{0L}(\zeta) \tag{5.11}$$

and

$$\varphi_0(\zeta) = \frac{1}{2}Dw_1 + \varphi_{0L}(\zeta), \tag{5.12}$$

where

$$\mathcal{X}_{0L}(\zeta) = -i(W_1)D\Pi + w_1c + w_2\Pi, \quad \varphi_{0L}(\zeta) = -i(W_1)\Pi + \frac{1}{2}W_2 \tag{5.13}$$

are zero order terms.

Denote by $\hat{\mathcal{E}}$ the strain tensor of the three-dimensional shell. By taking the first approximation from (7.61), (7.59) and (7.55) in [157], we obtain *Naghdi's displacement-strain relations* (7.65) in [157] as follows:

$$\begin{cases} \hat{\mathcal{E}}|_\Omega(p) = \Upsilon(\zeta) + z\mathcal{X}_0(\zeta), \\ i(N)\hat{\mathcal{E}}(p) = \varphi_0(\zeta) + \frac{z}{2}Dw_2, \quad \text{for} \quad p = x + xN(x) \in S \\ \hat{\mathcal{E}}(N,N)(p) = w_2, \end{cases} \tag{5.14}$$

where $\hat{\mathcal{E}}|_\Omega$ denotes the component of the three-dimensional shell strain $\hat{\mathcal{E}}$ on the middle surface.

We assume that the material of the shell is homogenous and isotropic. Let $x \in \Omega$ be given. Let E_1, E_2 be an orthonormal frame at x in the induced metric g of the surface M from the Euclidean metric of $\mathbb{R}^3$. Then E_1, E_2 and E_3 forms an orthonormal frame in the Euclidean metric of $\mathbb{R}^3$ where $E_3 = N$. The stress-strain relations of the three-dimensional shell on the middle surface are

$$\sigma_{ij} = \frac{E}{1+\mu}[\hat{\mathcal{E}}_{ij} + \frac{\mu}{1-2\mu}\operatorname{tr}\hat{\mathcal{E}}\delta_{ij}] \quad \text{for} \quad x \in \Omega \tag{5.15}$$

for $1 \leq i, j \leq 3$ where $\sigma_{ij} = \sigma(E_i, E_j)$, $\hat{\mathcal{E}}_{ij} = \hat{\mathcal{E}}(E_i, E_j)$, E is *Young's modulus*, and μ is *Poisson's ratio*. It follows from the formula (5.15) that

$$\begin{aligned} \sum_{ij}\sigma_{ij}\hat{\mathcal{E}}_{ij} &= \frac{E}{1+\mu}[|\hat{\mathcal{E}}|^2 + \frac{\mu}{1-2\mu}(\operatorname{tr}\hat{\mathcal{E}})^2] \\ &= \frac{E}{1+\mu}[\sum_{ij=1}^2 \hat{\mathcal{E}}_{ij}^2 + 2\sum_{i=1}^2 \hat{\mathcal{E}}_{i3}^2 + \hat{\mathcal{E}}_{33}^2 + \frac{\mu}{1-2\mu}(\sum_{i=1}^2 \hat{\mathcal{E}}_{ii} + \hat{\mathcal{E}}_{33})^2] \\ &= \frac{E}{1+\mu}[|\hat{\mathcal{E}}|_\Omega|^2 + 2|i(N)\hat{\mathcal{E}}|^2 + \hat{\mathcal{E}}^2(N,N)] \\ &\quad + \frac{\mu}{1-2\mu}(\operatorname{tr}\hat{\mathcal{E}}|_\Omega + \hat{\mathcal{E}}(N,N))^2] \end{aligned} \tag{5.16}$$

for $x \in \Omega$. Then *the strain energy of the three-dimensional shell* is obtained by integrating (5.16) over S:

$$I(\zeta) = \alpha \int_\Omega \int_{-h/2}^{h/2} J(x, z)\, dz dx \tag{5.17}$$

where

$$
\begin{aligned}
J(x, z) &= [|\hat{\mathcal{E}}|_\Omega|^2 + 2|i(N)\hat{\mathcal{E}}|^2 + \hat{\mathcal{E}}^2(N, N) \\
&\quad + \beta(\operatorname{tr}\hat{\mathcal{E}}|_\Omega + \hat{\mathcal{E}}(N, N))^2](1 + \operatorname{tr}\Pi z + \kappa z^2),
\end{aligned}
\tag{5.18}
$$

$$\alpha = E/(1 + \mu), \quad \beta = \mu/(1 - 2\mu), \tag{5.19}$$

where κ is the Gauss curvature of the middle surface.

By inserting the formula (5.14) into the formula (5.18), we have

$$
\begin{aligned}
J(x, z) &= \{|\Upsilon(\zeta)|^2 + 2|\varphi_0(\zeta)|^2 + w_2^2 + \beta(\operatorname{tr}\Upsilon(\zeta) + w_2)^2 \\
&\quad + [2\langle\Upsilon(\zeta), \mathcal{X}_0(\zeta)\rangle + \langle\varphi_0(\zeta), Dw_2\rangle + 2\beta(\operatorname{tr}\Upsilon(\zeta) + w_2)\operatorname{tr}\mathcal{X}_0(\zeta)]z \\
&\quad + [|\mathcal{X}_0(\zeta)|^2 + |Dw_2|^2/2 + \beta\overset{2}{\operatorname{tr}}\mathcal{X}_0(\zeta)]z^2\}(1 + \operatorname{tr}\Pi z + \kappa z^2) \\
&= |\Upsilon(\zeta)|^2 + 2|\varphi_0(\zeta)|^2 + w_2^2 + \beta(\operatorname{tr}\Upsilon(\zeta) + w_2)^2 \\
&\quad + \{[|\Upsilon(\zeta)|^2 + 2|\varphi_0(\zeta)|^2 + w_2^2 + \beta(\operatorname{tr}\Upsilon(\zeta) + w_2)^2]\operatorname{tr}\Pi \\
&\quad + 2\langle\Upsilon(\zeta), \mathcal{X}_0(\zeta)\rangle + \langle\varphi_0(\zeta), Dw_2\rangle + 2\beta(\operatorname{tr}\Upsilon(\zeta) + w_2)\operatorname{tr}\mathcal{X}_0(\zeta)\}z \\
&\quad + \{|\mathcal{X}_0(\zeta)|^2 + |Dw_2|^2/2 + \beta\overset{2}{\operatorname{tr}}\mathcal{X}_0(\zeta) \\
&\quad + [|\Upsilon(\zeta)|^2 + 2|\varphi_0(\zeta)|^2 + w_2^2 + \beta(\operatorname{tr}\Upsilon(\zeta) + w_2)^2]\kappa \\
&\quad + [2\langle\Upsilon(\zeta), \mathcal{X}_0(\zeta)\rangle + \langle\varphi_0(\zeta), Dw_2\rangle + 2\beta(\operatorname{tr}\Upsilon(\zeta) + w_2)\operatorname{tr}\mathcal{X}_0(\zeta)]\operatorname{tr}\Pi\}z^2 \\
&\quad + \{[2\langle\Upsilon(\zeta), \mathcal{X}_0(\zeta)\rangle + \langle\varphi_0(\zeta), Dw_2\rangle + 2\beta(\operatorname{tr}\Upsilon(\zeta) + w_2)\operatorname{tr}\mathcal{X}_0(\zeta)]\kappa \\
&\quad + [|\mathcal{X}_0(\zeta)|^2 + |Dw_2|^2/2 + \beta\overset{2}{\operatorname{tr}}\mathcal{X}_0(\zeta)]\operatorname{tr}\Pi\}z^3 \\
&\quad + [|\mathcal{X}_0(\zeta)|^2 + |Dw_2|^2/2 + \beta\overset{2}{\operatorname{tr}}\mathcal{X}_0(\zeta)]\kappa z^4.
\end{aligned}
\tag{5.20}
$$

Then it follows from the formula (5.20) that

$$
\begin{aligned}
\int_{-h/2}^{h/2} &J(x, z)\, dz \\
&= [|\Upsilon(\zeta)|^2 + 2|\varphi_0(\zeta)|^2 + w_2^2 + \beta(\operatorname{tr}\Upsilon(\zeta) + w_2)^2]h \\
&\quad + \{|\mathcal{X}_0(\zeta)|^2 + |Dw_2|^2/2 + \beta\overset{2}{\operatorname{tr}}\mathcal{X}_0(\zeta) + \\
&\quad + [|\Upsilon(\zeta)|^2 + 2|\varphi_0(\zeta)|^2 + w_2^2 + \beta(\operatorname{tr}\Upsilon(\zeta) + w_2)^2]\kappa + [2\langle\Upsilon(\zeta), \mathcal{X}_0(\zeta)\rangle \\
&\quad + \langle\varphi_0(\zeta), Dw_2\rangle + 2\beta(\operatorname{tr}\Upsilon(\zeta) + w_2)\operatorname{tr}\mathcal{X}_0(\zeta)]\operatorname{tr}\Pi\}h^3/12 \\
&\quad + [|\mathcal{X}_0(\zeta)|^2 + |Dw_2|^2/2 + \beta\overset{2}{\operatorname{tr}}\mathcal{X}_0(\zeta)]\kappa\, h^5/80.
\end{aligned}
\tag{5.21}
$$

As usual in the thin shell theory, we assume that

$$h/\mathbf{R} \ll 1 \tag{5.22}$$

where $\mathbf{R}$ is *the smallest principal radius of the curvature of the undeformed middle surface* is given by (4.25). By Lemma 4.2, the assumption (5.22), and the formula (5.21), the following approximation to the strain energy of the shell is justified

$$I(\zeta) = \alpha h \int_\Omega \{|\Upsilon(\zeta)|^2 + 2|\varphi_0(\zeta)|^2 + w_2^2 + \beta(\operatorname{tr}\Upsilon(\zeta) + w_2)^2$$
$$+\gamma[|\mathcal{X}_0(\zeta)|^2 + |Dw_2|^2/2 + \beta\overset{2}{\operatorname{tr}}\mathcal{X}_0(\zeta)]\} \, dx \tag{5.23}$$

where $\gamma = h^2/12$.

Now we are able to associate the following symmetric, bilinear form with the strain energy, directly defined on space $(H^1(\Omega, \Lambda))^2 \times (H^1(\Omega))^2$,

$$\mathcal{B}_0(\zeta, \eta) = \frac{\alpha h}{2} \int_\Omega B_0(\zeta, \eta) \, dx \tag{5.24}$$

where

$$B_0(\zeta, \eta) = 2\langle\Upsilon(\zeta), \Upsilon(\eta)\rangle + 4\langle\varphi_0(\zeta), \varphi_0(\eta)\rangle + 2w_2u_2$$
$$+2\beta(\operatorname{tr}\Upsilon(\zeta) + w_2)(\operatorname{tr}\Upsilon(\eta) + u_2) + 2\gamma\langle\mathcal{X}_0(\zeta), \mathcal{X}_0(\eta)\rangle$$
$$+\gamma\langle Dw_2, Du_2\rangle + 2\gamma\beta\operatorname{tr}\mathcal{X}_0(\zeta)\operatorname{tr}\mathcal{X}_0(\eta) \tag{5.25}$$

where $\zeta, \eta \in (H^1(\Omega, \Lambda))^2 \times (H^1(\Omega))^2$ are given by

$$\zeta = (W_1, W_2, w_1, w_2), \quad \eta = (U_1, U_2, u_1, u_2).$$

Let $\hat{\Gamma}$ be a relatively open portion of boundary Γ of the middle surface. Set

$$\mathcal{L}^2(\Omega) = (L^2(\Omega, \Lambda))^2 \times (L^2(\Omega))^2, \tag{5.26}$$

$$\mathcal{H}^1(\Omega) = (H^1(\Omega, \Lambda))^2 \times (H^1(\Omega))^2, \tag{5.27}$$

$$\mathcal{H}_{\hat{\Gamma}}^1(\Omega) = (H_{\hat{\Gamma}}^1(\Omega, \Lambda))^2 \times (H_{\hat{\Gamma}}^1(\Omega))^2. \tag{5.28}$$

Suppose that Load$(\cdot)$ is a linear functional which is associated with the external load, given by

$$\text{Load}(\zeta) = (\zeta, \mathcal{F})_{\mathcal{L}^2(\Omega)} \quad \text{for all} \quad \zeta \in \mathcal{L}^2(\Omega) \tag{5.29}$$

for some $\mathcal{F} \in \mathcal{L}^2(\Omega)$.

Then, for a Naghdi's shell which is clamped along Γ_0 and free on Γ_1, where $\Gamma_0 \cup \Gamma_1 = \Gamma$, we derive the following variational problem:

For $\mathcal{F} \in \mathcal{L}^2(\Omega)$, find $\zeta \in \mathcal{H}_{\Gamma_0}^1(\Omega)$ such that $\mathcal{B}(\zeta, \eta) = \text{Load}(\eta) \tag{5.30}$

for all $\eta \in \mathcal{H}_{\Gamma_0}^1(\Omega)$. Similarly, we may propose our problems for other kinds of boundary conditions by choosing different Sobolev spaces.

Let $\boldsymbol{\Delta}_\beta$ be an operator of the Hodge-Laplace type, given by

$$\boldsymbol{\Delta}_\beta = -[\delta d + 2(1+\beta)d\delta] \tag{5.31}$$

which will be applied to vector fields or equivalently to 1-forms, where d is the exterior derivative, δ is its formal adjoint, given by (1.109), and β is given in (5.19).

For $x \in \Omega$ and $G \in T^2(\Omega)$ given, $\langle G, i(X)D\Pi \rangle$ is a linear functional with respect to $X \in M_x$. Then there is a vector in M_x, denoted by $\mathcal{P}G$, such that

$$\langle G, i(X)D\Pi \rangle = \langle \mathcal{P}G, X \rangle \quad \text{for} \quad X \in M_x. \tag{5.32}$$

Therefore $\mathcal{P}\colon T^2(\Omega) \to \mathcal{X}(\Omega)$ is a linear operator.

The following Green formulas are the key to the boundary control problems which present the relations between the interior and the boundary of a displacement vector field.

Theorem 5.1 *Let the bilinear form $\mathcal{B}_0(\cdot,\cdot)$ be given by (5.24). For $\zeta = (W_1, W_2, w_1, w_2)$, $\eta = (U_1, U_2, u_1, u_2) \in \mathcal{H}^1(\Omega)$, we have*

$$\mathcal{B}_0(\zeta,\eta) = \frac{\alpha h}{2}(\mathcal{A}_0\zeta,\eta)_{\mathcal{L}^2(\Omega)} + \frac{\alpha h}{2}\int_\Gamma \partial(\mathcal{A}_0\zeta,\eta)\,d\Gamma \tag{5.33}$$

where

$$\partial(\mathcal{A}_0\zeta,\eta) = \langle B_{01}(\zeta), U_1 \rangle + \gamma\langle B_{02}(\zeta), U_2 \rangle + 2\langle\varphi_0(\zeta),\nu\rangle u_1 + \gamma\frac{\partial w_2}{\partial\nu}u_2, \tag{5.34}$$

ν, τ *are the normal and the tangential along the boundary Γ, $\mathcal{L}^2(\Omega)$, $\mathcal{H}^1(\Omega)$ are the Sobolev spaces given by (5.26) and (5.27), respectively,*

$$\begin{cases} B_{01}(\zeta) = 2\,\mathrm{i}\,(\nu)\Upsilon(\zeta) + 2\beta(\mathrm{tr}\,\Upsilon(\zeta) + w_2)\nu, \\ B_{02}(\zeta) = 2\,\mathrm{i}\,(\nu)\mathcal{X}_0(\zeta) + 2\beta(\mathrm{tr}\,\mathcal{X}_0(\zeta))\nu, \end{cases} \tag{5.35}$$

$$\begin{aligned} \mathcal{A}_0\zeta &= -(\boldsymbol{\Delta}_\beta W_1 + F_1(\zeta), \\ &\quad \gamma\boldsymbol{\Delta}_\beta W_2 + F_2(\zeta), \ \Delta w_1 + f_1(\zeta), \ \gamma\Delta w_2 + f_2(\zeta)), \end{aligned} \tag{5.36}$$

$F_i(\zeta)$, $f_i(\zeta)$ *are first order terms (≤ 1) for $i = 1, 2$, $\boldsymbol{\Delta}_\beta$ is given by (5.31), and Δ is the Laplacian of the Riemann manifold M with the induced metric from $\mathbb{R}^3$.*

Proof. Denote by $F(\zeta)$ lower order terms (≤ 1) which may be different from line to line and from term to term.

It follows from the formula (5.25) that

$$\begin{aligned} \int_\Omega \mathcal{B}_0(\zeta,\eta)\,dx &= 2(\Upsilon(\zeta),\Upsilon(\eta))_{L^2(\Omega,T^2)} + 4(\varphi_0(\zeta),\varphi_0(\eta))_{L^2(\Omega,\Lambda)} \\ &\quad + 2(w_2, u_2) + 2\beta(\mathrm{tr}\,\Upsilon(\zeta) + w_2, \ \mathrm{tr}\,\Upsilon(\eta) + u_2) \\ &\quad + 2\gamma(\mathcal{X}_0(\zeta),\mathcal{X}_0(\eta))_{L^2(\Omega,T^2)} + \gamma(Dw_2, Du_2)_{L^2(\Omega,\Lambda)} \\ &\quad + 2\gamma\beta(\mathrm{tr}\,\mathcal{X}_0(\zeta), \ \mathrm{tr}\,\mathcal{X}_0(\eta)). \end{aligned} \tag{5.37}$$

Let us compute each term in the right hand side of the formula (5.37), separately, as follows.

By the formula (5.6) and the formula (4.44) in Chapter 4, we obtain

$$2(\Upsilon(\zeta), \ \Upsilon(\eta))_{L^2(\Omega, T^2)} = ((\delta d + 2d\delta)W_1 + F(\zeta), \ U_1)_{L^2(\Omega, T^2)}$$

$$+(F(\zeta), u_1) + 2 \int_\Gamma \Upsilon(\zeta)(\nu, U_1) \, d\Gamma. \tag{5.38}$$

It follows from the divergence formula and the formulas (5.12) and (5.13) that

$$4(\varphi_0(\zeta), \varphi_0(\eta))_{L^2(\Omega, \Lambda)} = (2\varphi_0(\zeta), \ Du_1 + 2\varphi_{0L}(\eta))_{L^2(\Omega, \Lambda)}$$

$$= -(\Delta w_1 + F(\zeta), \ u_1) + 2 \int_\Gamma \langle \varphi_0(\zeta), \nu \rangle u_1 \, d\Gamma$$

$$+(F(\zeta), U_1)_{L^2(\Omega, \Lambda)} + (F(\zeta), U_2)_{L^2(\Omega, \Lambda)} \tag{5.39}$$

where the following computation is used (by (5.2))

$$(2\varphi_0(\zeta), \ -2\,\mathrm{i}\,(U_1)\Pi)_{L^2(\Omega, \Lambda)}$$

$$= -4 \int_\Omega \langle \varphi_0(\zeta), \ \mathrm{i}\,(U_1)\Pi \rangle \, dx = -4 \int_\Omega \Pi(\varphi_0(\zeta), U_1) \, dx$$

$$= -4(\mathcal{S}\varphi_0(\zeta), U_1)_{L^2(\Omega, \Lambda)} = (F(\zeta), U_1)_{L^2(\Omega, \Lambda)}.$$

Using the formula (4.46), we have

$$2\beta(\mathrm{tr}\,\Upsilon(\zeta) + w_2, \ \mathrm{tr}\,\Upsilon(\eta) + u_2)_{L^2(\Omega)}$$

$$= 2\beta(d\delta W_1 + F(\zeta), U_1)_{L^2(\Omega, \Lambda)} + 2\beta \int_\Gamma (\mathrm{tr}\,\Upsilon(\zeta) + w_2)\langle U_1, \nu \rangle \, d\Gamma$$

$$+(F(\zeta), u_1) + (F(\zeta), u_2). \tag{5.40}$$

Using the formulas (5.11), (1.155) and (1.156), we obtain

$$2\gamma(\mathcal{X}_0(\zeta), \mathcal{X}_0(\eta))_{L^2(\Omega, T^2)}$$

$$= \gamma(DW_2 + D^*W_2 + \mathcal{X}_{0L}(\zeta), \ \frac{1}{2}(DU_2 + D^*U_2) + \mathcal{X}_{0L}(\eta))_{L^2(\Omega, \Lambda)}$$

$$= \gamma(DW_2, DU_2)_{L^2(\Omega, \Lambda)} + \gamma(D^*W_2, DU_2)_{L^2(\Omega, \Lambda)} + (F(\zeta), U_2)_{L^2(\Omega, \Lambda)}$$

$$+(F(\zeta), U_1)_{L^2(\Omega, \Lambda)} + (F(\zeta), u_1) + (F(\zeta), u_2)$$

$$= \gamma((\delta d + 2d\delta)W_2, U_2)_{L^2(\Omega, \Lambda)} + \gamma \int_\Gamma (\langle D_\nu W_2, U_2 \rangle + \langle D_{U_2} W_2, \nu \rangle) \, d\Gamma$$

$$+(F(\zeta), U_2)_{L^2(\Omega, \Lambda)} + (F(\zeta), U_1)_{L^2(\Omega, \Lambda)}$$

$$+(F(\zeta), u_1) + (F(\zeta), u_2). \tag{5.41}$$

In addition,

$$\gamma \int_\Gamma \langle D_\nu W_2, U_2 \rangle + \langle D_{U_2} W_2, \nu \rangle) \, d\Gamma$$

$$= 2\gamma \int_\Gamma \mathcal{X}_0(\zeta)(\nu, U_2) d\Gamma + (F(\zeta), U_2)_{L^2(\Omega, \Lambda)} \tag{5.42}$$

and the Green formula for the Laplacian states that

$$\gamma(Dw_2, Du_2)_{L^2(\Omega,\Lambda)} = \gamma(-\Delta w_2, u_2) + \gamma \int_\Gamma u_2 \frac{\partial w_2}{\partial \nu}\, d\Gamma. \qquad (5.43)$$

Since

$$\text{tr}\, \mathcal{X}_0(\eta) = -\delta U_2 - \text{tr}\, i\,(U_1)D\Pi + u_1\, \text{tr}\, c + u_2\, \text{tr}\, \Pi,$$

by (1.137) in Chapter 1, we have

$$2\gamma\beta(\text{tr}\, \mathcal{X}_0(\zeta),\, \text{tr}\, \mathcal{X}_0(\eta))$$
$$= 2\gamma\beta(d\delta W_2 + F(\zeta), U_2)_{L^2(\Omega,\Lambda)} + 2\gamma\beta \int_\Gamma \langle U_2, \nu \rangle\, \text{tr}\, \mathcal{X}_0(\zeta)d\Gamma$$
$$+ (F(\zeta), U_1)_{L^2(\Omega,\Lambda)} + (F(\zeta), u_1) + (F(\zeta), u_2). \qquad (5.44)$$

Finally, we add up all the formulas from (5.38) to (5.44) to obtain the formula (5.33). $\square$

If the middle surface is flat, then $\Pi = c = 0$. The formulas (5.6)-(5.8) become

$$\begin{cases} 2\Upsilon(\zeta) = DW_1 + D^*W_1, \\ 2\mathcal{X}_0(\zeta) = DW_2 + D^*W_2, \\ 2\varphi_0(\zeta) = Dw_1 + W_2, \end{cases} \qquad (5.45)$$

where $W_2 = V$.

From the formulas (5.45) and the proof of Theorem 5.1, the following corollary and theorem are immediate.

Corollary 5.1 *For a plate, we have*

$$\mathcal{A}_0\zeta = - \begin{pmatrix} \mathbf{\Delta}_\beta W_1 + 2\beta dw_2 \\ \mathbf{\Delta}_\beta W_2 - W_2 - dw_1 \\ \Delta w_1 - \delta W_2 \\ \gamma\Delta w_2 - 2(1+\beta)w_2 + 2\beta\delta W_1 \end{pmatrix} \qquad (5.46)$$

where $\zeta = (W_1, W_2, w_1, w_2)$. *If* $\Phi = (\phi_1, \phi_2)$ *in the usual coordinate system* (x, y) *of* $\mathbb{R}^2$, *then*

$$\Delta\phi_1 = \frac{\partial^2 \phi_1}{\partial x^2} + \frac{\partial^2 \phi_1}{\partial y^2},$$

$$\mathbf{\Delta}_\beta\Phi = \begin{pmatrix} 2(1+\beta)\dfrac{\partial^2 \phi_1}{\partial x^2} + \dfrac{\partial^2 \phi_1}{\partial y^2} + 2(1+\beta)\dfrac{\partial^2 \phi_2}{\partial x \partial y} \\ 2(1+\beta)\dfrac{\partial^2 \phi_2}{\partial y^2} + \dfrac{\partial^2 \phi_2}{\partial x^2} + 2(1+\beta)\dfrac{\partial^2 \phi_1}{\partial x \partial y} \end{pmatrix}.$$

Theorem 5.2 *The variational problem (5.30) is equivalent to solving the following boundary value problem in unknown* $\zeta = (W_1, W_2, w_1, w_2)$

$$\frac{\alpha h}{2}\mathcal{A}_0\zeta = (F_1, F_2, f_1, f_2) \qquad (5.47)$$

subject to the boundary conditions

$$W_1|_{\Gamma_0} = W_2|_{\Gamma_0} = w_1|_{\Gamma_0} = w_2|_{\Gamma_0} = 0, \tag{5.48}$$

$$\frac{\alpha h}{2}B_{01}(\zeta)|_{\Gamma_1} = \frac{\alpha h \gamma}{2}B_{02}(\zeta)|_{\Gamma_1} = \alpha h \varphi_0(\zeta)|_{\Gamma_1} = \frac{\alpha h \gamma}{2}\frac{\partial w_2}{\partial \nu} = 0 \tag{5.49}$$

where $\mathcal{A}_0$ and B_{0i} are defined in (5.36) and (5.35), respectively, and

$$(F_1, F_2, f_1, f_2) = \mathcal{F}.$$

Ellipticity of the Strain Energy Now we consider ellipticity of the strain energy (5.24). A key issue is the following uniqueness results.

Lemma 5.1 *Let $\zeta = (W_1, W_2, w_1, w_2) \in \mathcal{H}^1_{\Gamma_0}(\Omega)$ be such that*

$$\Upsilon(\zeta) = 0, \quad \mathcal{X}_0(\zeta) = 0, \quad \varphi_0(\zeta) = 0, \quad and \quad w_2 = 0, \tag{5.50}$$

then $\zeta = 0$ for all $x \in \Omega$ where $\mathcal{H}^1_{\Gamma_0}(\Omega)$ is given by (5.28) and Γ_0 has a positive length.

Proof. By the formulas (5.6), (5.11), and (5.12), the assumptions (5.50) state that

$$DW_1 + D^*W_1 = -2w_1\Pi, \tag{5.51}$$

$$DW_2 + D^*W_2 = -2w_1c + 2\,\mathrm{i}\,(W_1)D\Pi, \tag{5.52}$$

$$Dw_1 = 2\,\mathrm{i}\,(W_1)\Pi - W_2. \tag{5.53}$$

Let $U = -2w_1\Pi$. It is easy to check that

$$DU = -2\Pi \otimes Dw_1 - 2w_1 D\Pi. \tag{5.54}$$

Using the formula (4.60) in Chapter 4 and the formula (5.54), we have

$$\begin{aligned}D^2W_1(X,Y,Z) &= -\Pi(X,Y)Z(w_1) - \Pi(Z,X)Y(w_1) + \Pi(Y,Z)X(w_1)\\&\quad +R(X,Y,Z,W_1) - w_1 D\Pi(X,Y,Z)\end{aligned} \tag{5.55}$$

for $X, Y, Z \in \mathcal{X}_0(\Omega)$, since $D\Pi$ is symmetric by Corollary 1.1, where $\mathbf{R}(\cdot,\cdot,\cdot,\cdot)$ is the curvature tensor.

Let $x \in \Omega$ be given. Let E_1, E_2 be a frame field normal at x such that $D_{E_i}E_j(x) = 0$ for $1 \le i, j \le 2$. By the formulas (5.55) and (1.114), we obtain at x

$$\begin{aligned}\Delta W_1 &= -\sum_{i=1}^{2}D^2_{E_iE_i}W_1 + \sum_{ij=1}^{2}R(E_i, E_j, W_1, E_j)E_i\\&= -\sum_{ij=1}^{2}D^2W_1(E_j, E_i, E_i)E_j + \kappa\,W_1\\&= 2\,\mathrm{i}\,(Dw_1)\Pi - (\mathrm{tr}\,\Pi)Dw_1 + w_1 D(\mathrm{tr}\,\Pi) + 2\kappa\,W_1 \end{aligned} \tag{5.56}$$

where Δ is the Hodge-Laplace operator and κ is the Gauss curvature. Moreover, it follows from the formulas (5.53) and (5.52) that

$$
\begin{aligned}
\Delta w_1 &= \sum_{i=1}^{2} D(2\,\mathrm{i}\,(W_1)\Pi - W_2)(E_i, E_i) \\
&= 2\sum_{i=1}^{2}[D\Pi(W_1, E_i, E_i) + \Pi(D_{E_i}W_1, E_i)] - \operatorname{div} W_2 \\
&= 2\langle W_1, D(\operatorname{tr}\Pi)\rangle + 2\operatorname{tr}\Pi(D.W_1, \cdot) - \operatorname{div} W_2 \\
&= w_1\operatorname{tr} c + \langle W_1, D(\operatorname{tr}\Pi)\rangle + 2\operatorname{tr}\Pi(D.W_1, \cdot). \quad (5.57)
\end{aligned}
$$

Next, we check the values of DW_1 and Dw_1 on Γ_0. Since $w_1|_{\Gamma_0} = W_1|_{\Gamma_0} = W_2|_{\Gamma_0} = 0$, it follows from the formulas (5.51)-(5.53) that

$$DW_1 + D^*W_1 = 0, \quad DW_2 + D^*W_2 = 0 \quad \text{and} \quad Dw_1 = 0 \quad (5.58)$$

for all $x \in \Gamma_0$. Moreover, the conditions $(DW_i + D^*W_i)|_{\Gamma_0} = 0$ and $W_i|_{\Gamma_0} = 0$ for $i = 1, 2$ imply that

$$DW_i = 0 \quad \text{on} \quad \Gamma_0 \quad (5.59)$$

Applying [5] to the elliptic system of the equations (5.56) and (5.57) yields that $W_1 = 0$ and $w_1 = 0$ on Ω. It follows from the formula (5.52) that

$$DW_2 + D^*W_2 = 0 \quad \text{on} \quad \Omega$$

which yields $W_2 = 0$ on Ω by Lemma 4.5. $\qquad\square$

Theorem 5.3 *Let* $\mathcal{B}_0(\cdot, \cdot)$ *be the bilinear form given by* (5.24). *Then there is* $c > 0$ *such that*

$$\mathcal{B}_0(\zeta, \zeta) \geq c\|\zeta\|^2_{\mathcal{H}^1_{\Gamma_0}(\Omega)} \quad \text{for} \quad \zeta \in \mathcal{H}^1_{\Gamma_0}(\Omega) \quad (5.60)$$

where $\mathcal{H}^1_{\Gamma_0}(\Omega)$ *is given by* (5.28) *with* $\Gamma_0 \subset \Gamma$ *having a positive length.*

Proof. Using the formulas (5.6), (5.11), and (5.12) in the expression (5.25), we obtain

$$
\begin{aligned}
\mathcal{B}_0(\zeta, \zeta) \geq{} &c_1(|DW_1 + D^*W_1|^2 + |DW_2 + D^*W_2|^2 + |Dw_1|^2 + |Dw_2|^2) \\
&- c_2(|W_1|^2 + |W_2|^2 + w_1^2 + w_2^2)
\end{aligned}
$$

which yields

$$\mathcal{B}_0(\zeta, \zeta) + c_2\|\zeta\|^2_{\mathcal{L}^2(\Omega)} \geq c_1\|\zeta\|^2_{\mathcal{H}^1_{\Gamma_0}(\Omega)}.$$

Then the inequality (5.60) follows from the above inequality and the uniqueness result, Lemma 5.1, by an argument as in Lemma 2.5. $\qquad\square$

By Theorem 5.3, the following results are immediate.

Theorem 5.4 *For $\mathcal{F} \in \mathcal{L}^2(\Omega)$ the problem (5.47)-(5.49) has a unique solution in $\mathcal{H}^1_{\Gamma_0}(\Omega)$.*

Equations of Motion In the last part of this section, we consider the equations of motion of the Naghdi model.

For this purpose we need to compute the kinetic energy. The relations (5.4), (5.5), and (5.9) yield the following displacement expression

$$\zeta(p) = W_1(x) + w_1(x)N + z\left[W_2(x) + w_2(x)N - i(W_1)\Pi\right] \tag{5.61}$$

for $p = x + zN \in S$.

Now we assume that the displacement vector field ζ depends on time t, and for simplicity, the mass density per unit of volume is one. Then the kinetic energy per unit area of the undeformed middle surface is obtained by integration with respect to thickness z

$$\int_{-h/2}^{h/2} |\zeta_t|^2 (1 + \operatorname{tr}\Pi z + \kappa z^2)\, dz \tag{5.62}$$

where κ is the Gauss curvature of the middle surface Ω. As for the formula (4.94), a computation in the thin shell theory with a relative error $h/\mathbf{R}$ yields that the total kinetic energy of the middle surface Ω is

$$\mathcal{K}(\zeta_t, \zeta_t) = h \int_\Omega [|W_{1t}|^2 + w_{1t}^2 + \gamma(|W_{2t}|^2 + w_{2t}^2)]\, dx \tag{5.63}$$

where the term $i(W_1)\Pi$ disappears because its error is $h/\mathbf{R}$.

The equations of motion for ζ are obtained by setting to zero the first variation of the Lagranian

$$\int_0^T \left[\mathcal{K}(\zeta_t, \zeta_t) + \operatorname{Load}(\zeta) - \mathcal{B}_0(\zeta, \zeta)\right] dt \tag{5.64}$$

(the "Principle of Virtual Work") where $\operatorname{Load}(\cdot)$ and $\mathcal{B}_0(\cdot, \cdot)$ are given by (5.29) and (5.24), respectively.

We assume that there is no external loading on the shell. Then the variation (5.64) is taken with respect to kinematically admissible displacements. We obtain, as a result of the calculation by (5.33) and (5.64),

Theorem 5.5 *Suppose that there are no external loads on the shell and the shell is clamped along a portion Γ_0 of the boundary Γ of the middle surface and free on Γ_1 where $\Gamma_0 \cup \Gamma_1 = \Gamma$. Then the displacement vector field $\zeta = (W_1, W_2, w_1, w_2)$ satisfies the following*

$$\begin{cases} h\mathbb{C}\zeta_{tt} + \dfrac{\alpha h}{2} \mathcal{A}_0\zeta = 0 & on \quad Q_\infty \\ \zeta(0, x) = \zeta_0(x), \quad \zeta_t(0, x) = \zeta_1(x) & on \quad \Omega, \end{cases} \tag{5.65}$$

subject to the boundary conditions

$$\zeta = 0 \quad on \quad \Sigma_{0\infty}, \tag{5.66}$$

$$\begin{cases} \dfrac{\alpha h}{2} B_{01}(\zeta) = \dfrac{\alpha h}{2} B_{02}(\zeta) = 0, \\ \dfrac{\alpha h}{2}\langle \varphi_0(\zeta), \nu \rangle = \dfrac{\alpha h}{2}\dfrac{\partial w_2}{\partial \nu} = 0, \end{cases} \quad on \quad \Sigma_{1\infty}, \tag{5.67}$$

where

$$\mathbb{C} = \begin{pmatrix} 1 & 0 & 0 & 0 \\ 0 & \gamma & 0 & 0 \\ 0 & 0 & 1 & 0 \\ 0 & 0 & 0 & \gamma \end{pmatrix}, \tag{5.68}$$

$A_0\zeta$, $B_{0i}(\zeta)$ *and* $\varphi_0(\zeta)$ *are given by Theorem 5.1, and*

$$Q_\infty = (0,\infty) \times \Omega, \quad \Sigma_{0\infty} = (0,\infty) \times \Gamma_0, \quad \Sigma_{1\infty} = (0,\infty) \times \Gamma_1.$$

By a change of variables by $\eta = \mathbb{C}^{1/2}\zeta$ and after changing t to t/c with $c^2\alpha = 2$ in the problem (5.65), we obtain the following system

$$\begin{cases} \eta_{tt} + \mathcal{A}\eta = 0 \quad on \quad Q_\infty \\ \eta(0) = \eta_0, \quad \eta_t(0) = \eta_1 \quad on \quad \Omega, \end{cases} \tag{5.69}$$

subject to the boundary conditions

$$\eta = 0 \quad on \quad \Sigma_{0\infty}, \tag{5.70}$$

$$\begin{cases} B_1(\eta) = B_2(\eta) = 0, \\ \langle \varphi(\zeta), \nu \rangle = \dfrac{\partial w_2}{\partial \nu} = 0, \end{cases} \quad on \quad \Sigma_{1\infty}, \tag{5.71}$$

where

$$\mathcal{A} = \mathbb{C}^{-1/2} A_0 \mathbb{C}^{-1/2}. \tag{5.72}$$

The bilinear form associated with $\mathcal{A}$ becomes

$$\begin{aligned} B(\zeta, \eta) &= B_0(\mathbb{C}^{-1/2}\zeta, \mathbb{C}^{-1/2}\eta) = 2\langle \Upsilon(\zeta), \Upsilon(\eta) \rangle + 4\langle \varphi(\zeta), \varphi(\eta) \rangle \gamma \\ &+ 2\beta(\text{tr } \Upsilon(\zeta) + w_2/\sqrt{\gamma})(\text{tr } \Upsilon(\eta) + u_2/\sqrt{\gamma}) + 2\langle \mathcal{X}(\zeta), \mathcal{X}(\eta) \rangle \\ &+ 2\beta \text{ tr } \mathcal{X}(\zeta) \text{ tr } \mathcal{X}(\eta) + \langle Dw_2, Du_2 \rangle + 2w_2 u_2/\gamma, \end{aligned} \tag{5.73}$$

$$\begin{cases} \Upsilon(\zeta) = \dfrac{1}{2}(DW_1 + D^*W_1) + w_1\Pi, \\ \mathcal{X}(\zeta) = \dfrac{1}{2}(DW_2 + D^*W_2) + w_2\Pi - \sqrt{\gamma}(\text{i}\,(W_1)D\Pi - w_1 c), \\ \varphi(\zeta) = \dfrac{1}{2}Dw_1 - \text{i}\,(W_1)\Pi + \dfrac{1}{\sqrt{\gamma}}W_2, \end{cases} \tag{5.74}$$

and this time the Green formula (5.33) reads

$$B(\zeta, \eta) = (\mathcal{A}\zeta, \eta)_{\mathcal{L}^2(\Omega)} + \int_\Gamma \partial(\mathcal{A}\zeta, \eta)\, d\Gamma \tag{5.75}$$

where

$$\mathcal{B}(\zeta, \eta) = \int_\Omega B(\zeta, \eta) \, dx, \tag{5.76}$$

$$\partial(\mathcal{A}\zeta, \eta) = \langle B_1(\zeta), U_1 \rangle + \langle B_2(\zeta), U_2 \rangle + 2\langle \varphi(\zeta), \nu \rangle u_1 + \frac{\partial w_2}{\partial \nu} u_2, \tag{5.77}$$

$$\begin{cases} B_1(\zeta) = 2\,\mathrm{i}\,(\nu)\Upsilon(\zeta) + 2\beta(\mathrm{tr}\,\Upsilon(\zeta) + w_2/\sqrt{\gamma})\nu, \\ B_2(\zeta) = 2\,\mathrm{i}\,(\nu)\mathcal{X}(\zeta) + 2\beta(\mathrm{tr}\,\mathcal{X}(\zeta))\nu, \end{cases} \tag{5.78}$$

$$\zeta = (W_1, W_2, w_1, w_2), \quad \text{and} \quad \eta = (U_1, U_2, u_1, u_2).$$

5.2 Observability Estimates from Boundary

In this section it is assumed that all the regularities we need hold since we are mainly concerned with inequalities for controllability/stabilization.

Let $b(\cdot, \cdot)$ be a bilinear form on $T^2(\Omega) \times T^2(\Omega)$, defined by

$$b(T_1, T_2) = \langle T_1, T_2 \rangle + \beta \,\mathrm{tr}\,T_1 \,\mathrm{tr}\,T_2 \quad \text{for} \quad T_1, \, T_2 \in T^2(\Omega) \tag{5.79}$$

where $\beta > 0$ is a constant, given in (5.19).

For $W \in H^1(\Omega, \Lambda)$, set

$$S(W) = \frac{1}{2}(DW + D^*W). \tag{5.80}$$

It is easy to check from Theorem 5.3 that there is a $\lambda_0 \geq 1$ such that

$$\lambda_0 \int_\Omega [b(S(W), S(W)) + |W|^2] \, dx \geq \|DW\|_{L^2(\Omega, T^2)}^2 \tag{5.81}$$

for $W \in H^1(\Omega, \Lambda)$.

Multiplier Identities For $W \in \mathcal{X}(\Omega)$ and $T \in T^2(\Omega)$, let $G(W, T) \in T^2(\Omega)$ be given by the formula (4.131) in Chapter 4.

Let V be a vector field on Ω such that there is a function ϑ on Ω that satisfies the relation

$$DV(X, X) = \vartheta(x)|X|^2 \quad \text{for all} \quad X \in M_x, \quad x \in \Omega. \tag{5.82}$$

For $\zeta = (W_1, W_2, w_1, w_2) \in \mathcal{H}^1(\Omega)$, set

$$m(\zeta) = (D_V W_1, D_V W_2, V(w_1), V(w_2)). \tag{5.83}$$

Lemma 5.2 *We have*

$$2\mathcal{B}(\zeta, m(\zeta)) = \int_\Gamma B(\zeta,\zeta)\langle V, \nu\rangle\, d\Gamma - 2\int_\Omega \vartheta B(\zeta,\zeta)dx$$
$$+ 2\int_\Omega e(\zeta,\zeta)\, dx + \mathrm{lo}\,(\zeta) \tag{5.84}$$

where $\mathcal{B}(\cdot,\cdot)$ is given in (5.76) and

$$e(\zeta,\zeta) = 2b(S(W_1), G(V, DW_1)) + 2b(S(W_2), G(V, DW_2))$$
$$+ 4\vartheta|\varphi(\zeta)|^2 + \vartheta|Dw_2|^2. \tag{5.85}$$

   Proof. Let us compute $\Upsilon(m(\zeta))$, $\mathcal{X}(m(\zeta))$, $\varphi(m(\zeta))$, and $\langle Dw_2, D(V(w_2))\rangle$, separately.
   From the identity (4.132) in Chapter 4 it follows that

$$\Upsilon\big(m(\zeta)\big) = D_V\Upsilon(\zeta) + G(V, DW_1) + \mathrm{lo}\,(\zeta). \tag{5.86}$$

   Let $x \in \Omega$ be given. Let E_1, E_2 be a frame field normal at x. We then obtain

$$D_V(\mathrm{i}\,(W_1)D\Pi)(E_i, E_j)$$
$$= V(D\Pi(W_1, E_i, E_j))$$
$$= D_V(D\Pi)(W_1, E_i, E_j) + D\Pi(D_V W_1, E_i, E_j) \quad \text{at} \quad x$$

which yields

$$D_V(\mathrm{i}\,(W_1)D\Pi) = \mathrm{i}\,(D_V W_1)D\Pi + \mathrm{i}\,(W_1)(D_V D\Pi). \tag{5.87}$$

Similarly, we have

$$D_V(\mathrm{i}\,(W_1)\Pi) = \mathrm{i}\,(D_V W_1)\Pi + \mathrm{i}\,(W_1)D_V\Pi \tag{5.88}$$

and

$$D(V(w_1)) = D_V Dw_1 + Dw_1(D.V). \tag{5.89}$$

   Using the identity (4.132) in Lemma 4.7 with $(W, w) = (W_2, w_2)$ and the formula (5.87), we obtain

$$\mathcal{X}(m(\zeta)) = D_V\mathcal{X}(\zeta) + G(V, DW_2) + \mathrm{lo}\,(\zeta). \tag{5.90}$$

   It follows from the formulas (5.88) and (5.89) that

$$\varphi(m(\zeta)) = D_V\varphi(\zeta) + \varphi(\zeta)(D.V) + \mathrm{lo}\,(\zeta)$$

which yields, via the relation (5.82),

$$2\langle\varphi(\zeta), \varphi(m(\zeta))\rangle = V(|\varphi(\zeta)|^2) + 2\vartheta|\varphi(\zeta)|^2 + \mathrm{lo}\,(\zeta). \tag{5.91}$$

Using the formula (2.6) in Chapter 2 and the relation (5.82), we obtain

$$2\langle Dw_2, D(V(w_2))\rangle = \operatorname{div}(|Dw_2|^2 V) = V(|Dw_2|^2) + 2\vartheta|Dw_2|^2. \quad (5.92)$$

Now we compute $\mathcal{B}(\zeta, m(\zeta))$. Noting that

$$\langle \Upsilon(\zeta), G(V, DW_1)\rangle + \beta \operatorname{tr} \Upsilon(\zeta) \operatorname{tr} G(V, DW_1)$$
$$= b(S(W_1), G(V, DW_1)) + \operatorname{lo}(\zeta)$$

and

$$\langle \mathcal{X}(\zeta), G(V, DW_1)\rangle + \beta \operatorname{tr} \mathcal{X}(\zeta) \operatorname{tr} G(V, DW_1)$$
$$= b(S(W_2), G(V, DW_2)) + \operatorname{lo}(\zeta),$$

it follows from the relations (5.86), (5.90), and (5.91) that

$$
\begin{aligned}
2\mathcal{B}(\zeta, m(\zeta)) &= \int_\Omega [V(B(\zeta, \zeta)) + 4b(S(W_1), G(V, DW_1)) \\
&\quad + 4b(S(W_2), G(V, DW_2)) + 2\vartheta|Dw_2|^2 \\
&\quad + 8\vartheta|\varphi(\zeta)|^2 + \operatorname{lo}(\zeta)]\, dx \\
&= \int_\Gamma B(\zeta, \zeta)\langle V, \nu\rangle\, d\Gamma - 2\int_\Omega \vartheta B(\zeta, \zeta)\, dx \\
&\quad + 2\int_\Omega e(\zeta, \zeta)\, dx + \operatorname{lo}(\zeta).
\end{aligned}
$$

$\square$

Let ζ be a displacement vector field of the shell. The total energy of the shell is defined by

$$2E(t) = \|\zeta_t\|^2_{\mathcal{L}^2(\Omega)} + \mathcal{B}(\zeta, \zeta) \quad (5.93)$$

where the bilinear form $\mathcal{B}(\cdot, \cdot)$ is given by (5.76).

Theorem 5.6 *Let $\zeta = (W_1, W_2, w_1, w_2) \in \mathcal{H}^1(\Omega)$ solve the problem*

$$\zeta_{tt} + \mathcal{A}\zeta = 0 \quad \text{for} \quad x \in Q = (0, T) \times \Omega. \quad (5.94)$$

Then

$$
\begin{aligned}
\int_\Sigma [2\partial(\mathcal{A}\zeta, m(\zeta)) &+ (|\zeta_t|^2 - B(\zeta, \zeta))\langle V, \nu\rangle]d\Sigma \\
&= 2(\zeta_t, m(\zeta))_{\mathcal{L}^2(\Omega)}\big|_0^T + 2\int_Q \vartheta[|\zeta_t|^2 - B(\zeta, \zeta)]\, dQ \\
&\quad + 2\int_Q e(\zeta, \zeta)\, dQ + L(\zeta)
\end{aligned}
\quad (5.95)
$$

where $\Sigma = (0, T) \times \Gamma$ and $e(\zeta, \zeta)$ is given by (5.85) and $L(\zeta)$ denotes lower *order terms relative to the energy* (5.93). *Moreover, if p is a function on Ω, then*

$$\int_\Sigma \partial(\mathcal{A}\zeta, p\zeta)\, dQ = \int_Q p[B(\zeta,\zeta) - |\zeta_t|^2]\, dQ + L(\zeta). \qquad (5.96)$$

Proof. We multiply the equation (5.94) by $2m(\zeta)$ and integrate over Q by parts.

By the relation (5.82), we have

$$
\begin{aligned}
(\zeta_{tt}, 2m(\zeta))_{\mathcal{L}^2(\Omega)} &= 2[(\zeta_t, m(\zeta))_{\mathcal{L}^2(\Omega)}]_t - 2(\zeta_t, m(\zeta_t))_{\mathcal{L}^2(\Omega)} \\
&= 2[(\zeta_t, m(\zeta))_{\mathcal{L}^2(\Omega)}]_t + 2\int_\Omega \vartheta |\zeta_t|^2 dx \\
&\quad - \int_\Gamma |\zeta_t|^2 \langle V, \nu \rangle\, d\Gamma.
\end{aligned}
\qquad (5.97)
$$

Using the formulas (5.75) and (5.84), we obtain

$$
\begin{aligned}
(\mathcal{A}\zeta, 2m(\zeta))_{\mathcal{L}^2(\Omega)} &= \int_\Gamma [B(\zeta,\zeta)\langle V, \nu \rangle - 2\partial(\mathcal{A}\zeta, m(\zeta))]\, d\Gamma \\
&\quad - 2\int_\Omega \vartheta B(\zeta,\zeta)\, dx + 2\int_\Omega e(\zeta,\zeta)\, dx + \mathrm{lo}\,(\zeta).
\end{aligned}
\qquad (5.98)
$$

The identity follows from the formulas (5.97), (5.98), and the equation (5.94).

We multiply the equation (5.94) by $p\zeta$ and integrate over Q by parts to obtain the identity (5.96). $\qquad \square$

Let us introduce the geometric assumptions for controllability/stabilization of the Naghdi shell.

Definition 5.1 *A vector field V on Ω is said to be an escape vector field for the Naghdi shell if the following assumptions hold:*
  (i)   There is a function ϑ on Ω such that the formula (5.82) holds true.
  (ii) Let

$$\iota(x) = \langle DV, \mathcal{E} \rangle / 2 \quad \text{for} \quad x \in \Omega \qquad (5.99)$$

where $\mathcal{E}$ denotes the volume element of the middle surface Ω. Suppose that the functions $\vartheta(x)$ and $\iota(x)$ satisfy the inequality

$$2\min_{x\in\Omega} \vartheta(x) > \lambda_0(1 + 2\beta)\max_{x\in\Omega} |\iota(x)| \qquad (5.100)$$

where $\lambda_0 \geq 1$ is given in (5.81) and β is given in (5.19).

Existence of such a vector field has been studied in Section 4.5 in Chapter 4 and many examples are given there.

Let V be an escape vector field for the Naghdi shell. Set

$$\sigma_0 = \max_{x\in\Omega} |V|, \quad \sigma_1 = \min_{x\in\Omega} \vartheta(x) - (1 + 2\beta)\max_{x\in\Omega} |\iota(x)|/2, \qquad (5.101)$$

$$\Gamma_0 = \{\, x \mid \langle V(x), \nu(x)\rangle > 0,\ x \in \Gamma \,\}, \quad \Gamma_1 = \Gamma/\Gamma_0. \tag{5.102}$$

Now we have the following observability estimates from boundary.

Theorem 5.7 *Let V be an escape vector field for the Naghdi shell. Let ζ solve the problem (5.94). Then for $T > 0$*

$$\int_\Sigma SB\, d\Sigma + \lambda_0\sigma_0[E(0) + E(T)] \geq 2\sigma_1 \int_0^T E(t)\, dt + L(\zeta) \tag{5.103}$$

where

$$SB = \partial(\mathcal{A}\zeta,\, 2m(\zeta) + \varrho\zeta) + [|\zeta_t|^2 - B(\zeta,\zeta)]\langle V, \nu\rangle,$$
$$m(\zeta) = (D_V W_1, D_V W_2, V(w_1), V(w_2)), \quad \varrho = 2\vartheta - \sigma_1. \tag{5.104}$$

Proof. We let $p = \varrho$ in the identity (5.96) and add it to the identity (5.95) to have

$$\int_\Sigma SB\, d\Sigma = 2(\zeta_t, m(\zeta))_{\mathcal{L}^2(\Omega)}|_0^T + \sigma_1 \int_Q [|\zeta_t|^2 - B(\zeta,\zeta)]dQ$$
$$+ 2\int_Q e(\zeta,\zeta)\, dQ + L(\zeta). \tag{5.105}$$

On the other hand, by the formulas (5.73), (5.79), and (5.80), it is easy to check that

$$B(\zeta,\zeta) = 2b(S(W_1), S(W_1)) + 2\beta b(S(W_2), S(W_2))$$
$$+ 4|\varphi(\zeta)|^2 + |Dw_2|^2 + \mathrm{lo}\,(\zeta). \tag{5.106}$$

By Lemma 5.3 below, we have

$$\int_Q e(\zeta,\zeta)\, dQ \geq \sigma_1 \int_Q B(\zeta,\zeta)\, dQ + L(\zeta). \tag{5.107}$$

Using the formula (5.106) and the inequality (5.81), we obtain

$$2|(\zeta_t, m(\zeta))_{\mathcal{L}^2(\Omega)}| \leq \sigma_0[\|\zeta_t\|_{\mathcal{L}^2(\Omega)}^2 + \sum_{i=1}^2 (\|DW_i\|_{L^2(\Omega,T^2)}^2 + \|Dw_i\|_{L^2(\Omega,\Lambda)}^2)]$$
$$\leq 2\lambda_0\sigma_0 E(t) + L(\zeta). \tag{5.108}$$

Now inserting the inequalities (5.108) and (5.107) into the formula (5.105) gives the inequality (5.103). $\qquad\square$

A similar argument as in Lemma 4.11 in Chapter 4 yields

Lemma 5.3 *Let V be an escape vector field for the Naghdi model and let $G(V, DW) \in T^2(\Omega)$ be given by (4.131) for $W \in H^1(\Omega, \Lambda)$. Then for $W \in H^1(\Omega, \Lambda)$*

$$\sigma_1 \int_\Omega b(S(W), S(W))\, dx \leq \int_\Omega b(S(W), G(V, DW))\, dx + \mathrm{lo}\,(\zeta). \tag{5.109}$$

Lemma 5.4 *Let $\Gamma_0 \subset \Gamma$ be a relative open set and let X be a vector field on Ω. Let $\zeta = (W_1, W_2, w_1, w_2)$, $\hat{\zeta} = (\hat{W}_1, \hat{W}_2, \hat{w}_1, \hat{w}_2) \in \mathcal{H}^1(\Omega)$ satisfy $\zeta = \hat{\zeta} = 0$ on Γ_0. Then*

$$B(\zeta, \hat{\zeta}) = \sum_{i=1}^{2}[2(1+\beta)DW_i(\nu, \nu)D\hat{W}_i(\nu, \nu)$$

$$+ DW_i(\tau, \nu)D\hat{W}_i(\tau, \nu) + \frac{\partial w_i}{\partial \nu}\frac{\partial \hat{w}_i}{\partial \nu}] \quad on \quad \Gamma_0 \qquad (5.110)$$

and

$$\partial(\mathcal{A}\zeta, D_X\hat{\zeta}) = B(\zeta, \hat{\zeta})\langle X, \nu\rangle \quad on \quad \Gamma_0 \qquad (5.111)$$

where $\partial(\mathcal{A}\cdot, \cdot)$ is given by (5.77) and $D_X\hat{\zeta} = (D_X\hat{W}_1, D_X\hat{W}_2, X(\hat{w}_1), X(\hat{w}_2))$.

Proof. The boundary conditions $\zeta|_{\Gamma_0} = 0$ yield

$$W_i = D_\tau W_i = 0, \quad w_i = \tau(w_i) = 0 \quad on \quad \hat{\Gamma} \qquad (5.112)$$

for $i = 1, 2$. By the formulas (5.74) and (5.112), we have

$$\Upsilon(\zeta)(\nu, \nu) = DW_1(\nu, \nu), \quad \Upsilon(\zeta)(\nu, \tau) = \frac{1}{2}DW_1(\tau, \nu), \quad \Upsilon(\zeta)(\tau, \tau) = 0,$$

$$\mathcal{X}(\zeta)(\nu, \nu) = DW_2(\nu, \nu), \quad \mathcal{X}(\zeta)(\nu, \tau) = \frac{1}{2}DW_2(\tau, \nu), \quad \mathcal{X}(\zeta)(\tau, \tau) = 0,$$

$$and \quad \varphi(\zeta) = \frac{1}{2}\frac{\partial w_1}{\partial \nu}\nu \quad on \quad \hat{\Gamma}. \qquad (5.113)$$

Moreover, the relations in (5.112) give

$$\begin{cases} D_X W_1 = \langle X, \nu\rangle(DW_1(\nu, \nu)\nu + DW_1(\tau, \nu)\tau), \\ D_X W_2 = \langle X, \nu\rangle(DW_2(\nu, \nu)\nu + DW_2(\tau, \nu)\tau), \\ X(w_1) = \langle X, \nu\rangle\frac{\partial w_1}{\partial \nu}, \\ X(w_2) = \langle X, \nu\rangle\frac{\partial w_2}{\partial \nu} \end{cases} \quad on \quad \Gamma_0. \qquad (5.114)$$

In particular,

$$\begin{cases} DW_1(\nu, X) = \langle X, \nu\rangle DW_1(\nu, \nu), \\ DW_1(\tau, X) = \langle X, \nu\rangle DW_1(\tau, \nu), \\ DW_2(\nu, X) = \langle X, \nu\rangle DW_2(\nu, \nu), \\ DW_2(\tau, X) = \langle X, \nu\rangle DW_2(\tau, \nu), \end{cases} \quad on \quad \Gamma_0. \qquad (5.115)$$

On the other hand, applying the relations in (5.112) and (5.113) to the formula (5.78), we obtain

$$\begin{cases} B_1(\zeta) = 2(1+\beta)DW_1(\nu, \nu)\nu + DW_1(\tau, \nu)\tau, \\ B_2(\zeta) = 2(1+\beta)DW_2(\nu, \nu)\nu + DW_2(\tau, \nu)\tau. \end{cases} \qquad (5.116)$$

Note that all the relations (5.112)-(5.116) are true if we replace ζ by $\hat{\zeta}$.

The formula (5.110) follows by using the relations in (5.113) in the formula (5.73) for ζ and $\hat{\zeta}$, respectively.

Applying the relations (5.113)-(5.116) to the formula (5.77) for ζ and $\hat{\zeta}$, respectively, we obtain (5.111). □

By a similar argument as in Proposition 4.3, we have

Proposition 5.1 *Let λ be a complex number and $\hat{\Gamma} \subset \Gamma$ be relatively open. Then the problem*

$$\begin{cases} \lambda\eta + \mathcal{A}\eta = 0 \quad on \quad \Omega, \\ W_i|_{\hat{\Gamma}} = D_\nu W_i|_{\hat{\Gamma}} = 0, \\ w_i|_{\hat{\Gamma}} = \dfrac{\partial w_i}{\partial \nu}|_{\hat{\Gamma}} = 0, \quad i = 1, 2 \end{cases} \tag{5.117}$$

has the unique zero solution.

Dirichlet Control First, we consider the Dirichlet mixed problem in unknown $\xi = (\Phi_1, \Phi_2, \phi_1, \phi_2)$ for $T > 0$

$$\begin{cases} \xi_{tt} + \mathcal{A}\xi = 0 \quad in \quad Q, \\ \xi(0) = \xi_0, \quad \xi_t(0) = \xi_1, \quad on \quad \Omega, \\ \xi|_{\Sigma_1} = 0, \quad \xi|_{\Sigma_0} = \varsigma \end{cases} \tag{5.118}$$

with control functions $\varsigma \in \mathcal{L}^2(\Sigma_0)$ where

$$\mathcal{L}^2(\Sigma_0) = (L^2(\Sigma_0, \Lambda))^2 \times (L^2(\Sigma_0))^2.$$

Its dual version in $\zeta = (W_1, W_2, w_1, w_2)$ is

$$\begin{cases} \zeta_{tt} + \mathcal{A}\zeta = 0 \quad in \quad Q, \\ \zeta = 0 \quad on \quad \Sigma, \\ \zeta(0) = \zeta_0, \quad \zeta_t(0) = \zeta_1 \quad on \quad \Omega. \end{cases} \tag{5.119}$$

Let ζ solve the Dirichlet problem (5.119). We solve the problem

$$\begin{cases} \eta_{tt} + \mathcal{A}\eta = 0 \quad in \quad Q, \\ \eta(T) = \eta_t(T) = 0 \quad on \quad \Omega, \\ \eta|_{\Sigma_1} = 0, \quad \eta|_{\Sigma_0} = D_\nu\zeta \end{cases} \tag{5.120}$$

where $D_\nu\zeta = (D_\nu W_1, D_\nu W_2, w_{1\nu}, w_{2\nu})$ and define

$$\Lambda(\zeta_0, \zeta_1) = (-\eta_t(0), \eta(0)).$$

Let ζ and $\hat{\zeta}$ solve the duality problem (5.119) with initial data (ζ_0, ζ_1) and $(\hat{\zeta}_0, \hat{\zeta}_1)$, respectively. It follows from (5.120), (5.119), and (5.111) that, for

$(\zeta_0, \zeta_1), (\hat{\zeta}_0, \hat{\zeta}_1) \in \mathcal{H}_0^1(\Omega) \times \mathcal{L}^2(\Omega)$,

$$(\Lambda(\zeta_0, \zeta_1), (\hat{\zeta}_0, \hat{\zeta}_1))_{\mathcal{L}^2(\Omega) \times \mathcal{L}^2(\Omega)} = -(\eta_t(0), \hat{\zeta}_0)_{\mathcal{L}^2(\Omega)} + (\eta(0), \hat{\zeta}_1)_{\mathcal{L}^2(\Omega)}$$

$$= \int_0^T [(\eta_t, \hat{\zeta})_{\mathcal{L}^2(\Omega)} - (\eta, \hat{\zeta}_t)_{\mathcal{L}^2(\Omega)}]_t dt$$

$$= \int_0^T [-(\mathcal{A}\eta, \hat{\zeta})_{\mathcal{L}^2(\Omega)} + (\eta, \mathcal{A}\hat{\zeta})_{\mathcal{L}^2(\Omega)}] dt$$

$$= \int_\Sigma [-\partial(\mathcal{A}\eta, \hat{\zeta}) + (\mathcal{A}\hat{\zeta}, \eta)] d\Sigma = \int_{\Sigma_0} \partial(\mathcal{A}\hat{\zeta}, D_\nu \zeta) d\Sigma$$

$$= \int_{\Sigma_0} B(\zeta, \hat{\zeta}) d\Sigma. \tag{5.121}$$

Remark 5.1 *By (5.121) and (5.110), if a solution ζ to the duality problem (5.119) satisfies*

$$(\Lambda(\zeta_0, \zeta_1), (\zeta_0, \zeta_1))_{\mathcal{L}^2(\Omega) \times \mathcal{L}^2(\Omega)} < \infty,$$

then the boundary control in (5.120) has a regularity

$$D_\nu \zeta \in \mathcal{L}^2(\Sigma_0).$$

Let $L = \mathcal{L}^2(\Omega) \times \mathcal{L}^2(\Omega)$ and $H = \mathcal{H}_0^1(\Omega) \times \mathcal{L}^2(\Omega)$. Applying Theorem 2.13, we obtain

Theorem 5.8 *The problem (5.118) is exactly $[\mathcal{L}^2(\Omega) \times \mathcal{H}_{\Gamma_1}^1(\Omega)]^*$ controllable by $(L^2(\Sigma_0, \Lambda))^2 \times (L^2(\Sigma_0))^2$ controls on $[0, T]$ if and only if there are $c_1 > 0$ and $c_2 > 0$ such that*

$$c_1 E(0) \leq \int_{\Sigma_0} B(\zeta, \zeta) \, d\Sigma \leq c_2 E(0) \tag{5.122}$$

for all solutions ζ of the problem (5.119).

Moreover, we have

Theorem 5.9 *If there is an escape vector field for the Naghdi shell on the middle surface, then the observability inequality (5.122) holds true for any $T > T_0$ where*

$$T_0 = \lambda_0 \sigma_0 / \sigma_1. \tag{5.123}$$

Proof. Let SB be given in (5.104). By (5.111),

$$\text{SB} = B(\zeta, \zeta) \langle V, \nu \rangle \quad \text{for} \quad (t, x) \in \Sigma. \tag{5.124}$$

Noting that $E(t) = E(0)$ for $t \geq 0$ we obtain by Theorem 5.7

$$\int_\Sigma \text{SB} \, d\Sigma \geq 2(\sigma_1 T - \lambda_0 \sigma_0) E(0) + L(\zeta). \tag{5.125}$$

In addition, by the definition Γ_0 in (5.102) and the formula (5.124), we have

$$\int_\Sigma SB\, d\Sigma \le \sigma_0 \int_{\Sigma_0} B(\varsigma, \varsigma)\, d\Sigma. \tag{5.126}$$

Thus the left hand side of the inequality (5.122) follows from the estimates (5.125) and (5.126) by a compactness/uniqueness argument in terms of Proposition 5.1, as in Lemma 2.5.

Finally, as usual, we take an vector field $H \in \mathcal{X}(\Omega)$ such that $H|_\Gamma = \nu$ and multiply the equation in (5.119) by $(D_H W_1, D_H W_2, H(w_1), H(w_2))$. As a result of the calculation, we obtain the right hand side of the inequality (5.122). □

Neumann Control Let $\overline{\Gamma_1} \neq \emptyset$ and $\overline{\Gamma_0} \cap \overline{\Gamma_1} = \emptyset$. We consider the problem in unknown $\xi = (\Phi_1, \Phi_2, \phi_1, \phi_2)$

$$\begin{cases} \xi_{tt} + \mathcal{A}\xi = 0 & \text{in} \quad Q, \\ \xi = 0 & \text{on} \quad \Sigma_1 \\ \partial \mathcal{A}\xi = \varsigma & \text{on} \quad \Sigma_0, \\ \xi(0) = \xi_0, \quad \xi_t(0) = \xi_1 & \text{on} \quad \Omega, \end{cases} \tag{5.127}$$

where

$$\partial \mathcal{A}\xi = (B_1(\xi), B_2(\xi), \varphi(\xi), \phi_2) \tag{5.128}$$

and ς is a boundary control. The dual problem for the above is the following in $\zeta = (W_1, W_2, w_1, w_2)$

$$\begin{cases} \zeta_{tt} + \mathcal{A}\zeta = 0 & \text{in} \quad Q, \\ \zeta = 0 & \text{on} \quad \Sigma_1 \\ \partial \mathcal{A}\zeta = 0 & \text{on} \quad \Sigma_0, \\ \zeta(0) = \zeta_0, \quad \zeta_t(0) = \zeta_1 & \text{on} \quad \Omega. \end{cases} \tag{5.129}$$

Let

$$\aleph: \quad H^1([0,T], \mathcal{L}^2(\Sigma_0)) \to [H^1([0,T], \mathcal{L}^2(\Sigma_0))]^*$$

be *the canonical map* with respect to $\mathcal{L}^2(\Sigma_0)$ where

$$\mathcal{L}^2(\Sigma_0) = (L^2(\Sigma_0, \Lambda))^2 \times (L^2(\Sigma_0))^2.$$

Then, for any $\psi \in H^1([0,T], \mathcal{L}^2(\Sigma_0))$, $\aleph\psi \in [H^1([0,T], \mathcal{L}^2(\Sigma_0))]^*$ satisfies

$$(\aleph\psi, \chi)_{\mathcal{L}^2(\Sigma_0)} = \int_{\Sigma_0} (\langle \psi_t, \chi_t \rangle + \langle \psi, \chi \rangle)\, d\Sigma \tag{5.130}$$

for all $\chi \in \mathcal{L}^2(\Sigma_0)$.

Let ζ solve the problem (5.129). We solve the problem

$$\begin{cases} \eta_{tt} + \mathcal{A}\eta = 0 & \text{in} \quad Q, \\ \eta(T) = \eta_t(T) = 0 & \text{on} \quad \Omega, \\ \eta|_{\Sigma_1} = 0, \quad \partial \mathcal{A}\eta|_{\Sigma_0} = -\aleph\zeta \end{cases} \tag{5.131}$$

and define
$$\Lambda(\zeta_0, \zeta_1) = (-\eta_t(0), \eta(0)).$$

Let ζ and $\hat{\zeta}$ solve the duality problem (5.129) with initial data (ζ_0, ζ_1) and $(\hat{\zeta}_0, \hat{\zeta}_1)$, respectively, such that $\zeta|_{\Sigma_0}$ and $\hat{\zeta}|_{\Sigma_0}$ are in the space $H^1([0,T], \mathcal{L}^2(\Sigma_0))$. As for (5.121), by (5.130), we have

$$(\Lambda(\zeta_0, \zeta_1), (\hat{\zeta}_0, \hat{\zeta}_1))_{\mathcal{L}^2(\Omega) \times \mathcal{L}^2(\Omega)} = -\int_{\Sigma_0} \partial(\mathcal{A}\eta, \hat{\zeta})d\Sigma$$

$$= -\int_{\Sigma_0} \langle \partial \mathcal{A}\eta, \hat{\zeta} \rangle \, d\Sigma = (\aleph\zeta, \hat{\zeta})_{\mathcal{L}^2(\Sigma_0)}$$

$$= \int_{\Sigma_0} (\langle \zeta_t, \hat{\zeta}_t \rangle + \langle \zeta, \hat{\zeta} \rangle)d\Sigma.$$

Then the estimate
$$(\Lambda(\zeta_0, \zeta_1), (\zeta_0, \zeta_1))_{\mathcal{L}^2(\Omega) \times \mathcal{L}^2(\Omega)} < \infty$$

implies that the boundary control in (5.131) has a regularity
$$-\aleph\zeta \in [H^1([0,T], \mathcal{L}^2(\Sigma_0))]^*.$$

Let $L = \mathcal{L}^2(\Omega) \times \mathcal{L}^2(\Omega)$ and $H = \mathcal{H}^1_{\Gamma_1}(\Omega) \times \mathcal{L}^2(\Omega)$. Applying Theorem 2.13, we obtain, via Theorem 5.11 below,

Theorem 5.10 *Let V be an escape vector field for the Naghdi shell. Let T_0 be given by (5.123). Then the problem (5.127) is exactly $[\mathcal{L}^2(\Omega) \times \mathcal{H}^1_{\Gamma_1}(\Omega)]^*$ controllable by $[H^1([0,T], \mathcal{L}^2(\Sigma_0))]^*$ controls on $[0,T]$ for $T > T_0$.*

A similar argument as for Theorem 5.9 yields

Theorem 5.11 *Let V be an escape vector field for the Naghdi shell. Let T_0 be given by (5.123). Then, for $T > T_0$, there is $c > 0$ such that*

$$\int_{\Sigma_0} (|\zeta_t|^2 + |\zeta|^2) \, dx \geq cE(0) \tag{5.132}$$

for all solutions to the problem (5.129) for which the left hand side of (5.132) is finite.

5.3   Stabilization by Boundary Feedback

Consider the following boundary feedback control problem in unknown $\zeta = (W_1, W_2, w_1, w_2)$

$$\begin{cases} \zeta_{tt} + \mathcal{A}\zeta = 0 & \text{in } Q_\infty, \\ \zeta(0) = \zeta_0, \quad \zeta_t(0) = \zeta_1 & \text{on } \Omega, \end{cases} \tag{5.133}$$

subject to the boundary conditions

$$\zeta = 0 \quad \text{on} \quad \Sigma_{1\infty} = (0,\infty) \times \Gamma_1 \tag{5.134}$$

and a boundary feedback

$$\begin{cases} B_1(\zeta) = J_1(\zeta_t)\nu + J_2(\zeta_t)\tau, \\ B_2(\zeta) = J_3(\zeta_t)\nu + J_4(\zeta_t)\tau, \\ 2\varphi(\zeta) = J_5(\zeta_t), \\ \dfrac{\partial w_2}{\partial \nu} = J_6(\zeta_t), \end{cases} \quad \text{on} \quad \Sigma_{0\infty} \tag{5.135}$$

where where J_i, $i = 1, \cdots, 6$ are feedback operators.

For simplicity, we denote

$$\check{\zeta} = (\langle W_1, \nu \rangle, \langle W_1, \tau \rangle, \langle W_2, \nu \rangle, \langle W_2, \tau \rangle, w_1, w_2)$$

for any $\zeta = (W_1, W_2, w_1, w_2)$. We will consider feedback laws to be defined by

$$J(\zeta) = -\check{\zeta}F \tag{5.136}$$

where $J(\zeta) = (J_1(\zeta), \cdots, J_6(\zeta))$ and $F = \left(f_{ij}(x) \right)$ are 6×6 matrices. We assume that

$$F \text{ is symmetric, positive semidefinite on } \overline{\Gamma}_0. \tag{5.137}$$

Define the total energy of the system (5.133)-(5.135) by

$$2E(t) = \int_\Omega [|\zeta_t|^2 + B(\zeta, \zeta)]\, dx \tag{5.138}$$

where the bilinear form $B(\cdot, \cdot)$ is given by the formula (5.73).

Using the Green formula (5.75) and the relations (5.133)-(5.136), we have

$$\frac{d}{dt}E(t) = \int_\Gamma \partial(\mathcal{A}\zeta, \zeta_t)\, d\Gamma = -\int_{\Gamma_0} \langle \check{\zeta_t}, F\check{\zeta_t} \rangle_{\mathbb{R}^6}\, d\Gamma. \leq 0 \tag{5.139}$$

Then the resulting closed-loop system under the feedback laws of (5.135) and (5.137) is dissipative in the sense that $E(t)$ is decreasing.

Variational Formula of the Closed-Loop System (5.133)-(5.137)
Let the Sobolev spaces $\mathcal{H}^1_{\Gamma_1}(\Omega)$ and $\mathcal{L}(\Omega)$ be given by (5.28) and (5.26), respectively. Introduce a bilinear form by

$$\alpha(\zeta, \eta) = \int_{\Gamma_0} \langle \check{\zeta}, F\check{\eta} \rangle_{\mathbb{R}^6}\, d\Gamma \tag{5.140}$$

where $\zeta = (W_1, W_2, w_1, w_2)$, $\eta = (U_1, U_2, u_1, u_2)$, and F is the feedback matrix. Then it follows from the Green formula (5.75) that an appropriate variational formulation for the system (5.133)-(5.137) is: Find $\zeta \in C([0,\infty); \mathcal{H}^1_{\Gamma_1}(\Omega)) \cap C^1([0,\infty), \mathcal{L}^2(\Omega))$ such that

$$\begin{cases} \frac{d}{dt}[(\zeta_t, \eta)_{\mathcal{L}^2(\Omega)} + \alpha(\zeta, \eta)] + \mathcal{B}(\zeta, \eta) = 0 \quad \forall \eta \in \mathcal{H}^1_{\Gamma_1}(\Omega), \\ \zeta(0) = \zeta_0 \in \mathcal{H}^1_{\Gamma_1}(\Omega), \quad \zeta_t(0) = \zeta_1 \in \mathcal{L}^2(\Omega). \end{cases} \tag{5.141}$$

Well-Posedness of the Closed-Loop System (5.133)-(5.137) Let the following assumption hold.

Assumption (H1) Γ_0 and Γ_1 satisfy the following conditions

$$\overline{\Gamma}_1 \neq \emptyset, \quad \overline{\Gamma}_0 \cap \overline{\Gamma}_1 = \emptyset, \quad \text{and} \quad \langle V, \nu \rangle \leq 0 \quad \text{on} \quad \Gamma_1. \tag{5.142}$$

By Theorem 5.3, the bilinear form $\mathcal{B}(\cdot, \cdot)$ introduces an equivalent inner product on $\mathcal{H}^1_{\Gamma_1}(\Omega)$. By the assumption (5.137), $\alpha(\cdot, \cdot)$ is continuous, symmetric, and non-negative. We identify $\mathcal{L}^2(\Omega)$ with its dual $\mathcal{L}^2(\Omega)^*$ so that we have the dense and continuous embeddings

$$\mathcal{H}^1_{\Gamma_1}(\Omega) \subset \mathcal{L}^2(\Omega) \subset \mathcal{H}^1_{\Gamma_1}(\Omega)^*. \tag{5.143}$$

Let $\aleph$ denote *the canonical isomorphism* of $\mathcal{H}^1_{\Gamma_1}(\Omega)$ endowed with the scalar product $\mathcal{B}(\cdot, \cdot)$ onto $\mathcal{H}^1_{\Gamma_1}(\Omega)$. Then

$$\mathcal{B}(\zeta, \eta) = (\aleph\zeta, \eta)_{\mathcal{L}^2(\Omega)} \quad \text{for} \quad \zeta, \eta \in \mathcal{H}^1_{\Gamma_1}(\Omega). \tag{5.144}$$

Furthermore, there is a non-negative operator $A \in L(\mathcal{H}^1_{\Gamma_1}(\Omega), \mathcal{H}^1_{\Gamma_1}(\Omega)^*)$ such that

$$\alpha(\zeta, \eta) = (A\zeta, \eta)_{\mathcal{L}^2(\Omega)} \quad \text{for} \quad \zeta, \eta \in \mathcal{H}^1_{\Gamma_1}(\Omega). \tag{5.145}$$

Then the equation in (5.141) can be written as

$$\frac{d}{dt}(\zeta_t + A\zeta) + \aleph\zeta = 0 \quad \text{in} \quad \mathcal{H}^1_{\Gamma_1}(\Omega)^*. \tag{5.146}$$

Let us formally rewrite (5.146) as the following system:

$$\mathbb{C}Y_t + \mathbb{N}Y = 0 \quad t \geq 0 \quad \text{in} \quad \mathcal{H}^1_{\Gamma_1}(\Omega)^* \times \mathcal{L}^2(\Omega) \tag{5.147}$$

where

$$\mathbb{C} = \begin{pmatrix} \aleph & 0 \\ 0 & I \end{pmatrix}, \quad \mathbb{N} = \begin{pmatrix} 0 & -\aleph \\ \aleph & A \end{pmatrix}, \quad Y = \begin{pmatrix} \zeta \\ \zeta_t \end{pmatrix}. \tag{5.148}$$

We shall solve the problem (5.147) in the space $\mathcal{H}^1_{\Gamma_1}(\Omega)^* \times \mathcal{L}^2(\Omega)$. In order to make sense of (5.147) in that space it is natural to introduce

$$D(\mathbb{N}) = \{ (\zeta, \eta) \in \mathcal{H}^1_{\Gamma_1}(\Omega) \times \mathcal{H}^1_{\Gamma_1}(\Omega) \,|\, \aleph\zeta + A\eta \in \mathcal{L}^2(\Omega) \}.$$

Since $\mathbb{C}$ is *the canonical isomorphism* of $\mathcal{H}^1_{\Gamma_1}(\Omega) \times \mathcal{L}^2(\Omega)$ onto $\mathcal{H}^1_{\Gamma_1}(\Omega)^* \times \mathcal{L}^2(\Omega)$, we rewrite (5.147) in the form

$$Y_t + \mathbb{C}^{-1}\mathbb{N}Y = 0 \quad \text{in} \quad \mathcal{H}^1_{\Gamma_1}(\Omega) \times \mathcal{L}^2(\Omega). \tag{5.149}$$

Then solutions of the system (5.133)-(5.137) are defined by (5.149).

We have

Theorem 5.12 $-\mathbb{C}^{-1}\mathbb{N}$ *is the infinitemal generator of a C_0-semigroup of contraction on* $\mathcal{H}^1_{\Gamma_1}(\Omega) \times \mathcal{L}^2(\Omega)$.

Proof. **Claim 1** $D(\mathbb{N})$ is dense in $\mathcal{H}^1_{\Gamma_1}(\Omega) \times \mathcal{L}^2(\Omega)$.
By the definitions of the operators $\aleph$ and A, we have

$$(\aleph\zeta + A\eta, \varsigma)_{\mathcal{L}^2(\Omega)} = \mathcal{B}(\zeta, \varsigma) + \alpha(\eta, \varsigma)$$
$$= (A\zeta, \varsigma) + \int_{\Gamma_0} [\partial(A\zeta, \varsigma) + \langle \breve{\eta}, F\breve{\varsigma} \rangle_{\mathbb{R}^6}] \, d\Gamma \qquad (5.150)$$

for $\varsigma \in \mathcal{H}^1_{\Gamma_1}(\Omega)$. The expression on the right-hand side of the formula (5.150) implies the relation

$$D(\mathbb{N}) \supset D_0 = \{ (\zeta, \eta) \,|\, \zeta \in \mathcal{H}^1_{\Gamma_1}(\Omega) \cap \mathcal{H}^2(\Omega), \ \eta \in \mathcal{H}^1_{\Gamma_1}(\Omega),$$
$$\partial(A\zeta, \varsigma) + \langle \breve{\eta}, F\breve{\varsigma} \rangle_{\mathbb{R}^6} = 0 \text{ on } \Gamma_0 \},$$

where

$$\mathcal{H}^2(\Omega) = (H^2(\Omega, \Lambda))^2 \times (H^2(\Omega))^2.$$

Indeed, if $(\zeta, \eta) \in D_0$, then

$$|(\aleph\zeta + A\eta, \varsigma)_{\mathcal{L}^2(\Omega)}| \leq c \|\zeta\|_{\mathcal{H}^2(\Omega)} \|\varsigma\|_{\mathcal{L}^2(\Omega)}$$

which yields $\aleph\zeta + A\eta \in \mathcal{L}^2(\Omega)$.

Since D_0 is dense in $\mathcal{H}^1_{\Gamma_1}(\Omega) \times \mathcal{L}^2(\Omega)$, Claim 1 follows.
Claim 2 $-\mathbb{C}^{-1}\mathbb{N}$ is dissipative. Noting that the inner product on $\mathcal{H}^1_{\Gamma_1}(\Omega)$ is $\mathcal{B}(\cdot, \cdot)$, it is shown by

$$(\mathbb{C}^{-1}\mathbb{N}(\zeta, \eta), (\zeta, \eta))_{\mathcal{H}^1_{\Gamma_1}(\Omega) \times \mathcal{L}^2(\Omega)} = ((-\eta, \aleph\zeta + A\eta), (\zeta, \eta))_{\mathcal{H}^1_{\Gamma_1}(\Omega) \times \mathcal{L}^2(\Omega)}$$
$$= -\mathcal{B}(\zeta, \eta) + \mathcal{B}(\zeta, \eta) + \alpha(\eta, \eta) = \alpha(\eta, \eta) \geq 0$$

for $(\zeta, \eta) \in D(\mathbb{N})$.
Claim 3 $\text{Range}(\lambda I + \mathbb{C}^{-1}\mathbb{N}) = \mathcal{H}^1_{\Gamma_1}(\Omega) \times \mathcal{L}^2(\Omega)$ for $\lambda > 0$. In fact, this is equivalent to

$$\text{Range}(\lambda^2 I + \lambda A + \aleph) = \mathcal{L}^2(\Omega).$$

But, by the Lax-Milgram theorem, it is actually true. $\qquad\square$

As a consequence of Theorem 5.12, we have the following result.

Theorem 5.13 *Let the assumption* $(H1)$ *in* (5.142) *hold. Then the problem* (5.133)-(5.137) *admits a unique solution with*

$$\zeta \in C([0, \infty); \mathcal{H}^1_{\Gamma_1}(\Omega)), \quad \zeta_t \in C([0, \infty); \mathcal{L}^2(\Omega)). \qquad (5.151)$$

Furthermore, if $(\zeta_0, \zeta_1) \in D(\mathcal{Q})$, *then the solution* ζ *satisfies*

$$\begin{cases} \zeta \in C^1([0, \infty); \mathcal{H}^1_{\Gamma_1}(\Omega)) \cap C^2([0, \infty); \mathcal{L}^2(\Omega)), \\ \aleph\zeta + A\zeta_t \in C([0, \infty); \mathcal{L}^2(\Omega)), \\ \zeta_{tt} + A\zeta_t + \aleph\zeta = 0 \quad t > 0, \\ \zeta(0) = \zeta_0, \quad \zeta_t(0) = \zeta_1. \end{cases} \qquad (5.152)$$

Solution ζ is called a *weak solution* if the conditions (5.151) are true. It is called a *strong solution* if the conditions (5.152) are true.

Trace Estimates on Boundary Temporarily, we forget the definition of β in (5.19) and assume that μ and β are just positive constants, independent of each other. By comparing the formula (5.31) in this chapter with the formula (4.37) in Chapter 4 it follows that $\boldsymbol{\Delta}_\beta = 2(1+\beta)\boldsymbol{\Delta}_\mu$ if $\mu = \dfrac{\beta}{1+\beta}$. Because of this relation the results in Proposition 4.6 in Chapter 4 hold true if replaced $\boldsymbol{\Delta}_\mu$ by $\boldsymbol{\Delta}_\beta$.

Consider the problem

$$\begin{cases} W_{tt} - \boldsymbol{\Delta}_\beta W = F & \text{in} \quad Q, \\ W|_{\Gamma_1} = 0 & \text{on} \quad \Sigma_1 \end{cases} \tag{5.153}$$

where $\boldsymbol{\Delta}_\beta$ is given by (5.31). Then

Proposition 5.2 *Let* $0 < \alpha < T/2$. *Let* W *be a solution to the problem* (5.153). *Then there is* $c_{\alpha T} > 0$ *such that*

$$\int_\alpha^{T-\alpha} \int_{\Gamma_0} |D_\tau W|^2 \, d\Gamma dt \le c_{\alpha T} \int_0^T \|F\|^2_{H^{-1/2}(\Omega,\Lambda)} \, dt$$

$$+ c_{\alpha T} \int_{\Sigma_0^T} (|W_t|^2 + |S(W)(\nu,\nu) - \beta\delta W|^2$$

$$+ |S(W)(\nu,\tau)|^2) \, d\Sigma + L(W) \tag{5.154}$$

where $S(W)$ *is given by the formula* (5.80).

Now we have

Theorem 5.14 *Let* $\zeta = (W_1, W_2, w_1, w_2)$ *be a solution of the closed-loop system* (5.133)-(5.137). *Let* $T/2 > \alpha > 0$ *be given. Then*

$$\sum_{i=1}^2 \int_\alpha^{T-\alpha} \int_{\Gamma_0} (|DW_i|^2 + |Dw_i|^2) \, d\Gamma dt \le c_{T\alpha} \int_{\Sigma_0} |\zeta_t|^2 d\Sigma + L(\zeta). \tag{5.155}$$

Proof. By Propositions 5.2 and Lemma 2.12, we need to estimate $D_\nu W_i$ and $\frac{\partial w_i}{\partial \nu}$. However, they are controlled by the feedback laws. It follows from the formulas (5.136), (5.135), (5.78), and (5.74) that

$$\begin{cases} \langle B_1(\zeta), \nu \rangle = J_1(\zeta_t), & \langle B_1(\zeta), \tau \rangle = J_2(\zeta_t), \\ \langle B_2(\zeta), \nu \rangle = J_3(\zeta_t), & \langle B_2(\zeta), \tau \rangle = J_4(\zeta_t), \\ 2\varphi(\zeta) = J_5(\zeta_t), & \dfrac{\partial w_2}{\partial \nu} = J_6(\zeta_t) \end{cases}$$

which yield

$$\begin{cases} DW_1(\nu,\nu) = \dfrac{1}{2(1+\beta)}[J_1(\zeta_t) - 2\beta DW_1(\tau,\tau)] + \mathrm{lo}_0(\zeta), \\ DW_1(\tau,\nu) = J_2(\zeta_t) - DW_1(\nu,\tau) + \mathrm{lo}_0(\zeta), \\ DW_2(\nu,\nu) = \dfrac{1}{2(1+\beta)}[J_3(\zeta_t) - 2\beta DW_2(\tau,\tau)] + \mathrm{lo}_0(\zeta), \\ DW_2(\tau,\nu) = J_4(\zeta_t) - DW_2(\nu,\tau) + \mathrm{lo}_0(\zeta), \\ \dfrac{\partial w_1}{\partial \nu} = J_5(\zeta_t) + \mathrm{lo}_0(\zeta), \\ \dfrac{\partial w_2}{\partial \nu} = J_6(\zeta_t) \end{cases} \tag{5.156}$$

where $\mathrm{lo}_0(\zeta)$ denote some zero order terms on ζ.

By the relations (5.156), we obtain

$$\begin{cases} |DW_i|^2 \le c(|\zeta_t|^2 + |D_\tau W_i|^2 + |\zeta|^2), \\ |Dw_i| \le c(|\zeta_t|^2 + |\dfrac{\partial w_i}{\partial \tau}|^2 + |\zeta|^2) \end{cases} \tag{5.157}$$

for $i = 1, 2$.

Finally, the inequality (5.155) follows the estimates (5.157) and Propositions 5.2 and Lemma 2.12. $\qquad\square$

Stabilization of the Closed-Loop System (5.133)-(5.137) For stabilization, we need a further assumption:

Assumption (H2) The 6×6 matrix F is positive on $\overline{\Gamma}_0$.

We have

Theorem 5.15 *Suppose that there exists an escape vector field for the Naghdi shell such that relations (5.99) and (5.100) hold. Let the assumptions* (**H1**) *and* (**H2**) *be true where the assumption* (**H1**) *is given by (5.142). Then there are* c_1, $c_2 > 0$ *such that*

$$E(t) \le c_1 E(0)e^{-c_2 t} \quad for \quad t \ge 0 \tag{5.158}$$

where $E(t)$ *is the total energy of the closed-loop system (5.133)-(5.137), given by (5.138).*

Proof. Let $\zeta = (W_1, W_2, w_1, w_2)$ be a solution of the closed loop system (5.133)-(5.137). Let the boundary terms SB be given by the formula in (5.104). It follows from Lemma 5.4 and the boundary conditions (5.134) that

$$\int_{\Sigma_1} \mathrm{SB}\, d\Sigma = \int_{\Sigma_1} B(\zeta,\zeta)\langle V,\nu \rangle\, d\Sigma \le 0, \tag{5.159}$$

via the assumption (**H1**). Moreover, using the formulas (5.140), (5.136), (5.135), and (5.77), we obtain

$$\int_{\Gamma_0} \partial(\mathcal{A}\zeta, m(\zeta) + \varrho\zeta)\, d\Gamma = -\alpha(\zeta_t, m(\zeta) + \varrho\zeta). \tag{5.160}$$

Let s_1, $s_2 > 0$ be such that

$$s_1 |X|^2 \leq \langle X, FX \rangle_{\mathbb{R}^6} \leq s_2 |X|^2 \quad \text{for} \quad X \in \mathbb{R}^6, \ x \in \Gamma_0.$$

Then the definition (5.140) of $\alpha(\cdot, \cdot)$ yields

$$s_1 \int_{\Gamma_0} |\zeta|^2 d\Gamma \leq \alpha(\zeta, \zeta) \leq s_2 \int_{\Gamma_0} |\zeta|^2 d\Gamma \tag{5.161}$$

which gives

$$|\alpha(\zeta_t, 2m(\zeta) + \varrho\zeta)| \leq \alpha^{1/2}(\zeta_t, \zeta_t)\alpha^{1/2}(2m(\zeta) + \varrho\zeta, 2m(\zeta) + \varrho\zeta)$$

$$\leq c(|\zeta_t|^2 + \sum_{i=1}^{2}(|DW_i|^2 + |Dw_i|^2) + |\zeta|^2). \tag{5.162}$$

Using the relations (5.162), (5.160), and (5.104), we have

$$\int_{\Sigma_0} \text{SB} \, d\Sigma = - \int_0^T \alpha(\zeta_t, 2m(\zeta) + \varrho\zeta) \, dt + \int_{\Sigma_0} [|\zeta_t|^2 - B(\zeta, \zeta)]\langle V, \nu \rangle d\Sigma$$

$$\leq c \int_{\Sigma_0} [|\zeta_t|^2 + \sum_{i=1}^{2}(|DW_i|^2 + |Dw_i|^2)]d\Sigma + L(\zeta). \tag{5.163}$$

Then applying the inequalities (5.163) and (5.159) to the inequality (5.103) in Theorem 5.7 yields

$$\sigma_1 \int_0^T E(t)dt \leq c \int_{\Sigma_0} [|\zeta_t|^2 + \sum_{i=1}^{2}(|DW_i|^2 + |Dw_i|^2)]d\Sigma$$

$$+ c[E(0) + E(T)] + L(\zeta). \tag{5.164}$$

Next, change the integral domain Σ_0 into $[\alpha, T - \alpha] \times \Gamma_0$ on the both sides of the inequalities (5.164) and use the inequality (5.155) to give

$$\sigma_1 \int_\alpha^{T-\alpha} E(t) \, dt \leq c[E(\alpha) + E(T - \alpha) + \int_{\Sigma_0} |\zeta_t|^2 d\Sigma] + L(\zeta). \tag{5.165}$$

Using the relations (5.139) and (5.140), we have for any $T \geq s \geq 0$

$$E(s) = E(T) + \int_s^T \alpha(\zeta_t, \zeta_t) \, dt \leq E(T) + c \int_{\Sigma_0} |\zeta_t|^2 \, d\Sigma$$

which yields

$$E(\alpha) \leq E(T) + \int_{\Sigma_0} |\zeta_t|^2 d\Sigma \tag{5.166}$$

and

$$\int_\alpha^{T-\alpha} E(t)dt \geq (T - 2\alpha)E(T) - cT \int_{\Sigma_0} |\zeta_t|^2 d\Sigma. \tag{5.167}$$

Using the inequalities (5.166) and (5.167) in the inequality (5.165) gives

$$E(T) \le c_T \int_{\Sigma_0} |\zeta_t|^2 d\Sigma \qquad (5.168)$$

for some $T > 0$ large where the lower order terms have been absorbed by Proposition 5.1 as in Lemma 2.5. Using the left hand side inequality in (5.161), the inequality (5.168), and the equations (5.139), we now have

$$E(T) \le c_T \int_0^T \alpha(\zeta_t, \zeta_t) \, dt = c_T(E(0) - E(T)),$$

that is,

$$E(T) \le \frac{c_T}{1 + c_T} E(0)$$

from which the exponential decay follows.   □

5.4   Stabilization of Transmission

Let a middle surface Ω be divided into two regions, Ω_1 and Ω_2, by a smooth curve Γ_3 on Ω. Set $\Gamma_i = \partial\Omega \cap \partial\Omega_i$ for $i = 1$ and 2. Denote by ν_i and τ_i the normal and the tangential on $\partial\Omega_i$ pointing outside Ω_i, respectively. Then

$$\nu_i = \nu, \quad \tau_i = \tau \quad \text{for} \quad x \in \Gamma_1 \cup \Gamma_2 = \partial\Omega,$$

$$\nu_1(x) = -\nu_2(x), \quad \tau_1(x) = -\tau_2(x) \quad \text{for} \quad x \in \Gamma_3. \qquad (5.169)$$

Suppose that the type of material in Ω_1 is different from that in Ω_2, where *Young's modulus* and *Poisson's coefficient* have a jump across the separatrix Γ_3.

First, let us clarify *transmission conditions for the Naghdi shell.* Let E_i and μ_i, respectively, denote Young's modulus and Poisson's coefficient of the material corresponding to the region Ω_i , $i = 1$, 2. Set

$$\alpha_i = E_i/(1 + \mu_i), \quad \beta_i = \mu_i/(1 - 2\mu_i). \qquad (5.170)$$

Let $\mathcal{A}_i$ be the Naghdi shell operator $\mathcal{A}_0$ on the portion Ω_i, given by the formula (5.36) in which α and β have been replaced with α_i and β_i, respectively, for $i = 1, 2$. Let $\zeta_i = (W_{i1}, W_{i2}, w_{i1}, w_{i2})$ be a displacement vector on Ω_i for $i = 1$ and 2, respectively. Similarly, let

$$\begin{cases} B_{i1}(\zeta_i) = 2\,\mathrm{i}\,(\nu_i)\Upsilon(\zeta_i) + 2\beta_i(\mathrm{tr}\,\Upsilon(\zeta_i) + w_{i2})\nu_i, \\ B_{i2}(\zeta_i) = 2\,\mathrm{i}\,(\nu_i)\mathcal{X}_0(\zeta_i) + 2\beta_i(\mathrm{tr}\,\mathcal{X}_0(\zeta_i))\nu_i, \end{cases} \qquad (5.171)$$

where $\mathrm{i}\,(\nu_i)$ denotes *the interior product.*

Definition 5.2 *A displacement vector field ζ is said to satisfy the transmission conditions for the Naghdi shell on the separatrix Γ_3 if*

$$
\begin{cases}
\zeta_1 = \zeta_2, \\
\alpha_1 B_{11}(\zeta_1) + \alpha_2 B_{21}(\zeta_2) = 0, \\
\alpha_1 B_{12}(\zeta_1) + \alpha_2 B_{22}(\zeta_2) = 0, \\
\alpha_1 \langle \varphi_0(\zeta_1), \nu_1 \rangle + \alpha_2 \langle \varphi_0(\zeta_2), \nu_2 \rangle = 0, \\
\alpha_1 \dfrac{\partial w_{12}}{\partial \nu_1} + \alpha_2 \dfrac{\partial w_{22}}{\partial \nu_2} = 0
\end{cases}
\qquad on \quad \Gamma_3. \qquad (5.172)
$$

where $\zeta_i = \zeta|_{\Omega_i}$ for $i = 1, 2$.

Consider the following boundary feedback problem in unknown $\zeta_i = (W_{i1}, W_{i2}, w_{i1}, w_{i2})$ for $x \in \Omega_i$

$$
\begin{cases}
\zeta_{itt} + \alpha_i \mathcal{A}_i \zeta_i = 0 & \text{in} \quad (0, \infty) \times \Omega_i, \\
\zeta_i(0) = \zeta_i^0, \quad \zeta_{it}(0) = \zeta_i^1 & \text{on} \quad \Omega_i, \quad i = 1, 2,
\end{cases}
\qquad (5.173)
$$

which is subject to the boundary conditions

$$
\zeta_1 = 0 \quad \text{on} \quad (0, \infty) \times \Gamma_1, \qquad (5.174)
$$

the transmission conditions (5.172) on $(0, \infty) \times \Gamma_3$, $\qquad (5.175)$

and a boundary feedback on $(0, \infty) \times \Gamma_2$

$$
\begin{cases}
B_{21}(\zeta_2) = J_1(\zeta_{2t})\nu + J_2(\zeta_{2t})\tau, \\
B_{22}(\zeta_2) = J_3(\zeta_{2t})\nu + J_4(\zeta_{2t})\tau, \\
2\varphi(\zeta_2) = J_5(\zeta_{2t}), \\
\dfrac{\partial w_{22}}{\partial \nu} = J_6(\zeta_{2t})
\end{cases}
\qquad \text{on} \quad (0, \infty) \times \Gamma_2 \qquad (5.176)
$$

where J_i, $i = 1, \cdots, 6$ are feedback operators. We introduce feedback laws by

$$
\left(J_1(\zeta), \cdots, J_6(\zeta) \right) = -F\check{\zeta} \quad \text{on} \quad (0, \infty) \times \Gamma_2 \qquad (5.177)
$$

where F are symmetric, nonnegative 6×6 matrices for $x \in \Gamma_2$ and $\check{\zeta} = (\langle W_1, \nu \rangle, \langle W_1, \tau \rangle, \langle W_2, \nu \rangle, \langle W_2, \tau \rangle, w_1, w_2)$ for $\zeta = (W_1, W_2, w_1, w_2)$.

For convenience, set

$$
\alpha(x) = \alpha_i, \quad \beta(x) = \beta_i, \quad \mathcal{A}_x = \mathcal{A}_i, \quad \text{and} \quad \zeta = \zeta_i
$$

as $x \in \Omega_i$ for $i = 1, 2$. Define the total energy of the closed loop system (5.173)-(5.177) by

$$
2E(t) = \int_\Omega [|\zeta_t|^2 + \alpha(x) B_x(\zeta, \zeta)]\, dx \qquad (5.178)
$$

where $B_x(\cdot, \cdot) = B_i(\cdot, \cdot)$ for $x \in \Omega_i$ where $B_i(\cdot, \cdot) = B_0(\cdot, \cdot)$ is given by the formula (5.25) in which E and μ are replaced by E_i and μ_i for $i = 1$ and 2, respectively. It follows from the Green formula (5.33) on Ω_i that

$$\int_{\Omega_i} B_i(\zeta, \eta)\, dx = \int_{\Omega_i} \langle \mathcal{A}_i \zeta, \eta \rangle\, dx + \int_{\Gamma_i \cup \Gamma_3} \partial(\mathcal{A}_i \zeta, \eta)\, d\Gamma \tag{5.179}$$

for $i = 1$ and 2.

By the formulas (5.172)-(5.179), we obtain

$$\frac{d}{dt} E(t) = \sum_{i=1}^{2} \int_{\Omega_i} [\langle \zeta_{it}, \zeta_{itt} \rangle + \alpha_i B_i(\zeta_i, \zeta_{it})]\, dx$$

$$= \sum_{i=1}^{2} \alpha_i \int_{\Gamma_i \cup \Gamma_3} \partial(\mathcal{A}_i \zeta_i, \zeta_{it})\, d\Gamma = \alpha_2 \int_{\Gamma_2} \partial(\mathcal{A}_2 \zeta_2, \zeta_{2t})\, d\Gamma$$

$$= -\alpha_2 \int_{\Gamma_2} \langle F \breve{\zeta}_{2t}, \breve{\zeta}_{2t} \rangle_{\mathbb{R}^6}\, d\Gamma \leq 0$$

which implies that the closed-loop system (5.173)-(5.177) is dissipative.

By similar arguments as in Section 5.3 , we obtain the existence, uniqueness and regularity of solutions to the closed-loop system (5.173)-(5.177) as follows. Set

$$\mathcal{L}^2(\Omega) = (L^2(\Omega, \Lambda))^2 \times (L^2(\Omega))^2, \quad \mathcal{H}^1_{\Gamma_1}(\Omega) = (H^1_{\Gamma_1}(\Omega, \Lambda))^2 \times (H^1_{\Gamma_1}(\Omega))^2,$$

$$\mathcal{W} = \{ \zeta \,|\, \zeta \in \mathcal{H}^1_{\Gamma_1}(\Omega),\ \zeta_i = \zeta|_{\Omega_i} \in (H^2_{\Gamma_1}(\Omega_i, \Lambda))^2 \times (H^2_{\Gamma_1}(\Omega_i))^2,$$
$$i = 1, 2;\ \zeta \text{ satisfies the transmission conditions } (5.172) \}.$$

Theorem 5.16 *For* $(\zeta^0, \zeta^1) \in \mathcal{H}^1_{\Gamma_1}(\Omega) \times \mathcal{L}^2(\Omega)$, *the closed-loop system* (5.173) $-$ (5.177) *admits a unique solution with* $\zeta \in C([0, \infty), \mathcal{H}^1_{\Gamma_1}(\Omega))$ *and* $\zeta_t \in C([0, \infty), \mathcal{L}^2(\Omega))$. *Moreover, if the initial data satisfy* $\zeta^0 \in \mathcal{W}$, $\zeta_1 \in \mathcal{H}^1_{\Gamma_1}(\Omega)$, *and* $\partial(\mathcal{A}_2 \zeta_2^0, \eta) = -\langle F \breve{\zeta}_2^0, \breve{\eta} \rangle_{\mathbb{R}^6}$ *on* Γ_2 *for all* $\eta \in \mathcal{H}^1_{\Gamma_1}(\Omega)$, *then* $\zeta \in C([0, \infty); \mathcal{W} \cap \mathcal{H}^1_{\Gamma_1}(\Omega))$.

Next, let us clarify some assumptions for the stabilization of the system (5.173)-(5.177), which will be based on the geometric properties and the materials of the Naghdi shell.

Assumption (H1) Let V be *an escape vector field V for the Naghdi shell.* Let $\vartheta(x)$ and $\iota(x)$ be defined by the formulas (5.82) and (5.99), respectively. Moreover, suppose that ϑ and ι meet the inequality

$$2 \min_{x \in \Omega} \vartheta(x) > \lambda_0 (1 + 2\beta_0) \max_{x \in \Omega} |\iota(x)| \tag{5.180}$$

where $\lambda_0 \geq 1$ is the elliptic number, given by the inequality (5.81) and $\beta_0 = \max\{\beta_1, \beta_2\}$. Set

$$\sigma_0 = \max_{x \in \Omega} |V|, \quad \sigma_1 = \min_{x \in \Omega} \vartheta(x) - \lambda_0 (1 + 2\beta_0) \max_{x \in \Omega} |\iota(x)|/2.$$

Assumption (H2) $\overline{\Gamma}_1 \neq \emptyset$ and $\langle V, \nu \rangle \leq 0$ for $x \in \Gamma_1$.

Assumption (H3) $\alpha_1 \neq \alpha_2$, $(\alpha_2 - \alpha_1)\langle V, \nu_1 \rangle \leq 0$, and $(\alpha_2\beta_2 - \alpha_1\beta_1)\langle V, \nu_1 \rangle \leq 0$ for $x \in \Gamma_3$.

Assumption (H4) $F \in C^1(\Gamma_2)$ is positive on $\overline{\Gamma}_2$.

For existence of an escape vector field with the relations (5.82) and (5.180), see Section 4.5 of Chapter 4.

Then the main results of this section are

Theorem 5.17 *Let the assumptions* (**H1**) $-$ (**H4**) *hold. Then there are* $c_1 > 0$ *and* $c_2 > 0$ *such that*

$$E(t) \leq c_1 e^{-c_2 t} \quad \forall \, t \geq 0. \tag{5.181}$$

The proof of this theorem will be given at the end of this section. For this stabilization, we need the following observability estimate, Theorem 5.18 and the boundary trace estimate, Theorem 5.19.

Using Theorem 5.7 on Ω_i with $\mathcal{A}$ replaced by $\alpha_i \mathcal{A}_i$ and $B(\cdot, \cdot)$ replaced by $\alpha_i B_i(\cdot, \cdot)$, respectively for $i = 1, 2$, and the transmission condition (5.172), we obtain

Theorem 5.18 *Let the assumption* (**H1**) *hold. Suppose that* $\zeta = (W_1, W_2, w_1, w_2)$ *solve the problem*

$$\zeta_{tt} + \alpha(x)\mathcal{A}_x \zeta = 0 \tag{5.182}$$

such that the right hand side of the estimate (5.183) *below is well defined. Let* $T > 0$ *be given. Then there is* $c > 0$ *independent of* ζ *such that*

$$2\sigma_1 \int_0^T E(t) \, dt \; \leq \int_\Sigma \text{SB}\,|_\Sigma d\Sigma + \int_{\Sigma_3} \text{SB}\,|_{\Sigma_3} d\Sigma_3$$
$$+ \sigma_0 \lambda_0 [E(0) + E(T)] + L(\zeta) \tag{5.183}$$

where

$$\text{SB}\,|_\Sigma = [|\zeta_t|^2 - \alpha(x)B_x(\zeta, \zeta)]\langle V, \nu \rangle + \alpha(x)\partial(\mathcal{A}_x\zeta, 2m(\zeta) + \varrho\zeta),$$
$$m(\zeta) = (D_V W_1, D_V W_2, v(w_1), V(w_2)), \quad \varrho = 2\vartheta - \sigma_1, \tag{5.184}$$

$$\text{SB}_{\Sigma_3} = 2[\alpha_1 \partial(\mathcal{A}_1 \zeta_1, m(\zeta_1)) + \alpha_2 \partial(\mathcal{A}_2 \zeta_2, m(\zeta_2))]$$
$$+ [\alpha_2 B_2(\zeta_2, \zeta_2) - \alpha_1 B_1(\zeta_1, \zeta_1)]\langle V, \nu_1 \rangle. \tag{5.185}$$

In addition, by similar methods in Theorem 5.14, we have

Theorem 5.19 *Let* ζ *solve the system* (5.173) $-$ (5.176). *Then for* $T/2 > \alpha > 0$ *there is* $c_{T\alpha} > 0$ *such that*

$$\sum_{i=1}^2 \int_\alpha^{T-\alpha} \int_{\Gamma_2} (|DW_{2i}|^2 + |Dw_{2i}|^2) \, d\Sigma \leq c_{T\alpha} \int_{\Sigma_2} |\zeta_t|^2 \, d\Sigma + L(\zeta). \tag{5.186}$$

The following lemma is a key to the transmission problem.

Lemma 5.5 *The transmission conditions (5.172) and the material assumption* (**H3**) *imply that*

$$\int_{\Sigma_3} SB\,|_{\Sigma_3} \leq L(\zeta).$$

Proof. We compute on the division Γ_3.

The transmission conditions (5.172) yield that on Γ_3

$$\begin{cases} D_{\tau_1}W_{1i} + D_{\tau_2}W_{2i} = 0, \\ \dfrac{\partial w_{1i}}{\partial \tau_1} + \dfrac{\partial w_{2i}}{\partial \tau_2} = 0 \qquad i = 1,\,2, \end{cases} \tag{5.187}$$

$$\begin{cases} \alpha_1(1+\beta_1)DW_{1i}(\nu_1,\nu_1) + \alpha_1\beta_1 DW_{1i}(\tau_1,\tau_1) \\ \quad = \alpha_2(1+\beta_2)DW_{2i}(\nu_2,\nu_2) + \alpha_2\beta_2 DW_{2i}(\tau_2,\tau_2) + \mathrm{lo}\,(\zeta), \\ \alpha_1\Upsilon(\zeta_1)(\nu_1,\tau_1) = \alpha_2\Upsilon(\zeta_2)(\nu_2,\tau_2), \\ \alpha_1\mathcal{X}_0(\zeta_1)(\nu_1,\tau_1) = \alpha_2\mathcal{X}_0(\zeta_2)(\nu_2,\tau_2) + \mathrm{lo}\,(\zeta), \\ \alpha_1\dfrac{\partial w_{1i}}{\partial \nu_1} + \alpha_2\dfrac{\partial w_{2i}}{\partial \nu_2} = 0, \\ i = 1,\,2. \end{cases} \tag{5.188}$$

Moreover, the relations (5.187) and (5.188) reach

$$\begin{cases} \alpha_1(1+\beta_1)DW_{1i}(\nu_1,\nu_1) = \alpha_2(1+\beta_2)DW_{2i}(\nu_2,\nu_2) \\ \quad +(\alpha_2\beta_2 - \alpha_1\beta_1)DW_{1i}(\tau_1,\tau_1) + \mathrm{lo}\,(\zeta), \\ \alpha_1 DW_{1i}(\tau,\nu_1) = \alpha_2 DW_{2i}(\tau_2,\nu_2) \\ \quad +(\alpha_2 - \alpha_1)DW_{1i}(\nu_1,\tau_1) + \mathrm{lo}\,(\zeta), \\ i = 1,\,2. \end{cases} \tag{5.189}$$

Note that $m(\zeta_i) = D_V\zeta_i = \langle V,\nu_i\rangle D_{\nu_i}\zeta_i + \langle V,\tau_i\rangle D_{\tau_i}\zeta_i$ for $i = 1$ and 2. Then the transmission conditions (5.172) imply that

$$\alpha_1\partial(\mathcal{A}_1\zeta_1, m(\zeta_1)) + \alpha_2\partial(\mathcal{A}_2\zeta_2, m(\zeta_2))$$
$$= [\alpha_1\partial(\mathcal{A}_1\zeta_1, D_{\nu_1}\zeta_1) - \alpha_2\partial(\mathcal{A}_2\zeta_2, D_{\nu_2}\zeta_2)]\langle V,\nu_1\rangle.$$

Then

$$\begin{aligned} SB\,|_{\Sigma_3} &= \{2\alpha_1\partial(\mathcal{A}_1\zeta_1, D_{\nu_1}\zeta_1) - 2\alpha_2\partial(\mathcal{A}_2\zeta_2, D_{\nu_2}\zeta_2) \\ &\quad +\alpha_2 B_2(\zeta_2,\zeta_2) - \alpha_1 B_1(\zeta_1,\zeta_1)\}\langle V,\nu_1\rangle. \end{aligned} \tag{5.190}$$

We shall compute the right hand side of (5.190).

The definitions of $\partial(\mathcal{A}_i\zeta_i, D_{\nu_i}\zeta_i)$ and $B_i(\zeta_i,\zeta_i)$ read that

$$\begin{aligned} \partial(\mathcal{A}_i\zeta_i, D_{\nu_i}\zeta_i) &= \langle B_{i1}(\zeta_i), D_{\nu_i}W_{i1}\rangle + \gamma\langle B_{i2}(\zeta_i), D_{\nu_i}W_{i2}\rangle \\ &\quad +2\langle\varphi_0(\zeta_i),\nu_i\rangle\dfrac{\partial w_{i1}}{\partial \nu_i} + \gamma(\dfrac{\partial w_{i2}}{\partial \nu_i})^2 \end{aligned} \tag{5.191}$$

and

$$B_i(\zeta_i, \zeta_i) = 2|\Upsilon(\zeta_i)|^2 + 2\beta_i[\text{tr } \Upsilon(\zeta_i)]^2 + 2\gamma|\mathcal{X}_0(\zeta_i)|^2 + 2\gamma\beta_i[\text{tr } \mathcal{X}_0(\zeta_i)]^2$$
$$+\gamma|Dw_{i2}|^2 + 4|\varphi_0(\zeta_i)|^2 + \text{lo}(\zeta), \tag{5.192}$$

respectively, on Σ_3 for $i = 1, 2$.

From the formulas (5.171), we obtain

$$2\alpha_i\langle B_{i1}(\zeta_i), D_{\nu_i}W_{i1}\rangle - 2\alpha_i\{|\Upsilon(\zeta_i)|^2 + \beta_i[\text{tr } \Upsilon(\zeta_i)]^2\}$$
$$= 2\alpha_i(1 + \beta_i)[DW_{i1}(\nu_i, \nu_i)]^2 - 2\alpha_i(1 + \beta_i)[DW_{i1}(\tau_i, \tau_i)]^2$$
$$+4\alpha_i[\Upsilon(\zeta_i)(\nu_i, \tau_i)]^2 - 4\alpha_i\Upsilon(\zeta_i)(\nu_i, \tau_i)DW_{i1}(\nu_i, \tau_i). \tag{5.193}$$

Let

$$q_i = \alpha_i(1 + \beta_i) \quad \text{for} \quad i = 1, 2, \quad p = q_2 - q_1.$$

It follows from the formulas (5.193) and (5.187)-(5.189) that

$$2\alpha_1\langle B_{11}(\zeta_1), D_{\nu_1}W_{11}\rangle - 2\alpha_1\{|\Upsilon(\zeta_1)|^2 + \beta_1[\text{tr } \Upsilon(\zeta_1)]^2\}$$
$$-2\alpha_2\langle B_{21}(\zeta_2), D_{\nu_2}W_{21}\rangle + 2\alpha_2\{|\Upsilon(\zeta_2)|^2 + \beta_2[\text{tr } \Upsilon(\zeta_2)]^2\}$$
$$= 2q_1[DW_{11}(\nu_1, \nu_1)]^2 - 2q_2[DW_{21}(\nu_2, \nu_2)]^2 + 2p[DW_{11}(\tau_1, \tau_1)]^2$$
$$+\frac{4\alpha_1}{\alpha_2}(\alpha_2 - \alpha_1)[\Upsilon(\zeta_1)(\nu_1, \tau_1)]^2 + \text{lo}(\zeta)$$
$$= \frac{2q_1p}{q_2}[DW_{11}(\nu_1, \nu_1)]^2 + \frac{2}{q_2}[q_2p - (\alpha_1\beta_1 - \alpha_2\beta_2)^2][DW_{11}(\tau_1, \tau_1)]^2$$
$$+\frac{4q_1}{q_2}(\alpha_2\beta_2 - \alpha_1\beta_1)DW_{11}(\nu_1, \nu_1)DW_{11}(\tau_1, \tau_1)$$
$$+\frac{4\alpha_1}{\alpha_2}(\alpha_2 - \alpha_1)[\Upsilon(\zeta_1)(\nu_1, \tau_1)]^2 + \text{lo}(\zeta). \tag{5.194}$$

Since

$$q_1p = q_1(\alpha_2 - \alpha_1) + q_1(\alpha_2\beta_2 - \alpha_1\beta_1),$$
$$q_2p - (\alpha_1\beta_1 - \alpha_2\beta_2)^2 = q_2(\alpha_2 - \alpha_1) + q_1(\alpha_2\beta_2 - \alpha_1\beta_1),$$

we have

the right hand side of (5.194)
$$= \frac{2}{q_2}\Big\{q_1(\alpha_2 - \alpha_1)[DW_{11}(\nu_1, \nu_1)]^2 + q_2(\alpha_2 - \alpha_1)[DW_{11}(\tau_1, \tau_1)]^2$$
$$+q_1(\alpha_2\beta_2 - \alpha_1\beta_1)[DW_{11}(\nu_1, \nu_1) + DW_{11}(\tau_1, \tau_1)]^2\Big\}$$
$$+\frac{4\alpha_1}{\alpha_2}(\alpha_2 - \alpha_1)[\Upsilon(\zeta_1)(\nu_1, \tau_1)]^2 + \text{lo}(\zeta). \tag{5.195}$$

From the relations (5.194) and (5.195) and the assumption **(H3)**, we obtain

$$\Big\{2\alpha_1\langle B_{11}(\zeta_1), D_{\nu_1}W_{11}\rangle - 2\alpha_1\{|\Upsilon(\zeta_1)|^2 + \beta_1[\text{tr } \Upsilon(\zeta_1)]^2\}$$
$$-2\alpha_2\langle B_{21}(\zeta_2), D_{\nu_2}W_{21}\rangle + 2\alpha_2\{|\Upsilon(\zeta_2)|^2 + \beta_2[\text{tr } \Upsilon(\zeta_2)]^2\}\Big\}\langle V, \nu_1\rangle$$
$$\le \text{lo}(\zeta). \tag{5.196}$$

A similar computation gives

$$\left\{2\alpha_1\langle B_{12}(\zeta_1), D_{\nu_1}W_{21}\rangle - 2\alpha_1\{|\mathcal{X}_0(\zeta_1)|^2 + \beta_1[\text{tr }\mathcal{X}_0(\zeta_1)]^2\}\right.$$

$$\left. -2\alpha_2\langle B_{22}(\zeta_2), D_{\nu_2}W_{22}\rangle + 2\alpha_2\{|\mathcal{X}_0(\zeta_2)|^2 + \beta_2[\text{tr }\mathcal{X}_0(\zeta_2)]^2\}\right\}\langle V, \nu_1\rangle$$

$$\leq \text{lo}(\zeta). \tag{5.197}$$

Moreover, we have

$$4\alpha_1\langle\varphi_0(\zeta_1), \nu_1\rangle\frac{\partial w_{11}}{\partial\nu_1} - 4\alpha_2\langle\varphi_0(\zeta_2), \nu_2\rangle\frac{\partial w_{21}}{\partial\nu_2}$$

$$+4\alpha_2|\varphi_0(\zeta_2)|^2 - 4\alpha_1|\varphi_0(\zeta_1)|^2 = \text{lo}(\zeta) \tag{5.198}$$

and

$$2\alpha_1\left(\frac{\partial w_{12}}{\partial\nu_1}\right)^2 - 2\alpha_2\left(\frac{\partial w_{22}}{\partial\nu_2}\right)^2 + \alpha_2|Dw_{22}|^2 - \alpha_1|Dw_{12}|^2$$

$$= \alpha_1\left(\frac{\partial w_{12}}{\partial\nu_1}\right)^2 - \alpha_1\left(\frac{\partial w_{12}}{\partial\tau_1}\right)^2 - \alpha_1\left(\frac{\partial w_{22}}{\partial\nu_2}\right)^2 + \alpha_2\left(\frac{\partial w_{22}}{\partial\tau_2}\right)^2$$

$$= \frac{\alpha_1}{\alpha_2}(\alpha_2 - \alpha_1)\left(\frac{\partial w_{12}}{\partial\nu_1}\right)^2 + (\alpha_2 - \alpha_1)\left(\frac{\partial w_{12}}{\partial\tau_1}\right)^2$$

which yields

$$\left\{2\alpha_1\left(\frac{\partial w_{12}}{\partial\nu_1}\right)^2 - 2\alpha_2\left(\frac{\partial w_{22}}{\partial\nu_2}\right)^2 + \alpha_2|Dw_{22}|^2 - \alpha_1|Dw_{12}|^2\right\}\langle V, \nu_1\rangle \leq 0. \tag{5.199}$$

Finally, we obtain the estimate

$$\int_{\Sigma_3} \text{SB}\, d\Sigma \leq \text{lo}(\zeta)$$

by combining (5.190), (5.191), (5.192), (5.196), (5.197), (5.198), and (5.199). □

Proof of Theorem 5.17 Using Lemma 5.5 in Theorem 5.18 yields

$$2\sigma_1\int_0^T E(t)\, dt \leq \int_\Sigma (\text{SB}_1 + \text{SB}_2)d\Sigma$$

$$+\sigma_0\lambda_0[E(0) + E(T)] + \text{lo}(\zeta) \tag{5.200}$$

where

$$\text{SB}_1 = [|\zeta_t|^2 - \alpha(x)B_x(\zeta, \zeta)]\langle V, \nu\rangle, \tag{5.201}$$

$$\text{SB}_2 = \alpha(x)\partial(\mathcal{A}_x\zeta, 2m(\zeta) + \varrho\zeta). \tag{5.202}$$

By the boundary condition (5.174) and the identity (5.111), we have

$$\text{SB}_1|_{\Sigma_1} = -\alpha_1 B_1(\zeta, \zeta)\langle V, \nu\rangle$$

and

$$SB_2|_{\Sigma_1} = 2\alpha_1 B_1(\zeta, \zeta)\langle V, \nu \rangle.$$

It follows from the assumption **(H2)** that

$$\int_{\Sigma_1} (SB_1 + SB_2)d\Sigma = \int_{\Sigma_1} \alpha_1 B_1(\zeta, \zeta)\langle V, \nu \rangle d\Sigma \leq 0. \tag{5.203}$$

Next, set

$$a_1(\zeta, \eta) = \int_{\Gamma_2} \alpha_2 \langle F\check{\zeta}, \check{\eta} \rangle_{\mathbb{R}^6} \, d\Gamma.$$

Then the assumption **(H4)** implies that there are $s_1, s_2 > 0$ such that for $\eta = (U_1, U_2, u_1, u_2)$

$$s_1 \int_{\Gamma_2} |\eta|^2 \, d\Gamma \leq a_1(\eta, \eta) \leq s_2 \int_{\Gamma_2} |\eta|^2 \, d\Gamma. \tag{5.204}$$

Noting that the relation $E'(t) = -a_1(\zeta_{2t}, \zeta_{2t})$, we have, for any $T \geq \alpha \geq 0$,

$$E(\alpha) = E(T) + \int_{\alpha}^{T} a_1(\zeta_{2t}, \zeta_{2t}) \, dt. \tag{5.205}$$

From the feedback laws (5.176) and (5.177), we obtain

$$\int_{\Sigma_2} SB_2 d\Sigma = -\int_0^T a_1(\zeta_{2t}, 2m(\zeta_2) + \varrho\zeta) \, dt$$

$$\leq c \int_{\Sigma_2} [|\zeta_{2t}|^2 + \sum_{i=1}^{2}(|DW_{2i}|^2 + |Dw_{2i}|^2)] \, d\Sigma + L(\zeta). \tag{5.206}$$

Moreover, clearly, the following estimate is true,

$$\int_{\Sigma_2} SB_1 d\Sigma \leq c \int_{\Sigma_2} [|\zeta_{2t}|^2 + \sum_{i=1}^{2}(|DW_{2i}|^2 + |Dw_{2i}|^2)] \, d\Sigma + L(\zeta). \tag{5.207}$$

Substituting the relations (5.203), (5.206), and (5.207) in the inequality (5.200) yields

$$2\sigma_1 \int_0^T E(t) \, dt \leq c \int_{\Sigma_2} [|\zeta_{2t}|^2 + \sum_{i=1}^{2}(|DW_{2i}|^2 + |Dw_{2i}|^2)] \, d\Sigma$$

$$+ \sigma_0 \lambda_0 [E(0) + E(T)] + L(\zeta). \tag{5.208}$$

Now, change the integral domain Σ_2 into $[\alpha, T - \alpha] \times \Gamma_2$ in both sides of the inequalities (5.208) and use the inequality (5.186) and the relation (5.205) to get

$$2\sigma_1 \int_{\alpha}^{T-\alpha} E(t) \, dt \leq c\left\{ E(\alpha) + E(T - \alpha) + \int_{\Sigma_2} |\zeta_{2t}|^2 \, d\Sigma \right\} + L(\zeta). \tag{5.209}$$

By the relation (5.205), we find, for any $T \geq s \geq 0$,

$$E(T) - c \int_{\Sigma_2} |\zeta_{2t}|^2 \, d\Sigma \leq E(s) \leq E(T) + c \int_{\Sigma_2} |\zeta_{2t}|^2 \, d\Sigma. \tag{5.210}$$

Using the inequality (5.210) in the inequality (5.209) and the uniqueness in Proposition 5.1, we obtain, for some T suitably large,

$$E(T) \leq c_T \int_{\Sigma_2} |\zeta_{2t}|^2 \, d\Sigma. \tag{5.211}$$

Finally, using the left hand side of the inequality (5.204) in the estimate (5.211), we have

$$E(T) \leq c_T \int_0^T a_1(\zeta_{2t}, \zeta_{2t}) \, dt = c_T(E(0) - E(T)),$$

that is,

$$E(T) \leq \frac{c_T}{1 + c_T} E(0). \tag{5.212}$$

Then the estimate (5.181) follows from the inequality (5.212). $\qquad\square$

Exercises

5.1 *Let the Sobolev space $\mathcal{H}^1(\Omega)$ be given by (5.27) and let the bilinear form $\mathcal{B}(\cdot, \cdot)$ be given by (5.76). Prove that there is $c > 0$ such that*

$$\mathcal{B}(\zeta, \zeta) \leq c\|\zeta\|_{\mathcal{H}^1(\Omega)}^2 \quad \text{for all} \quad \zeta \in \mathcal{H}^1(\Omega).$$

5.2 *Let the operator $\mathcal{A}$ of the Naghdi shell be given by the formula (5.72). Then for $\zeta = (W_1, W_2, w_1, w_2)$,*

$$\mathcal{A}\zeta = (\boldsymbol{\Delta}_\beta W_1, \boldsymbol{\Delta}_\beta W_2, \Delta w_1, \Delta w_2) + \mathrm{lo}\,(\zeta). \tag{5.213}$$

5.3 *Write out the proof of Lemma 5.3 in detail by following the proof of Lemma 4.11 in Chapter 4.*

5.4 *Prove Proposition 5.1.*

5.5 *Prove Theorem 5.18.*

5.6 *Prove Theorem 5.19.*

5.7 *Prove that the transmission conditions (5.172) imply the relations (5.188) and (5.189).*

5.5　Notes and References

Sections 5.1 and 5.2 are from [213]; Section 5.3 is from [25]; Section 5.4 is from [35].

The displacement-strain relations (5.14) come from [157] which is the foundation stone of the Naghdi shell. The Green formulas in Theorem 5.1 and the equations of motion in Theorem 5.5 are given in the form of coordinates free. Modeling of the Naghdi shell in the form of coordinates free is the key for a series of advances on this model.

In this direction, [35] is an excellent work. One of its main contributions clarifies the transmission assumption **(H3)**, by a complicated computation as in Lemma 5.5, which guarantees the transmission stabilization only by feedbacks on the portion Γ_2 of the boundary.

A recent advance on the Naghdi model is made by [30] which is not included in this chapter. [30] presents some well-posedness results and an explicit, simple formula of the feedthrough operator ([202], [203]), respectively, for the input-output system of the Naghdi shell. The well-posedness results in [30] can guarantee the equivalence between exact controllability and stabilization by Russell's principle. Moreover, the feedthrough operator of the Naghdi model in [30] shows that the speeds of propagation of the Naghdi shell are 1 and $1/\sqrt{2(1+\beta)}$ along the tangential direction and the normal, respectively, where β is given in (5.19).

Chapter 6

Koiter's Shells. Modeling and Controllability

The Koiter shell was introduced by [98]. In this model the second fundamental form of the middle surface goes into the change of curvature tensor which makes its analysis much more complicated than that of the shallow shell. We are then faced with new tasks: to derive the dynamic system starting from the strain tensor and the curvature tensor and to set up appropriate geometric conditions for controllability/stabilization. Multiplier identities are established and exact controllability from boundary is presented.

6.1 Equations of Equilibria. Equations of Motion

Let us assume that the middle surface of the shell occupies a bounded region Ω of the surface M in $\mathbb{R}^3$. The shell, a body in $\mathbb{R}^3$, is defined by

$$S = \{\, p \mid p = x + zN(x),\ x \in \Omega,\ -h/2 < z < h/2 \,\}$$

where h is the thickness of the shell, small.

As usual, denote by $\zeta(x)$ the displacement vector of point x of the middle surface. We decompose the displacement vector $\zeta(x)$ into a direct sum

$$\zeta(x) = W(x) + wN(x) \qquad W(x) \in M_x, \quad x \in \Omega,$$

that is, W and w are the components of ζ on the tangential plane and on the normal of the undeformed middle surface Ω, respectively. For convenience, we let $\zeta = (W, w)$. For the Koiter model, *the linearized strain tensor and the change of curvature tensor of the middle surface* Ω are given by

$$\Upsilon(\zeta) = \frac{1}{2}(DW + D^*W) + w\Pi \tag{6.1}$$

and
$$\rho(\zeta) = D^2 w - \Pi(\cdot, D.W) - \Pi(D.W, \cdot) - \mathrm{i}(W)D\Pi - wc, \qquad (6.2)$$

respectively, where Π is the second fundamental form, c the third fundamental form of the surface M given by (5.3), $D^2 w$ the Hessian of w, and $\cdot$ denotes the position of the variable. In (6.2), $\mathrm{i}(W)D\Pi$ is *the interior product of the tensor field $D\Pi$ by the vector field W*. The relations (6.1) and (6.2) are given by [98], also see [12] or [134].

Remark 6.1 *If we describe the strain tensors* (6.1) *and* (6.2) *by a coordinate, they look as follows. Let the middle surface be given by a coordinate*

$$\varphi(x_1, x_2) = (\varphi_1(x_1, x_2), \varphi_2(x_1, x_2), \varphi_3(x_1, x_2)) \quad for \quad (x_1, x_2) \in \mathbb{R}^2.$$

Set
$$\mathbf{a}_\alpha = \left(\frac{\partial \varphi_1}{\partial x_\alpha}, \frac{\partial \varphi_2}{\partial x_\alpha}, \frac{\partial \varphi_3}{\partial x_\alpha}\right), \quad 1 \le \alpha \le 2.$$

Define $\mathbf{a}^\alpha$ by
$$\langle \mathbf{a}_\beta, \mathbf{a}^\alpha \rangle = \delta_{\alpha\beta}$$

and let
$$W = w_1 \mathbf{a}^1 + w_2 \mathbf{a}^2.$$

Then in the classical notation the tensors (6.1) *and* (6.2) *become*

$$\Upsilon_{\alpha\beta} = \frac{1}{2}(w_{\alpha|\beta} + w_{\beta|\alpha}) - b_{\alpha\beta}w,$$

$$\rho_{\alpha\beta} = w_{|\alpha\beta} - c_{\alpha\beta}w + b_\alpha^\lambda w_{\lambda|\beta} + b_\beta^\lambda w_{\lambda|\alpha} + b_{\alpha|\beta}^\lambda w_\lambda$$

where
$$b_{\alpha\beta} = -\Pi(\mathbf{a}_\alpha, \mathbf{a}_\beta), \quad c_{\alpha\beta} = c(\mathbf{a}_\alpha, \mathbf{a}_\beta), \quad \partial_\alpha N = b_\alpha^\lambda \mathbf{a}_\lambda,$$

$$w_{|\alpha\beta} = D^2 w(\mathbf{a}_\alpha, \mathbf{a}_\beta), \quad w_{\alpha|\beta} = \langle D_{\mathbf{a}_\beta} W, \mathbf{a}_\alpha \rangle, \quad b_{\alpha|\beta}^\lambda = -D\Pi(\mathbf{a}^\lambda, \mathbf{a}_\alpha, \mathbf{a}_\beta).$$

The shell strain energy associated to the displacement vector field ζ of the middle surface Ω can be written as

$$\mathcal{B}_1(\zeta, \zeta) = \frac{Eh}{1 - \mu^2} \int_\Omega B(\zeta, \zeta)\, dx \qquad (6.3)$$

where
$$B(\zeta, \zeta) = a(\Upsilon(\zeta), \Upsilon(\zeta)) + \gamma a(\rho(\zeta), \rho(\zeta)), \quad \gamma = h^2/12, \qquad (6.4)$$

$$a(T, T) = (1 - \mu)\langle T, T \rangle + \mu(\operatorname{tr} T)^2, \quad T \in T^2(\overline{\Omega}),$$

for $x \in \Omega$, where E, μ respectively denote *Young's modulus* and *Poisson's coefficient* of the material. The expression (6.3) is an approximation to the shell strain energy. Its derivation from the three-dimensional elasticity theory is carried out by integration on the thickness of the shell, following the methods of asymptotic expressions. For a formal utilization of these methods in shell theory, we refer to [95], [157], [183], and the bibliographies of these papers.

For an isotropic material, with the expression (6.3), we are able to associate the following symmetric bilinear form, directly defined on the middle surface Ω:

$$\mathcal{B}(\zeta, \eta) = \int_\Omega B(\zeta, \eta) \, dx \tag{6.5}$$

where $\eta = (U, u)$.

The Ellipticity of the Strain Energy for the Koiter Shell The ellipticity is an indispensable result for all control problems. To establish it, we need

Lemma 6.1 *Let $\Gamma_0 \subset \Gamma$ have a positive length and let a displacement vector field $\zeta = (W, w)$ be such that*

$$W|_{\Gamma_0} = 0, \quad w|_{\Gamma_0} = \frac{\partial w}{\partial \nu}\Big|_{\Gamma_0} = 0. \tag{6.6}$$

If $\rho(\zeta) = \Upsilon(\zeta) = 0$ for all $x \in \Omega$, then $\zeta = 0$ for $x \in \Omega$.

Proof. Let $U = DW + D^*W$. Then $\Upsilon = 0$ gives $U = -2w\Pi$. It is easy to check that

$$DU = -2\Pi \otimes Dw - 2wD\Pi. \tag{6.7}$$

Let $x \in \Omega$ be given. Let E_1, E_2 be a frame field normal at x such that $D_{E_i} E_j(x) = 0$ for $1 \le i, j \le 2$. By the formula (4.60) in Chapter 4, we have

$$\sum_{i=1}^{2} D^2 W(E_j, E_i, E_i) = -2\Pi(Dw, E_j) + \operatorname{tr} \Pi \langle Dw, E_j \rangle$$
$$-\kappa \langle W, E_j \rangle - w \langle DH, E_j \rangle \quad \text{at} \quad x \tag{6.8}$$

where κ is the Gauss curvature. It follows from the formulas (6.8) and (1.114) that at x

$$\boldsymbol{\Delta} W = -\sum_{i=1}^{2} D_{E_i E_i}^2 W + \sum_{ij=1}^{2} \mathbf{R}(E_i, E_j, W, E_j) E_i$$
$$= -\sum_{ij=1}^{2} D^2 W(E_j, E_i, E_i) E_j + \kappa W$$
$$= 2 \, \mathrm{i}\,(Dw)\Pi - HDw - wDH + 2\kappa W. \tag{6.9}$$

On the other hand, by (6.2), the assumption $\rho(\zeta) = 0$ implies that

$$\Delta w = \operatorname{tr} D^2 w = 2 \operatorname{tr} \Pi(\cdot, D.W) + \operatorname{tr} \mathrm{i}\,(W) D\Pi + w \operatorname{tr} \mathrm{c} \tag{6.10}$$

for $x \in \Omega$.

Moreover, the conditions in (6.6) and the relation $DW + D^*W = -2w\Pi$ yield that $DW|_{\Gamma_0} = 0$.

Finally, applying the uniqueness theorem of [170] to the system (6.9) and (6.10), we obtain that $\zeta = 0$. □

Theorem 6.1 *Let $\Gamma_0 \subset \Gamma$ have a positive length. Then there is a constant $c > 0$ such that, for all $\zeta \in H^1_{\Gamma_0}(\Omega, \Lambda) \times H^2_{\Gamma_0}(\Omega)$,*

$$\mathcal{B}(\zeta, \zeta) \geq c\|\zeta\|^2_{H^1_{\Gamma_0}(\Omega,\Lambda) \times H^2_{\Gamma_0}(\Omega)}. \tag{6.11}$$

Proof. By Lemma 4.5 and the inequality (4.78) in Chapter 4, we have constants $c_1, c_2 > 0$ such that

$$\int_\Omega a(\Upsilon(\zeta), \Upsilon(\zeta))\, dx \geq c_1\|W\|^2_{H^1_{\Gamma_0}(\Omega,\Lambda)} - c_2\|w\|^2_{L^2(\Omega)}. \tag{6.12}$$

In addition, from (6.2), there is $c_3 > 0$ such that

$$|\rho(\zeta)| \geq |D^2 w|^2 - c_3(|DW|^2 + |w|^2) \quad \text{for} \quad x \in \Omega. \tag{6.13}$$

Let $0 < \varepsilon < 1$ be such that $\varepsilon(1-\mu)c_3 < c_1$. Then

$$\int_\Omega a(\rho(\zeta), \rho(\zeta))\, dx \geq \varepsilon(1-\mu) \int_\Omega |\rho(\zeta)|^2\, dx$$
$$\geq \varepsilon(1-\mu)\|w\|^2_{H^2_{\Gamma_0}(\Omega)} - \varepsilon(1-\mu)c_3\|W\|^2_{H^1_{\Gamma_0}(\Omega,\Lambda)}$$
$$- [\varepsilon(1-\mu) + c_3]\|w\|^2_{H^1(\Omega)}. \tag{6.14}$$

It follows from the inequalities (6.12) and (6.14) that there are constants $c_4, c_5 > 0$ such that

$$\mathcal{B}(\zeta, \zeta) + c_5\|w\|^2_{H^1(\Omega)} \geq c_4\|\zeta\|^2_{H^1(\Omega,\Lambda) \times H^2(\Omega)}. \tag{6.15}$$

Finally the inequality (6.11) follows from the inequality (6.15) and Lemma 6.1 by a compactness-uniqueness argument as in Lemma 2.5. □

We recall the following operators which are needed to describe the Koiter shell operator.

Operator $\mathcal{S}$ Let the operator $\mathcal{S}: L^2(\Omega, \Lambda) \to L^2(\Omega, \Lambda)$ be defined by (5.2), that is,

$$\Pi(X, Y) = \langle \mathcal{S}X, Y \rangle \quad \text{for} \quad X, Y \in M_x, \quad x \in \Omega. \tag{6.16}$$

Then for each $x \in M$, $\mathcal{S}: M_x \to M_x$ is a symmetric, linear operator. It is clear that for $X \in \mathcal{X}(M)$

$$\mathcal{S}X = \tilde{D}_X N \tag{6.17}$$

where $\tilde{D}$ is the covariant differential of the Euclidean metric of $\mathbb{R}^3$ and N is the normal of the surface M.

Operator $\mathcal{Q}$ Let the operator $\mathcal{Q}$: $L^2(\Omega, T^2) \to L^2(\Omega, \Lambda)$ be defined by

$$\langle X, \mathcal{Q}T \rangle = \operatorname{tr} \operatorname{i}(X)DT \quad \text{for} \quad X \in \mathcal{X}(M), \quad T \in T^2(\Omega). \tag{6.18}$$

$-\mathcal{Q}$ is the formal adjoint of the covariant differential operator D. For the further properties of $\mathcal{Q}$, see Section 1.4.

Operator $\mathcal{P}$ Let the operator $\mathcal{P}$ be defined by (5.32). Then

$$\langle T, \operatorname{i}(X)D\Pi \rangle = \langle \mathcal{P}T, X \rangle \quad \text{for} \quad T \in T^2(M), \quad X \in M_x, \quad x \in M. \tag{6.19}$$

Lemma 6.2 *We have*

$$\Pi(\cdot, D.W) + \Pi(D.W, \cdot) + \operatorname{i}(W)D\Pi = D(\mathcal{S}W) + D^*(\mathcal{S}W) - \operatorname{i}(W)D\Pi. \tag{6.20}$$

Moreover, let δ be, given by (1.109), the formal adjoint of the exterior derivative d. Then

$$\begin{aligned} \mathcal{Q}\rho &= (\delta d + 2d\delta)\mathcal{S}W + d\Delta w - 2\kappa\mathcal{S}W + \mathcal{Q}(\operatorname{i}(W)D\Pi) \\ &\quad + \kappa\,dw - \operatorname{i}(Dw)\operatorname{c} - w\mathcal{Q}\operatorname{c} \end{aligned} \tag{6.21}$$

where κ is the Gauss curvature function.

Proof. Let $x \in \Omega$ be given and let E_1, E_2 be a frame field normal at x. Using the relations $D_{E_i}E_j(x) = 0$ and the formula (6.16), we obtain at x

$$\begin{aligned} D(\mathcal{S}W)(E_i, E_j) &= E_j(\langle \mathcal{S}W, E_i \rangle) = E_j[\Pi(E_i, W)] \\ &= D\Pi(W, E_i, E_j) + \Pi(E_i, D_{E_j}W) \end{aligned}$$

which yield the formula (6.20) since $D\Pi$ is symmetric by Corollary 1.1.

Using the formulas (1.150) and (3.100), we have

$$\mathcal{Q}D^2 w = -\boldsymbol{\Delta}dw + \kappa dw = d\Delta w + \kappa dw. \tag{6.22}$$

In addition, it follows from the identity (6.20) and the formulas (1.150) and (1.151) that

$$\begin{aligned} &\mathcal{Q}[\Pi(\cdot, D.W) + \Pi(D.W, \cdot) + \operatorname{i}(W)D\Pi] \\ &= -(\delta d + 2d\delta)\mathcal{S}W + 2\kappa\mathcal{S}W - \mathcal{Q}(\operatorname{i}(W)D\Pi). \end{aligned} \tag{6.23}$$

Moreover, it is easy to check from (1.144) that

$$\mathcal{Q}(w\operatorname{c}) = \operatorname{i}(Dw)\operatorname{c} + w\mathcal{Q}\operatorname{c}. \tag{6.24}$$

Finally, we obtain the formula (6.21) by the formulas (6.2), (6.22), (6.23), and (6.24). $\qquad\square$

Denote by $\mathcal{O}_j(\zeta)$ some terms which contain at most the j-th order derivatives of the vector field ζ with respect to the spacial variables for $0 \leq j \leq 3$. We have the following Green formula for the Koiter shell.

Theorem 6.2 *Let the bilinear form* $\mathcal{B}(\cdot, \cdot)$ *be given by* (6.5). *Then, for* $\zeta = (W, w)$, $\eta = (U, u) \in H^1(\Omega, \Lambda) \times H^2(\Omega)$,

$$\mathcal{B}(\zeta, \eta) = (\mathcal{A}\eta, \zeta)_{L^2(\Omega,\Lambda) \times L^2(\Omega)} + \int_{\Gamma} \partial(\mathcal{A}\eta, \zeta) d\Gamma, \tag{6.25}$$

where

$$\partial(\mathcal{A}\zeta, \eta) = \langle V_1(\zeta), U \rangle + v_2(\zeta) \frac{\partial u}{\partial \nu} + v_3(\zeta) u, \tag{6.26}$$

ν *is the normal along the curve* Γ,

$$\mathcal{A}\zeta = \begin{pmatrix} (\mathcal{A}\zeta)_1 \\ (\mathcal{A}\zeta)_2 \end{pmatrix} \tag{6.27}$$

where

$$(\mathcal{A}\zeta)_1 = -(\boldsymbol{\Delta}_\mu + 4\gamma S \boldsymbol{\Delta}_\mu S)W - 2\gamma S d\delta dw + \gamma(1-\mu)\mathcal{P}D^2 w$$
$$+ \gamma\mu(\Delta w)\mathcal{Q}\Pi + \mathcal{O}_1(\zeta), \tag{6.28}$$

$$(\mathcal{A}\zeta)_2 = \gamma(\Delta^2 w - 2\delta d\delta SW) + \gamma\mu\Delta\langle W, \mathcal{Q}\Pi \rangle$$
$$+ \gamma(1-\mu)\{\delta[i(Dw)\mathbf{c} - \mathcal{Q}(i(W)\Pi)] - \langle D^2 w, \mathbf{c} \rangle\}$$
$$+ \gamma[(1-\mu)\kappa - 2\mu \operatorname{tr} \mathbf{c}]\Delta w + \mathcal{O}_1(\zeta), \tag{6.29}$$

$\boldsymbol{\Delta}_\mu$ *is of the Hodge-Laplacian type, applied to 1-forms (or equivalent vector fields), defined by*

$$\boldsymbol{\Delta}_\mu = -(\frac{1-\mu}{2}\delta d + d\delta), \tag{6.30}$$

d *the exterior differential,* δ *the formal adjoint of* d, Δ *the Laplacian on the manifold* M $(\Delta = -\delta d)$,

$$\begin{cases} V_1(\zeta) = B_1(\zeta) - 2\gamma S B_2(\zeta), \\ v_2(\zeta) = \gamma\langle B_2(\zeta), \nu \rangle, \\ v_3(\zeta) = -\gamma[\langle B_3(\zeta), \nu \rangle + (1-\mu)\frac{\partial}{\partial \tau}\rho(\zeta)(\nu, \tau)], \end{cases} \tag{6.31}$$

τ *is the tangential along the curve* Γ, *and*

$$\begin{cases} B_1(\zeta) = (1-\mu)i(\nu)\Upsilon(\zeta) + \mu(\operatorname{tr} \Upsilon(\zeta))\nu, \\ B_2(\zeta) = (1-\mu)i(\nu)\rho(\zeta) + \mu(\operatorname{tr} \rho(\zeta))\nu, \\ B_3(\zeta) = (1-\mu)\mathcal{Q}\rho(\zeta) + \mu d(\operatorname{tr} \rho(\zeta)). \end{cases} \tag{6.32}$$

Proof. We need to compute the integral $\int_\Omega a(\rho(\zeta), \rho(\eta)) \, dx$. To this end, let us compute $\int_\Omega \langle \rho(\zeta), \rho(\eta) \rangle \, dx$ and $\int_\Omega \operatorname{tr} \rho(\zeta) \operatorname{tr} \rho(\eta) \, dx$, separately.

Using the formulas (1.145) and (6.21), we have

$$(\rho(\zeta), D^2 u)_{L^2(\Omega, T^2)} = -(\mathcal{Q}\rho, du)_{L^2(\Omega, \Lambda)} + \int_\Gamma \rho(\nu, du) \, d\Gamma$$

$$= \left(\Delta^2 w - 2\delta d\delta SW - \delta\mathcal{Q}(i(W)D\Pi) + \kappa\Delta w + \delta(i(Dw)\mathbf{c}) + \mathcal{O}_1(\zeta), \; u \right)$$

$$+ \int_\Gamma [\rho(\nu, du) - u\langle \nu, \mathcal{Q}\rho \rangle] \, d\Gamma. \tag{6.33}$$

Similarly, we obtain

$$- \left(\rho(\varsigma),\ \Pi(\cdot, D.U) + \Pi(D.U, \cdot) + \mathrm{i}\,(U)D\Pi + u\,c \right)_{L^2(\Omega, T^2)}$$

$$= - \left(\rho(\varsigma),\ 2D(SU) - \mathrm{i}\,(U)D\Pi + u\,c \right)_{L^2(\Omega, T^2)}$$

$$= 2(Q\rho, SU)_{L^2(\Omega, \Lambda)} - 2\int_\Gamma \rho(\nu, SU)\,d\Gamma + (\mathcal{P}D^2w + \mathcal{O}_1(\varsigma),\ U)_{L^2(\Omega, \Lambda)}$$

$$+ (\langle D^2w,\ c\rangle + \mathcal{O}_1(\varsigma),\ u)$$

$$= \left(- 2Sd\delta dw + 2S(\delta d + 2d\delta)SW + \mathcal{P}D^2w + \mathcal{O}_1(\varsigma),\ U \right)_{L^2(\Omega, \Lambda)}$$

$$+ (-\langle D^2w,\ c\rangle + \mathcal{O}_1(\varsigma),\ u) - 2\int_\Gamma \rho(\nu, SU)\,d\Gamma. \tag{6.34}$$

It follows from the formulas (6.33) and (6.34) that

$$\int_\Omega \langle \rho(\varsigma), \rho(\eta) \rangle\,dx$$

$$= \left(- 2Sd\delta dw + 2S(\delta d + 2d\delta)SW + \mathcal{P}D^2w + \mathcal{O}_1(\varsigma),\ U \right)_{L^2(\Omega, \Lambda)}$$

$$+ \left(\Delta^2 w - 2\delta d\delta SW - \delta Q(\mathrm{i}\,(W)D\Pi) + \kappa\Delta w + \delta(\mathrm{i}\,(Dw)\,c) \right.$$

$$\left. - \langle D^2w,\ c\rangle + \mathcal{O}_1(\varsigma),\ u \right)$$

$$+ \int_\Gamma \{-2\rho(\nu, SU) + [\rho(\nu, du) - u\langle \nu, Q\rho\rangle]\}\,d\Gamma. \tag{6.35}$$

Next, we compute the integral $\int_\Omega \mathrm{tr}\,\rho(\varsigma)\,\mathrm{tr}\,\rho(\eta)\,dx$. Noting that $\mathrm{tr}\,DSU = -\delta SU$, we have from the formulas (6.2) and (6.20)

$$\mathrm{tr}\,\rho(\varsigma) = \Delta w + 2\delta SW + \mathrm{tr}\,\mathrm{i}\,(W)D\Pi - w\,\mathrm{tr}\,c\,; \tag{6.36}$$

$$\mathrm{tr}\,\rho(\eta) = \Delta u + 2\delta SU + \mathrm{tr}\,l_U\,D\Pi - u\,\mathrm{tr}\,c\,. \tag{6.37}$$

By Green's formula for the Laplacian,

$$(\mathrm{tr}\,\rho(\varsigma), \Delta u) = (\Delta\,\mathrm{tr}\,\rho(\varsigma),\ u) + \int_\Gamma [\mathrm{tr}\,\rho(\varsigma)\frac{\partial u}{\partial \nu} - u\frac{\partial\,\mathrm{tr}\,\rho(\varsigma)}{\partial \nu}]\,d\Gamma$$

$$= (\Delta^2 w - 2\delta d\delta SW + \Delta\langle W, Q\Pi\rangle - (\mathrm{tr}\,c)\Delta w + \mathcal{O}_1(\varsigma),\ u)$$

$$+ \int_\Gamma [\mathrm{tr}\,\rho(\varsigma)\frac{\partial u}{\partial \nu} - u\frac{\partial\,\mathrm{tr}\,\rho(\varsigma)}{\partial \nu}]\,d\Gamma \tag{6.38}$$

since $\Delta = -\delta d$ when it is applied to a function. Moreover,

$$(\operatorname{tr}\rho(\zeta),\ 2\delta SU + \operatorname{tr} i(U)D\Pi - u\operatorname{tr} \mathrm{c}\,)$$

$$= 2(d\operatorname{tr}\rho(\zeta),\ SU)_{L^2(\Omega,\Lambda)} - 2\int_\Gamma \langle SU,\nu\rangle \operatorname{tr}\rho(\zeta)\,d\Gamma$$

$$+((\Delta w)Q\Pi + \mathcal{O}_1(\zeta),\ U)_{L^2(\Omega,\Lambda)} + (-(\operatorname{tr}\mathrm{c}\,)\Delta w + \mathcal{O}_1(\zeta),\ u)$$

$$= (-2Sd\delta dw + 4Sd\delta SW + (\Delta w)Q\Pi + \mathcal{O}_1(\zeta),\ U)_{L^2(\Omega,\Lambda)}$$

$$+(-(\operatorname{tr}\mathrm{c}\,)\Delta w + \mathcal{O}_1(\zeta),\ u) - 2\int_\Gamma \langle SU,\nu\rangle \operatorname{tr}\rho(\zeta)\,d\Gamma. \qquad (6.39)$$

It follows from the formulas (6.36)-(6.39) that

$$\int_\Omega \operatorname{tr}\rho(\zeta)\operatorname{tr}\rho(\eta)\,dx$$

$$= (-2Sd\delta dw + 4Sd\delta SW + (\Delta w)Q\Pi + \mathcal{O}_1(\zeta),\ U)_{L^2(\Omega,\Lambda)}$$

$$+\left(\Delta^2 w - 2\delta d\delta SW + \Delta\langle W,Q\Pi\rangle - 2(\operatorname{tr}\mathrm{c}\,)\Delta w + \mathcal{O}_1(\zeta),\ u\right)$$

$$+\int_\Gamma [\operatorname{tr}\rho(\zeta)\frac{\partial u}{\partial\nu} - u\frac{\partial\operatorname{tr}\rho(\zeta)}{\partial\nu} - 2\langle SU,\nu\rangle \operatorname{tr}\rho(\zeta)]\,d\Gamma. \qquad (6.40)$$

Finally, using the formula (4.47), (6.35), and (6.40), we obtain the formula (6.25). □

Remark 6.2 *There are higher order coupling terms* $-2\gamma Sd\delta dw$ *and* $-2\gamma\delta d\delta W$ *in the principal parts of the Koiter shell* (6.28) *and* (6.29), *respectively, because the second fundamental form goes into the change of the curvature tensor (see* (6.2)). *It is a coupling relationship that makes analysis on the Koiter shell much more difficult than that on the shallow shell. If we set* $S = 0$ *in the formulas* (6.28) *and* (6.29), *then the formulas* (6.27) *become shallow shell ones where only the lower order terms are coupled.*

The detailed expressions (6.28) *and* (6.29) *are necessary when we consider a uniqueness problem in Section 6.2 (see the proof of Theorem 6.5).*

If the shell is flat, a plate, the equations in the formulas (6.25) *are uncoupled. The formulas on the components* W *and* w *are the same as in* [105], [109], *a Kirchhoff plate.*

Definition 6.1 *The operator* $\mathcal{A}$, *given by* (6.27), *is called the Koiter operator.*

The Kinetic Energy of the Koiter Shell   We will consider the kinetic energies of the Koiter shell under the classical Kirchhoff-Love assumption and the Koiter complementary hypotheses, respectively.

The Kinetic Energy of the Koiter Shell under the Classical Kirchhoff-Love Assumption   In this case, the kinetic energy is the same as for the shallow shell (see Section 4.3)

$$\mathcal{P} = \int_\Omega \mathcal{H}(t)\,dx = \int_\Omega (h|W_t|^2 + hw_t^2 + \frac{h^3}{12}|Dw_t|^2)\,dx. \qquad (6.41)$$

By the "Principle of Virtual Work" and the Green formula (6.25), we obtain the following displacement equations for the Koiter shell after changing t to λt with $\lambda^2 E/(1 - \mu^2) = 1$.

Theorem 6.3 *We assume that there are no external loads on the shell and the shell is clamped along a portion Γ_0 of Γ and free on Γ_1, where $\Gamma_0 \cup \Gamma_1 = \Gamma$ and $\Gamma_0 \cap \Gamma_1 = \emptyset$. Then the displacement vector $\zeta = (W, w)$ satisfies the following boundary value problem:*

$$\begin{cases} W_{tt} + (\mathcal{A}\zeta)_1 = 0, \\ w_{tt} - \gamma \Delta w_{tt} + (\mathcal{A}\zeta)_2 = 0, \\ \zeta(0) = \zeta^0, \quad \zeta_t(0) = \zeta^1 \quad in \quad Q_\infty, \end{cases} \tag{6.42}$$

$$\begin{cases} W = 0, \\ w = \dfrac{\partial w}{\partial \nu} = 0, \end{cases} \quad on \ \Sigma_{0\infty}, \tag{6.43}$$

$$\begin{cases} V_1(\zeta) = 0, \\ v_2(\zeta) = 0, \\ \gamma \dfrac{\partial w_{tt}}{\partial \nu} + v_3(\zeta) = 0, \end{cases} \quad on \ \Sigma_{1\infty}, \tag{6.44}$$

where $(\mathcal{A}\zeta)_1$ and $(\mathcal{A}\zeta)_2$ are given by the formulas (6.28) and (6.29), respectively, and

$$Q_\infty = \Omega \times (0, \infty), \quad \Sigma_{0\infty} = \Gamma_0 \times (0, \infty), \quad \Sigma_{1\infty} = \Gamma_1 \times (0, \infty). \tag{6.45}$$

The Kinetic Energies of the Koiter Shell in the Koiter Hypotheses In order to eliminate a contradiction in the classical Kirchhoff-Love assumptions, Koiter introduced complementary hypotheses which permit us to derive a satisfactory approximation of the displacement field of the shell from only the knowledge of the displacement field of the middle surface. These hypotheses are:

Koiter's Hypotheses ([95], pages 15-16)

(i) The normal to the undeformed middle surface remains normal to the deformed middle surface after the deformation;

(ii) During the deformation, the stresses are approximatively plane and parallel to the tangent plane to the middle surface.

Let $\sigma(\cdot, \cdot)$ be the stress tensor of the three-dimensional shell. Then the Koiter assumption (ii) means $i(N)\sigma = 0$, that is,

$$\sigma(X, N)(p) = 0 \quad for \quad X \in \mathcal{X}(M) \tag{6.46}$$

and

$$\sigma(N, N)(p) = 0 \tag{6.47}$$

for $p = x + zN \in S$.

Let $\overline{\Upsilon}(\cdot,\cdot)$ be the strain tensor of the three-dimensional shell. For an isotropic material, we have

$$\sigma(\cdot,\cdot) = \frac{E}{1+\mu}\overline{\Upsilon}(\cdot,\cdot) + \frac{\mu E}{(1+\mu)(1-2\mu)}(\tilde{\mathrm{tr}}\,\overline{\Upsilon})\tilde{g}(\cdot,\cdot) \tag{6.48}$$

where E, μ respectively denote *Young's modulus* and *Poisson's coefficient* of the material, respectively, and $\tilde{\mathrm{tr}}$ and $\tilde{g}(\cdot,\cdot)$ are the trace and the Euclidean metric in $\mathbb{R}^3$, respectively. It is easy to check that the relations (6.46)-(6.48) yield the following formulas

$$\overline{\Upsilon}(X,N) = 0 \quad \text{for} \quad X \in \mathcal{X}(M) \tag{6.49}$$

and

$$(1-\mu)\overline{\Upsilon}(N,N) + \mu\,\mathrm{tr}\,\overline{\Upsilon} = 0 \quad \text{for} \quad p = x + zN \in S \tag{6.50}$$

where tr is the trace of M in the induced metric from $\mathbb{R}^3$.

Let the shell occupy a region $\mathcal{F}(S)$ in $\mathbb{R}^3$ after deformation where

$$\mathcal{F}(S) = \{\,\mathcal{F}(p)\,|\,p = x + zN(x) \in S\,\}$$

and $\mathcal{F}\colon S \to \mathbb{R}^3$ is the deformation map. The Koiter assumption (i) implies

$$\mathcal{F}(p) = \mathcal{F}(x) + \overline{z}\overline{N}(\mathcal{F}(x)) \quad \text{for} \quad p = x + zN(x) \in S \tag{6.51}$$

where $\overline{N}$ is the normal to the deformed middle surface $\overline{\Omega} = \{\,\mathcal{F}(x)\,|\,x \in \Omega\,\}$ and $\overline{z} = \overline{z}(p)$ is the distance from $\mathcal{F}(p)$ to the surface $\overline{\Omega}$.

Lemma 6.3 *Under the Koiter assumptions* (i) *and* (ii), *we have the following approximation*

$$\overline{z}(p) = z[1 - \alpha\,\mathrm{tr}\,\Upsilon(\zeta)] - \frac{z^2}{2}\alpha\,\mathrm{tr}\,D(\,\mathrm{i}\,(W)\Pi - Dw) \tag{6.52}$$

for $p = x + zN \in S$ *where* $\alpha = \mu/(1-\mu)$ *and* $\Upsilon(\zeta)$ *is the strain tensor of the undeformed middle surface* Ω *and* $\zeta = (W,w)$.

Proof. Let $\overline{\zeta} = \mathcal{F}(p) - p$ be the displacement vector of the material point $p = x + zN$. Then

$$\overline{\zeta}(p) = W + (w - z)N + \overline{z}\overline{N} \quad \text{for} \quad p = x + zN.$$

Using the formula (4.87), we have

$$\tilde{D}_N\overline{\zeta} = \frac{d\overline{z}}{dz}\overline{N} - N = \frac{d\overline{z}}{dz}(\,\mathrm{i}\,(W)\Pi - Dw) + (\frac{d\overline{z}}{dz} - 1)N, \tag{6.53}$$

$$\tilde{D}_X\overline{\zeta} = \tilde{D}_X W + \overline{z}\tilde{D}_X(\,\mathrm{i}\,(W)\Pi - Dw) + X(\overline{z})(\,\mathrm{i}\,(W)\Pi - Dw)$$
$$+ (w + \overline{z} - z)\tilde{D}_X N + [X(w) + X(\overline{z})]]N \tag{6.54}$$

where $X \in \mathcal{X}(\Omega)$ and $\tilde{D}$ is the covariant differential of the Euclidean metric of $\mathbb{R}^3$. It follows from the formulas (6.53), (6.54), and the formula (4.87) that

$$\frac{d\bar{z}}{dz} = 1 + \overline{\Upsilon}(N, N), \tag{6.55}$$

$$\langle \tilde{D}_N \bar{\zeta}, X \rangle = \frac{d\bar{z}}{dz}[\Pi(W, X) - X(w)], \tag{6.56}$$

$$\begin{aligned}\langle \tilde{D}_X \bar{\zeta}, X \rangle &= DW(X, X) + \bar{z}D(\mathrm{i}(W)\Pi - Dw)(X, X) \\ &\quad + X(\bar{z})[\Pi(W, X) - X(w)] + (w + \bar{z} - z)\Pi(X, X)\end{aligned} \tag{6.57}$$

for $X \in \mathcal{X}(M)$.

Using the relations (6.49), (6.54), and (6.56), we obtain

$$\begin{aligned}0 &= 2\overline{\Upsilon}(N, X) = \langle \tilde{D}_X \bar{\zeta}, N \rangle + \langle \tilde{D}_N \bar{\zeta}, X \rangle \\ &= -\Pi(W, X) - \bar{z}\Pi(\mathrm{i}(W)\Pi - Dw, X) + X(w) + X(\bar{z}) \\ &\quad + \frac{d\bar{z}}{dz}[\Pi(W, X) - X(w)] \quad \text{for} \quad X \in \mathcal{X}(M)\end{aligned}$$

which yields

$$D\bar{z} = (\frac{d\bar{z}}{dz} - 1)(Dw - \mathrm{i}(W)\Pi) + \bar{z}\mathrm{i}(\mathrm{i}(W)\Pi - Dw)\Pi. \tag{6.58}$$

Let $\mathbf{R}$ be the smallest principal radius of curvature of the undeformed middle surface given by (4.25). Then the estimate

$$|\bar{z}\mathrm{i}(\mathrm{i}(W)\Pi - Dw)\Pi| \le c\frac{h}{\mathbf{R}}|\mathrm{i}(W)\Pi - Dw|$$

for some $c > 0$ implies that the formula (6.58) can be approximated by

$$D\bar{z} = (\frac{d\bar{z}}{dz} - 1)(Dw - \mathrm{i}(W)\Pi). \tag{6.59}$$

Combining the formulas (6.57) and (6.59) gives

$$\begin{aligned}\mathrm{tr}\,\overline{\Upsilon} &= \mathrm{tr}\,DW + \bar{z}\,\mathrm{tr}\,D(\mathrm{i}(W)\Pi - Dw) + \langle D\bar{z}, \mathrm{i}(W)\Pi - Dw \rangle \\ &\quad + (w + \bar{z} - z)\,\mathrm{tr}\,\Pi \\ &= \mathrm{tr}\,\Upsilon(\zeta) + \bar{z}\,\mathrm{tr}\,D(\mathrm{i}(W)\Pi - Dw) - (\frac{d\bar{z}}{dz} - 1)|\mathrm{i}(W)\Pi - Dw|^2 \\ &\quad + (\bar{z} - z)\,\mathrm{tr}\,\Pi\end{aligned}$$

from which we have, by dropping the nonlinear terms,

$$\mathrm{tr}\,\overline{\Upsilon} = \mathrm{tr}\,\Upsilon(\zeta) + \bar{z}[\mathrm{tr}\,D(\mathrm{i}(W)\Pi - Dw) + \mathrm{tr}\,\Pi] - z\,\mathrm{tr}\,\Pi. \tag{6.60}$$

Using the formulas (6.60) and (6.55) in the formula (6.50), we have

$$\frac{d\bar{z}}{dz} + \alpha[\operatorname{tr} D(\,\mathrm{i}\,(W)\Pi - Dw) + \operatorname{tr}\Pi]\bar{z} = \alpha\operatorname{tr}\Pi z + 1 - \alpha\operatorname{tr}\Upsilon(\zeta) \qquad (6.61)$$

with $\bar{z}(0) = 0$.

Since

$$\alpha|\operatorname{tr}\Pi z| \le 2\alpha\frac{h}{\mathbf{R}}, \quad \alpha|\operatorname{tr}\Pi\bar{z}| \le c\frac{h}{\mathbf{R}}$$

for some $c > 0$, the terms $\alpha\operatorname{tr}\Pi z$ and $\alpha\operatorname{tr}\Pi\bar{z}$ can be omitted compared with 1 from the equation (6.61) by the assumption (4.28). Thus

$$\frac{d\bar{z}}{dz} + \alpha\operatorname{tr} D(\,\mathrm{i}\,(W)\Pi - Dw)\bar{z} = 1 - \alpha\operatorname{tr}\Upsilon(\zeta), \quad \bar{z}(0) = 0. \qquad (6.62)$$

We solve the equation (6.62), then linearize the solution with respect to the displacement vector, and obtain the formula (6.52). $\qquad\square$

Using the formulas (6.52) and (4.87), we have, after a linearization,

$$\bar{\zeta}(p) = \zeta + zY_1 + z^2 Y_2 \qquad (6.63)$$

where

$$Y_1 = \mathrm{i}\,(W)\Pi - Dw - \alpha[\operatorname{tr}\Upsilon(\zeta)]N, \quad Y_2 = -\frac{1}{2}\alpha[\operatorname{tr} D(\,\mathrm{i}\,(W)\Pi - Dw)]N. \qquad (6.64)$$

Then the kinetic energy per unit area of the unformed middle surface is obtained by integration with respect to z

$$\mathcal{H}(t) = \int_{-h/2}^{h/2} |\bar{\zeta}_t|^2 (1 + \operatorname{tr}\Pi z + \kappa z^2)\, dz. \qquad (6.65)$$

Since

$$|\bar{\zeta}_t|^2 |\operatorname{tr}\Pi z + \kappa z^2| \le \frac{h}{\mathbf{R}}|\bar{\zeta}_t|^2 + \frac{1}{4}\left(\frac{h}{\mathbf{R}}\right)^2 |\bar{\zeta}_t|^2 \quad \text{for} \quad |z| \le \frac{h}{2},$$

we have, by the formula (6.63),

$$\mathcal{H}(t) = \int_{-h/2}^{h/2} |\bar{\zeta}_t|^2\, dz = (|W_t|^2 + w_t^2)h - \alpha w_t \operatorname{tr} D(\,\mathrm{i}\,(W_t)\Pi - Dw_t)\frac{h^3}{12}$$

$$+|Y_{1t}|^2 \frac{h^3}{12} + |Y_{2t}|^2 \frac{h^5}{80}. \qquad (6.66)$$

By a similar argument, the term $|Y_{1t}|^2 h^3/12$ in the right hand side of the formula (6.66) can be replaced with $[|Dw_t|^2 + \alpha^2(\operatorname{tr} DW_t)^2]h^3/12$. We then obtain

$$\mathcal{H}(t) = \{|W_t|^2 + [w_t - \frac{h^2\alpha}{24}\operatorname{tr} D(\,\mathrm{i}\,(W_t)\Pi - Dw_t)]^2\}h + |Dw_t|^2 h^3/12$$

$$+\alpha^2(\operatorname{tr} DW_t)^2 h^3/12 + \alpha^2[\operatorname{tr} D(\,\mathrm{i}\,(W_t)\Pi - Dw_t)]^2 h^5/720. \qquad (6.67)$$

Let the operator $\mathcal{S}$ be given by (6.16). We introduce a bilinear form on $L^2(\Omega, \Lambda) \times L^2(\Omega)$ by

$$
\begin{aligned}
H(\zeta, \eta) &= \int_\Omega \{\langle W, U \rangle + [w + \alpha\sigma_0\delta(\mathrm{i}\,(W)\Pi - Dw)][u + \alpha\sigma_0\delta(\mathrm{i}\,(U)\Pi - Du)] \\
&\quad + 2\sigma_0\langle Dw, Du \rangle + 2\alpha^2\sigma_0\delta W\delta U \\
&\quad + \alpha^2\sigma_1\delta(\mathcal{S}W - Dw)\delta(\mathcal{S}U - Du)\} \, dx
\end{aligned}
\tag{6.68}
$$

where

$$
\zeta = (W, w), \quad \eta = (U, u), \quad \sigma_0 = h^2/24, \quad \sigma_1 = h^4/720.
\tag{6.69}
$$

We need

Lemma 6.4 *We have*

$$
\begin{aligned}
H(\zeta, \eta) &= \Big(\Phi(\zeta), U\Big)_{L^2(\Omega, \Lambda)} + \Big(\Psi(\zeta), u\Big) \\
&\quad + \int_\Gamma [\langle \Gamma_1(\zeta), U \rangle + \Gamma_2(\zeta)\frac{\partial u}{\partial \nu} + \Gamma_3(\zeta)u] \, d\Gamma
\end{aligned}
\tag{6.70}
$$

where

$$
\begin{aligned}
\Phi(\zeta) &= W + 2\alpha^2\sigma_0 d\delta W + \alpha^2(\sigma_0^2 + \sigma_1)\mathcal{S}d\delta\mathcal{S}W \\
&\quad + \alpha\sigma_0\mathcal{S}dw + \alpha^2(\sigma_0^2 + \sigma_1)\mathcal{S}d\Delta w,
\end{aligned}
\tag{6.71}
$$

$$
\begin{aligned}
\Psi(\zeta) &= w - 2(1 - \alpha)\sigma_0\Delta w + \alpha^2(\sigma_0^2 + \sigma_1)\Delta^2 w \\
&\quad + \alpha\sigma_0\delta\mathcal{S}W - \alpha^2(\sigma_0^2 + \sigma_1)\delta d\delta\mathcal{S}W,
\end{aligned}
\tag{6.72}
$$

$$
\Gamma_1(\zeta) = -2\alpha^2\sigma_0\delta W\nu - \alpha[\sigma_0 w + \alpha(\sigma_0^2 + \sigma_1)(\delta\mathcal{S}W + \Delta w)]\mathcal{S}\nu,
\tag{6.73}
$$

$$
\Gamma_2(\zeta) = \alpha\sigma_0 w + \alpha^2(\sigma_0^2 + \sigma_1)[\delta(\mathcal{S}W) + \Delta w],
\tag{6.74}
$$

$$
\Gamma_3(\zeta) = \sigma_0(2 - \alpha)\frac{\partial w}{\partial \nu} - \alpha^2(\sigma_0^2 + \sigma_1)(\frac{\partial\delta\mathcal{S}W}{\partial \nu} + \frac{\partial\Delta w}{\partial \nu}).
\tag{6.75}
$$

Proof. We compute each term in the right hand side of (6.70) where the formulas (1.136) and (1.137) have been used many times.

Noting that $i\,(W)\Pi = SW$ and $D = d$ if applied to a function, we have

$$\left(w + \alpha\sigma_0\delta(i\,(W)\Pi - Dw),\ u + \alpha\sigma_0\delta(i\,(U)\Pi - Du)\right)$$

$$= \left(w + \alpha\sigma_0\delta(SW - Dw),\ u\right)$$

$$+\alpha\sigma_0\left(dw + \alpha\sigma_0 d\delta(SW - Dw),\ SU - Du\right)_{L^2(\Omega,\Lambda)}$$

$$-\alpha\sigma_0\int_\Gamma [w + \alpha\sigma_0\delta(SW - Dw)]\langle SU - Du,\ \nu\rangle\,d\Gamma$$

$$= \left(w + 2\alpha\sigma_0\Delta w + \alpha^2\sigma_0^2\Delta^2 w + \alpha\sigma_0\delta SW - \alpha^2\sigma_0^2\delta d\delta SW,\ u\right)$$

$$+\left(\alpha\sigma_0 Sdw + \alpha^2\sigma_0^2 Sd\delta(SW - Dw),\ U\right)_{L^2(\Omega,\Lambda)}$$

$$-\alpha\sigma_0\int_\Gamma [w + \alpha\sigma_0\delta(SW - Dw)]\langle S\nu, U\rangle\,d\Gamma$$

$$+\alpha\sigma_0\int_\Gamma [w + \alpha\sigma_0\delta(SW - Dw)]\frac{\partial u}{\partial\nu}\,d\Gamma$$

$$-\alpha\sigma_0\int_\Gamma \langle Dw + \alpha\sigma_0 d\delta(SW - Dw),\ \nu\rangle u\,d\Gamma; \tag{6.76}$$

$$\left(Dw, Du\right)_{L^2(\Omega,\Lambda)} = \left(\delta dw, u\right) + \int_\Gamma \langle Dw, \nu\rangle u\,d\Gamma; \tag{6.77}$$

$$\left(\delta W, \delta U\right) = \left(d\delta W, U\right)_{L^2(\Omega,\Lambda)} - \int_\Gamma (\delta W)\langle U, \nu\rangle\,d\Gamma; \tag{6.78}$$

$$\left(\delta(SW - Dw),\ \delta(SU - Du)\right) = \left(Sd\delta(SW - Dw),\ U\right)_{L^2(\Omega,\Lambda)}$$

$$-\left(\delta d\delta(SW - Dw),\ u\right) - \int_\Gamma \delta(SW - Dw)[\langle S\nu, U\rangle - \frac{\partial u}{\partial\nu}]\,d\Gamma$$

$$-\int_\Gamma \langle d\delta(SW - Dw),\ \nu\rangle u\,d\Gamma. \tag{6.79}$$

Inserting the formulas (6.76)-(6.79) in the formula (6.68), we obtain the formula (6.70). □

Equations of Motion of the Koiter Shell in Koiter's Hypotheses
By the "Principle of Virtual Work" and the Green formula (6.25), we obtain the following displacement equations for the Koiter shell after changing t to λt with $\lambda^2 E/(1 - \mu^2) = 1$:

Theorem 6.4 *We assume that there are no external loads on the shell and the shell is clamped along a portion Γ_0 of Γ and free on Γ_1, where $\Gamma_0 \cup \Gamma_1 = \Gamma$ and $\Gamma_0 \cap \Gamma_1 = \emptyset$. Let $\Phi(\zeta)$, $\Psi(\zeta)$, $\Gamma_1(\zeta)$, $\Gamma_2(\zeta)$ and $\Gamma_3(\zeta)$ be defined by Lemma*

6.4. *Then the displacement vector* $\zeta = (W, w)$ *satisfies the following boundary value problem:*

$$\begin{cases} \Phi(\zeta_{tt}) + (\mathcal{A}\zeta)_1 = 0, \\ \Psi(\zeta_{tt}) + (\mathcal{A}\zeta)_2 = 0, \\ \zeta(0) = \zeta^0, \quad \zeta_t(0) = \zeta^1 \quad in \quad Q_\infty, \end{cases} \qquad (6.80)$$

$$\begin{cases} W = 0, \\ w = \dfrac{\partial w}{\partial \nu} = 0, \end{cases} \quad on \; \Sigma_{0\infty}, \qquad (6.81)$$

$$\begin{cases} \Gamma_1(\zeta_{tt}) + V_1(\zeta) = 0, \\ \Gamma_2(\zeta_{tt}) + v_2(\zeta) = 0, \quad on \; \Sigma_{1\infty}, \\ \Gamma_3(\zeta_{tt}) + v_3(\zeta) = 0, \end{cases} \qquad (6.82)$$

where $(\mathcal{A}\zeta)_1$ *and* $(\mathcal{A}\zeta)_2$ *are given by the formulas* (6.28) *and* (6.29), *respectively, and*

$$Q_\infty = \Omega \times (0, \infty), \quad \Sigma_{0\infty} = \Gamma_0 \times (0, \infty), \quad \Sigma_{1\infty} = \Gamma_1 \times (0, \infty). \qquad (6.83)$$

6.2 Uniqueness for the Koiter Shell

We consider the uniqueness of the Cauchy problem for the Koiter shell which is needed to absorb the lower order terms in obtaining controllability/stabilization for the equations of motion in Theorem 6.3.

Let M be a smooth surface in $\mathbb{R}^3$ and let a middle surface of the shell occupy a bounded region Ω of the surface M whose boundary Γ is smooth. Let $\hat{\Gamma} \subset \Gamma$ have a positive length. Consider the following Cauchy problem in unknown $\zeta = (W, w)$:

$$\mathcal{A}\zeta = \begin{pmatrix} \mathcal{O}_1(\zeta) \\ \mathcal{O}_1(\zeta) + \mathcal{O}_2(w) \end{pmatrix} \quad on \quad \Omega \qquad (6.84)$$

subject to the boundary conditions

$$\begin{cases} W = DW = 0 \quad on \quad \hat{\Gamma}, \\ w = \dfrac{\partial w}{\partial \nu} = \Delta w = \dfrac{\partial \Delta w}{\partial \nu} = 0 \quad on \quad \hat{\Gamma} \end{cases} \qquad (6.85)$$

where $\mathcal{A}$ is the Koiter operator, given by the formula (6.27), and $\mathcal{O}_j$ denote some terms with the orders of the derivatives less than j.

Let $\mathbf{R}$ be the smallest principal radius of curvature of the undeformed middle surface given by (4.25). In general the condition $h/\mathbf{R} \ll 1$ is assumed in the thin shell. Here we need

$$h/\mathbf{R} < \sqrt{3}. \qquad (6.86)$$

In this section we will establish the following uniqueness results.

Theorem 6.5 *Let the condition (6.86) be true. If one of the following assumptions holds:*

(a) $\inf_{x\in\Omega}(|\Pi|^2 - 2\kappa) > 0$; *or*

(b) $|\Pi|^2 = 2\kappa$ *for every* $x \in \Omega$, *where* κ *is the Gauss curvature of the surface* M, *then the problem* (6.84) − (6.85) *has a unique zero solution.*

Remark 6.3 *The assumptions* (a) *and* (b) *are easy to check. Clearly,* (a) *is true when* $\Pi \neq 0$ *and* $\kappa \leq 0$. *Moreover, if* λ_1 *and* λ_2 *are the two eigenvalues of the second fundamental form* Π, *then*

$$|\Pi|^2 - 2\kappa = (\lambda_1 - \lambda_2)^2.$$

Then the assumption (a) *is equivalent to*

$$\lambda_1 \neq \lambda_2 \quad \forall x \in \overline{\Omega}$$

and the assumption (b) *means*

$$\lambda_1 = \lambda_2 \quad \forall x \in \overline{\Omega}.$$

Let us make a preparation for the proof of Theorem 6.5 which will be given at the end of this section.

We introduce some basic Carleman estimates of the first order differential operators from [223] where some of them were given by [191]. Those estimates will play an important role in our proof of Theorem 6.5. Consider a differential operator ∂ on $\mathbb{R}^2$, given by

$$\partial = \partial_{x_1} + \lambda\partial_{x_2} \tag{6.87}$$

where $\partial_{x_i} = \dfrac{\partial}{\partial x_i}$ and $\lambda \in C^1$ with $\mathrm{Im}\,\lambda \neq 0$ at the origin $(0,0)$. From Propositions 1.2 ([223, p. 3]) and 1.4 ([223, p. 45]), we have

Lemma 6.5 *There exist positive constants* C, k_0, T_0, *and* r *such that, for all* $k \geq k_0$ *and* $0 < T \leq T_0$,

$$\int_{F(T,r)} e^{k(x_1-T)^2}|w|^2\, dx_1 dx_2 \leq \frac{C}{k}\int_{F(T,r)} e^{k(x_1-T)^2}|\partial w|^2\, dx_1 dx_2 \tag{6.88}$$

and

$$\int_{F(T,r)} e^{k(x_1-T)^2}(|\partial_{x_1}w|^2 + |\partial_{x_2}w|^2)\, dx_1 dx_2$$
$$\leq C(1 + kT^2)\int_{F(T,r)} e^{k(x_1-T)^2}|\partial w|^2\, dx_1 dx_2 \tag{6.89}$$

for $w \in C^\infty$ *with* $\mathrm{supp}\, w \subset F(T,r)$ *where*

$$F(T,r) = \{\,(x_1, x_2)\,|\, 0 \leq x_1 \leq T,\, |x_2| \leq r\,\}. \tag{6.90}$$

Next, we move the estimates in Lemma 6.5 to the Riemannian manifold M with the induced metric from the Euclidean metric of $\mathbb{R}^3$. Let $x_0 \in M$ be given and let E_1, E_2 be a frame at x_0. Consider a differential operator Φ, given by

$$\Phi = E_1 + \lambda E_2 \tag{6.91}$$

where $\lambda \in C^1$ with $\operatorname{Im} \lambda \neq 0$.

Let $\varphi(x) = (y_1, y_2)$ be a coordinate system on M at x_0 with $\varphi(x_0) = (0, 0)$. We have

Lemma 6.6 *There exist positive constants C, k_0, T_0, and r such that, for all $k \geq k_0$ and $0 < T \leq T_0$,*

$$\int_{\varphi^{-1}(F(T,r))} e^{k(y_1(x)-T)^2} |w|^2 \, dg \leq \frac{C}{k} \int_{\varphi^{-1}(F(T,r))} e^{k(y_1(x)-T)^2} |\Phi w|^2 \, dg \tag{6.92}$$

where dg is the volume element of the induced metric g of M from $\mathbb{R}^3$ and

$$\int_{\varphi^{-1}(F(T,r))} e^{k(y_1(x)-T)^2} (|E_1 w|^2 + |E_2 w|^2) \, dg$$

$$\leq C(1 + kT^2) \int_{\varphi^{-1}(F(T,r))} e^{k(y_1(x)-T)^2} |\Phi w|^2 \, dg \tag{6.93}$$

for $w \in C^\infty$ with $\operatorname{supp} w \subset \varphi^{-1}(F(T,r))$.

Proof. Let

$$E_i = \alpha_{i1}\partial_{y_1} + \alpha_{i2}\partial_{y_2}$$

where $\partial_{y_i} = \dfrac{\partial}{\partial y_i}$ and $(\alpha_{ij})_{2\times 2}$ are real matrices with $\det(\alpha_{ij}) \neq 0$. Then $\operatorname{Im} \lambda \neq 0$ and $\det(\alpha_{ij}) \neq 0$ together imply that

$$\alpha_{11} + \lambda\alpha_{21} \neq 0, \quad \operatorname{Im}\left(\frac{\alpha_{12} + \lambda\alpha_{22}}{\alpha_{11} + \lambda\alpha_{21}}\right) \neq 0.$$

Then Lemma 6.6 follows by applying Lemma 6.5 to Φ since

$$\Phi = (\alpha_{11} + \lambda\alpha_{21})\left(\partial_{y_1} + \frac{\alpha_{12} + \lambda\alpha_{22}}{\alpha_{11} + \lambda\alpha_{21}}\partial_{y_2}\right).$$

$\square$

If we let $\Phi = E_1 + i E_2$, then $|\Phi w|^2 \leq 2|Dw|^2$. Applying Lemma 6.6 repeatedly, we have

Lemma 6.7 *For a positive integer m given, there exist positive constants C, k_0, T_0, and r such that, for all $k \geq k_0$ and $0 < T \leq T_0$,*

$$\sum_{j=0}^{m-1} \int_{\varphi^{-1}(F(T,r))} e^{k(y_1(x)-T)^2} |D^j w|^2 \, dg$$

$$\leq \frac{C}{k} \int_{\varphi^{-1}(F(T,r))} e^{k(y_1(x)-T)^2} |D^m w|^2 \, dg \tag{6.94}$$

for $w \in C^\infty$ with $\text{supp}\, w \subset \varphi^{-1}(F(T,r))$.

Let the operators $\mathbf{\Delta}_\mu$ and $\mathcal{S}$ be given by the formula (6.30) and (6.17), respectively. We introduce a fourth order differential operator $\mathbf{A}$ on $H^4(M, \Lambda)$ by

$$\mathbf{A}W = \mathbf{\Delta\Delta}_\mu W - \beta \mathcal{S}(\delta d)^2 \mathcal{S}W \tag{6.95}$$

where $\mathbf{\Delta} = \delta d + d\delta$ and β is a positive constant.

We need

Lemma 6.8 *We have*

$$[\mathcal{S}, \mathbf{\Delta}]W = \mathcal{O}_1(W) \quad for \quad W \in H^2(\Omega, \Lambda). \tag{6.96}$$

Furthermore, if $|\Pi|^2 = 2\kappa$ for all $x \in \Omega$, then

$$[\mathcal{S}, d\delta]W = \mathcal{O}_1(W), \quad [\mathcal{S}, \delta d]W = \mathcal{O}_1(W) \tag{6.97}$$

for $W \in H^2(\Omega, \Lambda)$.

Proof. Let $x \in \Omega$ be given. Since $\mathcal{S} \colon M_x \to M_x$ is a symmetric, linear operator, there exists an orthonormal basis e_1, e_2 of M_x and real numbers λ_1, λ_2 such that

$$\mathcal{S}e_i = \lambda_i e_i \quad for \quad i = 1, 2. \tag{6.98}$$

Let E_1, E_2 be a frame normal at x such that

$$E_i(x) = e_i, \quad D_{E_i} E_j(x) = 0 \tag{6.99}$$

for $1 \le i, j \le 2$. Then in a neighborhood of x, we have

$$\begin{aligned}
D(\mathcal{S}W)(E_i, E_j) &= \langle D_{E_j}(\mathcal{S}W), E_i \rangle = E_j(\Pi(W, E_i)) - \Pi(W, D_{E_j} E_i) \\
&= D\Pi(W, E_i, E_j) + \Pi(D_{E_j} W, E_i).
\end{aligned} \tag{6.100}$$

Using the formulas (6.98)-(6.100), we obtain at x

$$\begin{aligned}
D^2(\mathcal{S}W)(E_i, E_j, E_k) &= E_k(D(\mathcal{S}W)(E_i, E_j)) \\
&= D^2\Pi(W, E_i, E_j, E_k) + D\Pi(D_{E_k} W, E_i, E_j) + D\Pi(D_{E_j} W, E_i, E_k) \\
&\quad + \Pi(D_{E_k} D_{E_j} W, E_i) \\
&= \lambda_i \langle D_{E_k} D_{E_j} W, E_i \rangle + \mathcal{O}_1(W) = \lambda_i D^2 W(E_i, E_j, E_k) + \mathcal{O}_1(W) \tag{6.101}
\end{aligned}$$

for $1 \le i, j, k \le 2$.

It follows from the formulas (6.98), (6.99), (6.101), and (1.114), that at x

$$\begin{aligned}
\mathbf{\Delta}\mathcal{S}W(x) &= -\sum_j D_{E_j} D_{E_j}(\mathcal{S}W) + \sum_{ij} \mathbf{R}(E_i, E_j, \mathcal{S}W, E_j) E_i \\
&= -\sum_{ij} D^2(\mathcal{S}W)(E_i, E_j, E_j) E_i + \kappa \mathcal{S}W \\
&= \mathcal{S}\mathbf{\Delta}W + \mathcal{O}_1(W).
\end{aligned}$$

Next, let $|\Pi|^2 = 2\kappa$ for all $x \in \Omega$. Then $\lambda_1 = \lambda_2 = \lambda$ for all $x \in \Omega$. By the formulas (6.101) and (1.119), we have

$$
d\delta \mathcal{S}W = -\sum_{ij} \langle D_{E_i} D_{E_j}(\mathcal{S}W), E_j \rangle E_i = -\sum_{ij} D^2(\mathcal{S}W)(E_j, E_j, E_i) E_i
$$
$$
= \mathcal{S}d\delta W + \mathcal{O}_1(W).
$$

Similarly, we have $\delta d\mathcal{S}W = \mathcal{S}\delta dW + \mathcal{O}_1(W)$. □

We consider the structure of the operator $\mathbf{A}$ in a frame field. To this end, we assume that $x_0 \in M$ is given such that

$$
|\Pi|^2(x_0) > 2\kappa(x_0). \tag{6.102}
$$

Then there are two distinct eigenvalues λ_1 and λ_2 of the operators $\mathcal{S}$ for x in a neighborhood of x_0. Then there is a frame E_1, E_2 at x_0 such that $\mathcal{S}E_i = \lambda_i E_i$ for x in a neighborhood of x_0. We have the formula

$$
\Pi = \lambda_1 E_1 \otimes E_1 + \lambda_2 E_2 \otimes E_2. \tag{6.103}
$$

Lemma 6.9 *Let $x_0 \in M$ be such that the assumption (6.102) holds true. Let a frame E_1, E_2 at x_0 be such that the formula (6.103) is true. If*

$$
W = w_1 E_1 + w_2 E_2, \tag{6.104}
$$

then

$$
-2\mathbf{A}W = \Delta f_1 E_1 + \Delta f_2 E_2 + \mathcal{O}_3(W) \tag{6.105}
$$

where

$$
f_1 = 2E_1^2 w_1 + a_1 E_2^2 w_1 + \varsigma E_1 E_2 w_2, \tag{6.106}
$$

$$
f_2 = \varsigma E_1 E_2 w_1 + a_2 E_1^2 w_2 + 2E_2^2 w_2, \tag{6.107}
$$

$$
\begin{cases} a_i = 1 - \mu + 2\beta\lambda_i^2, \\ \varsigma = 1 + \mu - 2\beta\kappa, \end{cases} \quad for \quad i = 1, 2, \tag{6.108}
$$

where κ is the Gauss curvature.

Proof. Since $\mathbf{\Delta}\delta d = (\delta d)^2$, we have by the formula (6.96)

$$
\mathbf{A}W = \mathbf{\Delta}(\Delta_\mu W - \beta \mathcal{S}\delta d\mathcal{S}W) + \mathcal{O}_3(W). \tag{6.109}
$$

For the expression (6.105) it suffices to compute $\Delta_\mu W - \beta \mathcal{S}\delta d\mathcal{S}W$.

Using the formulas (6.104), (1.117), and (1.119), we have

$$
\delta dW = -\sum_i D_{E_i} D_{E_i} W + \sum_{ij} \langle D_{E_i} D_{E_j} W, E_i \rangle E_j + \mathcal{O}_1(W)
$$
$$
= (E_1 E_2 w_2 - E_2^2 w_1) E_1 + (E_1 E_2 w_1 - E_1^2 w_2) E_2 + \mathcal{O}_1(W) \tag{6.110}
$$

and

$$d\delta W = -\sum_{ij} \langle D_{E_i} D_{E_j} W, E_j \rangle E_i + \mathcal{O}_1(W)$$

$$= -(E_1^2 w_1 + E_1 E_2 w_2)E_1 - (E_2 E_1 w_1 + E_2^2 w_2)E_2 + \mathcal{O}_1(W). \quad (6.111)$$

It follows from the formulas (6.30), (6.110) and (6.111) that

$$2\Delta_\mu W = [2E_1^2 w_1 + (1-\mu)E_2^2 w_1 + (1+\mu)E_1 E_2 w_2]E_1$$

$$+[(1-\mu)E_1^2 w_2 + 2E_2^2 w_2 + (1+\mu)E_1 E_2 w_1]E_2 + \mathcal{O}_1(W). \quad (6.112)$$

In addition, replacing W with $\mathcal{S}W$ in the formula (6.110) and then applying the operator $\mathcal{S}$ to it, we obtain

$$\mathcal{S}\delta d\mathcal{S}W = \lambda_1(\lambda_2 E_1 E_2 w_2 - \lambda_1 E_2^2 w_1)E_1$$

$$+\lambda_2(\lambda_1 E_1 E_2 w_1 - \lambda_2 E_1^2 w_2)E_2 + \mathcal{O}_1(W). \quad (6.113)$$

Then it follows from the formulas (6.112) and (6.113) that

$$2\Delta_\mu W - 2\beta \mathcal{S}\delta d\mathcal{S}W = f_1 E_1 + f_2 E_2. \quad (6.114)$$

Finally, the formula (6.105) follows from the formulas (6.109) and (6.114). □

A simple computation yields

Lemma 6.10 *Let $x_0 \in \Omega$ be such that the condition (6.102) holds. Let $\beta > 0$ satisfy*

$$\beta|\Pi|^2(x_0) < 1 - \mu. \quad (6.115)$$

Then

$$(2+\varsigma)^2 > a_1 a_2 > (2-\varsigma)^2 \quad (6.116)$$

where a_i and ς are given in the formulas (6.108).

Now we establish a Carleman estimate for the operator **A**.

Theorem 6.6 *Let $x_0 \in \Omega$ be such that the condition (6.102) holds. Let $\beta > 0$ satisfy the inequality (6.115). Let $\varphi(x) = (y_1, y_2)$ be a coordinate system on M at x_0 with $\varphi(x_0) = (0,0)$. Then there exist positive constants C, k_0, T_0, and r such that, for all $k \geq k_0$ and $0 < T \leq T_0$,*

$$\sum_{j=0}^{3} \int_{\varphi^{-1}(F(T,r))} e^{k(y_1(x)-T)^2} |D^j W|^2 \, dg \quad (6.117)$$

$$\leq \frac{C}{k} \int_{\varphi^{-1}(F(T,r))} e^{k(y_1(x)-T)^2} |\mathbf{A}W|^2 \, dg \quad (6.118)$$

for every $W \in C^\infty(M, \Lambda)$ such that $\operatorname{supp} W \subset \varphi^{-1}(F(T,r))$ where $F(T,r)$ is given by the formula (6.90).

Proof. **Step 1** Denote $W = (w_1, w_2)^\tau$. We define an operator Ψ on $L^2(\Omega) \times L^2(\Omega)$ by

$$\Psi W = G_0 E_1^4 W + G_1 E_2^4 W + G_2 E_2^3 E_1 W$$
$$+ G_3 E_2^2 E_1^2 W + G_4 E_2 E_1^3 W \tag{6.119}$$

where

$$E_2^j E_1^{4-j} W = (E_2^j E_1^{4-j} w_1, \ E_2^j E_1^{4-j} w_2)^\tau, \quad 0 \le j \le 4, \tag{6.120}$$

$$G_0 = \begin{pmatrix} 2 & 0 \\ 0 & a_2 \end{pmatrix}, \quad G_1 = \begin{pmatrix} a_1 & 0 \\ 0 & 2 \end{pmatrix}, \tag{6.121}$$

$$G_2 = G_4 = \begin{pmatrix} 0 & \varsigma \\ \varsigma & 0 \end{pmatrix}, \quad G_3 = \begin{pmatrix} a_1 + 2 & 0 \\ 0 & a_2 + 2 \end{pmatrix} \tag{6.122}$$

where a_i and ς are given in the formulas (6.108).

Let f_1 and f_2 be given by the formulas (6.106) and (6.107), respectively. It is easy to check that

$$\Psi W = \begin{pmatrix} \Delta f_1 \\ \Delta f_2 \end{pmatrix} + \mathcal{O}_3(W). \tag{6.123}$$

Let

$$Z = (z_1, z_2, z_3, z_4)^\tau \quad \text{on} \quad \left(L^2(\Omega) \right)^8 \tag{6.124}$$

where

$$z_j = (E_2^{4-j} E_1^{j-1} w_1, \ E_2^{4-j} E_1^{j-1} w_2)^\tau, \quad 1 \le j \le 4.$$

It is easy to check that the following relation holds

$$E_1 Z + J E_2 Z = \left(\mathcal{O}_3(W), \mathcal{O}_3(W), \mathcal{O}_3(W), G_0^{-1} \Psi W \right)^\tau \tag{6.125}$$

where

$$J = \begin{pmatrix} 0 & -I_2 & 0 & 0 \\ 0 & 0 & -I_2 & 0 \\ 0 & 0 & 0 & -I_2 \\ G_0^{-1} G_1 & G_0^{-1} G_2 & G_0^{-1} G_3 & G_0^{-1} G_4 \end{pmatrix} \tag{6.126}$$

and I_2 is the unit matrix of 2×2.

Step 2 We diagonalize the matrix J. Noting that

$$2\lambda^4 + (a_1 + 2)\lambda^2 + a_1 = (\lambda^2 + 1)(2\lambda^2 + a_1) \quad \text{and}$$
$$a_2 \lambda^4 + (a_2 + 2)\lambda^2 + 2 = (\lambda^2 + 1)(a_2 \lambda^2 + 2),$$

we have

$$\det(\lambda I - J) = \frac{1}{2a_2} \det(G_0 \lambda^4 - G_4 \lambda^3 + G_3 \lambda^2 - G_2 \lambda + G_1)$$

$$= \frac{1}{2a_2} \{ [2\lambda^4 + (a_1 + 2)\lambda^2 + a_1][a_2 \lambda^4 + (a_2 + 2)\lambda^2 + 2]$$
$$- \varsigma^2 (\lambda^2 + 1)^2 \lambda^2 \}$$

$$= \frac{1}{2a_2} (\lambda^2 + 1)^2 f(\lambda^2) \tag{6.127}$$

where
$$f(\lambda^2) = 2a_2\lambda^4 + (4 + a_1a_2 - \varsigma^2)\lambda^2 + 2a_1. \tag{6.128}$$

Consider solutions to the equation $f(\lambda^2) = 0$. Let

$$2a_2h^2 + (4 + a_1a_2 - \varsigma^2)h + 2a_1 = 0. \tag{6.129}$$

The inequality (6.116) implies that

$$(4 + a_1a_2 - \varsigma^2)^2 - 16a_1a_2$$
$$= (2 - \varsigma - \sqrt{a_1a_2})(2 + \varsigma - \sqrt{a_1a_2})[(2 + \sqrt{a_1a_2})^2 - \varsigma^2] < 0.$$

Then for a solution of the equation (6.129) and by (6.116) again

$$\operatorname{Re} h = -\frac{4 + a_1a_2 - \varsigma^2}{4a_2} < -\frac{4 - \varsigma}{2a_2} < 0. \tag{6.130}$$

Then the equation $f(\lambda^2) = 0$ has four roots h_1, h_2, h_3, and h_4 to satisfy
(a) $h_j \neq \pm i$ and $\operatorname{Im} h_j \neq 0$ for $1 \leq j \leq 4$;
(b) $h_j \neq h_l$ for $j \neq l$, $1 \leq j, l \leq 4$.
Then the matrix J has four eigenfunctions corresponding to eigenvalues h_j, respectively. On the other hand, a simple computation shows that the eigenvalues i and $-i$ of the matrix J have linearly independent eigenvectors

$$(e_{2j}, -i\,e_{2j}, -e_{2j}, i\,e_{2j}) \quad \text{for} \quad j = 1, 2$$

where $e_{21} = (1, 0)$ and $e_{22} = (0, 1)$ and

$$(e_{2j}, i\,e_{2j}, -e_{2j}, -i\,e_{2j}) \quad \text{for} \quad j = 1, 2,$$

respectively. Therefore, there are eight linearly independent eigenvectors of the matrix J and we have a differentiable, invertible matrix $I\!N$ such that

$$I\!N^{-1}JI\!N = \operatorname{diagonal}\{i, i, -i, -i, h_1, h_2, h_3, h_4\}. \tag{6.131}$$

Step 3 Let $Y = I\!N^{-1}Z$ where Z is given in (6.124). By Step 2, each component of $E_1Y + I\!N^{-1}JE_2Y$ has a form $E_1Y_j + \lambda_jE_2Y_j$ with $\operatorname{Im} \lambda_j \neq 0$. We apply Lemma 6.6 to $E_1Y_j + \lambda_jE_2Y_j$ and obtain, via the formula (6.125),

$$\int_{\varphi^{-1}(F(T,r))} e^{k(y_1(x)-T)^2}|Z|^2\,dg$$
$$\leq \frac{C}{k}\int_{\varphi^{-1}(F(T,r))} e^{k(y_1(x)-T)^2}(|\Psi W|^2 + |\mathcal{O}_3(W)|^2)\,dg. \tag{6.132}$$

Finally, the estimate (6.118) follows from the relations (6.132), (6.124), (6.105), and (6.94). $\qquad\square$

We are ready to give a proof of Theorem 6.5.

Proof of Theorem 6.5.

(i) Let the assumption (a) hold in Theorem 6.5.

By the formulas (6.27)-(6.29), the system (6.84) consists of the following two equations

$$\mathbf{\Delta}_\mu W + 4\gamma\mathcal{S}\mathbf{\Delta}_\mu\mathcal{S}W + 2\gamma\mathcal{S}d\delta dw$$
$$= \gamma(1-\mu)\mathcal{P}D^2w + \gamma\mu(\Delta w)\mathcal{Q}\Pi + \mathcal{O}_1(\zeta); \tag{6.133}$$

$$\gamma\Delta^2 w = 2\gamma\delta d\delta\mathcal{S}W - \gamma\mu\Delta\langle W, \mathcal{Q}\Pi\rangle$$
$$-\gamma(1-\mu)\{\delta[\mathrm{i}(Dw)\mathrm{c} - \mathcal{Q}(\mathrm{i}(W)\Pi)] - \langle D^2w, \mathrm{c}\rangle\}$$
$$-\gamma[(1-\mu)\kappa - 2\mu\operatorname{tr}\mathrm{c}]\Delta w + \mathcal{O}_1(\zeta) + \mathcal{O}_2(w). \tag{6.134}$$

Let $x_0 \in \Omega$ be given and let E_1, E_2 be a frame field at x_0 such that the formula (6.103) holds. Noting that $\delta dw = -\Delta w$ and $\mathcal{S}E_i = \lambda_i E_i$, we have, from the equation (6.133),

$$\lambda_1 E_1(\Delta w)E_1 + \lambda_2 E_2(\Delta w)E_2 = \frac{1}{2\gamma}[\mathbf{\Delta}_\mu W + 4\gamma\mathcal{S}\mathbf{\Delta}_\mu\mathcal{S}W$$
$$-\gamma(1-\mu)\mathcal{P}D^2w - \gamma\mu(\Delta w)\mathcal{Q}\Pi + \mathcal{O}_1(\zeta)]. \tag{6.135}$$

The condition (6.102) implies that $|\Pi|^2 > 0$ in a neighborhood of x_0. We may assume that $\lambda_1 \neq 0$. Applying the operator E_j to both the sides of the equation (6.135) yields

$$E_j E_1(\Delta w) = \mathcal{O}_3(\zeta) \quad \text{for} \quad j = 1,\, 2. \tag{6.136}$$

In addition, using the equation (6.134), we have

$$E_2 E_2(\Delta w) = \Delta^2 w - E_1 E_1(\Delta w) = \mathcal{O}_3(\zeta). \tag{6.137}$$

It follows from the relations (6.135) and (6.137) that

$$D^2(\Delta w) = \mathcal{O}_3(\zeta). \tag{6.138}$$

Moreover, the equation (6.138) yields

$$\mathbf{\Delta}(D^2 w(E_i, E_j)) = \mathbf{\Delta}E_i E_j(w) + \mathcal{O}_3(\zeta) = -E_i E_j(\Delta w) + \mathcal{O}_3(\zeta) = \mathcal{O}_3(\zeta)$$

for $1 \leq i,\, j \leq 2$, that is,

$$\mathbf{\Delta}D^2 w = \mathcal{O}_3(\zeta). \tag{6.139}$$

Applying the operator d to the equation (6.134) gives

$$\gamma d\Delta^2 w = 2\gamma(d\delta)^2\mathcal{S}W + \mathcal{O}_3(\zeta). \tag{6.140}$$

Using the relations (6.96), (6.138), and (6.139), we obtain

$$\mathbf{\Delta}\mathcal{S}d\delta dw = -\mathbf{\Delta}\mathcal{S}d\Delta w = -\mathcal{S}\mathbf{\Delta}d\Delta w - [\mathbf{\Delta}, \mathcal{S}]d\Delta w$$
$$= -\mathcal{S}\mathbf{\Delta}d\Delta w + \mathcal{O}_3(\zeta)$$
$$= \mathcal{S}d(\Delta^2 w) + \mathcal{O}_3(\zeta). \tag{6.141}$$

Applying the operator $\mathbf{\Delta} = d\delta + \delta d$ to both the sides of the equation (6.133), we obtain, via the formulas (6.96), (6.30), (6.138), (6.139), (6.140), and (6.141)

$$
\begin{aligned}
\mathbf{A}W &= \mathbf{\Delta}\mathbf{\Delta}_\mu W - \beta \mathcal{S}(\delta d)^2 \mathcal{S}W \\
&= \mathbf{\Delta}[\mathbf{\Delta}_\mu W + 4\gamma \mathcal{S}\mathbf{\Delta}_\mu \mathcal{S}W + 2\gamma \mathcal{S}d\delta dw] \\
&\quad + 4\gamma \mathcal{S}(d\delta)^2 \mathcal{S}W - 2\gamma \mathbf{\Delta}\mathcal{S}d\delta dw + \mathcal{O}_3(\zeta) \\
&= 2\gamma[\mathcal{S}d(\mathbf{\Delta}^2 w) - \mathbf{\Delta}\mathcal{S}d\delta dw] + \mathcal{O}_3(\zeta) = \mathcal{O}_3(\zeta)
\end{aligned}
\tag{6.142}
$$

where $\beta = 2\gamma(1 - \mu)$.

On the other hand, a similar argument as in the proof of Proposition 4.3 yields

$$
\begin{cases}
W|_{\hat{\Gamma}} = DW|_{\hat{\Gamma}} = D^2 W|_{\hat{\Gamma}} = D^3 W|_{\hat{\Gamma}} = 0, \\
w|_{\hat{\Gamma}} = \dfrac{\partial w}{\partial \nu}|_{\hat{\Gamma}} = \Delta w|_{\hat{\Gamma}} = \dfrac{\partial \Delta w}{\partial \nu}|_{\hat{\Gamma}} = 0.
\end{cases}
\tag{6.143}
$$

Moreover, by the condition (6.86), we have

$$
\beta |\Pi|^2(x_0) = \frac{h^2(1 - \mu)}{6}(\lambda_1^2 + \lambda_2^2) \leq \frac{h^2}{3\mathbf{R}^2}(1 - \mu) < 1 - \mu.
$$

Then the Carleman estimate in Theorem 6.6 holds true for the operator $\mathbf{A}$. By the conditions (6.143), we have that $\zeta = 0$ on Ω after we apply Theorem 6.6 of this section to the component $\mathbf{A}W$ and Theorem 1.1 of Chapter Two of [223] to the component $\mathbf{\Delta}^2 w$, respectively.

(ii) Next, we assume that $|\Pi|^2 = 2\kappa$ for all $x \in \Omega$. It is easy to check that there is a function $\varphi \in C^\infty$ such that

$$
\Pi = \varphi g, \quad \mathcal{S}W = \varphi W \quad \text{for} \quad x \in \Omega
\tag{6.144}
$$

where g is the induced metric of the surface M from the Euclidean metric of $\mathbb{R}^3$. Moreover, we have $D\Pi = g \otimes D\varphi$. Then the symmetry of $D\Pi$ (Corollary 1.1) implies that $\varphi = \lambda$ is a constant. Then $D\Pi = 0$, $\mathcal{P} = 0$, and $\mathcal{Q}\Pi = 0$. Using those results in the formulas (6.28) and (6.29), the system (6.84) becomes

$$
(1 + 4\gamma\lambda^2)\mathbf{\Delta}_\mu W - 2\gamma\lambda d(\Delta w) = \mathcal{O}_1(\zeta);
\tag{6.145}
$$

$$
\gamma(\mathbf{\Delta}^2 w - 2\lambda\delta d\delta W) = \mathcal{O}_2(\zeta).
\tag{6.146}
$$

Applying the differential operator d to the equation (6.146), we have

$$
d(\mathbf{\Delta}^2 w) = 2\lambda(d\delta)^2 W + \mathcal{O}_2(\zeta).
\tag{6.147}
$$

Applying the differential operator $d\delta$ to the equation (6.145) and using the formula (6.147), we obtain

$$
(d\delta)^2 W = \frac{2\gamma\lambda}{1 + 4\gamma\lambda^2} d(\mathbf{\Delta}^2 w) + \mathcal{O}_3(\zeta) = \frac{4\gamma\lambda^2}{1 + 4\gamma\lambda^2}(d\delta)^2 W + \mathcal{O}_3(\zeta)
$$

which yields

$$(d\delta)^2 W = \mathcal{O}_3(\zeta). \tag{6.148}$$

Furthermore, applying the differential operator δd to the equation (6.145) gives

$$(\delta d)^2 W = \mathcal{O}_3(\zeta). \tag{6.149}$$

It follows from the formulas (6.148) and (6.149) that

$$\mathbf{\Delta}^2 W = \mathcal{O}_3(\zeta). \tag{6.150}$$

Finally, we apply [170] to the system (6.146), (6.150) and (6.143) to have $\zeta = 0$. $\qquad\square$

6.3 Multiplier Identities

In this section we shall establish some multiplier identities for the Koiter model. The multipliers we will need are

$$f\zeta_1, \quad f\zeta_2, \quad \text{and} \quad m(\zeta) = (D_V W, V(w))$$

where $\zeta = (W, w)$, $\zeta_1 = (W, 0)$, $\zeta_2 = (0, w)$, f is a function, and V is a vector field which satisfies the following assumption (6.151).

Let V be a vector field on Ω such that there is a function ϑ on Ω that satisfies the relation

$$DV(X, X) = \vartheta(x)|X|^2 \quad \text{for all} \quad X \in \Omega_x \quad x \in \Omega. \tag{6.151}$$

For the existence of such a vector field, see Section 4.5.

For $W \in H^1(\Omega, \Lambda)$, we define

$$\begin{aligned} F(V, W)(X, Y) &= \Pi(D_X W, D_Y V) + \Pi(D_X V, D_Y W) \\ &\quad + D_V \Pi(D_X W, Y) + D_V \Pi(X, D_Y W) \end{aligned} \tag{6.152}$$

for all $X, Y \in M_x$, $x \in \Omega$.

We need

Lemma 6.11 *Let a vector field V be such that the relation (6.151) holds and let $m(\zeta) = (D_V W, V(w))$. Then*

$$\begin{aligned} \rho(m(\zeta)) &= D_V \rho(\zeta) + \rho(\zeta)(D.V, \cdot) + \rho(\zeta)(\cdot, D.V) \\ &\quad + F(V, W) + \mathcal{O}_0(W) + \mathcal{O}_1(w) \end{aligned} \tag{6.153}$$

where the change of the curvature tensor ρ is defined by the formula (6.2).

Proof. It follows from the formula (4.133) that

$$D^2(V(w)) = D_V D^2 w + 2G(V, D^2 w) + \mathcal{O}_1(w) \qquad (6.154)$$

where the tensor $G(V, D^2 w)$ is defined by the formula (4.131) in Chapter 4.

Let $x \in \Omega$ be given and let E_1, E_2 be a frame normal at x. Then

$$[V, E_j] = D_V E_j - D_{E_j} V = -D_{E_j} V \quad \text{at } x$$

for $j = 1,2$. Then

$$D_{E_j} D_V W = D_{E_j} D_V W + D_{D_{E_j} V} W + \mathbf{R}_{V E_j} W \quad \text{at} \quad x \qquad (6.155)$$

where $\mathbf{R}_{V E_j}$ is *the curvature operator*.

By the formulas (6.16) and (6.155), we have at x

$$\Pi(E_i, D_{E_j} D_V W) = \Pi(E_i, D_V D_{E_j} W) + \Pi(E_i, D_{D_{E_j} V} W) + \Pi(E_i, R_{V E_j} W)$$
$$= V(\Pi(E_i, D_{E_j} W)) - D_V \Pi(E_i, D_{E_j} W) + \Pi(E_i, D_{D_{E_j} V} W)$$
$$+ \mathbf{R}(V, E_j, W, \mathcal{S} E_i) \qquad (6.156)$$

for $1 \le i, j \le 2$.

Using the formulas (6.154) and (6.156), we obtain at x

$$\rho(m(\zeta))(E_i, E_j) = D^2(V(w))(E_i, E_j) - \Pi(E_i, D_{E_j} D_V W)$$
$$- \Pi(E_j, D_{E_i} D_V W) - \mathrm{i}\,(D_V W) D \Pi(E_i, E_j) - V(w) \,\mathrm{c}\,(E_i, E_j)$$
$$= V(\rho(E_i, E_j)) + \rho(E_i, D_{E_j} V) + \rho(E_j, D_{E_i} V)$$
$$+ F(V, W)(E_i, E_j) + \mathcal{O}_0(W) + \mathcal{O}_1(w)$$

which yields the formula (6.153).   □

Lemma 6.12 *We have*

$$2\mathcal{B}(\zeta, m(\zeta)) = \int_\Gamma B(\zeta, \zeta)\langle V, \nu \rangle \, d\Gamma + 2\wp(V, \zeta) + \mathrm{lo}\,(\zeta)$$
$$+ 2 \int_\Omega \vartheta[\gamma a(\rho(\zeta), \rho(\zeta)) - a(\Upsilon(\zeta), \Upsilon(\zeta))] \, dx \qquad (6.157)$$

where $\mathcal{B}(\cdot, \cdot)$ and $B(\cdot, \cdot)$ are given by the formulas (6.5) and (6.4), respectively, and

$$\wp(V, \zeta) = \int_\Omega [a(\Upsilon(\zeta), G(V, DW)) + \gamma a(\rho(\zeta), F(V, W))] \, dx. \qquad (6.158)$$

Proof. The expression for $a(\Upsilon(\zeta), \Upsilon(m(\zeta)))$ has been given in the formula (4.143) in Chapter 4. Here we need to compute $a(\rho(\zeta), \rho(m(\zeta)))$.

Using Theorem 3.4, we have

$$a(\rho(\zeta), \rho(\zeta)(\cdot, D.V)) = \vartheta(x) a(\rho(\zeta), \rho(\zeta)). \qquad (6.159)$$

It follows from the formulas (6.153), (6.159), and (6.151) that

$$
\begin{aligned}
2a(\rho(\zeta), \rho(m(\zeta))) &= V(a(\rho(\zeta), \rho(\zeta))) + 4\vartheta a(\rho(\zeta), \rho(\zeta)) \\
&\quad + 2a(\rho(\zeta), F(V, W)) + \mathrm{lo}\,(\zeta) \\
&= \mathrm{div}\,[a(\rho(\zeta), \rho(\zeta))V] + 2\vartheta a(\rho(\zeta), \rho(\zeta)) \\
&\quad + 2a(\rho(\zeta), F(V, W)) + \mathrm{lo}\,(\zeta)
\end{aligned} \tag{6.160}
$$

Using the formula (4.143) in Chapter 4 and the formula (6.160), we obtain

$$
\begin{aligned}
2B(\zeta, m(\zeta)) &= \mathrm{div}\,[B(\zeta, \zeta)V] + 2\vartheta[\gamma a(\rho(\zeta), \rho(\zeta)) - a(\Upsilon(\zeta), \Upsilon(\zeta))] \\
&\quad + 2a(\Upsilon(\zeta), G(V, DW)) + 2\gamma a(\rho(\zeta), F(V, W)) + \mathrm{lo}\,(\zeta).
\end{aligned} \tag{6.161}
$$

Integrating the identity (6.161) over Ω yields the formula (6.157). $\qquad\square$

Let $\zeta = (W, w)$ solve the problem

$$
\zeta_{tt} + \gamma(0, -\Delta w_{tt}) + \mathcal{A}\zeta = 0 \tag{6.162}
$$

where $\mathcal{A}$ is *the Koiter operator*, given by the formula (6.27), and $\gamma = h^2/12$. We have

Theorem 6.7 *Let $\zeta = (W, w)$ solve the problem (6.162) and let p be a function. Then*

$$
\int_{\Sigma} \partial(\mathcal{A}\zeta, p\zeta_1)\, d\Sigma = \int_{Q} p[B(\zeta, \zeta_1) - |W_t|^2]\, dQ + L(\zeta); \tag{6.163}
$$

$$
\begin{aligned}
\int_{\Sigma} \left[\partial(\mathcal{A}\zeta, p\zeta_2) + \gamma pw \frac{\partial w_{tt}}{\partial \nu}\right] d\Sigma \\
= \int_{Q} p[B(\zeta, \zeta_2) - w_t^2 - \gamma|Dw_t|^2]\, dQ + L(\zeta)
\end{aligned} \tag{6.164}
$$

where $\zeta_1 = (W, 0)$, $\zeta_2 = (0, w)$, $\Sigma = (0, T) \times \Gamma$ and $Q = (0, T) \times \Omega$.

Proof. Clearly, we have

$$
\Upsilon(p\zeta_i) = p\Upsilon(\zeta_i) + \mathrm{lo}\,(\zeta), \quad \rho(p\zeta_i) = p\rho(\zeta_i) + \mathrm{lo}\,(\zeta)
$$

for $i = 1, 2$, which yield, via the formula (6.4),

$$
B(\zeta, p\zeta_i) = pB(\zeta, \zeta_i) + \mathrm{lo}\,(\zeta). \tag{6.165}
$$

We multiply the equation (6.162) by $p\zeta_1$, integrate by parts, and obtain, by the Green formula (6.25) and the formula (6.165),

$$
\begin{aligned}
0 &= \int_0^T \left(\zeta_{tt} + \gamma(0, -\Delta w_{tt}) + \mathcal{A}\zeta,\ p\zeta_1\right)_{L^2(\Omega, \Lambda) \times L^2(\Omega)} dt \\
&= \int_Q [\langle W_{tt}, pW\rangle + B(\zeta, p\zeta_1)]\, dQ - \int_{\Sigma} \partial(\mathcal{A}\zeta, p\zeta_1)\, d\Sigma \\
&= \int_Q p[B(\zeta, \zeta_1) - |W_t|^2]\, dQ - \int_{\Sigma} \partial(\mathcal{A}\zeta, p\zeta_1)\, d\Sigma + L(\zeta)
\end{aligned}
$$

which is the formula (6.163).

Similarly, we multiply the equation (6.162) by $p\zeta_2$ and obtain the identity (6.164). □

Theorem 6.8 *Let $\zeta = (W, w)$ solve the problem (6.162). Then*

$$\int_\Sigma [|\zeta_t|^2 + \gamma|Dw_t|^2 - B(\zeta, \zeta)]\langle V, \nu\rangle d\Sigma$$

$$+2\int_\Sigma [\partial(A\zeta, m(\zeta)) + \gamma V(w)\frac{\partial w_{tt}}{\partial \nu}]d\Sigma$$

$$= 2Z|_0^T + 2\int_0^T \wp(V, \zeta)\, dt + L(\zeta)$$

$$+2\int_Q \vartheta[|\zeta_t|^2 + \gamma a(\rho(\zeta), \rho(\zeta)) - a(\Upsilon(\zeta), \Upsilon(\zeta))]\, dQ \qquad (6.166)$$

where

$$Z = (\zeta_t, m(\zeta))_{L^2(\Omega, \Lambda) \times L^2(\Omega)} + \gamma(Dw_t, D(V(w)))_{L^2(\Omega, \Lambda)}. \qquad (6.167)$$

Proof. We multiply the equation (6.162) by $2m(\zeta) = 2(D_V W, V(w))$ and integrate by parts. We have

$$\int_Q \langle \zeta_{tt}, 2m(\zeta)\rangle\, dQ = (\zeta_t, 2m(\zeta))_{L^2(\Omega, \Lambda) \times L^2(\Omega)}\Big|_0^T$$

$$+2\int_Q \vartheta|\zeta_t|^2\, dQ - \int_\Sigma |\zeta_t|^2\langle V, \nu\rangle\, d\Sigma; \qquad (6.168)$$

$$\int_Q \langle(0, \Delta w_{tt}), 2m(\zeta)\rangle\, dQ - 2\int_\Sigma V(w)\frac{\partial w_{tt}}{\partial \nu}\, d\Sigma = -2\int_Q \langle Dw_{tt}, D(V(w))\rangle\, dQ$$

$$= -2(Dw_t, D(V(w))_{L^2(\Omega, \Lambda)}\Big|_0^T + 2\int_Q \langle Dw_t, D(V(w_t))\rangle\, dQ$$

$$= -2(Dw_t, D(V(w))_{L^2(\Omega, \Lambda)}\Big|_0^T + \int_\Sigma |Dw_t|^2\langle V, \nu\rangle\, d\Sigma \qquad (6.169)$$

where the following formula is used:

$$2\langle Dw_t, D(V(w_t))\rangle = 2DV(Dw_t, Dw_t) + \text{div}\,(|Dw_t|^2 V) - |Dw_t|^2\,\text{div}\,V$$

$$= \text{div}\,(|Dw_t|^2 V).$$

By using the formulas (6.168), (6.169), and the Green formula (6.25) in the equation (6.162), we obtain the identity (6.166). □

6.4   Observability Estimates from Boundary

We consider boundary controllability for the Koiter model. For that purpose, we introduce some assumptions. Let $\Gamma_1 \subset \Gamma$ and $\Gamma_1 \neq \emptyset$. By the ellipticity of the Koiter shell, Theorem 6.1, there is $\lambda_0 > 0$ such that

$$\lambda_0 \mathcal{B}(\zeta, \zeta) \geq \|\zeta\|^2_{H^1_{\Gamma_1}(\Omega, \Lambda) \times H^2_{\Gamma_1}(\Omega)} \tag{6.170}$$

for all $\zeta \in H^1_{\Gamma_1}(\Omega, \Lambda) \times H^2_{\Gamma_1}(\Omega)$.

Definition 6.2 *Let a vector field V on $\overline{\Omega}$ be such that the relation (6.151) holds. V is said to be an escape vector field for the Koiter shell if the following inequality holds:*

$$\min_{x \in \Omega} \vartheta(x) > \max_{x \in \Omega} \upsilon(x) \tag{6.171}$$

where

$$\upsilon(x) = 1 + [2 + \lambda_0(1 + \mu)(1 + 2\gamma|\Pi|^2)]|\iota(x)| + 16\lambda_0(1 + \mu)\gamma|D_V\Pi|^2,$$

and the functions $\vartheta(x)$ and $\iota(x)$ are given in the formulas (6.151) and (5.99), respectively.

Remark 6.4 *By Theorem 4.8, there always exists a vector field V satisfying the assumption (6.151). Then the key to being an escape vector field for the Koiter shell is the assumption (6.171).*

For M being of constant curvature or of revolution, then there exists a vector field V on the whole M with $\iota(x) = 0$ for all $x \in M$, see Theorems 4.9 and 4.10 in Chapter 4. In those cases the condition (6.171) becomes

$$\min_{x \in \Omega} \vartheta(x) > 1 + 16\lambda_0(1 + \mu)\gamma \max_{x \in \Omega} |D_V\Pi|^2.$$

Let us see several examples.

Example 6.1 (*Circular Cylindrical Shells*)   *Consider a circular cylindroid to be defined by*

$$M = \{ (x_1, x_2, x_3) \,|\, x_2^2 + x_3^2 = 1 \}.$$

Then M is of zero curvature with the induced metric g in $\mathbb{R}^3$.

Given $x^0 = (x_1^0, x_2^0, x_3^0) \in M$. Let $A(x^0) = \{ (x_1, -x_2^0, -x_3^0) \,|\, x_1 \in (-\infty, \infty) \}$. Let $\rho(x) = d(x^0, x)$ be the distance function on M from x^0 to $x \in M$. Let

$$V = 2\rho D\rho \quad \text{for} \quad x \in M/A(x^0). \tag{6.172}$$

It is easy to check that

$$D_X V = 2X \quad \forall\, X \in M_x, \ x \in M/A(x^0); \tag{6.173}$$

$$\vartheta(x) = 2, \quad \iota(x) = 0, \quad D_V\Pi = 0 \quad \text{on} \quad M/A(x^0). \tag{6.174}$$

From the relations (6.173) and (6.174) we have reached the following con-clusion: For any $\Omega \subset M/A(x^0)$, V, given by (6.172), is an escape vector field for the Koiter shell.

Example 6.2 (*Spherical Shells*)   Let M be the sphere of the constant curva-ture $\alpha = c^2$. Then

$$\Pi = cg, \quad D\Pi = cDg = 0$$

where g is the induced metric of M in $\mathbb{R}^3$.

Given $x^0 \in M$. Let $\rho(x) = d(x^0, x)$ be the distance function on M from x^0 to $x \in M$. Set $h(x) = -\cos(c\rho(x))$ and define

$$H = Dh = c\sin c\rho D\rho \quad \text{for} \quad x \in M/\{-x^0\}.$$

We obtain

$$DH = c^2 \cos c\rho(x)\, g \quad \text{for} \quad x \in M/\{-x^0\}.$$

We then arrive at the following conclusion: Let

$$\overline{\Omega} \subset \{x \,|\, x \in M,\, d(x^0, x) < \frac{\pi}{2c}\}.$$

Let

$$V = \alpha H \quad \text{for} \quad x \in \Omega$$

where $\alpha = (c^2 \min_{x\in\Omega} \cos c\rho(x) + 1)^{-1}$. Then

$$\vartheta(x) = \alpha c^2 \cos c\rho(x), \quad \iota(x) = 0, \quad D_V\Pi = 0 \quad \text{for} \quad x \in \Omega,$$

that is, V is an escape vector field for the Koiter shell.

Example 6.3 (*Helical Shells*)   Consider a helical surface, defined by

$$M = \{\alpha(t, s) | (t, s) \in \mathbb{R}^2, \quad t > 0\},$$

where

$$\alpha(t, s) = (t\cos s, t\sin s, c_0 s), \quad c_0 > 0.$$

Set

$$E_1 = \frac{\partial}{\partial t} = (\cos s, \sin s, 0), \quad E_2 = \frac{1}{\sqrt{t^2 + c_0^2}}\frac{\partial}{\partial s} = \frac{1}{\sqrt{t^2 + c_0^2}}(-t\sin s, t\cos s, c_0),$$

$$N = \frac{1}{\sqrt{t^2 + c_0^2}}(c_0\sin s, -c_0\cos s, t).$$

Then E_1, E_2 makes up a frame field and N is unit normal vector on the whole surface M.

Let $\Omega \subset M$ be a bounded open set such that $\overline{\Omega} \subset M$. Then

$$c_1 = \inf_{\alpha(t,s)\in\Omega} \frac{t}{\sqrt{t^2+c_0^2}} > 0, \quad c_2 = \sup_{\alpha(t,s)\in\Omega} \frac{t|s|}{\sqrt{t^2+c_0^2}} < \infty.$$

We take

$$V = f_1 E_1 + f_2 E_2 \qquad (6.175)$$

where

$$f_1 = \sqrt{t^2+c_0^2}\Big(\int_0^t \frac{dt}{\sqrt{t^2+c_0^2}} + \frac{1+2c_2+\lambda_0(1+\mu)c_2}{c_1}\Big), \quad f_2 = \sqrt{t^2+c_0^2}s.$$

We obtain

$$DV = \vartheta g + \iota \mathcal{E},$$

$$\vartheta = 1 + \frac{t}{t^2+c_0^2}f_1, \quad \iota = -\frac{st}{\sqrt{t^2+c_0^2}}.$$

It is easy to check that

$$\min_{x\in\Omega} \vartheta(x) \geq 2 + 2c_2 + \lambda_0(1+\mu)c_2;$$

$$\upsilon(x) \leq 1 + 2c_2 + \lambda_0(1+\mu)c_2 + c_3\gamma|\Pi|^2$$

where

$$c_3 = 2\lambda_0(1+\mu) \sup_{\alpha(t,s)\in\Omega} \Big[c_2 + \frac{32t^2}{t^2+c_0^2}\Big(s^2 + \frac{1}{t^2+c_0^2}f_1^2\Big)\Big].$$

To make the error of the thin shell small, h/R has to be small. If h/R is so small that

$$c_3\gamma|\Pi|^2 \leq \frac{c_3}{6}\Big(\frac{h}{R}\Big)^2 < 1, \qquad (6.176)$$

we then have

$$\min_{x\in\Omega} \vartheta(x) > \max_{x\in\Omega} \upsilon(x),$$

that is, under the assumption (6.176), V, given by (6.175), is an escape vector field for the Koiter shell.

Lemma 6.13 *Let V be an escape vector field for the Koiter shell. Then for $\zeta = (W, w) \in H^1(\Omega, \Lambda) \times H^2(\Omega)$*

$$\wp(V, \zeta) \geq \int_\Omega \vartheta(x)[a(\Upsilon(\zeta), \Upsilon(\zeta)) - \gamma a(\rho(\zeta), \rho(\zeta_1))]dx$$

$$- \frac{\max_{x\in\overline{\Omega}} \upsilon(x)}{2} \int_\Omega B(\zeta, \zeta)\,dx + \mathrm{lo}\,(\zeta) \qquad (6.177)$$

where $\wp(V, \zeta)$ is defined by the formula (6.158) and $\zeta_1 = (W, 0)$.

Proof. Note that $a(\Upsilon(\zeta_1), \Upsilon(\zeta_1)) = a(\Upsilon(\zeta), \Upsilon(\zeta)) + \text{lo}(\zeta)$ and the inequality (4.235) in Chapter 4 yields

$$
\begin{aligned}
a(\Upsilon(\zeta), G(V, DW)) \geq{}& \vartheta(x) a(\Upsilon(\zeta), \Upsilon(\zeta)) \\
& -\frac{1+\mu}{2} |\iota(x)| |DW|^2 + \text{lo}(\zeta).
\end{aligned} \tag{6.178}
$$

Let us estimate $a(\rho(\zeta), F(V, W))$. Set

$$
\ae(\cdot, \cdot) = \Pi(D.V, D.W) + \Pi(D.W, D.V); \tag{6.179}
$$

$$
\text{Im}\,(\cdot, \cdot) = D_V \Pi(\cdot,\ D.W) + D_V \Pi(D.W,\ \cdot). \tag{6.180}
$$

Let $x \in \Omega$ be given. Since Π is symmetric, we may take e_1, e_2 to be an orthonormal basis of M_x, with the positive orientation such that

$$
\Pi(e_i, e_i) = \lambda_i, \quad i = 1, 2, \quad \Pi(e_1, e_2) = \Pi(e_2, e_1) = 0. \tag{6.181}
$$

Then the formula (4.176) of Lemma 4.9 in Chapter 4 implies

$$
D_{e_1} V = \vartheta(x) e_1 - \iota(x) e_2 \quad \text{and} \quad D_{e_2} V = \iota(x) e_1 + \vartheta(x) e_2. \tag{6.182}
$$

Denote $W_{ij} = DW(e_i, e_j)$ for $1 \leq i, j \leq 2$. Then use of the relations (6.181) and (6.182) gives

$$
\begin{aligned}
\ae(e_1, e_1) ={}& \Pi(D_{e_1} V, D_{e_1} W) + \Pi(D_{e_1} W, D_{e_1} V) \\
={}& \vartheta(x)[\Pi(e_1, D_{e_1} W) + \Pi(D_{e_1} W, e_1)] - 2\lambda_2 \iota(x) W_{21} \\
={}& -\vartheta(x)\rho(\zeta_1)(e_1, e_1) - 2\lambda_2 \iota(x) W_{21} + \text{lo}(\zeta); \tag{6.183} \\
\ae(e_1, e_2) ={}& -\vartheta(x)\rho(\zeta_1)(e_1, e_2) + \iota(x)[\lambda_1 W_{11} - \lambda_2 W_{22}] + \text{lo}(\zeta); \tag{6.184} \\
\ae(e_2, e_2) ={}& -\vartheta(x)\rho(\zeta_1)(e_2, e_2) + 2\lambda_1 \iota(x) W_{12} + \text{lo}(\zeta). \tag{6.185}
\end{aligned}
$$

From the formulas (6.183)-(6.185), one obtains

$$
\begin{aligned}
a(\rho(\zeta), \ae) ={}& -\vartheta(x) a(\rho(\zeta), \rho(\zeta_1)) + 2\iota(x)\{(1 - \mu)[-\lambda_2 W_{21} \rho(\zeta)(e_1, e_1) \\
& +\lambda_1 W_{12} \rho(e_2, e_2) + (\lambda_1 W_{11} - \lambda_2 W_{22})\rho(\zeta)(e_2, e_1)] \\
& +\mu \operatorname{tr} \rho(\zeta)[\lambda_1 W_{12} - \lambda_2 W_{21}]\} + \text{lo}(\zeta). \tag{6.186}
\end{aligned}
$$

We apply Schwartz's inequality to the second term in the right hand side of the inequality (6.186) to give

$$
\begin{aligned}
a(\rho(\zeta),\ \ae) \geq{}& -\vartheta(x) a(\rho(\zeta), \rho(\zeta_1)) - |\iota(x)| a(\rho(\zeta), \rho(\zeta)) \\
& -|\iota(x)| |\Pi|^2 |DW|^2 + \text{lo}(\zeta). \tag{6.187}
\end{aligned}
$$

On the other hand, we apply Schwartz's inequality again to obtain

$$
\begin{aligned}
2a(\rho(\zeta),\ \text{Im}\,) \geq{}& -a(\rho(\zeta), \rho(\zeta)) - a(\text{Im}\,, \text{Im}\,) \\
\geq{}& -a(\rho(\zeta), \rho(\zeta)) - 16(1 + \mu)|D_V \Pi|^2 |DW|^2. \tag{6.188}
\end{aligned}
$$

Finally, we combine the inequalities (6.178), (6.187), and (6.188) to get the inequality (6.177). □

Let $\zeta = (W, w)$ solve the problem (6.162). Define the total energy of the shell by

$$E(t) = \frac{1}{2}[\|W_t\|^2_{L^2(\Omega,\Lambda)} + \|w_t\|^2 + \gamma\|Dw_t\|^2_{L^2(\Omega,\Lambda)} + \mathcal{B}(\zeta, \zeta)], \qquad (6.189)$$

and we set

$$\sigma_1 = \min_{x\in\Omega} \vartheta(x) - \max_{x\in\Omega} \upsilon(x), \qquad (6.190)$$

$$Q = \Omega \times (0, T), \quad \Sigma = \Gamma \times (0, T).$$

We say that $L(\zeta)$ are *lower order terms* if, any $\varepsilon > 0$, there is $C_\varepsilon > 0$ such that

$$L(\zeta) \le \varepsilon \int_0^T E(\tau)d\tau + C_\varepsilon\{\|W(0)\|^2_{L^2(\Omega,\Lambda)} + \|w(0)\|^2 + \gamma\|w_t(0)\|^2$$

$$+ \|Dw(0)\|^2_{L^2(\Omega,\Lambda)} + \|W(T)\|^2_{L^2(\Omega,\Lambda)} + \|w(T)\|^2$$

$$+ \gamma\|w_t(T)\|^2 + \|Dw(T)\|^2_{L^2(\Omega,\Lambda)}$$

$$+ \int_0^T [\|W\|^2_{L^2(\Omega,\Lambda)} + \|w\|^2 + \gamma\|w_t\|^2 + \|Dw\|^2_{L^2(\Omega,\Lambda)}]dt\}. \quad (6.191)$$

First, we have the following observability estimates.

Theorem 6.9 *Let V be an escape vector field for the Koiter shell and let $\zeta = (W, w)$ solve the problem (6.162). Given $T > 0$. Then there is $C_T > 0$, independent of ζ, such that*

$$2\sigma_1 \int_0^T E(t)\, dt \le \int_\Sigma SB\, d\Sigma + \sigma_0\lambda_0[E(0) + E(T)] + C_T L(\zeta) \quad (6.192)$$

where

$$SB = [|\zeta_t|^2 + \gamma|Dw_t|^2 - B(\zeta, \zeta)]\langle V, \nu\rangle + \partial(\mathcal{A}\zeta, 2m(\zeta))$$

$$+ \vartheta\partial(\mathcal{A}\zeta, \zeta_1 - \zeta_2) + \gamma(2V(w) - \vartheta w)\frac{\partial w_{tt}}{\partial\nu} \qquad (6.193)$$

where

$$m(\zeta) = (D_V W, V(w)), \quad \zeta_1 = (W, 0), \quad \zeta_2 = (0, w).$$

Proof. We take $p = \vartheta(x)$ in the identity (6.163) and $p = -\vartheta(x)$ in the identity (6.164), respectively, and then add them up to

$$\int_\Sigma \left[\partial\left(\mathcal{A}\zeta, \vartheta(\zeta_1 - \zeta_2)\right) - \gamma\vartheta w\frac{\partial w_{tt}}{\partial\nu}\right] d\Sigma = \int_Q \vartheta B(\zeta, \zeta_1 - \zeta_2)\, dQ$$

$$+ \int_Q \vartheta(w_t^2 + \gamma|Dw_t|^2 - |W_t|^2)\, dQ + L(\zeta). \qquad (6.194)$$

Adding the identities (6.194) and (6.166) up and using the inequality (6.177), we further have

$$\int_{\Sigma} [|\zeta_t|^2 + \gamma|Dw_t|^2 - B(\zeta, \zeta)] \langle V, \nu \rangle \, d\Sigma$$

$$+ \int_{\Sigma} [\vartheta\left(\mathcal{A}\zeta, 2m(\zeta) + \vartheta(\zeta_1 - \zeta_2)\right) + \gamma\left(2V(w) - \vartheta w\right) \frac{\partial w_{tt}}{\partial \nu}] \, d\Sigma$$

$$= \int_Q \vartheta[|\zeta_t|^2 + \gamma|Dw_t|^2 + B(\zeta, \zeta)] \, dQ$$

$$+ 2\int_Q \vartheta w_t^2 \, dQ + 2\int_Q \vartheta[\gamma a(\rho(\zeta), \rho(\zeta_1)) - a(\Upsilon(\zeta), \Upsilon(\zeta))] \, dQ$$

$$+ 2Z|_0^T + 2\int_0^T \wp(V, \zeta) \, dt + L(\zeta)$$

$$\geq 2\sigma_1 \int_0^T E(t) \, dt + 2Z|_0^T + L(\zeta) \tag{6.195}$$

where the following relations have been used

$$B(\zeta, \zeta_2) = \gamma a(\rho(\zeta), \rho(\zeta_2)) + \text{lo}(\zeta),$$

$$\gamma a(\rho(\zeta), \rho(\zeta)) - B(\zeta, \zeta_2) = \gamma a(\rho(\zeta), \rho(\zeta_1)) + \text{lo}(\zeta).$$

Finally, we obtain the inequality (6.192) by Lemma 4.10 for the term $2Z|_0^T$. $\quad\square$

Dirichlet Action We consider the boundary control problem in unknown $\varsigma = (U, u)$

$$\begin{cases} \varsigma_{tt} - \gamma(0, \Delta u_{tt}) + \mathcal{A}\varsigma = 0 & \text{in} \quad Q, \\ \varsigma(0) = \varsigma^0, \quad \varsigma_t(0) = \varsigma^1 & \text{in} \quad \Omega, \\ U|_{\Gamma_1} = 0, \quad U|_{\Gamma_0} = \Phi & \text{for} \quad 0 < t < T, \\ u|_{\Gamma_1} = 0, \quad \frac{\partial u}{\partial \nu}|_{\Gamma_1} = 0 & \text{for} \quad 0 < t < T, \\ u|_{\Gamma_0} = \phi, \quad \frac{\partial u}{\partial \nu}|_{\Gamma_0} = \varphi & \text{for} \quad 0 < t < T \end{cases} \tag{6.196}$$

where Φ, ϕ and φ are boundary controls. It's dual version in $\zeta = (W, w)$ is given by

$$\begin{cases} \zeta_{tt} - \gamma(0, \Delta w_{tt}) + \mathcal{A}\zeta = 0 & \text{in} \quad Q, \\ \zeta(0) = \zeta^0, \quad \zeta_t(0) = \zeta^1 & \text{in} \quad \Omega, \\ W|_{\Gamma} = 0 & \text{for} \quad 0 < t < T, \\ w|_{\Gamma} = \frac{\partial w}{\partial \nu}|_{\Gamma} = 0 & \text{for} \quad 0 < t < T. \end{cases} \tag{6.197}$$

We have

Lemma 6.14 *Let $\hat{\Gamma} \subset \Gamma$ be given and let $\zeta_i = (W_i, w_i) \in H^1(\Omega, \Lambda) \times H^2(\Omega)$ be such that*

$$W_i = 0, \quad w_i = \frac{\partial w_i}{\partial \nu} = 0 \quad for \quad x \in \hat{\Gamma} \tag{6.198}$$

for $i = 1, 2$. Let $X \in \mathcal{X}(\Omega)$ be a vector field. Then on $\hat{\Gamma}$

$$B(\zeta_1, \zeta_2) = DW_1(\nu, \nu)DW_2(\nu, \nu) + (1 - \mu)DW_1(\tau, \nu)DW_2(\tau, \nu)/2$$
$$+\gamma\{\rho(\zeta_1)(\nu, \nu)\rho(\zeta_2)(\nu, \nu) + 2(1 - \mu)DW_1(\mathcal{S}\tau, \nu)DW_2(\mathcal{S}\tau, \nu)\}; \quad (6.199)$$

$$\partial(\mathcal{A}\zeta_1, D_X\zeta_2) = B(\zeta_1, \zeta_2)\langle X, \nu\rangle. \quad (6.200)$$

Proof. The proof will be complete by a careful computation.

It follows from the boundary conditions (6.198) that

$$\begin{cases} D_\tau W_i = Dw_i = D_\tau Dw_i = 0, \\ \dfrac{\partial X(w_i)}{\partial \nu} = \langle X, \nu\rangle D^2 w_i(\nu, \nu) \end{cases} \quad \text{for} \quad x \in \hat{\Gamma} \quad (6.201)$$

where τ is the unit tangential along Γ for $i = 1, 2$. Then the formulas (6.201) imply that on $\hat{\Gamma}$

$$\begin{cases} \Upsilon(\zeta_i)(\nu, \nu) = DW_i(\nu, \nu), \quad \Upsilon(\zeta_i)(\tau, \tau) = 0, \\ \Upsilon(\zeta_i)(\nu, \tau) = DW_i(\tau, \nu)/2, \end{cases} \quad (6.202)$$

$$\begin{cases} \rho(\zeta_i)(\nu, \nu) = D^2 w_i(\nu, \nu) - 2DW_i(\mathcal{S}\nu, \nu), \quad \rho(\zeta_i)(\tau, \tau) = 0, \\ \rho(\zeta_i)(\nu, \tau) = -DW_i(\mathcal{S}\tau, \nu), \end{cases} \quad (6.203)$$

$$\begin{cases} \langle B_1(\zeta_i), \nu\rangle = DW_i(\nu, \nu), \quad \langle B_1(\zeta_i), \tau\rangle = (1 - \mu)DW_i(\tau, \nu)/2, \\ \langle B_2(\zeta_i), \nu\rangle = \rho(\zeta_i)(\nu, \nu), \quad \langle B_2(\zeta_i), \tau\rangle = -(1 - \mu)DW_i(\mathcal{S}\tau, \nu), \end{cases} \quad (6.204)$$

for $i = 1, 2$.

Inserting the formulas (6.202) and (6.203) into the formula (6.4), we obtain the formula (6.199).

On the other hand, using (6.201)-(6.204) in the formulas (6.26), we obtain

$$\partial(\mathcal{A}\zeta_1, D_X\zeta_2) = \langle V_1(\zeta_1), D_X W_2\rangle + v_2(\zeta_1)\frac{\partial X(w_2)}{\partial \nu}$$
$$= \langle X, \nu\rangle[\langle B_1(\zeta_1), D_\nu W_2\rangle - 2\gamma\langle B_2(\zeta_1), \mathcal{S}D_\nu W_2\rangle + \gamma\langle B_2(\zeta_1), \nu\rangle D^2 w_2(\nu, \nu)]$$
$$= \langle X, \nu\rangle\{\langle B_1(\zeta_1), \nu\rangle DW_2(\nu, \nu) + \langle B_1(\zeta_1), \tau\rangle DW_2(\tau, \nu)$$
$$+\gamma\langle B_2(\zeta_1), \nu\rangle[D^2 w_2(\nu, \nu) - 2DW_2(\mathcal{S}\nu, \nu)] - 2\gamma\langle B_2(\zeta_1), \tau\rangle DW_2(\mathcal{S}\tau, \nu)\}$$
$$= \langle X, \nu\rangle\{DW_1(\nu, \nu) + (1 - \mu)DW_1(\tau, \nu)DW_2(\tau, \nu)/2$$
$$+\gamma[\rho(\zeta_1)(\nu, \nu)\rho(\zeta_2)(\nu, \nu) + 2(1 - \mu)DW_1(\mathcal{S}\tau, \nu)DW_2(\mathcal{S}\tau, \nu)]\}$$
$$= \langle X, \nu\rangle B(\zeta_1, \zeta_2) \quad \text{for} \quad x \in \hat{\Gamma}.$$

$\square$

Let $\zeta = (W, w)$ solve the problem (6.197). Then we solve the problem in $\eta = (\Phi, \phi)$

$$\begin{cases} \eta_{tt} - \gamma(0, \Delta\phi_{tt}) + \mathcal{A}\eta = 0 \quad \text{in} \quad Q, \\ \eta(T) = \eta_t(T) = 0 \quad \text{in} \quad \Omega, \\ \Phi = \phi = \dfrac{\partial \phi}{\partial \nu} = 0 \quad \text{on} \quad \Sigma_1, \\ \Phi = -D_\nu W, \quad \phi = -v_3(\zeta), \quad \dfrac{\partial \phi}{\partial \nu} = -D^2 w(\nu, \nu) \quad \text{on} \quad \Sigma_0 \end{cases} \quad (6.205)$$

where $v_3(\zeta)$ is given in (6.31).

We define

$$\Lambda(\zeta_0, \zeta_1) = (-\eta_t(0) + \gamma(0, \Delta\phi_t(0)), \quad \eta(0) - \gamma(0, \Delta\phi(0))).$$

Let ζ and $\hat{\zeta}$ solve the problem (6.197) with initial data (ζ_0, ζ_1) and $(\hat{\zeta}_0, \hat{\zeta}_1)$, respectively. It follows from (6.197), (6.225), and (6.25) that

$$\begin{aligned}
&((-\eta_t(0), \eta(0)), (\hat{\zeta}_0, \hat{\zeta}_1))_{[L^2(\Omega,\Lambda) \times L^2(\Omega)]^2} \\
&= -(\eta_t(0), \hat{\zeta}_0)_{L^2(\Omega,\Lambda) \times L^2(\Omega)} + (\eta(0), \hat{\zeta}_1)_{L^2(\Omega,\Lambda) \times L^2(\Omega)} \\
&= \int_0^T \frac{d}{dt}[(\eta_t, \hat{\zeta})_{L^2(\Omega,\Lambda) \times L^2(\Omega)} - (\eta, \hat{\zeta}_t)_{L^2(\Omega,\Lambda) \times L^2(\Omega)}]dt \\
&= \gamma \int_0^T [(\Delta\phi_{tt}, \hat{w}) - (\phi, \Delta\hat{w}_{tt})]dt \\
&\quad + \int_0^T [(\mathcal{A}\eta, \hat{\zeta})_{L^2(\Omega,\Lambda) \times L^2(\Omega)} - (\eta, \mathcal{A}\hat{\zeta})_{L^2(\Omega,\Lambda) \times L^2(\Omega)}]dt \\
&= \gamma[(\Delta\phi(0), \hat{w}_1) - (\Delta\phi_t(0), \hat{w}_0)] - \int_{\Sigma_0} \partial(\mathcal{A}\hat{\zeta}, \eta)d\Sigma. \quad (6.206)
\end{aligned}$$

From Lemma 6.14, we have

$$\begin{aligned}
&(\Lambda(\zeta_0, \zeta_1), (\hat{\zeta}_0, \hat{\zeta}_1))_{[L^2(\Omega,\Lambda) \times L^2(\Omega)]^2} \\
&= \int_{\Sigma_0} [B(\zeta, \hat{\zeta}) + v_3(\zeta)v_3(\hat{\zeta})]\langle V, \nu\rangle d\Sigma.
\end{aligned}$$

Let $L = [L^2(\Omega, \Lambda) \times L^2(\Omega)]^2$ and $H = [H_0^1(\Omega, \Lambda) \times H_0^2(\Omega)] \times [L^2(\Omega, \Lambda) \times H_0^1(\Omega)]$. Since $-\Delta: H_0^1(\Omega) \to H^{-1}(\Omega)$ and $-\Delta: L^2(\Omega) \to H^{-2}(\Omega)$ are isomorphisms, respectively, applying Theorem 2.13 yields

Theorem 6.10 *The problem* (6.196) *is exactly* $[L^2(\Omega, \Lambda) \times H_0^1(\Omega)] \times [H^{-1}(\Omega, \Lambda) \times L^2(\Omega)]$ *controllable by* $L^2(\Sigma_0, \Lambda) \times L^2(\Sigma_0) \times L^2(\Sigma_0)$ *controls on* $[0, T]$ *if the following observability inequality is true: There is* $c_T > 0$ *satisfying*

$$\int_{\Sigma_0} [B(\zeta, \zeta) + v_3^2(\zeta)]d\Sigma \geq c_T E(0) \quad (6.207)$$

for all solutions ζ *to the problem* (6.197).

We have taken one more control $\phi = v_1(\zeta)$ in the problem (6.225) in order to have

Lemma 6.15 *Let* $\zeta = (W, w)$ *solve the problem* (6.197). *If*

$$B(\zeta, \zeta) = v_3(\zeta) = 0 \quad for \quad x \in \Gamma_0, \quad (6.208)$$

then

$$W = DW = 0, \quad w = \frac{\partial w}{\partial \nu} = \Delta w = \frac{\partial \Delta w}{\partial \nu} = 0 \quad for \quad x \in \Gamma_0. \quad (6.209)$$

Proof. It follows from (6.199) and (6.201) that

$$DW = 0, \quad \rho(\zeta)(\nu, \nu) = 0 \quad \text{for} \quad x \in \Gamma_0,$$

which yields, via (6.203),

$$\rho(\zeta) = 0, \quad \Delta w = D^2 w(\nu, \nu) = \operatorname{tr} \rho(\zeta) = 0 \quad \text{for} \quad x \in \Gamma_0.$$

In particular, $D^2 w = 0$ for $x \in \Gamma_0$.

It remains to prove $\dfrac{\partial \Delta w}{\partial \nu} = 0$ for $x \in \Gamma_0$. Let $x \in \Gamma_0$ be given. Let E_1, E_2 be a frame field on M normal at x such that

$$E_1(x) = \tau, \quad E_2(x) = \nu,$$

and

$$D_{E_i} E_j(x) = 0, \quad [E_i, E_j](x) = 0$$

for all i, j. Then

$$D_{E_j} D_{E_i} W = D_{E_i} D_{E_j} W + \mathbf{R}_{E_i E_j} W \quad \text{at} \quad x$$

for all i, j, where $\mathbf{R}_{E_i E_j}$ is the curvature operator. By $W(x) = Dw(x) = 0$, we have

$$\begin{aligned}
\Pi(D_{E_j} D_{E_i} W, E_k) &= \langle D_{E_j} D_{E_i} W, \mathcal{S} E_k \rangle \\
&= \Pi(D_{E_i} D_{E_j} W, E_k) + \mathbf{R}(E_i, E_j, W, \mathcal{S} E_k) \\
&= \Pi(D_{E_i} D_{E_j} W, E_k) \quad \text{at} \quad x
\end{aligned}$$

and, by (1.22),

$$\begin{aligned}
D^3 w(E_i, \nu, E_i) &= D^3 w(E_i, E_i, \nu) + \mathbf{R}(\nu, E_i, Dw, E_i) \\
&= D^3 w(E_i, E_i, \nu) \quad \text{at} \quad x,
\end{aligned}$$

for all i, j, k. We obtain, at x,

$$\begin{aligned}
D\rho(\zeta)(\nu, E_i, E_i) &= D\rho(E_i, \nu, E_i) = E_i(\rho(\zeta)(E_i, E_2)) \\
&= D^3 w(E_i, \nu, E_i) - \Pi(D_{E_i} D_{E_2} W, E_i) - \Pi(E_i, D_{E_i} D_{E_2} W) \\
&= D\rho(\zeta)(E_i, E_i, \nu) \quad \text{at} \quad x,
\end{aligned}$$

for $i = 1, 2$, which give

$$\operatorname{tr} i(\nu) D\rho = \frac{\partial \operatorname{tr} \rho(\zeta)}{\partial \nu} \quad \text{for} \quad x \in \Gamma_0. \tag{6.210}$$

Since

$$\frac{\partial \rho(\zeta)}{\partial \tau} = 0 \quad \text{for} \quad x \in \Gamma_0,$$

we obtain, via (6.30), (6.31), (6.18), and (6.210),

$$0 = v_3(\zeta) = -\gamma\langle B_3(\zeta), \nu\rangle = -\gamma\frac{\partial\,\mathrm{tr}\,\rho}{\partial\nu} = -\gamma\frac{\partial\Delta w}{\partial\nu} \quad \text{for} \quad x \in \Gamma_0.$$

□

Now, let us consider the inequality (6.207). We make the following assumption for a uniqueness result:

Assumption (H) Let one of the following conditions hold

$$\inf_{x\in\Omega}(|\Pi|^2 - 2\kappa) > 0, \quad \text{or} \quad |\Pi|^2 = 2\kappa \quad \text{for all } x \in \Omega \tag{6.211}$$

where κ is the Gauss curvature of the surface M. See Remark 6.3.

We have

Theorem 6.11 *Let V be an escape vector field for the Koiter shell such that the inequality (6.171) holds. Moreover, we assume that the assumption (H) is true. Then for any $T > T_0$, there exists $c_T > 0$ such that the inequality (6.207) holds where*

$$T_0 = \lambda_0\sigma_0/\sigma_1, \quad \Gamma_0 = \{\, x \,|\, x \in \Gamma, \langle V(x), \nu(x)\rangle > 0\,\} \tag{6.212}$$

where σ_1 is given by (6.190) and

$$\sigma_0 = \max_{x\in\Omega}|V(x)|. \tag{6.213}$$

Proof. It follows from the boundary conditions in (6.197), Lemma 6.14, and the formula (6.193) that

$$SB = B(\zeta,\zeta)\langle V,\nu\rangle \quad \text{for} \quad (t,x) \in \Sigma. \tag{6.214}$$

It follows from the estimate (6.192) and the formula (6.214) that for any $\varepsilon > 0$ small and for $T > 0$

$$2(\sigma_1 T - \sigma_0\lambda_0)E(0) \le \int_{\Sigma_0} B(\zeta,\zeta)\,d\Sigma + L(\zeta)$$

where ζ solves the problem (6.197). Then the inequality (6.207) holds for $T > T_0$ which differs by a lower order term. To get the inequality (6.207) by absorbing the lower order terms, as in Lemma 2.5, we need the following uniqueness: If $\eta = (U, u)$ solves the problem

$$\lambda^2(U, u - \Delta u) + \mathcal{A}\eta = 0$$

where λ is a complex number and η satisfies the Dirichlet boundary conditions on Γ_0 such that

$$B(\eta,\eta) + v_3^2(\eta) = 0, \tag{6.215}$$

then

$$\eta = 0 \quad \text{on} \quad \Omega. \tag{6.216}$$

However, by Lemma 6.15, the above result is given by Theorem 6.5. $\qquad\square$

Neumann Control Let $\Gamma_1 \neq \emptyset$ and $\Gamma_0 \cap \Gamma_1 = \emptyset$. We turn to the control problem in $\varsigma = (U, u)$

$$\begin{cases} \varsigma_{tt} - \gamma(0, \Delta u) + \mathcal{A}\varsigma = 0 & \text{in} \quad Q, \\ \varsigma(0) = \varsigma^0, \quad \varsigma_t(0) = \varsigma^1 & \text{on} \quad \Omega, \end{cases} \tag{6.217}$$

where we act on $\Sigma_1 = (0, T) \times \Gamma_1$ by

$$U = 0, \quad u = \frac{\partial u}{\partial \nu} = 0 \tag{6.218}$$

and on $\Sigma_0 = (0, T) \times \Gamma_0$ by

$$V_1(\varsigma) = \Phi, \quad v_2(\varsigma) = \phi, \quad \gamma \frac{\partial u_{tt}}{\partial \nu} + v_3(\varsigma) = \varphi, \tag{6.219}$$

where Φ, ϕ and φ are control functions. The dual problem for the above is the following in $\zeta = (W, w)$

$$\begin{cases} \zeta_{tt} - \gamma(0, \Delta w_{tt}) + \mathcal{A}\zeta = 0 & \text{in} \quad Q, \\ \zeta(0) = \zeta^0, \quad \zeta_t(0) = \zeta^1 & \text{on} \quad \Omega, \end{cases} \tag{6.220}$$

subject to the boundary condition

$$W = 0, \quad w = \frac{\partial w}{\partial \nu} = 0 \quad \text{on} \quad \Sigma_1, \tag{6.221}$$

$$V_1(\zeta) = 0, \quad v_2(\zeta) = 0, \quad \gamma \frac{\partial w_{tt}}{\partial \nu} + v_3(\zeta) = 0 \quad \text{on} \quad \Sigma_0. \tag{6.222}$$

We consider the spaces

$$L^0 = L^2(\Sigma_0, \Lambda) \times L^2(\Sigma_0) \times L^2(\Sigma_0),$$

$$H^0 = H^1([0, T], L^2(\Gamma_0, \Lambda) \times L^2(\Gamma_0) \times H^1(\Gamma_0)).$$

Denote by H^{0*} the conjugate space of H^0 with respect to the space L^0. Let $\aleph \colon H^0 \to H^{0*}$ be *the canonical map*.

Let $\zeta = (W, w)$ solve the problem (6.220)-(6.222) such that

$$\int_{\Sigma_0} [|W_t|^2 + \gamma(\frac{\partial w_t}{\partial \nu})^2 + |w_t|^2 + \gamma(\frac{\partial w_t}{\partial \tau})^2] d\Sigma < \infty. \tag{6.223}$$

We define

$$\Gamma\zeta = (W, \frac{\partial w}{\partial \nu}, w) \quad \text{for} \quad x \in \Gamma.$$

By Theorem 2.12, the condition (6.223) means that $\aleph\Gamma\zeta \in H^{0*}$ satisfies

$$
\begin{aligned}
(\aleph\Gamma\zeta, \varpi)_{L^0} &= (\Gamma\zeta, \varpi)_{H^0} \\
&= \int_{\Sigma_0} (\langle W_t, A_t \rangle + \gamma\frac{\partial w_t}{\partial \nu} a_t + \gamma\frac{\partial w_t}{\partial \tau}\frac{b_t}{\partial \tau})d\Sigma
\end{aligned}
\tag{6.224}
$$

for all $\varpi = (A, a, b) \in H^0$. Then we solve the problem in $\eta = (\Phi, \phi)$

$$
\begin{cases}
\eta_{tt} - \gamma(0, \Delta\phi_{tt}) + \mathcal{A}\eta = 0 & \text{in} \quad Q, \\
\eta(T) = \eta_t(T) = 0 & \text{in} \quad \Omega, \\
\Phi = \phi = \dfrac{\partial\phi}{\partial\nu} = 0 & \text{on} \quad \Sigma_1, \\
(V_1(\eta), v_2(\eta), \gamma\dfrac{\phi_{tt}}{\partial\nu} + v_3(\eta)) = \aleph\Gamma\zeta. & \text{on} \quad \Sigma_0.
\end{cases}
\tag{6.225}
$$

We define Λ by

$$
\Lambda(\zeta_0, \zeta_1) = (-\eta_t(0), \eta(0)).
\tag{6.226}
$$

This time, we use the space

$$
L = L^2(\Omega, \Lambda) \times H^1_{\Gamma_1}(\Omega),
$$

and consider an equivalent inner product on L, given by

$$
((W, w), (U, u))_L = \int_\Omega (\langle W, U \rangle + wu + \gamma\langle Dw, Du \rangle)dx.
$$

Let ζ and $\hat{\zeta}$ solve the problem (6.220)-(6.222) with initial data (ζ_0, ζ_1) and $(\hat{\zeta}_0, \hat{\zeta}_1)$ such that (6.223) are true, respectively. Using (6.220)-(6.222),(6.225),

(6.25), and (6.224), we obtain

$$(\Lambda(\zeta_0, \zeta_1), \ (\hat{\zeta}_0, \hat{\zeta}_1))_{L \times L} = -(\eta_t(0), \hat{\zeta}_0)_L + (\eta(0), \hat{\zeta}_1)_L$$

$$= \int_0^T [(\eta_{tt}, \hat{\zeta})_L - (\eta, \hat{\zeta}_{tt})_L] dt$$

$$= \int_Q [\langle \Phi_{tt}, \hat{W} \rangle + \phi_{tt} \hat{w} + \gamma \langle D\phi_{tt}, D\hat{w} \rangle dQ$$

$$- \int_Q [\langle \Phi, \hat{W}_{tt} + \phi \hat{w}_{tt} + \gamma \langle D\phi, D\hat{w}_{tt} \rangle] dQ$$

$$= \int_Q [\langle \eta_{tt} - \gamma(0, \Delta \phi_{tt}), \ \hat{\zeta} \rangle - \langle \eta, \ \hat{\zeta}_{tt} - \gamma(0, \Delta \hat{w}_{tt}) \rangle] dQ$$

$$+ \gamma \int_\Sigma [\hat{w} \frac{\phi_{tt}}{\partial \nu} - \phi \frac{\partial \hat{w}_{tt}}{\partial \nu}] d\Sigma$$

$$= \int_\Sigma \{\partial(A\eta, \hat{\zeta}) - \partial(A\hat{\zeta}, \eta) + \gamma[\hat{w} \frac{\phi_{tt}}{\partial \nu} - \phi \frac{\partial \hat{w}_{tt}}{\partial \nu}]\} d\Sigma$$

$$= \int_{\Sigma_0} \{[\langle V_1(\eta), \hat{W} \rangle + v_2(\eta) \frac{\partial \hat{w}}{\partial \nu} + (v_3(\eta) + \gamma \frac{\partial \phi_{tt}}{\partial \nu}) \hat{w}] - [v_3(\hat{\zeta}) + \gamma \frac{\partial \hat{w}_{tt}}{\partial \nu}] \phi\} d\Sigma$$

$$= (\aleph \Gamma(\zeta), \Gamma(\hat{\zeta}))_{L^2(\Sigma_0, \Lambda) \times L^2(\Sigma_0) \times L^2(\Sigma_0)}$$

$$= \int_{\Sigma_0} (\langle W_t, \hat{W}_t \rangle + w_t \hat{w}_t + \gamma \langle Dw_t, D\hat{w}_t \rangle) d\Sigma. \tag{6.227}$$

Let

$$H = [H^1_{\Gamma_1}(\Omega) \times H^2_{\Gamma_1}(\Omega)] \times L^2(\Omega, \Lambda) \times H^1_{\Gamma_1}(\Omega).$$

Denote by H^* the conjugate space of H with respect to L. Then applying Theorem 2.13 gives

Theorem 6.12 *The problem* (6.217)−(6.219) *is exact* H^* *controllable by* H^{0*} *controls on* $[0, T]$ *if the following inequality is true: There is* $c_T > 0$ *such that*

$$\int_{\Sigma_0} [|W_t|^2 + |w_t|^2 + \gamma |Dw_t|^2] d\Sigma \geq c_T E(0) \tag{6.228}$$

for all solutions ζ *to the problem* (6.220) − (6.222) *for which the left hand side of* (6.228) *is finite.*

Furthermore, we have

Theorem 6.13 *Let* V *be an escape vector field for the Koiter shell and let the assumption* (H) *in* (6.211) *hold. Then, for any* $T > T_0$, *there is* $c_T > 0$ *such that the inequality* (6.228) *holds where* T_0 *and* Γ_0 *are given in* (6.212).

Proof. It follows from the boundary conditions in (6.221)-(6.222) and Lemma 6.14 that

$$\text{SB} = \langle V, \nu \rangle \begin{cases} |\zeta_t|^2 + \gamma |Dw_t|^2 - B(\zeta, \zeta)] & \text{on} \quad \Sigma_0, \\ B(\zeta, \zeta) \langle V, \nu \rangle & \text{on} \quad \Sigma_1. \end{cases} \tag{6.229}$$

Applying Theorem 6.9, we obtain (6.228). □

Exercises

6.1 *Let M be a surface of $\mathbb{R}^3$ with the induced metric g from $\mathbb{R}^3$ and let Π be the second fundamental form of M. If $|\Pi|^2 = 2\kappa$ for all $x \in M$, then there is a function φ such that*

$$\Pi = \varphi g, \quad \mathcal{S}W = \varphi W \quad for \quad x \in M, \ W \in \mathcal{X}(M)$$

where the operator $\mathcal{S}$ is defined by the formula (6.17).

6.2 *Prove Lemma 6.10.*

6.3 *Let $\zeta = (W, w)$ be a smooth solution to the problem (6.84)-(6.85). Prove that the formula (6.143) holds.*

6.5   Notes and References

Sections 6.1, 6.3, and 6.4 are from [37]; Section 6.2 is from [26].

The motion equations for the Koiter shell in Theorem 6.3 were given by [37] where the kinetic energy of the shell was based on to the classical Kirchhoff-Love Assumptions. However, under Koiter's Hypotheses [95] the kinetic energy of the displacement vector field for the Koiter shell is very complicated, see the formula (6.67), which leads to different motion equations in Theorem 6.4. The equations in Theorem 6.4 are new and first published here.

The ellipticity of the strain energy for the Koiter shell was proved by [13] where the authors used the classical differential geometry where the middle surface of the shell was given by one coordinate path. Here we present a proof of the ellipticity (Theorem 6.1) that is very simple due to the Bochner technique.

In the excellent work [40] the Bochner technique again plays an important role in establishing the corresponding Carleman estimates to obtain the uniqueness for the Koiter shell.

Chapter 7

Control of the Quasilinear Wave Equation in Higher Dimensions

This chapter introduces the latest advances on control of the quasilinear wave equation in higher dimensions by combining the Riemannian geometrical approach with some knowledge from nonlinear partial differential equations where the quasilinearity arises in the principal part of the systems.

Section 7.1 presents some necessary estimates from the nonlinear partial differential equations.

In Section 7.2 we prove the locally exact controllability around the equilibrium under existence of *an escape vector field for a metric*. We then establish the globally exact controllability in such a way that the state of the quasilinear wave equation moves from an equilibrium in one location to an equilibrium in another location under some geometrical conditions. The Dirichlet action and the Neumann action are studied, respectively. Some examples are presented to verify the globally exact controllability.

In Section 7.3 we consider the stabilization of the quasilinear wave equation with a structure of an input-output in the boundary when initial data and boundary inputs are near a given equilibrium of the system. We show that the stabilization of solutions depends not only on this dissipation structure but also on existence of *an escape vector field for the Riemannian metric*, given by the coefficients and the equilibrium of the system. In particular, we prove that the norm of the state of the system decays exponentially if the input stops after a finite time and if there exists an escape vector field in the metric, which implies the exponential stabilization of the system by boundary feedback.

In Section 7.4 we study the stabilization of smooth solutions for the quasilinear wave equation by an internal local damping when initial data are close to a given equilibrium. We show that *an escape region for the metric* can guarantee the global solutions and their exponential stabilization. These escape regions depend not only on the sectional curvature of a Riemannian metric but also on the equilibrium of the system.

7.1   Boundary Traces and Energy Estimates

We shall treat boundary traces and energy estimates of the quasilinear wave equation in three types of boundary data: Dirichlet data, Neumann data, and Robin data, respectively.

Let $\Omega \subset \mathbb{R}^n$ be an open, bounded set with the smooth boundary Γ. Suppose that Γ consists of two parts, Γ_0 and Γ_1. If $\Gamma_0 \neq \emptyset$, we further assume that $\overline{\Gamma_1} \cap \overline{\Gamma_0} = \emptyset$.

Let $A(x,y) = (a_{ij}(x,y))$ be symmetric, positive on $\overline{\Omega} \times \mathbb{R}^n$ where a_{ij} are smooth functions. Let b be a smooth function on $\Omega \times \mathbb{R}^n \times \mathbb{R}$.

Definition 7.1 $w \in H^2(\Omega)$ *is said to be an equilibrium if*

$$\sum_{ij=1}^{n} a_{ij}(x, \nabla w) w_{x_i x_j} + b(x, \nabla w, 0) = 0 \quad for \quad x \in \Omega. \tag{7.1}$$

Let w be an equilibrium and let $T > 0$ be given. We consider the problem

$$\begin{cases} u_{tt} = \sum_{ij=1}^{n} a_{ij}(x, \nabla u) u_{x_i x_j} + b(x, \nabla u, u_t) & \text{for} \quad (t,x) \in Q, \\ u = w & \text{for} \quad (t,x) \in \Sigma_1, \end{cases} \tag{7.2}$$

where

$$Q = (0, T) \times \Omega, \quad \Sigma_1 = \Sigma_1.$$

In this section we establish some energy and boundary trace estimates of the problem (7.2) in preparation for controllability/stabilization. The target space is $H^m(\Omega) \times H^{m-1}(\Omega)$ where we need the integer m to satisfy

$$m \geq [n/2] + 3. \tag{7.3}$$

Remark 7.1 *We assume the condition (7.3) in order to have the existence of small time solutions, for example, by [91] or [48], if initial data on Ω and boundary value on the portion Γ_0 are specified, respectively. The target space will be larger if m is smaller. In the linear case, the number m can be any real number in $(-\infty, \infty)$, see [117], [192], and [193]. In [114], the authors were able to establish global solutions in $H^m(\Omega) \times H^{m-1}(\Omega)$ for $m \geq 1$ where m may be much smaller than in the condition (7.3) because they had a lower regularity for their model. Here we have to assume that the condition (7.3) holds since we are short of the corresponding local regularities when $m < [n/2] + 3$.*

In general, solutions of the problem (7.2) may blow up in a finite time even if the initial data and the boundary value are smooth. We here assume that the system (7.2) has small time solutions and study it's energy estimates and boundary traces.

Let $u \in \cap_{k=0}^{m} C^k \left([0,T], H^{m-k}(\Omega)\right)$ be a solution of the problem (7.2) for some $T > 0$. Let $w \in H^m(\Omega)$ be an equilibrium of the problem (7.2). We can make a transform by $u = w + \phi$ and consider the ϕ-problem

$$\begin{cases} \phi_{tt} = \sum_{ij} a_{ij}(x, \nabla w + \nabla \phi)\phi_{x_i x_j} + b_0(x, \nabla \phi, \phi_t) & \text{for} \quad (t, x) \in Q, \\ \phi = 0 & \text{for} \quad (t, x) \in \Sigma_1 \end{cases} \tag{7.4}$$

where

$$b_0(x, y, s) = b(x, \nabla w + y, s) \quad \text{for} \quad (x, y, s) \in \Omega \times \mathbb{R}^n \times \mathbb{R}.$$

We introduce a linear operator $\mathcal{A}(t)$ by

$$\mathcal{A}(t)v = \sum_{ij=1}^{n} a_{ij}(x, \nabla w + \nabla \phi)v_{x_i x_j} \quad \text{for} \quad v \in H^2(\Omega). \tag{7.5}$$

Then

$$(-\mathcal{A}(t)v_1, v_2) = -\int_\Gamma v_2 \langle A\nabla v_1, \nu \rangle d\Gamma + (A\nabla v_1, \nabla v_2) + (\mathbb{C}v_1, v_2) \tag{7.6}$$

for $v_1, v_2 \in H^1(\Omega)$ where $A = (a_{ij}(x, \nabla w + \nabla \phi))$, ν is the outside normal of Γ in the Euclidean metric, and

$$\mathbb{C}v_1 = \sum_{ij=1}^{n} (a_{ij}(x, \nabla w + \nabla \phi))_{x_j} v_{1 x_i}.$$

Then the problem (7.4) can be rewritten as

$$\begin{cases} \phi_{tt} = \mathcal{A}(t)\phi(t) + b_0(x, \nabla \phi, \phi_t), & \text{for} \quad (t, x) \in Q, \\ \phi = 0 & \text{for} \quad (t, x) \in \Sigma_1. \end{cases} \tag{7.7}$$

Boundary Traces and Energy Estimates with Dirichlet Data on Γ_0 Let $\phi \in \cap_{k=0}^{m} C^k\left([0, T], H^{m-k}(\Omega)\right)$ be a solution of the problem (7.7) for some $T > 0$. We introduce

$$\mathcal{E}(t) = \sum_{k=0}^{m} \|\phi^{(k)}(t)\|_{m-k}^2, \quad \mathcal{E}_{\Gamma_0, D}(t) = \sum_{k=0}^{m-1} \|\phi^{(k)}(t)\|_{m-k-1/2, \Gamma_0}^2, \tag{7.8}$$

$$Q(t) = \sum_{k=1}^{m} \left(\|\phi^{(k)}(t)\|^2 + \|\nabla \phi^{(k-1)}(t)\|^2\right), \quad L(t) = \sum_{k=2}^{m} \mathcal{E}^k(t), \tag{7.9}$$

$$Q_{\Gamma_0, D}(t) = \sum_{k=1}^{m} \left(\|\phi^{(k)}(t)\|_{\Gamma_0}^2 + \|\phi^{(k-1)}(t)\|_{1, \Gamma_0}^2\right), \tag{7.10}$$

where $\|\cdot\|$, $\|\cdot\|_j$, $\|\cdot\|_{\Gamma_0}$, and $\|\cdot\|_{j, \Gamma_0}$ are norms of $L^2(\Omega)$, $H^j(\Omega)$, $L^2(\Gamma_0)$, and $H^j(\Gamma_0)$, respectively, for $1 \leq j \leq m$.

We shall establish

Theorem 7.1 *Let $\gamma > 0$ be given. Let ϕ be a solution of the problem (7.7) on the interval $[0, T]$ for some $T > 0$ such that*

$$\sup_{0 \le t \le T} \|\phi(t)\|_m \le \gamma \quad and \quad \sup_{0 \le t \le T} \|\dot{\phi}(t)\|_{m-1} \le \gamma. \tag{7.11}$$

Then there is $c_\gamma > 0$, which depends on the number γ but is independent of solution ϕ, such that

$$Q(t) \le \mathcal{E}(t) \le c_\gamma Q(t) + c_\gamma \mathcal{E}_{\Gamma_0, D}(t) + c_\gamma L(t) \quad for \quad 0 \le t \le T \tag{7.12}$$

and

$$Q(t) \le c_\gamma Q(0) + c_\gamma \int_0^t [(1 + \mathcal{E}^{1/2}(t))Q(t) + Q_{\Gamma_0, D}(t) + L(t)]dt \tag{7.13}$$

for $t \in [0, T]$.

Remark 7.2 *We observe that the conditions (7.3) and (7.11) imply that*

$$\phi_t, \ \partial_{x_i}\phi \in C^1(\overline{\Omega}) \quad for \quad 1 \le i \le n \tag{7.14}$$

which is also one of the reasons why we need the condition (7.3).

The proof of Theorem 7.1 will be given after Lemma 7.5.

First, we collect here a few basic properties of Sobolev spaces in the following lemma which play an important role in our estimates in the sequel. Let $\|\cdot\|_s$ be the norm of $H^s(\Omega)$ for $s \ge 0$. In particular, denote by $\|\cdot\| = \|\cdot\|_0$ the norm of $L^2(\Omega)$.

Lemma 7.1 *Let $\Omega \subset \mathbb{R}^n$ be a bounded, open set. Then*
(i) *Let $s_1 > s_2$ be given. For any $\varepsilon > 0$ there is $c_\varepsilon > 0$ such that*

$$\|w\|_{s_2}^2 \le \varepsilon \|w\|_{s_1}^2 + c_\varepsilon \|w\|^2 \quad for \quad w \in H^{s_1}(\Omega). \tag{7.15}$$

(ii) *Let $s_i \ge 0$ be given for $i = 1, 2$. If $0 \le r \le \min\{s_1, s_2, s_1 + s_2 - [n/2] - 1\}$, then there is a constant $c > 0$ such that*

$$\|fg\|_r \le c\|f\|_{s_1}\|g\|_{s_2} \quad for \quad f \in H^{s_1}(\Omega), \ g \in H^{s_2}(\Omega). \tag{7.16}$$

(iii) *Let $s_j \ge 0$ be given for $j = 1, \cdots, k$. Let*

$$0 \le r \le \min_{1 \le i \le k} \min_{j_1 \le \cdots \le j_i} \{s_{j_1} + \cdots + s_{j_i} - (i-1)([n/2] + 1)\}.$$

Then there is a constant $c > 0$ such that

$$\|f_1 \cdots f_k\|_r \le c\|f_1\|_{s_1} \cdots \|f_k\|_{s_k} \quad for \quad f_j \in H^{s_j}(\Omega), \ 1 \le j \le k. \tag{7.17}$$

For the proofs of (ii) and (iii) in the above lemma, for example, see books [1], or [195].

Lemma 7.2 (i) *Let $f(x, y, s)$ be a smooth function on $\overline{\Omega} \times \mathbb{R}^n \times \mathbb{R}$. Set $F(x) = f(x, \nabla \phi, \phi_t)$. For $0 \le k \le m-1$, there is $c = c(\sup_{x \in \Omega}(|\nabla \phi|, |\phi_t|)) > 0$ such that*

$$\|F\|_k \le c \sum_{j=0}^{k} (1 + \|\phi\|_m + \|\phi_t\|_{m-1})^j. \qquad (7.18)$$

(ii) *Let ϕ be a solution of the problem (7.7) and let $\gamma > 0$ be given. Suppose that the condition (7.11) holds true. Then there is $c_\gamma > 0$, which depends on the γ, such that*

$$\|v\|_{k+1}^2 \le c_\gamma \left(\|\mathcal{A}(t)v\|_{k-1}^2 + \|v\|_{k+1/2, \Gamma_0}^2 + \|v\|_k^2 \right) \qquad (7.19)$$

for $v \in H^k(\Omega) \cap H^1_{\Gamma_1}(\Omega)$ and $0 \le k \le m-1$.

Proof. (i) By induction. The inequality (7.18) is clearly true for $k = 0$. Suppose that it holds for $0 \le k < m-1$. Since

$$F_{x_i} = f_{x_i}(x, \nabla \phi, \phi_t) + \sum_{j=1}^{n} f_{y_j}(x, \nabla \phi, \phi_t)\phi_{x_i x_j} + f_s(x, \nabla \phi, \phi_t)\phi_{t x_i}$$

for $1 \le i \le n$, by using the inequality (7.16) and the induction assumption for $f_{x_i}(x, \nabla \phi, \phi_t)$ and for $f_{y_j}(x, \nabla \phi, \phi_t)$, respectively, we obtain

$$
\begin{aligned}
\|F\|_{k+1} &= \left(\|F\|^2 + \sum_{i=1}^{n} \|F_{x_i}\|_k^2 \right)^{1/2} \\
&\le c + c \sum_{i=1}^{n} \|f_{x_i}(\cdot, \nabla \phi, \phi_t)\|_k + c \sum_{ij=1}^{n} \|f_{y_j}(\cdot, \nabla \phi, \phi_t)\|_k \|\phi_{x_i x_j}\|_{m-2} \\
&\quad + c\|f_s(\cdot, \nabla \phi, \phi_t)\|_k \|\phi_{t x_i}\|_{m-2} \\
&\le c + c \sum_{j=0}^{k} (1 + \|\phi\|_m + \|\dot{\phi}\|_{m-1})^j \\
&\le c \sum_{j=0}^{k+1} (1 + \|\phi\|_m + \|\dot{\phi}\|_{m-1})^j.
\end{aligned}
$$

(ii) A standard method as to the linearly elliptic problem can give the inequality (7.19); for example see [68]. $\qquad \square$

Lemma 7.3 *Let b be a smooth function on $\Omega \times \mathbb{R}^n \times \mathbb{R}$. Let $\gamma > 0$ be given and let ϕ be a solution of the problem (7.7) on the interval $[0, T]$ for some $T > 0$ such that the condition (7.11) holds true. Then there is $c_\gamma > 0$, which depends on the number γ, such that*

$$
\begin{aligned}
\|b^{(k)}(x, \nabla \phi, \phi_t)\|_{m-k-2}^2 \le{}& c_\gamma \|\phi^{(k)}(t)\|_{m-k-1}^2 \\
&+ c_\gamma \|\phi^{(k+1)}(t)\|_{m-k-2}^2 + c_\gamma p_k
\end{aligned} \qquad (7.20)
$$

where

$$p_1 = 0; \quad p_k = \sum_{i=2}^{k} \mathcal{E}^i(t), \quad k \geq 2$$

for $1 \leq k \leq m - 2$ and

$$\|\mathcal{A}^{(j)}(t)\phi^{(k-j)}(t)\|_{m-k-2}^2 \leq c_\gamma \sum_{i=1}^{j} \mathcal{E}^{1+i}(t) \qquad (7.21)$$

for $1 \leq j \leq k \leq m - 2$ where the operator $\mathcal{A}(t)$ is defined by (7.5).

Proof. Denote by $\hat{D} = (D, \partial)$ the covariant differential on the space $\mathbb{R}^n \times \mathbb{R}$ with the Euclidean metric. Then it is easy to check that

$$b^{(k)}(x, \nabla\phi, \phi_t) = \sum_{i=1}^{k} \sum_{r_1 + \cdots + r_i = k} \hat{D}^i b((\nabla\phi^{(r_1)}, \phi_t^{(r_1)}), \cdots, (\nabla\phi^{(r_i)}, \phi_t^{(r_i)}))$$

where $\hat{D}^i b_0$ denotes the covariant differential of the function $b_0(x, y, s)$ of order i with respect to the variable (y, s) in the Euclidean metric of $\mathbb{R}^n \times \mathbb{R}$.

Let us estimate the term $\hat{D}b((\nabla\phi^{(k)}, \phi_t^{(k)}))$ first. We have

$$\hat{D}b((\nabla\phi^{(k)}, \phi_t^{(k)})) = \sum_{l=1}^{n} b_{y_l}(x, \nabla\phi, \phi_t)\phi_{x_l}^{(k)}(t) + b_s(x, \nabla\phi, \phi_t)\phi_t^{(k)}.$$

It follows from the inequalities (7.16), (7.18) and (7.11) that

$$\|\hat{D}b((\nabla\phi^{(k)}, \phi_t^{(k)}))\|_{m-k-2} \leq c \sum_{l} \|b_{y_l}\|_{m-1}\|\phi^{(k)}(t)\|_{m-k-1}$$

$$+ \|b_s\|_{m-1}\|\phi^{(k+1)}\|_{m-k-2} \leq c_\gamma(\|\phi^{(k)}(t)\|_{m-k-1} + \|\phi^{(k+1)}(t)\|_{m-k-2}). \quad (7.22)$$

Denote $x_0 = t$. For $2 \leq i \leq k$, we observe that

$$\hat{D}^i b((\nabla\phi^{(r_1)}, \phi_t^{(r_1)}), \cdots, (\nabla\phi^{(r_i)}, \phi_t^{(r_i)}))$$

are sums of terms such as

$$f(x, \nabla\phi, \phi_t)\phi_{x_{l_1}}^{(r_1)}(t) \cdots \phi_{x_{l_i}}^{(r_i)}(t)$$

where $r_1 + \cdots + r_i = k$ for $0 \leq l_j \leq n$ and $1 \leq j \leq i$. Using the inequalities (7.17) and (7.18), we have

$$\|f\phi_{x_{l_1}}^{(r_1)} \cdots \phi_{x_{l_i}}^{(r_i)}\|_{m-k-2} \leq c_\gamma \|\phi^{(r_1)}(t)\|_{m-r_1} \cdots \|\phi^{(r_i)}(t)\|_{m-r_i}$$

$$\leq c_\gamma \mathcal{E}^{i/2}(t). \qquad (7.23)$$

The inequality (7.20) follows from the inequalities (7.22) and (7.23).

Moreover, by the inequalities (7.16) and (7.20), we obtain

$$\|a_{pq}^{(j)}(\cdot, \nabla w + \nabla \phi)\phi_{x_p x_q}^{(k-j)}\|_{m-k-2}$$

$$\leq c\|a_{pq}^{(j)}(\cdot, \nabla w + \nabla \phi)\|_{m-j-2}\|\phi_{x_p x_q}^{(k-j)}\|_{m+j-k-2} \leq c_\gamma \sum_{i=1}^{j} \mathcal{E}^{(i+1)/2}(t)$$

for $1 \leq p, q \leq n$ which implies that the inequality (7.21) is true. $\qquad\square$

Lemma 7.4 *Let $\gamma > 0$ be given. Let ϕ be a solution of the problem (7.7) on the interval $[0, T]$ for some $T > 0$ such that the condition (7.11) holds true. Then there is $c_\gamma > 0$ such that the inequality (7.12) is true.*

Proof. It is clear that

$$\|\phi^{(m)}(t)\|^2 + \|\phi^{(m-1)}(t)\|_1^2 = \|\phi^{(m)}(t)\|^2 + \|\nabla\phi^{(m-1)}(t)\|^2$$
$$+\|\phi^{(m-1)}(t)\|^2 \leq \mathcal{Q}(t). \qquad (7.24)$$

Proceeding by induction, we assume that for some $1 \leq j \leq m - 1$

$$\|\phi^{(j)}(t)\|_{m-j}^2 \leq c_\gamma \mathcal{Q}(t) + c_\gamma \mathcal{E}_{\Gamma_0, D}(t) + c_\gamma L(t) \qquad (7.25)$$

which, as shown above, is true for $j = m$ and $j = m - 1$. For $m \geq j \geq 3$, we assume that the inequality (7.25) is true for j and $j - 1$. We need to prove that it is true for $j - 2$. Formal differentiation of the first equation in (7.7) by $j - 2$ times with respect to t yields

$$\phi^{(j)}(t) = \mathcal{A}(t)\phi^{(j-2)}(t) + b_0^{(j-2)}(x, \nabla\phi, \phi_t)$$
$$+ \sum_{i=1}^{j-2} C_i^{j-2} \mathcal{A}^{(i)}(t)\phi^{(j-2-i)}(t) \qquad (7.26)$$

where C_k^{j-2} are the coefficients of Leibniz's rule for differentiating a product. Using the inequalities (7.19) and (7.20), we obtain

$$\|\phi^{(j-2)}(t)\|_{m-j+2}^2 \leq c_\gamma\|\mathcal{A}(t)\phi^{(j-2)}(t)\|_{m-j}^2 + c_\gamma\|\phi^{(j-2)}(t)\|_{m-j+3/2, \Gamma_0}^2$$
$$+c_\gamma\|\phi^{(j-2)}(t)\|_{m-j+1}^2$$
$$\leq c_\gamma\|\phi^{(j)}(t)\|_{m-j}^2 + c_\gamma\|\phi^{(j-1)}(t)\|_{m-j+1}^2 + c_\gamma\|\phi^{(j-2)}(t)\|_{m-j+1}^2$$
$$+c_\gamma\mathcal{E}_{\Gamma_0, D}(t) + c_\gamma \sum_{i=2}^{j-1} \mathcal{E}^i(t) \qquad (7.27)$$

where the relation $\|\phi^{(j-1)}(t)\|_{m-j}^2 \leq \|\phi^{(j-1)}(t)\|_{m-j+1}^2$ has been used. Then the inequality (7.12) follows by induction where the following inequality has also been used to absorb the term $\|\phi^{(j-2)}(t)\|_{m-j+1}^2$ in the right hand side of the inequality (7.27),

$$\|\phi^{(j-2)}(t)\|_{m-j+1}^2 \leq \varepsilon\|\phi^{(j-2)}(t)\|_{m-j+2}^2 + c_{\gamma,\varepsilon}\|\phi^{(j-2)}(t)\|^2$$

for $\varepsilon > 0$ small, since $\|\phi^{(j-2)}(t)\|^2 \leq \mathcal{Q}(t)$ for $m \geq j \geq 3$. $\qquad\square$

Lemma 7.5 *Let $\gamma > 0$ be given and let ϕ be a solution of the problem (7.7) on the interval $[0, T]$ for some $T > 0$ such that the condition (7.11) holds true. Let $\varphi \in H^1(Q)$ solve the linear problem*

$$\begin{cases} \varphi_{tt}(t) = \mathcal{A}(t)\varphi(t) + F(t) & \text{for} \quad (t, x) \in Q, \\ \varphi = 0 & \text{for} \quad (t, x) \in \Sigma_1. \end{cases} \tag{7.28}$$

Set
$$\Upsilon(t) = \|\dot\varphi(t)\|^2 + \|\nabla\varphi(t)\|^2, \quad \Upsilon_{\Gamma_0, D}(t) = \|\dot\varphi(t)\|^2_{\Gamma_0} + \|\varphi\|^2_{1,\Gamma_0}.$$

Then there is $c_\gamma > 0$ such that

$$\Upsilon(t) \leq c_\gamma \Upsilon(0)$$
$$+ c_\gamma \int_0^t \left[(1 + \|\dot\phi(\tau)\|_{m-1})\Upsilon(\tau) + \Upsilon_{\Gamma_0, D}(\tau) + \|F(\tau)\|^2 \right] d\tau \tag{7.29}$$

for $0 \leq t \leq T$.

Proof. Let
$$P(t) = \|\dot\varphi(t)\|^2 + (A\nabla\varphi, \nabla\varphi). \tag{7.30}$$

Using the formula (7.6), we obtain

$$\dot{P}(t) = 2(\ddot\varphi(t), \dot\varphi(t)) + 2(A\nabla\varphi(t), \nabla\dot\varphi(t)) + \left(\dot{A}\nabla\varphi, \nabla\varphi\right)$$
$$= 2(F + C\varphi, \dot\varphi) + \left(\dot{A}\nabla\varphi, \nabla\varphi\right) + 2\int_{\Gamma_0} \dot\varphi\varphi_{\nu_A} d\Gamma$$

where $\varphi_{\nu_A} = \langle A(x, \nabla\phi)\nabla\varphi, \nu\rangle$. It follows from the positiveness of $A(x, \nabla\phi)$ that

$$\Upsilon(t) \leq c_\gamma P(t) \leq c_\gamma P(0)$$
$$+ c_\gamma \int_0^t \left[(\|\nabla\varphi\| + \|F(t)\|)\|\dot\varphi\| + \|\dot\phi\|_{m-1}\|\nabla\varphi\|^2 \right] dt$$
$$+ \varepsilon \int_0^t \int_{\Gamma_0} \varphi_{\nu_A}^2 d\Gamma dt + c_{\gamma, \varepsilon} \int_0^t \int_{\Gamma_0} \dot\varphi^2 d\Gamma dt \tag{7.31}$$

for $0 \leq t \leq T$ where $\varepsilon > 0$ is given small.

To obtain the inequality (7.29) from the inequality (7.31), we have to estimate the term $\int_0^t \int_\Gamma \varphi_{\nu_A}^2 d\Gamma dt$. We now introduce a Riemannian metric

$$g = A^{-1}(x, \nabla\phi)$$

on $\overline{\Omega}$ for $0 \leq t \leq T$ so that the couple $(\overline{\Omega}, g)$ are Riemannian manifolds for $t \in [0, T]$.

If $\Gamma_0 = \emptyset$, the inequality (7.29) is clearly true due to the inequality (7.31). We assume that $\Gamma_0 \neq \emptyset$. Since $\overline{\Gamma_1} \cap \overline{\Gamma_0} = \emptyset$, there is a vector field H on $\overline{\Omega}$ such that

$$H|_{\Gamma_1} = 0 \quad \text{in a neigborhood of } \Gamma_1; \quad H|_{\Gamma_0} = \nu_A. \tag{7.32}$$

Applying the identity (2.13) in Theorem 2.1 to the problem (7.28) yields

$$\operatorname{div}\{2H(\varphi)A(x,\nabla\phi)\nabla\varphi-(|\nabla_g\varphi|_g^2-u_t^2)H\}+2[C\varphi+F(t)]H(\varphi)$$
$$=2[\varphi_t H(\varphi)]_t+2DH(\nabla_g\varphi,\nabla_g\varphi)+(\varphi_t^2-|\nabla_g\varphi|_g^2)\operatorname{div}H.$$

We integrate the above equation over Ω to obtain

$$\int_{\Gamma_0}\left[\varphi_{\nu_A}^2+\frac{1}{2}(\dot{\varphi}^2-|\nabla_g\varphi|_g^2)|\nu_A|_g^2\right]d\Gamma$$
$$=\frac{d}{dt}(\dot{\varphi},H(\varphi))+\int_\Omega\left[D_gH(\nabla_g\varphi,\nabla_g\varphi)+\frac{1}{2}(\dot{\varphi}^2-|\nabla_g\varphi|_g^2)\operatorname{div}H\right]dx$$
$$-\int_\Omega(C\varphi+F(t))\,H(\varphi)dx. \tag{7.33}$$

Using the relation

$$|\nabla_g\varphi|_g^2=\frac{\varphi_{\nu_A}^2}{|\nu_A|_g^2}+|\nabla_{\Gamma_g}\varphi|_g^2\quad\text{for}\quad x\in\Gamma$$

in the formula (7.33) where ∇_{Γ_g} is the gradient in the induced metric on Γ from the Riemannian metric g, we have

$$\int_0^t\int_{\Gamma_0}\varphi_{\nu_A}^2\,d\Gamma dt\le c_\gamma\left[\Upsilon(t)+\Upsilon(0)\right]$$
$$+c_\gamma\int_0^t\int_\Omega\left[\Upsilon(t)+\Upsilon_{\Gamma_0,D}(t)\right]dxdt. \tag{7.34}$$

Finally, we insert the inequality (7.34) into the inequality (7.31), and choose $\varepsilon>0$ so small that the term $\varepsilon c_\gamma\Upsilon(t)$ can be absorbed by the left hand side of the inequality (7.31) to obtain the inequality (7.29). □

Proof of Theorem 7.1 The inequality (7.12) has been proved in Lemma 7.4. Now we prove the inequality (7.13).

We let $\varphi=\phi^{(j-2)}(t)$ for $2\le j\le m+1$ in the equation (7.28) and apply Lemma 7.5 to obtain

$$\|\phi^{(j-1)}(t)\|^2+\|\nabla\phi^{(j-2)}(t)\|^2$$
$$\le c_\gamma Q(0)+c_\gamma\int_0^t\left[\left(1+\mathcal{E}^{1/2}(t)\right)Q(t)+Q_{\Gamma_0,D}(t)\right]dt$$
$$+c_\gamma\int_0^t(\|b_0^{(j-2)}\|^2+\sum_{k=1}^{j-2}\|\mathcal{A}^{(k)}(t)\phi^{(j-2-k)}(t)\|^2)dt. \tag{7.35}$$

In addition, a similar computation as in Lemma 7.3 yields

$$\|b_0^{(j-2)}\|^2\le c_\gamma\|\nabla\phi^{(j-2)}\|^2+c_\gamma\mathcal{L}(t)\le c_\gamma Q(t)+c_\gamma L(t), \tag{7.36}$$

and

$$\|\mathcal{A}^{(k)}(t)\phi^{(j-2-k)}(t)\|^2 \le c_\gamma L(t) \quad \text{for} \quad 1 \le k \le j-2. \tag{7.37}$$

The inequality (7.13) follows from (7.35)-(7.37).   □

Boundary Traces and Energy Estimates with Neumann data on
Γ_0 Let $\phi \in \cap_{k=0}^m C^k\left([0,T], H^{m-k}(\Omega)\right)$ be a solution of the problem (7.7) for some $T > 0$. We define

$$v_{\nu_A} = \langle A(x, \nabla\phi)\nabla v, \; \nu\rangle \quad \text{for} \quad v \in H^2(\Omega)$$

and let

$$\mathcal{E}_{\Gamma_0,N}(t) = \sum_{k=0}^{m-2} \|\phi_{\nu_A}^{(k)}(t)\|_{m-k-3/2,\Gamma_0}^2, \quad Q_{\Gamma_0,N} = \sum_{k=0}^{m-1} \|\phi_{\nu_A}^{(k)}\|_{1/2,\Gamma_0}^2.$$

We have

Theorem 7.2 *Let $\gamma > 0$ be given and let ϕ be a solution of the problem (7.7) on the interval $[0,T]$ for some $T > 0$ such that the conditions (7.11) hold. Then there is $c_\gamma > 0$, which only depends on the γ, such that*

$$Q(t) \le \mathcal{E}(t) \le c_\gamma Q(t) + c_\gamma \mathcal{E}_{\Gamma_0,N}(t) + c_\gamma L(t) \tag{7.38}$$

for $0 \le t \le T$ and

$$Q(t) \le c_\gamma Q(0) + c_\gamma \int_0^t [(1 + \mathcal{E}^{1/2}(t))Q(t) + Q_{\Gamma_0,N}(t) + L(t)]dt \tag{7.39}$$

for $t \in [0,T]$ where $\mathcal{E}(t)$ and $Q(t)$ are given in (7.8) and (7.9), respectively.

Proof. It will suffice to make some revisions in the proofs of Lemmas 7.4 and 7.5, respectively.

Using the ellipticity that there is $c_\gamma > 0$ such that

$$\|v\|_{k+1}^2 \le c_\gamma \left(\|\mathcal{A}(t)v\|_{k-1}^2 + \|v_{\nu_A}\|_{k-1/2,\Gamma_0}^2 + \|v\|_k^2\right) \tag{7.40}$$

for $v \in H^m(\Omega) \cap H_{\Gamma_1}^1(\Omega)$ and $0 \le k \le m-1$ in the proof of Lemma 7.4 yields the inequality (7.38).

Moreover, the second inequality (7.39) is based on the following

Lemma 7.6 *Let $\gamma > 0$ be given and let ϕ be a solution of the problem (7.7) on the interval $[0,T]$ for some $T > 0$ such that the conditions (7.11) hold true. Let $\varphi \in H^1(Q)$ solve the linear problem*

$$\begin{cases} \ddot{\varphi}(t) = A(t)\varphi + F(t) & \text{for} \quad (t,x) \in Q, \\ \varphi = 0 & \text{for} \quad (t,x) \in \Sigma_1. \end{cases} \tag{7.41}$$

Set

$$\Upsilon(t) = \|\dot{\varphi}(t)\|^2 + \|\nabla\varphi(t)\|^2, \quad \Upsilon_{\Gamma_0,N}(t) = \|\varphi_{\nu_A}\|^2_{H^{1/2}(\Gamma_0)}.$$

Then there is $c_\gamma > 0$ such that

$$\Upsilon(t) \le c_\gamma \Upsilon(0)$$
$$+c_\gamma \int_0^t \left[(1 + \|\dot{\phi}(t)\|_{m-1})\Upsilon(\tau) + \Upsilon_{\Gamma_0,N}(\tau) + \|F(\tau)\|^2 \right] d\tau \tag{7.42}$$

for $0 \le t \le T$.

Proof. Let

$$P(t) = \|\dot{\varphi}\|^2 + (A\nabla\varphi, \nabla\varphi).$$

Then

$$\dot{P}(t) = 2(F, \dot{\varphi}) + \left(\dot{A}\nabla\varphi, \nabla\varphi \right) + 2\int_{\Gamma_0} \varphi_{\nu_A} \dot{\varphi} d\Gamma. \tag{7.43}$$

Using the estimate

$$|(\varphi_{\nu_A}, \dot{\varphi})_{L^2(\Gamma_0)}| \le \|\dot{\varphi}\|_{H^{-1/2}(\Gamma_0)} \|\varphi_{\nu_A}\|_{H^{1/2}(\Gamma_0)} \le c\Upsilon(t) + c\Upsilon_{\Gamma_0,N}(t)$$

in (7.43) gives the inequality (7.42). □

Boundary Traces and Energy Estimates with Robin Data on Γ_0 This time, we consider the wave equation in a divergence form:

$$\begin{cases} \phi_{tt} = \text{div } \mathbf{a}(x, \nabla w + \nabla\phi) & \text{for} \quad (t, x) \in Q, \\ \phi = 0 & \text{for} \quad (t, x) \in \Sigma_1 \end{cases} \tag{7.44}$$

where $\mathbf{a}(\cdot, \cdot) = (a_1(\cdot, \cdot), \cdots, a_n(\cdot, \cdot)): \Omega \times \mathbb{R}^n \to \mathbb{R}^n$ is smooth nonlinear map such that $\left(a_{iy_j}(x, y) \right)$ are symmetric and positive for all $(x, y) \in \Omega \times \mathbb{R}^n$ and $w \in H^m(\Omega)$ is *an equilibrium* which is defined by

$$\text{div } \mathbf{a}(x, \nabla w) = 0 \quad \text{for} \quad x \in \Omega.$$

Let $\phi \in \cap_{k=0}^m C^k \left([0, T], H^{m-k}(\Omega) \right)$ be a solution to the problem (7.44) for some $T > 0$. Let $\lambda > 0$ be given. We define

$$\begin{cases} \mathbf{I}(t) = \phi_t + \lambda\langle\mathbf{a}(x, \nabla w + \nabla\phi), \nu\rangle, \\ \mathbf{O}(t) = \phi_t - \lambda\langle\mathbf{a}(x, \nabla w + \nabla\phi), \nu\rangle. \end{cases} \tag{7.45}$$

Let

$$\mathcal{E}_{\Gamma_0,R}(t) = \sum_{k=0}^{m-2} \|\mathbf{I}^{(k)}(t)\|^2_{m-k-3/2,\Gamma_0}, \quad \mathcal{L}(t) = \sum_{k=3}^{2m} \mathcal{E}^{k/2}(t),$$

$$\mathcal{Q}_{\Gamma_0,R,\mathbf{I}} = \sum_{k=0}^{m-1} \|\mathbf{I}^{(k)}(t)\|^2_{\Gamma_0}, \quad \mathcal{Q}_{\Gamma_0,R,\mathbf{O}} = \sum_{k=0}^{m-1} \|\mathbf{O}^{(k)}(t)\|^2_{\Gamma_0},$$

$$\mathcal{Q}_\lambda(t) = 2\lambda[\|\phi_t(t)\|^2 + (B_\phi(t)\nabla\phi, \nabla\phi)]$$
$$+2\lambda \sum_{k=1}^{m-1} [\|\phi^{(k+1)}(t)\|^2 + (A_\phi(t)\nabla\phi^{(k)}, \nabla\phi^{(k)})], \tag{7.46}$$

$$\mathcal{P}(t) = 4\lambda(\|\phi_t\|^2 + (B_\phi(t)\nabla\phi, \nabla\phi) + \|\phi\|^2) \tag{7.47}$$

where

$$B_\phi(t) = \Big(\int_0^1 a_{iy_j}(x, \nabla w + s\nabla\phi)ds \Big), \tag{7.48}$$

$$A_\phi(t) = \Big(a_{iy_j}(x, \nabla w + \nabla\phi) \Big). \tag{7.49}$$

We have

Theorem 7.3 *Let $\gamma > 0$ be given and let ϕ be a solution of the problem (7.44) on the interval $[0, T]$ for some $T > 0$ such that the conditions (7.11) hold. Then there are $c_{0\gamma}$, $c_\gamma > 0$, which only depend on the γ, such that*

$$c_{0,\gamma}\mathcal{Q}_\lambda(t) \leq \mathcal{E}(t) \leq c_\gamma \mathcal{Q}_\lambda(t) + c_\gamma \mathcal{E}_{\Gamma_0,R}(t) + c_\gamma \|\phi(t)\|^2 + c_\gamma \mathcal{L}(t) \tag{7.50}$$

for $0 \leq t \leq T$,

$$\dot{\mathcal{Q}}_\lambda(t) \leq 4\mathcal{Q}_{\Gamma_0,R,\mathbf{I}}(t) - \mathcal{Q}_{\Gamma_0,R,\mathbf{0}}(t) + c_\gamma \mathcal{L}(t), \tag{7.51}$$

$$-\dot{\mathcal{Q}}_\lambda(t) \leq 2\mathcal{Q}_{\Gamma_0,R,\mathbf{I}}(t) + 2\mathcal{Q}_{\Gamma_0,R,\mathbf{0}}(t) + c_\gamma \mathcal{L}(t), \tag{7.52}$$

$$\mathcal{P}(t) \leq \{\mathcal{P}(0) + c_\gamma \int_0^T [\mathcal{E}_{\Gamma_0,R}(\tau) + \mathcal{L}(\tau)]\, d\tau\}e^t \tag{7.53}$$

for $0 \leq s \leq t \leq T$ where $\mathcal{E}(t)$ is given in (7.8).

To prove Theorem 7.3, we need

Lemma 7.7 *Let $\gamma > 0$ be given and let ϕ solve the system (7.44). Let*

$$r_j(t) = \sum_{k=1}^{j} C_k^j \operatorname{div} A_\phi^{(k)} \nabla\phi^{(j-k)}, \quad r_{\Gamma_0 j}(t) = \sum_{k=1}^{j} C_k^j \langle A_\phi^{(k)} \nabla\phi^{(j-k)}, \nu \rangle$$

for $1 \leq j \leq m - 2$. Then

$$\|r_j\|_{m-2-j}^2, \quad \|r_{\Gamma_0,j}\|_{m-3/2-j}^2 \leq c_\gamma \sum_{k=2}^{m-1} \mathcal{E}^k(t) \tag{7.54}$$

for $1 \leq k \leq m - 2$.

Proof. Let b be a smooth function on $\Omega \times \mathbb{R}^n$. We have

$$b^{(k)}(x, \nabla\phi) = \sum_{i=1}^{k} \sum_{r_1+\cdots+r_i=k} D_y^i b(\nabla\phi^{(r_1)}, \cdots, \nabla\phi^{(r_i)}) \tag{7.55}$$

for $1 \leq k \leq j$ where $D_y^i b$ are the differentials of b of order i with respect to variable y. Thus, $r_j(t)$ is a sum of some functions in the form

$$(f(x, \nabla\phi)\phi_{x_{j_1}}^{(r_1)} \cdots \phi_{x_{j_i}}^{(r_i)} \phi_{x_p}^{(j-k)})_{x_q}$$

with $r_1 + \cdots + r_i = k$ for $1 \leq i \leq k$. Using the inequalities (7.17) and (7.18), we have

$$\|(f(x, \nabla\phi)\phi_{x_{j_1}}^{(r_1)} \cdots \phi_{x_{j_i}}^{(r_i)} \phi_{x_p}^{(j-k)})_{x_q}\|_{m-2-j}^2$$
$$\leq \|f(x, \nabla\phi)\phi_{x_{j_1}}^{(r_1)} \cdots \phi_{x_{j_i}}^{(r_i)} \phi_{x_p}^{(j-k)}\|_{m-1-j}^2 \leq c_\gamma \mathcal{E}^{i+1}(t) \tag{7.56}$$

which yield the inequality (7.54) on $r_j(t)$ for all $1 \leq j \leq m-2$. An similar argument gives the inequality (7.54) on $r_{\Gamma_0 j}(t)$ for all j. □

Proof of Theorem 7.3. The left-hand side of the inequality (7.50) is clearly true. We prove the right-hand side of it. We have

$$\|\phi^{(m)}(t)\|^2 + \|\phi^{(m-1)}(t)\|_1^2$$
$$\leq \|\phi^{(m)}(t)\|^2 + c_\gamma(A_\phi(t)\nabla\phi^{(m-1)}, \nabla\phi^{(m-1)}) + \|\phi^{(m-1)}(t)\|^2$$
$$\leq c_\gamma \mathcal{Q}_\lambda(t). \tag{7.57}$$

Proceeding by induction, we assume that for $3 \leq l \leq m$, when $j = l$ and $j = l-1$,

$$\|\phi^{(j)}(t)\|_{m-j}^2 \leq c_\gamma \mathcal{Q}(t) + c_\gamma \mathcal{E}_{\Gamma_0, R}(t) + c_\gamma \|\phi(t)\|^2 + c_\gamma \mathcal{L}(t) \tag{7.58}$$

which, as shown in (7.57), is true for $l = m$.

Formal differentiation of the equations in (7.44) and (7.45) for $l - 2$ times with respect to t yields

$$\begin{cases} \phi^{(l)} = \operatorname{div} A_\phi(t)\nabla\phi^{(l-2)} + r_{l-3}(t) & \text{for } (t, x) \in Q, \\ \phi^{(l-2)} = 0 & \text{for } (t, x) \in \Sigma_1 \\ \phi^{(l-1)} + \lambda\phi_{\nu_A}^{(l-2)} + \lambda r_{\Gamma_0, l-3} = \mathbf{I}^{(l-2)}(t) & \text{for } (t, x) \in \Sigma_0, \\ \phi^{(l-1)} - \lambda\phi_{\nu_A}^{(l-2)} - \lambda r_{\Gamma_0, l-3} = \mathbf{O}^{(l-2)}(t) & \text{for } (t, x) \in \Sigma_0 \end{cases} \tag{7.59}$$

where r_{l-3} and $r_{\Gamma_0, l-3}$ are given in Lemma 7.7 if $l \geq 4$ and they are zero when $l = 3$. Using the ellipticity (7.40), the equations in (7.59), and the estimates in Lemma 7.7, we obtain

$$\|\phi^{(l-2)}(t)\|_{m-l+2}^2$$
$$\leq c_\gamma(\|\operatorname{div} A_\phi(t)\nabla\phi^{(l-2)}\|_{m-l}^2 + \|\phi_{\nu_A}^{(l-2)}\|_{\Gamma_0, m-l+1/2}^2 + \|\phi^{(l-2)}(t)\|_{m-l+1}^2)$$
$$\leq c_\gamma \|\phi^{(l)}(t)\|_{m-l}^2 + c_\gamma \|\mathbf{I}^{(l-2)}(t)\|_{\Gamma_0, m-l+1/2}^2 + c_\gamma \|\phi^{(l-1)}(t)\|_{m-l}^2$$
$$+ \varepsilon\|\phi^{(l-2)}(t)\|_{m-l+2}^2 + c_{\gamma, \varepsilon}\|\phi^{(l-2)}(t)\|^2 + c_\gamma \mathcal{L}(t) \tag{7.60}$$

which implies that the inequality (7.58) is also true for $l - 2$. Then the inequality (7.58) holds for $1 \leq j \leq m$ by induction. To get the estimate (7.50), we need to prove that the inequality (7.58) is true for $j = 0$.

The system (7.44)-(7.45) can be rewritten as

$$\begin{cases} \phi_{tt} = \operatorname{div} B_\phi(t)\nabla\phi & \text{for} \quad (t,x) \in Q, \\ \phi = 0 & \text{for} \quad (t,x) \in \Sigma_1, \\ \phi_t + \lambda\phi_{\nu_B} = \mathbf{I}(t) & (t,x) \in \Sigma_0, \\ \phi_t - \lambda\phi_{\nu_B} = \mathbf{O}(t) & (t,x) \in \Sigma_0. \end{cases} \tag{7.61}$$

Then similar estimates as in (7.60) yield the inequality (7.58) for $j = 0$ where this time instead of the formulas in (7.59), we use the formulas in (7.61).

Let

$$\vartheta_0(t) = \|\phi_t(t)\|^2 + (B_\phi(t)\nabla\phi, \nabla\phi) \tag{7.62}$$

and, for $1 \leq j \leq m - 1$, let

$$\vartheta_j(t) = \|\phi^{(j+1)}(t)\|^2 + (A_\phi(t)\nabla\phi^{(j)}(t), \nabla\phi^{(j)}(t)). \tag{7.63}$$

For $1 \leq j \leq m - 1$ using the formulas (7.59) for $l = j + 2$ and the Green formula, we obtain

$$\begin{aligned} \dot{\vartheta}_j(t) &= 2(r_{j-1}(t), \phi^{(j+1)}(t)) + 2(\phi^{(j+1)}(t), (\phi^{(j)})_{\nu_A}(t)) \\ &\quad + (\dot{A}_\phi(t)\nabla\phi^{(j)}(t), \nabla\phi^{(j)}(t)) \\ &= (2\lambda)^{-1}(\|\mathbf{I}^{(j)}(t) - \lambda r_{\Gamma_0,j-1}(t)\|^2_{\Gamma_0} - \|\mathbf{O}^{(j)}(t) + \lambda r_{\Gamma_0,j-1}(t)\|^2_{\Gamma_0}) \\ &\quad + 2(r_{j-1}(t), \phi^{(j+1)}(t)) + (\dot{A}_\phi(t)\nabla\phi^{(j)}(t), \nabla\phi^{(j)}(t)). \end{aligned} \tag{7.64}$$

In the case $j = 0$ we use the formula (7.61) to obtain

$$\dot{\vartheta}_0(t) = (2\lambda)^{-1}(\|\mathbf{I}(t)\|^2_{\Gamma_0} - \|\mathbf{O}(t)\|^2_{\Gamma_0}). \tag{7.65}$$

The inequalities (7.51) and (7.52) follow from (7.64) and (7.65) since $\mathcal{Q}_\lambda(t) = 4\lambda \sum_{j=0}^{m-1} \vartheta_j(t)$ where the following estimates have been used

$$|(\dot{A}_\phi(t)\nabla\phi^{(j)}(t), \nabla\phi^{(j)}(t))| \leq c_\gamma \mathcal{E}^{3/2}(t) \quad \text{for} \quad 1 \leq j \leq m - 1.$$

Similarly, we have

$$\dot{\mathcal{P}}(t) \leq \|\mathbf{I}(t)\|^2_{\Gamma_0} + c_\gamma \mathcal{E}^{3/2}(t) + \mathcal{P}(t), \quad 0 \leq t \leq T,$$

which yields

$$\mathcal{P}(t) \leq \mathcal{P}(0) + c_\gamma \int_9^T [\mathcal{E}_{\Gamma_0,R}(\tau) + \mathcal{E}^{3/2}(\tau)]\, d\tau] + \int_0^t \mathcal{P}(\tau)\, d\tau \tag{7.66}$$

for $0 \leq t \leq T$. The inequality (7.53) follows from (7.66) by *Gronwall's inequality*. $\square$

7.2 Locally and Globally Boundary Exact Controllability

In this section we study the boundary exact controllability for the quasilinear wave equation. We derive the existence of long time solutions near an equilibrium, prove the local exact controllability around the equilibrium under existence of *an escape vector field for the metric*. We then establish the globally exact controllability in such a way that the state of the quasilinear wave equation moves from an equilibrium in one location to an equilibrium in another location under some geometrical conditions. The Dirichlet action and the Neumann action are studied, respectively. Escape vector fields here are determined by a Riemannian metric, given by the coefficients and equilibria of the quasilinear wave equation. Then the Riemmannian curvature theory provides a unique verifiable tool for this assumption. Some examples are presented to verify the globally exact controllability.

Boundary Exact Controllability with Dirichlet Action Let $\Omega \subset \mathbb{R}^n$ be an open, bounded set with the smooth boundary Γ. Suppose that Γ consists of two disjoint parts, Γ_0 and Γ_1, where $\Gamma_0 \neq \emptyset$. Let w be an equilibrium, given by (7.1), and let $T > 0$ be given. We consider a controllability problem around the equilibrium w

$$\begin{cases} u_{tt} = \sum_{ij=1}^{n} a_{ij}(x, \nabla u) u_{x_i x_j} + b(x, \nabla u) & \text{for} \quad (t, x) \in Q, \\ u = w \quad \text{for} \quad (t, x) \in \Sigma_1, \\ u = w + \varphi \quad \text{for} \quad (t, x) \in \Sigma_0, \\ u(0) = u_0, \quad u_t(0) = u_1 \end{cases} \tag{7.67}$$

where $a_{ij}(x, y)$, $b(x, y)$ are smooth functions on $\overline{\Omega} \times \mathbb{R}^n$ such that $A(x, y) = (a_{ij}(x, y)) > 0$ for $(x, y) \in \overline{\Omega} \times \mathbb{R}^n$ and $b(x, 0) = 0$ for $x \in \overline{\Omega}$.

Let $m \geq [n/2] + 3$ be a given positive integer.

Definition 7.2 *We say that* $(u_0, u_1) \in H^m(\Omega) \times H^{m-1}(\Omega)$ *and*

$$\varphi \in \cap_{k=0}^{m-2} C^k \left([0, T], H^{m-k-1/2}(\Gamma_0) \right)$$

satisfy the compatibility conditions of order m *with the Dirichlet data if*

$$u_0 = w, \quad u_k \in H^{m-k}(\Omega), \quad u_k = 0 \quad \text{for} \quad x \in \Gamma_1,$$

$$\varphi = 0, \quad \varphi^{(k)}(0) = u_k, \quad \text{for} \quad x \in \Gamma_0, \tag{7.68}$$

for $1 \leq k \leq m-1$ *where for* $k \geq 2$, $u_k = u^{(k)}(0)$, *as computed formally (and recursively) in terms of* u_0 *and* u_1, *using the equation in* (7.67).

Definition 7.3 *Let* u_0, u_1, $\hat{u}_0$, *and* $\hat{u}_1$ *be given functions on* $\overline{\Omega}$ *and* $T > 0$ *be* *given. If there is a boundary function φ on Σ_0 such that the solution of the problem (7.67) satisfies*

$$u(T) = \hat{u}_0, \quad \dot{u}(T) = \hat{u}_1 \quad for \quad x \in \Omega,$$

we say the system (7.67) is exactly controllable from (u_0, u_1) to $(\hat{u}_0, \hat{u}_1)$ at time T by boundary with the Dirichlet action.

We assume that short time solutions to the problem (7.67) exist for $(u_0, u_1) \in H^m(\Omega) \times H^{m-1}(\Omega)$ and study the possibility of moving it to another state in $H^m(\Omega) \times H^{m-1}(\Omega)$ at time T via a boundary control

$$\varphi \in \cap_{k=0}^{m-2} C^k \left([0, T], H^{m-k-1/2}(\Gamma_0) \right).$$

Solutions in Long Time In general, solutions of the system (7.67) may blow up in a finite time even if the initial data and the boundary control are smooth. However, in order to move one state to another, the control time must be larger than the wave length of the system. For those purposes, we shall establish the existence of long time solutions of the system around an equilibrium.

Theorem 7.4 *Let $w \in H^m(\Omega)$ be an equilibrium of the problem (7.67). Let $T > 0$ be arbitrarily given. Then there is $\varepsilon_T > 0$, which depends on the time T, such that, if $(u_0, u_1) \in H^m(\Omega) \times H^{m-1}(\Omega)$ satisfy*

$$\|u_0 - w\|_m < \varepsilon_T, \quad \|u_1\|_{m-1} < \varepsilon_T,$$

and if $\varphi \in \cap_{k=0}^{m-2} C^k \left([0, T], H^{m-k-1/2}(\Gamma_0) \right)$ with $\varphi^{(k)} \in H^1(\Sigma_0)$ for $0 \le k \le m - 1$ satisfies the compatibility conditions with (u_0, u_1) of order m such that

$$\sum_{k=0}^{m-1} \|\varphi^{(k)}\|^2_{C([0,T],H^{m-1/2-k}(\Gamma_0))} + \sum_{k=0}^{m-1} \|\varphi^{(k)}\|^2_{H^1(\Sigma_0)} < \varepsilon_T,$$

then the system (7.67) has a solution $u \in \cap_{k=0}^{m} C^k \left([0, T], H^{m-k}(\Omega) \right)$.

Proof. Let u be a short time solution to the problem (7.67) and let $\phi = u - w$. Then

$$\phi(0) = u_0 - w, \quad \phi_t(0) = u_1, \quad \phi|_{\Gamma_0} = \varphi,$$

and ϕ is a short time solution to the problem (7.7) with $b_0(x, \nabla\phi, \phi_t) = b(x, \nabla\phi)$ for $x \in \Omega$. We take $\gamma = 1$. Let

$$c_1 = c_\gamma \ge 1 \tag{7.69}$$

be fixed such that the corresponding inequalities (7.12) and (7.13) in Theorem 7.1 hold for t in the existence interval of the solution u, respectively.

Let $T_1 > 0$ be arbitrarily given. We shall prove that, if initial data (u_0, u_1) and boundary value φ are compatible of order m such that

$$\mathcal{E}(0) + \max_{0 \leq t \leq T_1} \mathcal{E}_{\Gamma_0, D}(t) + \int_0^{T_1} Q_{\Gamma_0, D}(t) dt \leq \frac{1}{48c_1^3} e^{-6c_1^2 T_1}, \tag{7.70}$$

then solutions to the problem (7.67) exist at least on the interval $[0, T_1]$.

We set

$$\eta = \frac{1}{4c_1} \leq \frac{1}{4} < \frac{1}{2}. \tag{7.71}$$

Since $\mathcal{E}(0) \leq \eta/4$, the solution of short time must satisfy

$$\mathcal{E}(t) \leq \eta \leq 1/2 < 1 \tag{7.72}$$

for some interval $[0, \delta]$.

Let δ_0 be the largest number such that (7.72) is true for $t \in [0, \delta_0)$. It follows from the inequalities (7.71) and (7.72) that

$$\sum_{i=2}^m \mathcal{E}^i(t) \leq 2\mathcal{E}^2(t) \leq \frac{1}{2c_1} \mathcal{E}(t) \quad \text{for} \quad t \in [0, \delta_0). \tag{7.73}$$

We shall prove $\delta_0 \geq T_1$ by contradiction.

Suppose that $\delta_0 < T_1$. Then in this interval $[0, \delta_0]$ the conditions (7.11) hold true. We apply Theorem 7.1. Using the inequalities (7.12) and (7.73), we obtain

$$\mathcal{E}(t) \leq 2c_1[Q(t) + \mathcal{E}_{\Gamma_0, D}(t)] \quad \text{for} \quad t \in [0, \delta_0). \tag{7.74}$$

Moreover, it follows from the inequalities (7.13), (7.12), (7.72), (7.73), and (7.69) that

$$Q(t) \leq c_1 \mathcal{E}(0) + 3c_1 \int_0^t \mathcal{E}(\tau) \, d\tau + c_1 \int_0^t Q_{\Gamma_0, D}(\tau) \, d\tau \tag{7.75}$$

for $t \in [0, \delta_0)$. Inserting the inequality (7.75) into the inequality (7.74), we have

$$\mathcal{E}(t) \leq 6c_1^2[\mathcal{E}(0) + \int_0^{T_1} Q_{\Gamma_0, N}(\tau) \, d\tau + \max_{0 \leq t \leq T_1} \mathcal{E}_{\Gamma_0, N}(t)] + 6c_1^2 \int_0^t \mathcal{E}(\tau) \, d\tau$$

which yields, via the condition (7.70) and the Gronwall inequality,

$$\mathcal{E}(\delta_0) \leq \eta/2 < \eta.$$

This contradicts the definition of δ_0. $\square$

In order to move one state to another in $H^m(\Omega) \times H^{m-1}(\Omega)$, we recall a boundary control space in (2.94) as follows, which is a Banach space. Let $B_D^m(\Sigma_0)$ consist of all the functions

$$\varphi \in \cap_{k=0}^{m-1} C^k \left([0,T], H^{m-1/2-k}(\Gamma_0)\right),\tag{7.76}$$

$$\varphi^{(k)} \in H^1(\Sigma_0)), \quad \varphi^{(k)}(0)|_{\Gamma_0} = 0\tag{7.77}$$

for $0 \le k \le m-1$ with the norm

$$\|\varphi\|_{B_D^m}^2 = \sum_{k=0}^{m-1} \|\varphi^{(k)}\|_{C\left([0,T],H^{m-k-1/2}(\Gamma_0)\right)}^2 + \sum_{k=0}^{m-1} \|\varphi^{(k)}\|_{H^1(\Sigma_0)}^2.$$

Clearly, the controllability of the problem (7.67) depends on the geometrical properties of not only the coefficients but also the equilibrium. We introduce the metric

$$g = A^{-1}(x, \nabla w)\tag{7.78}$$

on Ω and consider the couple (Ω, g) as a Riemannian manifold with a boundary Γ.

Definition 7.4 *An equilibrium w is said to be exactly controllable if there is an escape vector field on $\overline{\Omega}$ for the metric (7.78).*

Remark 7.3 *Existence of escape vector fields has been studied in Section 2.3. Some more examples will be given in the end of this section.*

Let $w \in H^m(\Omega)$ be exactly controllable and let H be the strictly convex function on $\overline{\Omega}$ in the metric g. Then there is a $\varrho_0 > 0$ such that

$$DH(X,X) \ge \varrho_0 |X|_g^2 \quad \text{for} \quad X \in \mathbb{R}_x^n, \quad x \in \overline{\Omega}.\tag{7.79}$$

We have

Theorem 7.5 *Let $w \in H^m(\Omega)$ be an equilibrium which is exactly controllable. Let*

$$T_0 = \frac{2}{\varrho_0} \max_{x \in \Omega} |H|_g.\tag{7.80}$$

Furthermore, if $\Gamma_1 \ne \emptyset$, we assume that

$$\langle H, \nu \rangle \le 0 \quad \text{for} \quad x \in \Gamma_1.\tag{7.81}$$

Then, for $T > T_0$ given, there is a neighborhood of the point $(w, 0)$ in $H^m(\Omega) \times H^{m-1}(\Omega)$ where we can move one state of the problem (7.67) to another by a boundary control function in the space $B_D^m(\Sigma_0)$.

Proof. We invoke Theorem 7.4 to define a map for $\varphi \in B_D^m(\Sigma_0)$ by setting

$$\Phi(\varphi) = (u(T), u_t(T)) \tag{7.82}$$

where u is the solution of the following problem

$$\begin{cases} u_{tt} = \sum_{ij=1}^n a_{ij}(x, \nabla u)u_{x_i x_j} + b(x, \nabla u) & \text{for} \quad (t, x) \in Q, \\ u = w & \text{for} \quad (t, x) \in \Sigma_1, \\ u = w + \varphi & \text{for} \quad (t, x) \in \Sigma_0, \\ u(0) = w, \quad u_t(0) = 0. \end{cases} \tag{7.83}$$

Let $\varepsilon_T > 0$ be given by Theorem 7.4. Then

$$\Phi : B(0, \varepsilon_T) \to H^m(\Omega) \times H^{m-1}(\Omega)$$

where $B(0, \varepsilon_T) \subset B_D^m(\Sigma_0)$ is the ball with the radius ε_T centered at 0.

We observe that $\Phi(0) = (w, 0)$. If we can prove that there is a neighborhood $\mathcal{U}$ of the point $(w, 0)$ in $H^m(\Omega) \times H^{m-1}(\Omega)$ such that $\mathcal{U} \subset$ the image $\Phi(B(0, \varepsilon_T))$, then for each state in $\mathcal{U}$ there will be a boundary control $\varphi \in B(0, \varepsilon_T)$ which steers the state to $(w, 0)$ by the time reversibility of the problem (7.83) and the proof will be complete.

For the above purpose, we need to evaluate

$$\Phi'(0)\varphi = \frac{\partial}{\partial \sigma}\Phi(\sigma\varphi)|_{\sigma=0} \quad \text{for} \quad \varphi \in B_D^m(\Sigma_0).$$

It is easy to check that

$$\Phi'(0)\varphi = (\phi(T), \phi_t(T)) \quad \text{for} \quad \varphi \in B_D^m(\Sigma_0) \tag{7.84}$$

where $\phi(t, x)$ is the solution of the linear system with variable coefficients in the space variable

$$\begin{cases} \phi_{tt} = \mathcal{A}\phi + F(\phi) & \text{for} \quad (t, x) \in Q, \\ \phi = 0 & \text{for} \quad (t, x) \in \Sigma_1, \\ \phi = \varphi & \text{for} \quad (t, x) \in \Sigma_0, \\ \phi(0) = \phi_t(0) = 0 & \text{for} \quad x \in \Omega \end{cases} \tag{7.85}$$

where

$$\mathcal{A}\phi = \sum_{ij=1}^n a_{ij}(x, \nabla w)\phi_{x_i x_j}, \quad F = (F_1, \cdots, F_n),$$

$$F_i = \sum_{lj} a_{ljy_i}(x, \nabla w)w_{x_l x_j} + b_{y_i}(x, \nabla w).$$

From [154], an escape vector field for the metric (7.78) guarantees that the observability estimate (2.122) is true for $T > T_0$. By Theorem 2.21, the linear problem (7.85) is exactly $H^m(\Omega) \times H^{m-1}(\Omega)$ controllable on $[0, T]$ with boundary control in $B_D^m(\Sigma_0)$. Then the linear map, given by (7.84),

$$\Phi'(0) : B_D^m(\Sigma_0) \to H^m(\Omega) \times H^{m-1}(\Omega)$$

is surjective. Then the nonlinear map Φ is locally surjective at $\varphi = 0$ by Lemma 7.8 below. The proof is complete.   □

The proof of the following lemma is classical; for example, see Theorem 3.1.19 in [11].

Lemma 7.8 *Let* $\mathbf{X}$ *and* $\mathbf{Y}$ *be Banach spaces. Let a map* $\Phi: \mathcal{O} \to \mathbf{Y}$, *where* $\mathcal{O}$ *is an open subset of* $\mathbf{X}$, *be Frechét differentiable. Let* $x_0 \in \mathcal{O}$. *If* $\Phi'(x_0): \mathbf{X} \to \mathbf{Y}$ *is surjective, then there is an open neighborhood of* $y_0 = \Phi(x_0)$ *contained in the image* $\Phi(\mathcal{O})$.

Remark 7.4 *The results in Theorem 7.5 are local. However, if we have enough equilibria exactly controllable, we can move the quasilinear wave state along a curve of equilibria, moving in successive small steps from one equilibrium to another nearby equilibrium until the target equilibrium is reached, see Theorem 7.6 below. This procedure can be done by the theorem of finite covering through finite steps. This idea was used by [186] for the quasilinear string.*

Globally Exact Controllability from One Equilibrium to Another with the Dirichlet Action Let $w \in H^m(\Omega)$ be a given equilibrium. For $\alpha \in [0,1]$, we assume that $w_\alpha \in H^m(\Omega)$ are the solutions of the Dirichlet problem

$$\begin{cases} \sum_{ij=1}^n a_{ij}(x, \nabla w_\alpha) w_{\alpha x_i x_j} + b(x, \nabla w_\alpha) = 0 & \text{for } x \in \Omega, \\ w_\alpha|_\Gamma = \alpha w|_\Gamma \end{cases} \tag{7.86}$$

such that

$$\sup_{\alpha \in [0,1]} \|w_\alpha\|_m < \infty. \tag{7.87}$$

For the existence of the classical solution to the Dirchlet problem (7.86), for example, see [68].
    We have

Theorem 7.6 *Let* $w \in H^m(\Omega)$ *be an equilibrium which is exactly controllable. Let* $w_\alpha \in H^m(\Omega)$, *given by (7.86), be also exactly controllable for all* $\alpha \in [0,1]$ *such that the condition (7.87) holds. Then, there are* $T > 0$ *and* $\varphi \in \mathrm{B}_D^m(\Sigma_0)$ *that is compatible with the initial data* $(w, 0)$ *such that the solution of the system (7.67) with* $(u_0, u_1) = (w, 0)$ *satisfies*

$$u(T) = \dot{u}(T) = 0.$$

Proof. By Theorem 7.5 and the theorem of finite covering it will suffice to prove that $w_\alpha: [0,1] \to H^m(\Omega)$ is continuous in $\alpha \in [0,1]$ because the interval $[0,1]$ is compact.

It is readily seen that $v_\alpha = \frac{\partial}{\partial\alpha}w_\alpha$ is the solution of the following linear, elliptic problem

$$\begin{cases} \sum_{ij=1}^n a_{ij}(x, \nabla w_\alpha)v_{\alpha x_i x_j} + \\ \sum_l^n [\sum_{ij}^n a_{ijy_l}(x, \nabla w_\alpha)w_{\alpha x_i x_j} + b_{y_l}(x, \nabla w_\alpha)]v_{\alpha x_l} = 0, \\ v_\alpha|_\Gamma = w|_\Gamma \end{cases} \quad (7.88)$$

for each $\alpha \in [0,1]$. Moreover, by the maximum principle for the above problem (7.88),

$$\sup_{x\in\Omega} |\frac{\partial}{\partial\alpha}w_\alpha| \leq \sup_{x\in\Gamma} |w|. \quad (7.89)$$

Let

$$\mathcal{A}(\alpha)v = \sum_{ij=1}^n a_{ij}(x, \nabla w_\alpha)v_{x_i x_j} \quad \text{for} \quad v \in H^2(\Omega), \ \alpha \in [0,1]. \quad (7.90)$$

Using the uniform bound (7.87), the ellipticity of the operator $\mathcal{A}(\alpha_0)$, and the estimate (7.89), we have

$$\begin{aligned}
&\|w_\alpha - w_{\alpha_0}\|_m \\
&\leq c\|\mathcal{A}(\alpha_0)(w_\alpha - w_{\alpha_0})\|_{m-2} + c|\alpha - \alpha_0|\|w\|_{m-1/2,\Gamma} + c\|w_\alpha - w_{\alpha_0}\| \\
&\leq c\|\mathcal{A}(\alpha_0)(w_\alpha - w_{\alpha_0})\|_{m-2} + c|\alpha - \alpha_0|(\|w\|_{m-1/2,\Gamma} + \sup_{x\in\Gamma}|w|). \quad (7.91)
\end{aligned}$$

Next, let us estimate $\|\mathcal{A}(\alpha_0)(w_\alpha - w_{\alpha_0})\|_{m-2}$.
$[\mathcal{A}(\alpha_0) - \mathcal{A}(\alpha)]w_\alpha$ and $b(x, \nabla w_\alpha) - b(x, \nabla w_{\alpha_0})$ can be written as sums of some terms, respectively, in the form

$$f(x, \nabla w_\alpha, \nabla w_{\alpha_0})(w_{\alpha_0 x_l} - w_{\alpha x_l})w_{\alpha x_i x_j}.$$

Applying the inequality (7.17) to the above products gives, via the bound (7.87) and the estimate (7.89),

$$\begin{aligned}
&\|\mathcal{A}(\alpha_0)(w_\alpha - w_{\alpha_0})\|_{m-2} \\
&\leq \|(\mathcal{A}(\alpha_0) - \mathcal{A}(\alpha))w_\alpha\|_{m-2} + \|b(x, \nabla w_a) - b(x, \nabla w_{\alpha_0})\|_{m-2} \\
&\leq c\|w_\alpha - w_{\alpha_0}\|_{m-1} \\
&\leq \varepsilon\|w_\alpha - w_{\alpha_0}\|_m + c_\varepsilon|\alpha - \alpha_0| \sup_{x\in\Gamma}|w|. \quad (7.92)
\end{aligned}$$

We obtain the desired result after substituting (7.92) into (7.91). $\qquad\square$

Remark 7.5 *Since the quasilinear wave equation is time-reversible, an equilibrium can be moved to another if they can both be moved to zero. However, this result only gives the existence of the control time T. We do not know how large the T is because it is given by the theorem of finite covering.*

Boundary Exact Controllability with the Neumann Action Next, we turn to the boundary control with the Neumann action. Let $\Gamma = \Gamma_0 \cup \Gamma_1$ and $\overline{\Gamma}_0 \cap \overline{\Gamma}_1 = \emptyset$ with Γ_1 being nonempty. Let

$$\mathbf{a}(\cdot, \cdot) = (a_1(\cdot, \cdot), \cdots, a_n(\cdot, \cdot))$$

where $a_i(\cdot, \cdot)$ are smooth functions on $\overline{\Omega} \times \mathbb{R}^n$ such that $\mathbf{a}(x, 0) = 0$ for $x \in \overline{\Omega}$ and $A(x, y) = \left(a_{iy_j}(x, y)\right) > 0$ for $(x, y) \in \overline{\Omega} \times \mathbb{R}^n$.

This time we assume that the quasilinear part of the system is in the divergence form. Let $T > 0$ be given. We consider a controllability problem

$$\begin{cases} \ddot{u} = \text{div}\,\mathbf{a}(x, \nabla u) & \text{on}\quad Q, \\ u = w & \text{on}\quad \Sigma_1, \\ \langle \mathbf{a}(x, \nabla u), \nu \rangle = \langle \mathbf{a}(x, \nabla w), \nu \rangle + \varphi & \text{on}\quad \Sigma_0, \\ u(0) = u_0, \quad \dot{u}(0) = u_1 \end{cases} \tag{7.93}$$

where ν is the normal of the boundary Γ in the Euclidean metric of $\mathbb{R}^n$ and w is an equilibrium, defined by

$$\text{div}\,\mathbf{a}(x, \nabla w) = 0 \quad \text{on}\quad \Omega. \tag{7.94}$$

We say that $(u_0, u_1) \in H^m(\Omega) \times H^{m-1}(\Omega)$ and

$$\varphi \in \cap_{k=0}^{m-2} C^k\left([0, T], H^{m-k-3/2}(\Gamma_0)\right)$$

satisfy the *compatibility conditions* of order m with the Neumann boundary data on Γ_0 and the Dirichlet data on Γ_1 if the relations (7.68) hold and

$$\varphi^{(k)}(0) = \begin{cases} 0 & \text{for}\quad x \in \Gamma_0, \quad k = 0, \\ \langle A(x, \nabla w)\nabla u_k, \nu \rangle & x \in \Gamma_0 \quad 1 \le k \le m-1, \end{cases}$$

where $u_k = u^{(k)}(0)$ are given by the equation in (7.93) for $k \ge 2$.

Solutions in Long Time Using the estimates in Theorem 7.2, similar arguments as in the proof of Theorem 7.4 yield

Theorem 7.7 *Let $w \in H^m(\Omega)$ be an equilibrium of the problem (7.93). Let $T > 0$ be arbitrarily given. Then there is $\varepsilon_T > 0$, which depends on the time T, such that, if $(u_0, u_1) \in H^m(\Omega) \times H^{m-1}(\Omega)$ satisfy*

$$\|u_0 - w\|_m < \varepsilon_T, \quad \|u_1\|_{m-1} < \varepsilon_T,$$

and $\varphi \in \cap_{k=0}^{m-2} C^k\left([0, T], H^{m-k-3/2}(\Gamma_0)\right)$ is such that $\varphi^{(k)} \in H^1(\Sigma_0)$ for $0 \le k \le m - 1$, which satisfies the compatibility conditions with (u_0, u_1) of order m and

$$\sum_{k=0}^{m-2} \|\varphi^{(k)}\|^2_{C([0,T], H^{m-k-3/2}(\Gamma_0))} + \sum_{k=0}^{m-2} \|\varphi^{(k)}\|^2_{H^1(\Sigma_0)} < \varepsilon_T,$$

then the system (7.93) has a unique solution $u \in \cap_{k=0}^{m} C^k\left([0, T], H^{m-k}(\Omega)\right)$.

Let $T > 0$ be given. We recall a Banach space $\mathrm{B}_N^m(\Sigma_0)$ by (2.144) as follows. Let $\mathrm{B}_N^m(\Sigma_0)$ consist of all the functions

$$\varphi \in \cap_{k=0}^{m-2} C^k([0,T], H^{m-k-3/2}(\Gamma_0)), \quad \varphi^{(k)} \in H^1(\Sigma_0),$$

$$\varphi^{(k)}(0) = 0, \quad x \in \Gamma_0, \quad 0 \le k \le m-1$$

with the norm

$$\|\varphi\|_{\mathrm{B}_N^m(\Sigma_0)}^2 = \sum_{k=0}^{m-2} \|\varphi^{(k)}\|_{C([0,T], H^{m-3/2-k}(\Gamma_0))}^2 + \sum_{k=0}^{m-2} \|\varphi^{(k)}\|_{H^1(\Sigma_0)}^2.$$

Geometry Condition for Local Controllability with the Neumann Action We introduce a Riemannian metric on Ω by

$$g = A^{-1}(x, \nabla w)$$

and consider the couple (Ω, g) as a Riemannian manifold with a boundary Γ where w is an equilibrium, given by the problem (7.94).

We have

Theorem 7.8 *Let* $\Gamma_1 \neq \emptyset$. *Let* $w \in H^{m+1}(\Omega)$ *be an equilibrium which is exactly controllable. Let* H *be an escape vector field on* $\overline{\Omega}$ *such that* $\langle H, \nu \rangle \le 0$ *for* $x \in \Gamma_1$. *Then there exists a time* $T_0 > 0$ *such that for* $T > T_0$ *given, there is a neighborhood of the point* $(w, 0)$ *in* $H^{m+1}(\Omega) \times H^m(\Omega)$ *where we can move one state of the problem* (7.93) *to another by a boundary control function in the space* $\mathrm{B}_N^m(\Sigma_0)$.

Proof. We use Theorem 7.7 to define a map for $\varphi \in \mathrm{B}_N^m(\Sigma_0)$ by setting

$$\Phi_N(\varphi) = (u(T), \dot{u}(T)) \tag{7.95}$$

where u is the solution of the following problem

$$\begin{cases} u_{tt} = \operatorname{div} \mathbf{a}(x, \nabla u) & \text{for} \quad (t, x) \in Q, \\ u|_{\Gamma_1} = w & \text{for} \quad t \in (0, T), \\ \langle \mathbf{a}(x, \nabla u), \nu \rangle|_{\Gamma_0} = \langle \mathbf{a}(x, \nabla w), \nu \rangle + \varphi & \text{for} \quad t \in (0, T), \\ u(0) = w, \quad u_t(0) = 0. \end{cases} \tag{7.96}$$

Let $\varepsilon_T > 0$ be given by Theorem 7.7. Then the map $\Phi_N \colon B(0, \varepsilon_T) \to H^m(\Omega) \times H^{m-1}(\Omega)$ is well defined where $B(0, \varepsilon_T) \subset \mathrm{B}_N^m(\Sigma_0)$ is the ball with the radius ε_T centered at 0. We observe that, since $w \in H^{m+1}(\Omega)$, $\Phi_N(0) = (w, 0) \in H^{m+1}(\Omega) \times H^m(\Omega)$. Then Theorem 7.8 is equivalent to the following claim: For some $T > 0$ there are $\varepsilon_1 > 0$ and $\varepsilon_2 > 0$ with $\varepsilon_T \ge \varepsilon_2$ such that

$$B_{H^{m+1}(\Omega) \times H^m(\Omega)}\left((w, 0), \varepsilon_1\right) \subset \Phi_N\left(B(0, \varepsilon_2)\right) \tag{7.97}$$

where $B_{H^{m+1}(\Omega) \times H^m(\Omega)}\left((w, 0), \varepsilon_1\right)$ is the ball with the radius ε_1 centered at 0 in the space $H^{m+1}(\Omega) \times H^m(\Omega)$.

It is easy to check that the map Φ_N is *Fréchet* differentiable on $B(0, \varepsilon_T)$. In particular,

$$\Phi'_N(0)\varphi = (v(T), \dot{v}(T)) \quad \text{for} \quad \varphi \in \mathrm{B}_N^m(\Sigma_0) \tag{7.98}$$

where $v(t, x)$ is the solution of the linear system with variable coefficients in the space variable

$$\begin{cases} v_{tt} = \operatorname{div} A(x, \nabla w) \nabla v & \text{for} \quad (t, x) \in Q, \\ v|_{\Gamma_1} = 0 & \text{for} \quad t \in (0, T), \\ v_{\nu_A}|_{\Gamma_0} = \varphi & \text{for} \quad t \in (0, T), \\ v(0) = \dot{v}(0) = 0 \end{cases} \tag{7.99}$$

where $v_{\nu_A} = \langle A(x, \nabla w) \nabla v, \nu \rangle$.

For $i = 1, 2$, let

$$\mathbf{X}_i = \mathrm{B}_N^{m+i-1}(\Sigma_0),$$

$$\mathbf{Y}_i = H^{m+i-1}(\Omega) \times H^{m+i-2}(\Omega).$$

It is easy to check by Theorem 7.2 that the mappings Φ_N, given by (7.95), are of C^1 from $\mathrm{B}_{\mathbf{X}_i}(0, r)$ to $\mathbf{Y}_i$ for $i = 1, 2$, and for some $r > 0$. By Lemma 7.9 below, to prove the relation (7.97) is to establish the exact controllability of the linear system (7.99) on the space $[H^{m+1}(\Omega) \cap H_{\Gamma_1}^1(\Omega)] \times [H^m(\Omega) \cap H_{\Gamma_1}^1(\Omega)]$, which has been given by Theorem 2.24. □

A similar argument as in the proof of Theorem (3.1.19) in [11] gives

Lemma 7.9 *Let $\mathbf{X}_1$, $\mathbf{X}_2$, $\mathbf{Y}_1$, and $\mathbf{Y}_2$ be Banach spaces with $\mathbf{X}_2 \subset \mathbf{X}_1$, $\mathbf{Y}_2 \subset \mathbf{Y}_1$, $\overline{\mathbf{X}_2} = \mathbf{X}_1$, and $\overline{\mathbf{Y}_2} = \mathbf{Y}_1$ which satisfy the relations*

$$\|x\|_{\mathbf{X}_1} \leq \|x\|_{\mathbf{X}_2} \quad \text{and} \quad \|y\|_{\mathbf{Y}_1} \leq \|y\|_{\mathbf{Y}_2}.$$

Suppose that $\Phi : \mathrm{B}_{\mathbf{X}_i}(0, r) \to \mathbf{Y}_i$ are mappings of C^1 for $i = 1, 2$ such that

$$\mathbf{Y}_2 \subset \Phi'(0)\mathbf{X}_1. \tag{7.100}$$

Then there is $\varepsilon > 0$ such that

$$\mathrm{B}_{\mathbf{Y}_2}(\Phi(0), \varepsilon) \subset \Phi(\mathrm{B}_{\mathbf{X}_1}(0, r)). \tag{7.101}$$

Remark 7.6 *Here we lose an explicit formula of T_0.*

Remark 7.7 *Unlike the control with the Dirichlet action, we only have the exact controllability results in the space $H^{m+1}(\Omega) \times H^m(\Omega)$ by a control $\varphi \in \mathrm{B}_N^m(\Sigma_0)$. This is because the Neumann action loses a regularity of order 1(actually, order $1/2$). For this point, we may see [117]. In addition, although we can move one state to another in the space $H^{m+1}(\Omega) \times H^m(\Omega)$, we can not guarantee the solution $(u(t), \dot{u}(t))$ of the problem (7.93) always stays in $H^{m+1}(\Omega) \times H^m(\Omega)$ in the process of the control for $0 \leq t \leq T$ where they are actually in the space $H^m(\Omega) \times H^{m-1}(\Omega)$ for all $t \in [0, T]$ by Theorem 7.7. The same things happen to the globally exact controllability results in Theorem 7.9 below.*

Globally Exact Controllability from One Equilibrium to Another with the Neumann Action Let $w \in H^{m+1}(\Omega)$ be given an equilibrium. For $\alpha \in [0,1]$, we assume that $w_\alpha \in H^{m+1}(\Omega)$ are solutions to the problem

$$\begin{cases} \operatorname{div} \mathbf{a}(x, \nabla w_\alpha) = 0 & \text{for} \quad x \in \Omega, \\ w_\alpha = \alpha w & \text{for} \quad x \in \Gamma \end{cases} \tag{7.102}$$

with, this time, a uniform bound

$$\sup_{\alpha \in [0,1]} \|w_\alpha\|_{m+1} < \infty. \tag{7.103}$$

Similar arguments, as in the proof of Theorem 7.6, yield

Theorem 7.9 *Let an equilibrium $w \in H^{m+1}(\Omega)$ be exactly controllable. Let $w_\alpha \in H^{m+1}(\Omega)$, given by (7.102), be also exactly controllable for all $\alpha \in [0,1]$ such that the condition (7.103) holds. Then, there are $T > 0$ and $\varphi \in B_N^m(\Sigma_0)$ which is compatible with the initial data $(w, 0)$ of order m such that the solution of the system (7.93) with $(u_0, u_1) = (w, 0)$ satisfies $u(T) = \dot{u}(T) = 0$.*

Finally, let us see some examples to verify the globally exact controllability in Theorem 7.6.

Example 7.1 *Let $n = 2$ and $m = 4$. Consider the control problem*

$$\begin{cases} u_{tt} = (|\nabla u|^2 + 1)\Delta u & \text{for} \quad (t, x) \in (0, T) \times \Gamma, \\ u|_\Gamma = \varphi & \text{for} \quad 0 \le t \le T, \\ u(0) = w \quad u_t(0) = 0 & \text{for} \quad x \in \Omega, \end{cases} \tag{7.104}$$

where $\Delta = \dfrac{\partial^2}{\partial x^2} + \dfrac{\partial^2}{\partial y^2}$.

Then $w \in H^4(\Omega)$ is an equilibrium if and only if w is harmonic on Ω, i.e.,

$$\Delta w = 0 \quad \text{for} \quad x \in \Omega. \tag{7.105}$$

Let $w \in H^4(\Omega)$ be an equilibrium. Then metric (7.78) is given by

$$g = A^{-1}(x, \nabla w), \quad A(x, \nabla w) = \begin{pmatrix} |\nabla w|^2 + 1 & 0 \\ 0 & |\nabla w|^2 + 1 \end{pmatrix}.$$

Then the condition (7.105) implies that

$$|\nabla(|\nabla w|^2 + 1)|^2 = 2|\nabla w|^2 |D^2 w|^2$$

where $D^2 w$ is the Hessian of w in the Euclidean metric of $\mathbb{R}^2$.

By the formulas (2.27) and (7.105), the Gauss curvature of the Riemannian manifold $(\overline{\Omega}, g)$ is

$$\kappa(x) = \frac{|D^2 w|^2}{|\nabla w|^2 + 1} \quad \text{for} \quad x \in \overline{\Omega}. \tag{7.106}$$

Then the zero equilibrium, $w = 0$, is exactly controllable for any $\Omega \subset \mathbb{R}^2$. In addition, by Corollary 2.1 and Theorem 7.6, we have the conclusion: If two equilibria $w_i \neq 0$ in $H^4(\Omega)$ are such that there are $x_i \in \overline{\Omega}$ satisfying

$$\overline{\Omega} \subset \{ x \in \mathbb{R}^2 \,|\, |x - x_i| < \gamma_i \} \tag{7.107}$$

where

$$\gamma_i = \frac{\pi(1 + \inf_{x \in \Omega} |\nabla w_i|^2)}{2 \sup_{x \in \Omega} |D^2 w_i|}, \tag{7.108}$$

for $i = 1$, 2, then there are a control time $T > 0$ and a control function $\varphi \in B_D^4(\Sigma_0)$ such that the solution of the problem (7.104) with the initial $(w_1, 0)$ satisfies

$$u(T) = w_2, \quad \dot{u}(T) = 0.$$

In particular, let

$$w_1 = a(x_1^2 - x_2^2), \quad w_2 = bx_1x_2, \quad \Omega = \text{ the unit disc}, \tag{7.109}$$

where $0 < a < \pi/(4\sqrt{2})$ and $0 < b < \pi/(2\sqrt{2})$. It is easy to check that w_i meet the conditions (7.107) for $i = 1$, 2. Then the state of the system (7.104) can be moved from $(w_1, 0)$ to $(w_2, 0)$ at some time $T > 0$.

Example 7.2 *Let $n = 2$. Consider the control problem*

$$\begin{cases} u_{tt} = (|\nabla u|^2 + 1)^{-1} \Delta u & \text{for} \quad (t, x) \in (0, T) \times \Gamma, \\ u|_\Gamma = \varphi & \text{for} \quad 0 \le t \le T, \\ u(0) = w \quad u_t(0) = 0 & \text{for} \quad x \in \Omega. \end{cases} \tag{7.110}$$

Let $w \in H^4(\Omega)$ be an equilibrium. The metric is

$$g = \begin{pmatrix} |\nabla w|^2 + 1 & 0 \\ 0 & |\nabla w|^2 + 1 \end{pmatrix}.$$

By (2.27), the Gauss curvature of $(\overline{\Omega}, g)$ is

$$k(x) = -\frac{|D^2 w|^2}{(|\nabla w|^2 + 1)^3} \le 0, \quad \forall\, x \in \overline{\Omega}.$$

By Theorems 2.6 and 7.6, for any two equilibria w_1, $w_2 \in H^4(\Omega)$ and any $\Omega \subset \mathbb{R}^n$, there are a control time $T > 0$ and a control function $\varphi \in B_D^4(\Sigma_0)$ such that the state of the system (7.110) is moved from $(w_1, 0)$ to $(w_2, 0)$.

Example 7.3 *Let $n = 2$ and $\Omega \subset \mathbb{R}^2$ be a bounded, open set. Consider a quasilinear wave equation*

$$u_{tt}(t, x) = \frac{u_{x_1x_1}(t, x)}{1 + u_{x_1}^2(t, x)} + \frac{u_{x_2x_2}(t, x)}{1 + u_{x_1}^2(t, x)} \quad \text{on} \quad Q. \tag{7.111}$$

Consider an equilibrium given by

$$w(x) = \lambda x_1 x_2 \quad for \quad x = (x_1, x_2) \in \mathbb{R}^2 \qquad (7.112)$$

where $\lambda \neq 0$ *is constant. Let*

$$w_\alpha(x) = \alpha w(x) \quad for \quad \alpha \in [0, 1]. \qquad (7.113)$$

Then

$$A(x, \nabla w_\alpha) = \begin{pmatrix} \dfrac{1}{1 + \alpha^2 \lambda^2 x_2^2} & 0 \\ 0 & \dfrac{1}{1 + \alpha^2 \lambda^2 x_1^2} \end{pmatrix}$$

and the metrics are

$$g_\alpha = \begin{pmatrix} 1 + \alpha^2 \lambda^2 x_2^2 & 0 \\ 0 & 1 + \alpha^2 \lambda^2 x_1^2 \end{pmatrix} \quad for \quad \alpha \in [0, 1].$$

By Lemma 2.3, the Gauss curvature is

$$k_\alpha(x) = -\frac{2 + \alpha^2 \lambda^2 |x|^2}{(1 + \alpha^2 \lambda^2 x_2^2)(1 + \alpha^2 \lambda^2 x_1^2)} \leq 0 \quad for \quad x \in \mathbb{R}^2$$

and for $\alpha \in [0, 1]$. *Then* $w_\alpha(x)$ *is exactly controllable for each* $\alpha \in [0, 1]$. *By Theorem 7.6, the equilibrium* w, *given by* (7.112), *can be steered to rest at a time* T *by a boundary control.*

7.3 Boundary Feedback Stabilization

We consider the existence of global solutions of the quasilinear wave equation in the divergence form with a boundary dissipation structure of an input-output when initial data and boundary inputs are near a given equilibrium of the system in a certain sense. We show that the existence of global solutions depends not only on this dissipation structure but also on a Riemannian metric, given by the coefficients and the equilibrium of the system, where *an escape vector field for the metric* will guarantee global solutions in time. In particular, we have exponential decays of the norm of the state of the system if the input stops after a finite time, which implies the exponential stabilization of the system by boundary feedback.

Let $\Omega \subset \mathbb{R}^n$ be a bounded, open set with the smooth boundary $\Gamma = \Gamma_0 \cup \Gamma_1$ and $\overline{\Gamma}_0 \cap \overline{\Gamma}_1 = \emptyset$. Let

$$\mathbf{a}(x, y) = (a_1(x, y), \cdots, a_n(x, y)) : \quad \overline{\Omega} \times \mathbb{R}^n \to \mathbb{R}^n$$

be a smooth mapping with $\mathbf{a}(x,0) = 0$ for $x \in \overline{\Omega}$ such that $A(x,y) = \left(a_{iy_j}(x,y)\right)$ are symmetrical and positive for $(x,y) \in \overline{\Omega} \times \mathbb{R}^n$. Let w be an equilibrium, defined by

$$\text{div } \mathbf{a}(x, \nabla w) = 0 \quad \text{for} \quad x \in \Omega.$$

Let

$$Q^\infty = (0, \infty) \times \Omega, \quad \Sigma_1^\infty = (0, \infty) \times \Gamma_1, \quad \Sigma_0^\infty = (0, \infty) \times \Gamma_0.$$

We consider the following problem

$$\begin{cases} u_{tt} = \text{div } \mathbf{a}(x, \nabla w + \nabla u) & \text{for} \quad (t,x) \in Q^\infty, \\ u = 0 & \text{for} \quad (t,x) \in \Sigma_1^\infty, \\ u(0,x) = u_0, \quad u_t(0,x) = u_1 & \text{for} \quad x \in \Omega \end{cases} \tag{7.114}$$

with an input $\mathbf{I}(t)$ and an output $\mathbf{O}(t)$ on the portion Γ_0 of the boundary Γ

$$\mathbf{I}(t) = u_t + \lambda \langle \mathbf{a}(x, \nabla w + \nabla u), \nu \rangle \quad \text{for} \quad (t,x) \in \Sigma_0^\infty, \tag{7.115}$$

$$\mathbf{O}(t) = u_t - \lambda \langle \mathbf{a}(x, \nabla w + \nabla u), \nu \rangle \quad \text{for} \quad (t,x) \in \Sigma_0^\infty, \tag{7.116}$$

where $\langle \cdot, \cdot \rangle$ is the Euclidean metric of $\mathbb{R}^n$, ν is the outside unit normal of Γ, and $\lambda > 0$ is a constant.

Remark 7.8 *If $\sigma(x,y)$ is a smooth function on $\overline{\Omega} \times \mathbb{R}^n$ such that*

$$\mathbf{a}(x,y) = (\sigma_{y_1}(x,y), \cdots, \sigma_{y_n}(x,y)),$$

we define the total energy of the quasilinear system (7.114) − (7.116) *at time t in the zero equilibrium by*

$$E(t) = 4\lambda \int_\Omega [u_t^2/2 + \sigma(x, \nabla u)] \, dx \tag{7.117}$$

and obtain by a simple computation

$$\frac{dE(t)}{dt} = \int_{\Gamma_0} |\mathbf{I}(t)|^2 \, d\Gamma - \int_{\Gamma_0} |\mathbf{O}(t)|^2 \, d\Gamma. \tag{7.118}$$

This formula expresses that the rate of change of the energy is equal to the power supplied to the system by the input minus the power taken out by the output on Γ_0. The balance equation (7.118) *means that the boundary structure* (7.115) *is dissipative.*

Remark 7.9 *It is well known that smooth solutions of quasilinear hyperbolic systems usually develop singularities after some time, see [93] and [133]. Since the structure of the boundary dissipation* (7.115) *makes the energy dissipative, we expect that the introduction of the boundary structure* (7.115) *assures the existence of a globally smooth solution. The aim of this section is to seek* *general geometrical conditions for a quasilinear part as in the problem (7.114) and for a general equilibrium w to assure the system (7.114) − (7.115) to have global solutions when initial data and inputs are small. The results in Theorems 7.10, 7.11 later show that an escape vector field for a metric, given by (7.119) below, is one of such geometrical conditions.*

Let $m \geq [n/2] + 3$ be a given positive integer. We introduce a Banach space for inputs. Let $\mathcal{I}^m(\Sigma_0^\infty)$ consist of all the functions $\mathbf{I}(t) = \mathbf{I}(t, x)$ on $\Sigma^\infty = (0, \infty) \times \Gamma_0$ such that

$$\mathbf{I}^{(k)}(t) \in L^2((0, \infty), H^{m-k-3/2}(\Gamma_0)) \cap C[0, \infty; H^{m-k-3/2}(\Gamma_0)]$$

for $0 \leq k \leq m - 2$ and $\mathbf{I}^{(m-1)}(t) \in L^2((0, \infty), L^2(\Gamma_0))$ with a norm

$$\|\mathbf{I}\|^2_{\mathcal{I}^m((0,\infty),\Gamma_0)} = \max_{0 \leq t < \infty} \mathcal{E}_{\Gamma_0, R}(t) + \int_0^\infty \mathcal{E}_{\Gamma_0, R}(\tau)\, d\tau$$

$$+ \int_0^\infty \|\mathbf{I}^{(m-1)}(\tau)\|^2_{\Gamma_0}\, d\tau$$

where $\mathcal{E}_{\Gamma_0, R}(t) = \sum_{k=0}^{m-2} \|\mathbf{I}^{(k)}(t)\|^2_{\Gamma_0, m-k-3/2}$.

Let $w \in H^m(\Omega)$ be an equilibrium of the system (7.114). We hope to find solutions $u(t, x)$ in $\cap_{k=0}^m C^k((0, \infty), H^{m-k}(\Omega))$ if $(u_0, u_1) \in H^m(\Omega) \times H^{m-1}(\Omega)$ is small in $H^m(\Omega) \times H^{m-1}(\Omega)$ and $\mathbf{I}$ is small in $\mathcal{I}^m((0, \infty), \Gamma_0)$ which satisfy the following *compatibility conditions* of order m

$$u_k|_{\Gamma_1} = 0, \quad 0 \leq k \leq m - 1; \quad \mathbf{I}^{(k)}(0) = \mathbf{I}_k, \quad 0 \leq k \leq m - 2,$$

where for $k \geq 2$, $u_k = u^{(k)}(0)$, as computed formally (and recursively) in terms of u_0 and u_1, using the equation in (7.114); for $0 \leq k \leq m - 2$, $\mathbf{I}_k = [u_t + \lambda \langle \mathbf{a}(x, \nabla w + \nabla u), \nu \rangle]^{(k)}(0)$ for $x \in \Gamma_0$.

We define

$$g = A^{-1}(x, \nabla w) \quad \text{for} \quad x \in \Omega \tag{7.119}$$

where $A(x, y) = (a_{iy_j}(x, y))$ for $(x, y) \in \overline{\Omega} \times \mathbb{R}^n$, as a Riemannian metric on $\overline{\Omega}$ and consider the couple $(\overline{\Omega}, g)$ as a Riemannian manifold with a boundary Γ. Here the metric g depends not only on the coefficients $a_{ij}(\cdot, \cdot)$ and but also on the equilibrium w. We denote by $\langle \cdot, \cdot \rangle_g$ the inner product induced by g.

We make the following geometrical assumptions.

(H1) There exists *an escape vector field H for the metric g on $\overline{\Omega}$* such that

$$DH(X, X) \geq \varrho_0 |X|^2_g \quad X \in \mathbb{R}^n_x, \quad x \in \overline{\Omega} \tag{7.120}$$

where D is the Levi-Civita connection of the metric g, DH is the covariant differential of H of the metric g, and $\varrho_0 > 0$ is a constant.

(H2) The boundary portions Γ_1 and Γ_0 satisfy

$$\langle H, \nu \rangle \leq 0 \quad \text{for} \quad x \in \Gamma_1, \tag{7.121}$$

$$\langle H, \nu \rangle \geq 0 \quad \text{for} \quad x \in \Gamma_0. \tag{7.122}$$

(H3) Let $\Gamma_1 \neq \emptyset$. If $\Gamma_1 = \emptyset$, we further assume that the initial value u_0 satisfies

$$\int_\Gamma u_0 \, d\Gamma = 0. \tag{7.123}$$

We assume that solutions of short time exist for the system (7.114)-(7.115) with appropriate initial data and boundary input. In fact, there are standard approaches to obtain solutions of short time, for example, see [91] and [48].

We shall establish the following

Theorem 7.10 *Let an equilibrium $w \in H^m(\Omega)$ be such that the assumptions* **(H1)**, **(H2)**, *and* **(H3)** *hold. Let (u_0, u_1) and* **I** *satisfy the compatibility conditions of m order. Then for any (u_0, u_1) in $H^m(\Omega) \times H^{m-1}(\Omega)$ small and any* **I** *in $\mathcal{I}^m((0, \infty), \Gamma_0)$ small, the system (7.114)-(7.115) has a global solution u in $\cap_{k=0}^m C^k((0, \infty), H^{m-k}(\Omega))$.*

If the input $\mathbf{I}(t) = 0$ after a finite time $T_0 > 0$, the energy will decay exponentially, which is the stabilization by feedback from boundary.

Theorem 7.11 *Let all the assumptions of Theorem 7.10 hold. If there is $T_0 > 0$ such that $\mathbf{I}(t) = 0$ for $t \geq T_0$, then there are $c_1 > 0$, $c_2 > 0$, and $\hat{T} \geq T_0$ such that*

$$\mathcal{E}(t) \leq c_1 e^{-c_2 t} \quad \text{for} \quad t \geq \hat{T} \tag{7.124}$$

where $\mathcal{E}(t) = \sum_{k=0}^m \|u^{(k)}\|_{m-k}^2$.

The proofs of Theorems 7.10 and 7.11 will be given at the end of this section.

Remark 7.10 *In the linear case the structure of dissipation, as above, has been studied thoroughly, for example, see [202] and references there.*

Remark 7.11 *The assumption (7.122) can be removed if an estimate of boundary trace can be established for the quasilinear wave equation as that for the linear wave equation in [122].*

Remark 7.12 *We need the assumption* **(H3)** *for a uniqueness result to get rid of some lower order terms, see Lemma 7.11. Instead of (7.123) there are other options.*

Example 7.4 *Let*

$$\mathbf{a}(y) = (a_1(y_1), a_2(y_2)) : \ \mathbb{R}^2 \to \mathbb{R}^2$$

where a_i are smooth functions on $\mathbb{R}$ for $i = 1, 2$ such that $a_i'(s) > 0$ for $s \in \mathbb{R}$. Let

$$w(x) = x_1 x_2 \quad \text{for} \quad x = (x_1, x_2) \in \mathbb{R}^2.$$

Then

$$\operatorname{div} \mathbf{a}(\nabla w) = 0 \quad for \quad x \in \mathbb{R}^2.$$

The metric is

$$g = A^{-1}(x, \nabla w) = \begin{pmatrix} 1/a_1'(x_2) & 0 \\ 0 & 1/a_2'(x_1) \end{pmatrix}.$$

Let $a_1(s) = a_2(s) = \arctan s$ for $s \in \mathbb{R}$. Then

$$g = \begin{pmatrix} 1 + x_2^2 & 0 \\ 0 & 1 + x_1^2 \end{pmatrix}.$$

By Lemma 2.3, the Gauss curvature of $(\mathbb{R}^2, g)$ is

$$\kappa(x) = -\frac{2 + |x|^2}{(1 + x_1^2)^2(1 + x_2^2)^2} \leq 0 \quad for \quad x = (x_1, x_2) \in \mathbb{R}^2.$$

*By Theorem 2.6, the assumption (**H1**) holds true for any $\Omega \subset \mathbb{R}^2$.* □

We shall make preparations for the proofs of Theorems 7.10 and 7.11.

Let $\gamma > 0$ be given and let u satisfy the problem (7.114) on the interval $[0, T]$ for some $T > 0$ such that

$$\sup_{0 \leq t \leq T} \|u\|_m \leq \gamma, \quad \sup_{0 \leq t \leq T} \|\dot{u}\|_{m-1} \leq \gamma. \tag{7.125}$$

For $t \in [0, T]$, we introduce a metric g_u on $\overline{\Omega}$ by

$$g_u = A_u^{-1}(t) = A^{-1}(x, \nabla w + \nabla u). \tag{7.126}$$

Consider the couple $(\overline{\Omega}, g_u)$ as a Riemannian manifold for each fixed t. Let X, Y be vector fields on $\overline{\Omega}$ and let f be a function. Then

$$\langle X, Y \rangle_{g_u} = \langle A_u^{-1}(t)X, Y \rangle, \quad \nabla_{g_u} f = A_u(t)\nabla f \tag{7.127}$$

where $\langle \cdot, \cdot \rangle$ and $\langle \cdot, \cdot \rangle_{g_u}$ are the inner products of the Euclidean metric and the metric g_u, respectively, and ∇ and ∇_{g_u} are the gradients of the Euclidean metric and the metric g_u, respectively. By (7.119) and (7.126)

$$g_0 = A^{-1}(x, \nabla w) = g. \tag{7.128}$$

In addition, it is easy to check from the assumption (7.125) that there are $c_{0,\gamma} > 0$ and $c_\gamma > 0$ such that

$$c_{0,\gamma} |\nabla_g f|_g^2 \leq |\nabla_{g_u} f|_{g_u}^2 \leq c_\gamma |\nabla_g f|_g^2 \quad for \quad t \in [0, T], \quad f \in C^\infty(\Omega). \tag{7.129}$$

Let D_{g_u} and D be the Levi-Civita connections of the Riemannian metrics g_u and g, respectively. Let H be a vector field on $\overline{\Omega}$. We denote by $D_{g_u}H$ and by DH the covariant differentials of the metric g_u and g, respectively. They are tensor fields of rank 2 on $\overline{\Omega}$. We define

$$\eta = D_{g_u}H - DH. \tag{7.130}$$

Lemma 7.10 *Let H be a vector field on $\overline{\Omega}$. Suppose that the tensor field of rank 2 $\eta = \eta(\cdot, \cdot)$ is given by the formula (7.130). Let $\gamma > 0$ be given and u be such that $\sup_{x \in \overline{\Omega}} |\nabla u| \leq \gamma$. Then there is $c_\gamma > 0$ such that*

$$|\eta(X, Y)| \leq c_\gamma (|\nabla u| + |\nabla^2 u|)|X||Y| \quad \text{for} \quad X, Y \in \mathbb{R}^n_x, \ x \in \overline{\Omega} \quad (7.131)$$

where ∇ is the covariant differential of the Euclidean metric of $\mathbb{R}^n$.

Proof. Using the relations (7.128) and (7.127), we have

$$D_{g_u} H\left(\frac{\partial}{\partial x_i}, \frac{\partial}{\partial x_j}\right) = \frac{\partial}{\partial x_j} \langle H, \frac{\partial}{\partial x_i} \rangle_{g_u} - \langle H, (D_{g_u})_{\frac{\partial}{\partial x_j}} \frac{\partial}{\partial x_i} \rangle_{g_u}$$

$$= \frac{\partial}{\partial x_j} \langle A_u^{-1}(t)H, \frac{\partial}{\partial x_i} \rangle - \sum_{k=1}^n \Gamma^k_{g_u ij} \langle A_u^{-1}(t)H, \frac{\partial}{\partial x_k} \rangle$$

$$= DH\left(\frac{\partial}{\partial x_i}, \frac{\partial}{\partial x_j}\right) + \eta\left(\frac{\partial}{\partial x_i}, \frac{\partial}{\partial x_j}\right) \quad (7.132)$$

where

$$\eta\left(\frac{\partial}{\partial x_i}, \frac{\partial}{\partial x_j}\right) = \frac{\partial}{\partial x_j} \langle [A_u^{-1}(t) - A^{-1}(x, \nabla w)]H, \frac{\partial}{\partial x_i} \rangle$$

$$- \sum_{k=1}^n \Gamma^k_{g_u ij} \langle A_u^{-1}(t)H, \frac{\partial}{\partial x_k} \rangle$$

$$+ \sum_{k=1}^n \Gamma^k_{gij} \langle A^{-1}(x, \nabla w)H, \frac{\partial}{\partial x_k} \rangle, \quad (7.133)$$

$\Gamma^k_{g_u ij}$ and Γ^k_{gij} are the coefficients of the connections D_{g_u} and D, respectively. Let $A_u^{-1}(t) = (a^{ij}(x, \nabla w + \nabla u))$. Then we have

$$\Gamma^k_{g_u ij} = \frac{1}{2} \sum_{l=1}^n a_{lk}(x, \nabla w + \nabla u)[(a^{lj}(x, \nabla w + \nabla u))_{\partial x_i}$$

$$+ (a^{li}(x, \nabla w + \nabla u))_{\partial x_j} - (a^{ij}(x, \nabla w + \nabla u))_{\partial x_l}]$$

$$= f_{ijk}(x, \nabla u) + p_{ijk}(u) \quad (7.134)$$

where

$$f_{ijk}(x, \nabla u) = \frac{1}{2} \sum_{l=1}^n a_{lk}(x, \nabla w + \nabla u)[a^{il}_{x_j}(x, \nabla w + \nabla u) + a^{jl}_{x_i}(x, \nabla w + \nabla u)$$

$$- a^{ij}_{x_l}(x, \nabla w + \nabla u)], \quad (7.135)$$

$$p_{ijk}(u) = \frac{1}{2} \sum_{lh=1}^n a_{lk}(x, \nabla w + \nabla u)[a^{il}_{y_h}(x, \nabla w + \nabla u)u_{x_h x_j}$$

$$+ a^{jl}_{y_h}(x, \nabla w + \nabla u)u_{x_h x_i} - a^{ij}_{y_h}(x, \nabla w + \nabla u)u_{x_h x_l}]. \quad (7.136)$$

The formula (7.136) yields

$$|p_{ijk}(u)| \le c_\gamma |\nabla^2 u|. \tag{7.137}$$

In addition, we have, by (7.135),

$$\Gamma_{g_{ij}}^k = f_{ijk}(x, 0). \tag{7.138}$$

Thus by (7.133)-(7.138),

$$\sum_{k=1}^n \Gamma_{g_u ij}^k \langle A_u^{-1}(t)H, \frac{\partial}{\partial x_k} \rangle = \sum_{k=1}^n \Gamma_{g_{ij}}^k \langle A^{-1}(x, \nabla w)H, \frac{\partial}{\partial x_k} \rangle$$

$$+ \sum_{k=1}^n [f_{ijk}(x, \nabla u) - f_{ijk}(x, 0)] \langle A_u^{-1}(t)H, \frac{\partial}{\partial x_k} \rangle$$

$$+ \sum_{k=1}^n \Gamma_{g_{ij}}^k \langle [A_u^{-1}(t) - A^{-1}(x, \nabla w)]H, \frac{\partial}{\partial x_k} \rangle$$

$$+ \sum_{k=1}^n p_{ijk}(u) \langle A_u^{-1}(t)H, \frac{\partial}{\partial x_k} \rangle. \tag{7.139}$$

Moreover, the relations

$$A_u^{-1}(t) - A^{-1}(x, \nabla w) = \left(\sum_{k=1}^n \int_0^1 a_{y_k}^{ij}(x, \nabla w + \tau \nabla u) d\tau u_{x_k} \right),$$

$$f_{ijk}(x, \nabla u) - f_{ijk}(x, 0) = \sum_{h=1}^n \int_0^1 f_{ijkx_h}(x, \tau \nabla u) d\tau u_{x_h},$$

imply that

$$|\langle [A_u^{-1}(t) - A^{-1}(x, \nabla w)]H, \frac{\partial}{\partial x_k} \rangle| \le c_\gamma |\nabla u|, \tag{7.140}$$

$$|\frac{\partial}{\partial x_j} \langle [A_u^{-1}(t) - A^{-1}(x, \nabla w)]H, \frac{\partial}{\partial x_i} \rangle| \le c_\gamma (|\nabla u| + |\nabla^2 u|), \tag{7.141}$$

$$|f_{ijk}(x, \nabla u) - f_{ijk}(x, 0)| \le c_\gamma |\nabla u|. \tag{7.142}$$

The inequality (7.131) follows from the formulas (7.133) and the estimates in (7.140)-(7.142). $\qquad \square$

Theorem 7.12 *Let the assumptions* **(H1)**, **(H2)**, *and* **(H3)** *hold. Let* $\gamma > 0$ *be given. Let* u *be a solution to the problem* (7.114) *on the interval* $[0, T]$ *for some* $T > 0$ *such that the estimates in* (7.125) *hold true. Then there are* $c_\gamma > 0$ *and* $T_\gamma > 3 \sup_{x \in \bar{\Omega}} |H|_g / \varrho_0$, *where* $\varrho_0 > 0$ *is given by* (7.120), *such that, if* $0 \le s \le t \le T$ *are such that* $t - s \ge T_\gamma$, *then*

$$\int_s^t \mathcal{Q}_\lambda(\tau) d\tau \le c_\gamma \int_s^t [\mathcal{Q}_{\Gamma_0, R, \mathbf{I}}(\tau) + \mathcal{Q}_{\Gamma_0, R, \mathbf{O}}(\tau) + \mathcal{L}(\tau)] d\tau \tag{7.143}$$

where $\mathcal{Q}_\lambda(t)$, $\mathcal{Q}_{\Gamma_0,R,\mathbf{I}}(t)$, $\mathcal{Q}_{\Gamma_0,R,\mathbf{O}}(t)$, and $\mathcal{L}(t)$ are defined in Theorem 7.3, respectively.

Proof. Let H be a vector field and let f be a function. We denote by $H(f)$ the directional derivative of the function f along the vector field H. Then

$$H(f) = \langle H, \nabla f \rangle = \langle H, \nabla_{g_u} f \rangle_{g_u}. \tag{7.144}$$

We assume that solutions to the problem (7.114)-(7.116) exist on $[0,T]$ for some $T > 0$. Let $0 \le s \le t \le T$ be given.

Step 1 We assume $1 \le j \le m-1$.

We differentiate the equations in (7.114)-(7.116) by j times with respect to the variable t (see (7.59)) to obtain

$$\begin{cases} u_{tt}^{(j)} = \operatorname{div} A_u(t)\nabla u^{(j)} + r_{j-1} & \text{for} \quad (t,x) \in Q, \\ u^{(j)} = 0 & \text{for} \quad (t,x) \in \Sigma_1 \\ u^{(j+1)} + \lambda u_{\nu_A}^{(j)} + \lambda r_{\Gamma_0,j-1} = \mathbf{I}^{(j)}(t) & \text{for} \quad (t,x) \in \Sigma_0, \\ u^{(j+1)} - \lambda u_{\nu_A}^{(j)} - \lambda r_{\Gamma_0,j-1} = \mathbf{O}^{(j)}(t) & \text{for} \quad (t,x) \in \Sigma_0 \end{cases} \tag{7.145}$$

where $r_{j-1}(t)$ and $r_{\Gamma_0,j-1}(t)$ are given in Lemma 7.7 for $j \ge 2$ and they are zero if $j = 1$.

For $0 \le j \le m-1$, let

$$\varphi = u^{(j)}.$$

We multiply the equation in (7.145) by $2H(\varphi)$, integrate by parts over $Q_s^t = (s,t) \times \Omega$, and obtain (see the identity (2.13))

$$\int_{\Sigma_s^t} [2H(\varphi)\varphi_{\nu_A} + (\dot{\varphi}^2 - |\nabla_{g_u}\varphi|_{g_u}^2)\langle H, \nu\rangle]d\Sigma$$

$$= 2(\dot{\varphi}, H(\varphi))\Big|_s^t - 2\int_{Q_s^t} [\dot{\varphi}\dot{H}(\varphi) + r_{j-1}(t)H(\varphi)]dQ$$

$$+ \int_{Q_s^t} [2D_{g_u}H(\nabla_{g_u}\varphi, \nabla_{g_u}\varphi) + (\dot{\varphi}^2 - |\nabla_{g_u}\varphi|_{g_u}^2)\operatorname{div} H]dQ \tag{7.146}$$

where $\Sigma_s^t = (s,t) \times \Gamma$. Let $f \in C^2(\Omega)$. We multiply the equation in (7.145) by $2f\varphi$, integrate by parts over Q_s^t, and obtain

$$2\int_{Q_s^t} f(\dot{\varphi}^2 - |\nabla_{g_u}\varphi|_{g_u}^2)dQ = 2(\dot{\varphi}, f\varphi)_{L^2(\Omega)}\Big|_s^t + \int_{\Sigma_s^t} [\varphi^2 f_{\nu_A} - 2f\varphi\varphi_{\nu_A}]d\Sigma$$

$$- \int_{Q_s^t} [\varphi^2 \operatorname{div} A_u(t)\nabla f + 2r_{j-1}(t)f\varphi]dQ. \tag{7.147}$$

Next, we assume that the vector field H on Ω satisfies the assumptions **(H1)** and **(H2)**. Noting that H does not depend on time t, we have from the identity (7.146),

$$\uplus_{\Sigma_s^t} = \uplus_{Q_s^t} \tag{7.148}$$

where

$$\uplus_{\Sigma_s^t} = \int_{\Sigma_s^t} [2H(\varphi)\varphi_{\nu_A} + (\dot{\varphi}^2 - |\nabla_{g_u}\varphi|_{g_u}^2)\langle H, \nu\rangle] d\Sigma \tag{7.149}$$

$$\uplus_{Q_s^t} = 2\,(\dot{\varphi}, H(\varphi))_{L^2(\Omega)}\Big|_s^t - 2\int_{Q_s^t} r_{j-1}(t)H(\varphi)\,dQ$$

$$+ \int_{Q_s^t} [2D_{g_u}H(\nabla_{g_u}\varphi, \nabla_{g_u}\varphi) + (\dot{\varphi}^2 - |\nabla_{g_u}\varphi|_{g_u}^2)\,\mathrm{div}\,H]dQ. \tag{7.150}$$

To obtain the inequality (7.143), we shall estimate $\uplus_{\Sigma_s^t}$ and $\uplus_{Q_s^t}$, respectively.

Estimate on $\uplus_{\Sigma_s^t}$ Decompose $\uplus_{\Sigma_s^t}$ as

$$\uplus_{\Sigma_s^t} = \uplus_{\Sigma_{1_s}^t} + \uplus_{\Sigma_{0_s}^t}$$

where $\Sigma_{1_s}^t = (s,t) \times \Gamma_1$ and $\Sigma_{0_s}^t = (s,t) \times \Gamma_0$. On $\Sigma_s^t = (s,t) \times \Gamma$ we have a decomposition of the direct sum

$$H = H_{\Gamma_{g_u}} + \langle H, \frac{A_u(t)\nu}{|A_u(t)\nu|_{g_u}}\rangle_{g_u} \frac{A_u(t)\nu}{|A_u(t)\nu|_{g_u}}. \tag{7.151}$$

Moreover, since the boundary condition $\varphi = u^{(j)} = 0$ on Γ_1 implies

$$H(\varphi) = \frac{\varphi_{\nu_A}}{\langle A_u(t)\nu, \nu\rangle}\langle H, \nu\rangle, \quad \nabla_{g_u}\varphi = \varphi_{\nu_A}\frac{A_u(t)\nu}{\langle A_u(t)\nu, \nu\rangle} \tag{7.152}$$

for $x \in \Gamma_1$, we obtain

$$\uplus_{\Sigma_{1_s}^t} = \int_{\Sigma_{1_s}^t} \frac{(\varphi_{\nu_A})^2}{\langle A_u(t)\nu, \nu\rangle}\langle H, \nu\rangle\, d\Sigma \leq 0, \tag{7.153}$$

via the assumption (7.121). On $\Sigma_{0_s}^t = (s,t) \times \Gamma_0$, the assumption (7.122) and the boundary formulas in Theorem 7.3 yield, via the estimates in Lemma 7.7, respectively,

$$\uplus_{\Sigma_{0_s}^t} \leq \int_{\Sigma_{0_s}^t} [2H(\varphi)\varphi_{\nu_A} + \dot{\varphi}^2\langle H, \nu\rangle]\, d\Sigma$$

$$\leq \varepsilon\|\varphi\|_{L^2((s,t),H^1(\Gamma_0))}^2 + c_{\gamma,\varepsilon}(\|\dot{\varphi}\|_{L^2(\Sigma_{0_s}^t)}^2 + \|\varphi_{\nu_A}\|_{L^2(\Sigma_{0_s}^t)}^2)$$

$$\leq c_{\gamma,\varepsilon}\int_s^t [\mathcal{Q}_{\Gamma_0,R,\mathbf{I}}(\tau) + \mathcal{Q}_{\Gamma_0,R,\mathbf{O}}(\tau)]\, d\tau$$

$$+ \varepsilon\|\varphi\|_{L^2((s,t),H^1(\Gamma_0))}^2 + c_{\gamma,\varepsilon}\int_s^t \mathcal{L}(\tau)\, d\tau \tag{7.154}$$

where $\varepsilon > 0$ is small.

We now estimate the term $\|\varphi\|^2_{L^2((s,t),H^1(\Gamma_0))}$. Let H_1 be a vector field on $\overline{\Omega}$ such that

$$H_1 = 0 \quad \text{for} \quad x \in \Gamma_1; \quad H_1 = A_u(t)\nu \quad \text{for} \quad x \in \Gamma_0. \tag{7.155}$$

Since the vector field $A_u(t)\nu/\langle A_u(t)\nu, \nu\rangle^{1/2}$ is the unit normal of the metric g_u along the boundary Γ, we have a directional decomposition

$$\nabla_{g_u}\varphi = \nabla_{\Gamma_{g_u}}\varphi + \varphi_{\nu_A}\frac{A_u(t)\nu}{\langle A_u(t)\nu, \nu\rangle}$$

which yields

$$|\nabla_{g_u}\varphi|^2_{g_u} = |\nabla_{\Gamma_{g_u}}\varphi|^2_{g_u} + \varphi^2_{\nu_A}/\langle A_u(t)\nu, \nu\rangle. \tag{7.156}$$

We replace the vector field H in the identity (7.146) with the above $-H_1$, and, by (7.156), we obtain

the left hand side of (7.146)

$$= \int_{\Sigma 0^t_s} [(|\nabla_{\Gamma_{g_u}}\varphi|^2_{g_u}\langle A_u(t)\nu, \nu\rangle - (\dot{\varphi}^2\langle A_u(t)\nu, \nu\rangle + \varphi^2_{\nu_A})]d\Sigma \tag{7.157}$$

and

the right hand side of (7.146)

$$\leq c_\gamma \mathcal{Q}_\lambda(t) + c_\gamma \mathcal{Q}_\lambda(s) + c_\gamma \int_s^t [\mathcal{Q}_\lambda(\tau) + \mathcal{L}(\tau)]d\tau. \tag{7.158}$$

Since

$$\int_{\Sigma 0^t_s} (\dot{\varphi}^2\langle A_u(t)\nu, \nu\rangle + \varphi^2_{\nu_A})]d\Sigma \leq c_\gamma \int_s^t [\mathcal{Q}_{\Gamma_0,R,\mathbf{I}}(\tau) + \mathcal{Q}_{\Gamma_0,R,\mathbf{O}}(\tau)]\, d\tau,$$

it follows from (7.157) and (7.158) that

$$\|\varphi\|^2_{L^2((s,t),H^1(\Gamma_0))} \leq c_\gamma \mathcal{Q}_\lambda(s) + c_\gamma \mathcal{Q}_\lambda(t)$$

$$+ c_\gamma \int_s^t [\mathcal{Q}_\lambda(\tau) + \mathcal{Q}_{\Gamma_0,R,\mathbf{I}}(\tau) + \mathcal{Q}_{\Gamma_0,R,\mathbf{O}}(\tau) + \mathcal{L}(\tau)]\, d\tau. \tag{7.159}$$

Moreover, the inequalities (7.51) and (7.52) in Theorem 7.3 imply that

$$\max\{\mathcal{Q}_\lambda(t), \mathcal{Q}_\lambda(s)\} \leq \frac{1}{t-s}\int_s^t \mathcal{Q}_\lambda(\tau)\, d\tau$$

$$+ c_\gamma \int_s^t [\mathcal{Q}_{\Gamma_0,R,\mathbf{I}}(\tau) + \mathcal{Q}_{\Gamma_0,R,\mathbf{O}}(\tau) + \mathcal{L}(\tau)]\, d\tau. \tag{7.160}$$

Inserting (7.160) into (7.159), and then inserting (7.159) into (7.154) yield

$$\biguplus_{\Sigma^t_s} \leq \varepsilon c_\gamma[1 + 1/(t-s)]\int_s^t \mathcal{Q}_\lambda(\tau)\, d\tau$$

$$+ c_{\gamma,\varepsilon}\int_s^t [\mathcal{Q}_{\Gamma_0,R,\mathbf{I}}(\tau) + \mathcal{Q}_{\Gamma_0,R,\mathbf{O}}(\tau) + \mathcal{L}(\tau)]\, d\tau \tag{7.161}$$

where $\varepsilon > 0$ can be small and $0 \le s \le t \le T$.

Estimate on $\uplus_{\mathbf{Q}_s^t}$ It follows from the identities (7.120) and (7.131) that

$$
\begin{aligned}
D_{g_u} H(\nabla_{g_u}\varphi, \nabla_{g_u}\varphi) &= DH(\nabla_{g_u}\varphi, \nabla_{g_u}\varphi) + \eta(\nabla_{g_u}\varphi, \nabla_{g_u}\varphi) \\
&\ge \varrho_0 c_{0,\gamma} |\nabla_{g_u}\varphi|^2_{g_u} - c_\gamma \mathcal{L}(t).
\end{aligned}
\tag{7.162}
$$

Noting that

$$
\int_\Omega |\nabla_{g_u}\varphi|^2_{g_u}\, dx = (A_u(t)\nabla\varphi, \nabla\varphi),
$$

from (7.150) and (7.162), we have

$$
\begin{aligned}
\uplus_{Q_s^t} &\ge 2\varrho_0 c_{0,\gamma} \int_{Q_s^t} |\nabla_{g_u}\varphi|^2_{g_u}\, dQ + \int_{Q_s^t} (\dot\varphi^2 - |\nabla_{g_u}\varphi|^2_{g_u})\,\mathrm{div}\,H\,dQ \\
&\quad - c_\gamma \mathcal{Q}_\lambda(s) - c_\gamma \mathcal{Q}_\lambda(t) - c_\gamma \int_s^t \mathcal{L}(\tau)\, d\tau \\
&\ge \varrho_0 c_{0,\gamma} \int_s^t \vartheta_j\, d\tau + 2 \int_{Q_s^t} f(\dot\varphi^2 - |\nabla_{g_u}\varphi|^2_{g_u})dQ \\
&\quad - c_\gamma \mathcal{Q}_\lambda(s) - c_\gamma \mathcal{Q}_\lambda(t) - c_\gamma \int_s^t \mathcal{L}(\tau)\, d\tau
\end{aligned}
\tag{7.163}
$$

where

$$
\vartheta_j(t) = \|\dot\varphi(t)\|^2_{L^2(\Omega)} + (A_u(t)\nabla\varphi, \nabla\varphi), \quad f = \frac{\mathrm{div}\, H - \varrho_0 c_{0,\gamma}}{2}.
$$

On the other hand, using the identity (7.147), we obtain

$$
\begin{aligned}
2 &\left| \int_{Q_s^t} f(\dot\varphi^2 - |\nabla_{g_u}\varphi|^2_{g_u})\, dQ \right| \\
&\le c_\gamma [\mathcal{Q}_\lambda(s) + \mathcal{Q}_\lambda(t)] + c_\gamma (\|\varphi(s)\|^2 + \|\varphi(t)\|^2) + c_\gamma \int_s^t \|\varphi(\tau)\|^2\, d\tau \\
&\quad + c_\gamma \int_s^t (\|\varphi\|^2_{\Gamma_0} + \|\varphi_{\nu_A}\|^2_{\Gamma_0})\, d\tau + c_\gamma \int_s^t \mathcal{L}(\tau)\, d\tau.
\end{aligned}
\tag{7.164}
$$

Inserting (7.164) into (7.163) and using the relations (7.148) and (7.161) give

$$
\begin{aligned}
\int_s^t \vartheta_j(\tau)\, d\tau &\le \varepsilon c_\gamma [1 + 1/(t-s)] \int_s^t \mathcal{Q}_\lambda(\tau) d\tau \\
&\quad + c_{\varepsilon,\gamma} \int_s^t [\mathcal{Q}_{\Gamma_0, R, \mathbf{I}}(\tau) + \mathcal{Q}_{\Gamma_0, R, \mathbf{O}}(\tau) + \mathcal{L}(\tau)] d\tau \\
&\quad + c_{\varepsilon,\gamma} \Upsilon(s) + c_{\varepsilon,\gamma} \Upsilon(t) + c_{\varepsilon,\gamma} \int_s^t \Upsilon(\tau) d\tau
\end{aligned}
\tag{7.165}
$$

for $1 \leq j \leq m-1$ where

$$\Upsilon(t) = \sum_{j=0}^{m-1} \|u^{(j)}(t)\|^2 \qquad (7.166)$$

is a lower order term with respect to $\mathcal{E}(t)$.

Furthermore, from (7.166), we have

$$\left| \dot{\Upsilon}(t) \right| = 2 \left| \sum_{k=0}^{m-1} (u^{(k+1)}, u^{(k)}) \right| \leq \varepsilon \mathcal{Q}_\lambda(t) + c_\varepsilon \Upsilon(t),$$

which implies

$$\max\{\Upsilon(s), \Upsilon(t)\} \leq \varepsilon \int_s^t \mathcal{Q}_\lambda(\tau)\,d\tau + [c_\varepsilon + 1/(t-s)] \int_s^t \Upsilon(\tau)\,d\tau \qquad (7.167)$$

for $0 \leq s < t \leq T$.

Inserting (7.167) into (7.165), we obtain constants $c_\gamma > 0$ and $T_\gamma > 3\sup_{x\in\bar{\Omega}} |H|_g/\varrho_0$ such that, if $0 \leq s < t \leq T$ and $t - s \geq T_\gamma$, then

$$\int_s^t \vartheta_j(\tau)\,d\tau \leq \varepsilon c_\gamma \int_s^t \mathcal{Q}_\lambda(\tau)\,d\tau$$

$$+ c_\gamma \int_s^t [\mathcal{Q}_{\Gamma_0,R,\mathbf{I}}(\tau) + \mathcal{Q}_{\Gamma_0,R,\mathbf{O}}(\tau) + \mathcal{L}(\tau) + \Upsilon(\tau)]\,d\tau \qquad (7.168)$$

for $1 \leq j \leq m-1$ and $\varepsilon > 0$ small.

Step 2 This time, we consider a metric

$$h_u = B_u^{-1}(t) = \left(\int_0^1 a_{iy_j}(x, \nabla w + \tau \nabla u) d\tau \right)^{-1}$$

to replace the metric in (7.126), and use the system

$$\begin{cases} u_{tt} = \operatorname{div} B_u(t)\nabla u & (t,x) \in Q, \\ u|_{\Gamma_1} = 0 & t \in (0,T), \\ u_t + \lambda u_{v_B} = \mathbf{I}(t) & (t,x) \in \Sigma_0, \\ u_t - \lambda u_{v_B} = \mathbf{O}(t) & (t,x) \in \Sigma_0, \\ u(0,x) = u_0, \quad u_t(0,x) = u_1 & x \in \Omega \end{cases}$$

instead of the system (7.145). By repeating the procedure of Step 1, we obtain that the inequality (7.168) holds when $\vartheta_j(t)$ is replaced with $\vartheta_0(t)$ where $\vartheta_0(t) = \|u_t(t)\|^2 + (B_u(t)\nabla u, \nabla u)$.

Finally, the inequality (7.143) follows from (7.168) and Lemma 7.11 below.

□

Lemma 7.11 *Let all the assumptions of Theorem 7.10 hold. Suppose that $c_\gamma > 0$ and $T_\gamma > 3\sup_{x\in\overline{\Omega}}|H|_g/\varrho_0$ such that the inequality (7.168) holds. If $0 \le s < t \le T$ and $t - s \ge T_\gamma$, then*

$$\int_s^t \Upsilon(\tau)\,d\tau \le c_\gamma \int_s^t [\mathcal{Q}_{\Gamma_0,R,\mathbf{I}}(\tau) + \mathcal{Q}_{\Gamma_0,R,\mathbf{O}}(\tau) + \mathcal{L}(\tau)]\,d\tau \qquad (7.169)$$

where $\Upsilon(t)$ is given by (7.166).

Proof. We use an idea from [66]. It suffices to prove that for any solution u to the problem (7.114)-(7.115),

$$\int_0^{T_\gamma} \Upsilon(\tau)\,d\tau \le c_\gamma \int_0^{T_\gamma} [\mathcal{Q}_{\Gamma_0,R,\mathbf{I}}(\tau) + \mathcal{Q}_{\Gamma_0,R,\mathbf{O}}(\tau) + \mathcal{L}(\tau)]\,d\tau. \qquad (7.170)$$

In fact, if the inequality (7.170) is satisfied, then for any $h > 0$, we have

$$\begin{aligned}
\int_h^{T_\gamma+h} \Upsilon(\tau)\,d\tau &= \int_0^{T_\gamma} \Upsilon(s+h)\,ds \\
&\le c_\gamma \int_0^{T_\gamma} [\mathcal{Q}_{\Gamma_0,R,\mathbf{I}}(s+h) + \mathcal{Q}_{\Gamma_0,R,\mathbf{O}}(s+h) + \mathcal{L}(s+h)]\,ds \\
&\le c_\gamma \int_h^{T_\gamma+h} [\mathcal{Q}_{\Gamma_0,R,\mathbf{I}}(\tau) + \mathcal{Q}_{\Gamma_0,R,\mathbf{O}}(\tau) + \mathcal{L}(\tau)]\,d\tau. \qquad (7.171)
\end{aligned}$$

So for $s \ge 0$, $s + mT_\gamma \le t \le s + (m+1)T_\gamma$ where m is an integer, we have

$$\begin{aligned}
\int_s^t \Upsilon(\tau)\,d\tau &= \int_s^{s+T_\gamma} \Upsilon(\tau)\,d\tau + \cdots + \int_{s+(m-1)T_\gamma}^{s+mT_\gamma} \Upsilon(\tau)\,d\tau + \int_{s+mT_\gamma}^t \Upsilon(\tau)\,d\tau \\
&\le c_\gamma \left(\int_s^{s+T_\gamma} + \cdots + \int_{s+(m-1)T_\gamma}^{s+mT_\gamma} + \int_{t-T_\gamma}^t \right) [\mathcal{Q}_{\Gamma_0,R,\mathbf{I}}(\tau) + \mathcal{Q}_{\Gamma_0,R,\mathbf{O}}(\tau) + \mathcal{L}(\tau)]\,d\tau \\
&\le 2c_\gamma \int_s^t [\mathcal{Q}_{\Gamma_0,R,\mathbf{I}}(\tau) + \mathcal{Q}_{\Gamma_0,R,\mathbf{O}}(\tau) + \mathcal{L}(\tau)]\,d\tau.
\end{aligned}$$

We prove the inequality (7.170) by contradiction. Suppose that (7.170) does not hold for some $\gamma_0 > 0$. Then, there exist initial data (ϕ_0^k, ϕ_1^k), inputs $\mathbf{I}_k$, and the corresponding solutions ϕ^k of the problem (7.114)-(7.115) over $[0, T]$ such that

$$\sup_{0\le t\le T} \|\phi^k\|_{H^m(\Omega)} \le \gamma_0,$$

$$\int_0^{T_\gamma} \Upsilon_k(\tau)\,d\tau \ge k \int_0^{T_\gamma} [\mathcal{Q}_{\mathbf{I}_k,\Gamma_0}(\tau) + \mathcal{Q}_{\mathbf{O}_k,\Gamma_0}(\tau) + \mathcal{L}_k(\tau)]\,d\tau \qquad (7.172)$$

where $\Upsilon_k(t) = \sum_{j=0}^{m-1} \|(\phi^k)^{(j)}(t)\|_{L^2(\Omega)}^2$, $\mathcal{L}_k(t) = \sum_{j=3}^{2m} \mathcal{E}_j^{j/2}(t)$, and

$$\mathcal{E}_k(t) = \sum_{j=0}^m \|(\phi^k)^{(j)}\|_{H^{m-j}(\Omega)}^2 \quad \text{for} \quad k \ge 1.$$

Set

$$c_k^2 = \int_0^{T_\gamma} \Upsilon_k(\tau)\, d\tau, \quad \psi_k = \phi^k/c_k, \quad \theta_k = \mathbf{I}_k/c_k, \quad \vartheta_k = \mathcal{O}_k/c_k. \quad (7.173)$$

Then

$$\sum_{j=0}^{m-1} \int_0^{T_\gamma} \|\psi_k^{(j)}(\tau)\|_{L^2(\Omega)}^2\, d\tau = 1 \quad (7.174)$$

and by (7.172)

$$\sum_{j=0}^{m-1} \int_0^{T_\gamma} \left(\|\theta_k^{(j)}\|_{L^2(\Gamma_0)}^2 + \|\vartheta_k^{(j)}\|_{L^2(\Gamma_0)}^2 \right) d\tau + \frac{1}{c_k^2} \sum_{j=3}^{2m} \int_0^{T_\gamma} \mathcal{E}_k^{j/2}(\tau)\, d\tau \to 0$$

as $k \to \infty$.

Since

$$\int_0^{T_\gamma} \Upsilon_k(\tau)\, d\tau \leq \int_0^{T_\gamma} \mathcal{E}_k(\tau)\, d\tau \leq T_\gamma^{1/2} \left(\int_0^{T_\gamma} \mathcal{L}_k(\tau)\, d\tau \right)^{1/2},$$

the inequality (7.172) implies that $\int_0^{T_\gamma} \mathcal{E}_k(\tau)\, d\tau$ goes to zero as k goes to infinity, which, in turn, shows that

$$c_k \to 0. \quad (7.175)$$

We divide both sides of the inequality (7.168) by c_k^2 where we have set $s = 0$, $t = T_\gamma$, and $\varphi = (\phi^k)^{(j)}$ to see that

$$\sum_{j=0}^{m-1} \int_0^{T_\gamma} [\|\psi_k^{(j+1)}(\tau)\|_{L^2(\Omega)}^2 + (A_{\phi^k}(\tau)\nabla\psi_k^{(j)}(\tau), \nabla\psi_k^{(j)}(\tau))_{L^2(\Omega)}]\, d\tau$$

are bounded for all $k \geq 1$. Then there is $p \in H^1\left((0, T_\gamma) \times \Omega\right)$ such that

$$\psi_k^{(j)} \rightharpoonup p^{(j)} \quad \text{weakly in} \quad H^1\left((0, T_\gamma) \times \Omega\right), \quad (7.176)$$

$$\psi_k^{(j)} \to p^{(j)} \quad \text{strongly in} \quad L^2\left((0, T_\gamma) \times \Omega\right) \quad (7.177)$$

for $0 \leq j \leq m - 1$ as k goes to infinity.

Next, we divide the system (7.145) by c_k^2 where we have set $u = \phi^k$ to show that $p^{(j)} \in H^1\left((0, T_\gamma) \times \Gamma\right)$ satisfy

$$\begin{cases} p^{(j)} = Ap^{(j)}, & (t, x) \in (0, T_\gamma) \times \Omega, \\ p^{(j)} = 0 & (t, x) \in (0, T_\gamma) \times \Gamma_1, \\ p^{(j)} + \lambda p_{\nu_A}^{(j)} = 0 & (t, x) \in (0, T_\gamma) \times \Gamma_0, \\ p^{(j)} - \lambda p_{\nu_A}^{(j)} = 0 & (t, x) \in (0, T_\gamma) \times \Gamma_0, \end{cases} \quad (7.178)$$

for $0 \le j \le m - 1$ where

$$\mathcal{A}p^{(j)} = \operatorname{div} A(x, \nabla w)\nabla p^{(j)}, \quad p_{\nu_A}^{(j)} = \langle A(x, \nabla w)\nabla p^{(j)}, \nu \rangle.$$

By the observability inequality for the linear wave equation with variable coefficients, Theorem 2.15, the formulas (7.178) imply that

$$p^{(j)} = 0 \quad \text{for} \quad 1 \le j \le m - 1.$$

Then p is a constant. Moreover, the assumption **H3** implies that $p = 0$. These contradict the relations (7.174), (7.176), and (7.177).   □

By a similar argument as in Lemma 7.11 the right hand side of the inequality (7.50) of Theorem 7.3 can be improved into

Lemma 7.12 *Let all the assumptions in Theorem 7.10 hold. Let $\gamma > 0$ be given and ϕ satisfy the problem (7.114)-(7.115) on the interval $[0, T]$ for $T > 3 \sup_{x \in \Omega} |H|_g / \varrho_0$ such that (7.125) are true. Then there is $c_\gamma > 0$ such that*

$$\mathcal{E}(t) \le c_\gamma \mathcal{Q}_\lambda(t) + c_\gamma \mathcal{E}_{\Gamma_0, R}(t) + c_\gamma \mathcal{L}(t) \tag{7.179}$$

for $T \ge t \ge 3 \sup_{x \in \Omega} |H|_g / \varrho_0$.

Now we are ready to give proofs for Theorems 7.10 and 7.11.

Proof of Theorem 7.10 Let $(\phi_0, \phi_1) \in H^m(\Omega) \times H^{m-1}(\Omega)$ and $\mathbf{I} \in \mathcal{I}^m((0, \infty), \Gamma_0)$ be given such that the problem (7.114)-(7.115) has a short time solution. We shall look for a constant $\eta_0 > 0$ such that, if

$$\mathcal{E}(0) + \|\mathbf{I}\|^2_{\mathcal{I}^m((0,\infty),\Gamma_0)} \le \eta_0,$$

then the solution of the problem (7.114)-(7.115) is global in time.

In the rest of this section we take

$$\gamma = 1.$$

Let

$$c_{0,1} > 0, \quad c_1 \ge 1, \quad T_1 > 3 \sup_{x \in \Omega} |H|_g / \varrho_0$$

be given such that the estimates in Theorems 7.3 and 7.12 hold true. Let

$$\Xi(\eta) = \sum_{k=0}^{2m-3} \eta^{k/2}, \quad \Theta(\eta) = 2e^\eta \max\{c_1 c_{0,1}^{-1} + 2c_1\lambda, 5c_1^2, 3c_1^2 \Xi(1)\eta\}$$

for $\eta \in (0, \infty)$.

Let

$$0 < \eta < \min\{1, \Theta^{-2}(2T_1)/4\} \tag{7.180}$$

be given. We assume that

$$\mathcal{E}(0) + \|\mathbf{I}\|^2_{\mathcal{I}^m((0,\infty),\Gamma_0)} \le \eta^{3/2} < \eta. \tag{7.181}$$

Then there is some $\delta > 0$ such that

$$\mathcal{E}(t) < \eta \tag{7.182}$$

for $t \in [0, \delta)$. Let $\delta_0 > 0$ be the largest number such that the estimate (7.182) holds true for $t \in [0, \delta_0)$. We shall prove that $\delta_0 = \infty$.

Step 1 We have

$$\delta_0 > 2T_1. \tag{7.183}$$

Indeed, letting (7.183) be not true, i.e., $\delta_0 \leq 2T_1$, we shall have a contradiction as follows. Using (7.47), (7.46), and (7.50), we have

$$\mathcal{P}(0) \leq 2(c_{0,1}^{-1} + 2\lambda)\mathcal{E}(0), \tag{7.184}$$

which gives, by (7.53),

$$\|\phi(t)\|^2 \leq 2\{(c_{0,1}^{-1} + 2\lambda)\mathcal{E}(0) + c_1\|\mathbf{I}\|_{\mathcal{I}^m((0,\infty),\Gamma_0)}^2 + 2c_1T_1\eta^{3/2}\Xi(1)\}e^{2T_1}$$
$$\leq 2e^{2T_1}\max\{c_{0,1}^{-1} + 2\lambda, c_1, 2c_1T_1\Xi(1)\}[\mathcal{E}(0) + \|\mathbf{I}\|_{\mathcal{I}^m}^2 + \eta^{3/2}] \tag{7.185}$$

since $L(t) \leq \eta^{3/2}\Xi(\eta) \leq \eta^{3/2}\Xi(1)$ for $0 \in [0, \delta_0)$ by (7.182). In addition, we multiply (7.51) by 2, add it to (7.52), then integrate it over $(0, t)$, and obtain

$$\mathcal{Q}_\lambda(t) \leq \max\{c_{0,1}^{-1}, 10, 6c_1T_1\Xi(1)\}[\mathcal{E}(0) + \|\mathbf{I}\|_{\mathcal{I}^m_{((0,\infty),\Gamma_0)}}^2 + \eta^{3/2}]. \tag{7.186}$$

Inserting the estimates (7.185) and (7.186) into the right hand side of the inequality (7.50) yields, via (7.180) and (7.181),

$$\mathcal{E}(t) \leq \Theta(2T_1)[\mathcal{E}(0) + \|\mathbf{I}\|_{\mathcal{I}^m_{((0,\infty),\Gamma_0)}}^2 + \eta^{3/2}] \leq 2\Theta(2T_1)\eta^{1/2}\eta < \eta \tag{7.187}$$

for $0 \leq t \leq \delta_0$, which contradicts the definition of δ_0.

Step 2 Let $T_1 \leq s < t < \delta_0$ with $t - s > T_1$. Integrating (7.179) over (s, t) yields

$$\int_s^t \mathcal{E}(\tau)d\tau \leq c_1 \int_s^t \mathcal{Q}_\lambda(\tau)d\tau + c_1\|\mathbf{I}\|_{\mathcal{I}^m((s,t),\Gamma_0)}^2 + c_1\Xi(1)\eta^{1/2} \int_s^t \mathcal{E}(\tau)d\tau,$$

that is,

$$\int_s^t \mathcal{E}(\tau)d\tau \leq c_1[1 - c_1\Xi(1)\eta^{1/2}]^{-1}\left(\int_s^t \mathcal{Q}_\lambda(\tau)d\tau + \|\mathbf{I}\|_{\mathcal{I}^m((s,t),\Gamma_0)}^2\right),$$

which, in turn, gives

$$\int_s^t L(\tau)d\tau \leq f(\eta^{1/2})\left(\int_s^t \mathcal{Q}_\lambda(\tau)d\tau + \|\mathbf{I}\|_{\mathcal{I}^m((s,t),\Gamma_0)}^2\right) \tag{7.188}$$

where

$$f(x) = c_1\Xi(1)x[1 - c_1\Xi(1)x]^{-1} \quad \text{for} \quad 0 < x < 1,$$

$$\|\mathbf{I}\|^2_{\mathcal{I}^m((s,t),\Gamma_0)} = \max_{s \le \tau < t} \mathcal{E}_{\Gamma_0,R}(\tau) + \int_s^t \mathcal{E}_{\Gamma_0,R}(\tau)\, d\tau + \int_s^t \|\mathbf{I}^{(m-1)}(\tau)\|^2_{\Gamma_0}\, d\tau.$$

Furthermore, by combining the inequalities (7.143) and (7.188), we obtain

$$\int_s^t \mathcal{Q}_\lambda(\tau) d\tau \; \le c_1 [1 - c_1 f(\eta^{1/2})]^{-1}[1 + f(\eta^{1/2})]\|\mathbf{I}\|^2_{\mathcal{I}^m((s,t),\Gamma_0)}$$

$$+ c_1 [1 - c_1 f(\eta^{1/2})]^{-1}\int_s^t \mathcal{Q}_{\Gamma_0,R,\mathbf{O}}(\tau) d\tau. \qquad (7.189)$$

On the other hand, we integrate the inequality (7.51) over (s,t) and obtain, via (7.188),

$$\mathcal{Q}_\lambda(t) \; \le \mathcal{Q}_\lambda(s) + [4 + c_1 f(\eta^{1/2})]\|\mathbf{I}\|^2_{\mathcal{I}^m((s,t),\Gamma_0)}$$

$$+ c_1 f(\eta^{1/2}) \int_s^t \mathcal{Q}_\lambda(\tau) d\tau - \int_s^t \mathcal{Q}_{\Gamma_0,R,\mathbf{O}}(\tau) d\tau. \qquad (7.190)$$

Let $\kappa > 0$ be given. We multiply (7.189) by κ, then add it to (7.190), and have

$$\mathcal{Q}_\lambda(t) + [\kappa - c_1 f(\eta^{1/2})] \int_s^t \mathcal{Q}_\lambda(\tau) d\tau \le \mathcal{Q}_\lambda(s)$$

$$+ \{\kappa c_1 [1 - c_1 f(\eta^{1/2})]^{-1}[1 + f(\eta^{1/2})] + 4 + c_1 f(\eta^{1/2})\}\|\mathbf{I}\|^2_{\mathcal{I}^m((s,t),\Gamma_0)}$$

$$+ \{\kappa c_1 [1 - c_1 f(\eta^{1/2})]^{-1} - 1\} \int_s^t \mathcal{Q}_{\Gamma_0,R,\mathbf{O}}(\tau) d\tau \qquad (7.191)$$

for $T_1 \le s \le t < \delta_0$ with $t - s \ge T_1$. Letting $\kappa = [1 - c_1 f(\eta^{1/2})]/c_1$ in (7.191) gives

$$\mathcal{Q}_\lambda(t) + \omega(\eta) \int_s^t \mathcal{Q}_\lambda(\tau) d\tau \le \mathcal{Q}_\lambda(s)$$

$$+ [5 + (1 + c_1)f(\eta^{1/2})]\|\mathbf{I}\|^2_{\mathcal{I}^m((s,t),\Gamma_0)} \qquad (7.192)$$

for $T_1 \le s \le t < \delta_0$ with $t - s \ge T_1$, where $\omega(\eta) = c_1^{-1} - (1 + c_1)f(\eta^{1/2})$.
  Let

$$\tau_0 = \max\{c_1 c_{0,1}^{-1}\Theta(2T_1), c_1(7 + c_1), c_1\Xi(1)\}.$$

We fix $\eta > 0$ such that

$$\eta^{1/2} < \frac{1}{2\tau_0}, \quad \omega(\eta) > 0. \qquad (7.193)$$

Then when the condition (7.181) holds, the inequalities (7.179), (7.187), and (7.192) imply that the estimate (7.182) holds with $\delta_0 = \infty$. Indeed, if $\delta_0 < \infty$, then we will have a contradiction as follows. The condition $2c_1\Xi(1)\eta < 1$ implies $f(\eta^{1/2}) \le 1$. Then using (7.192) with $s = T_1$ in (7.179), we have

$$\mathcal{E}(t) \le c_1 c_{0,1}^{-1}\mathcal{E}(T_1) + c_1(7 + c_1)\|\mathbf{I}\|^2_{\mathcal{I}^m((s,t),\Gamma_0)} + c_1\Xi(1)\eta^{3/2} \qquad (7.194)$$

for $0 \leq t \leq \delta_0$. Finally, we use the inequality (7.187) with $t = T_1$ in (7.194) and obtain, by (7.181) and (7.193),

$$\mathcal{E}(t) \leq \tau_0[\mathcal{E}(0) + \|\mathbf{I}\|_{\mathcal{I}^m((s,t),\Gamma_0)}^2 + \eta^{3/2}] \leq 2\tau_0 \eta^{1/2} \eta < \eta$$

for all $T_1 \leq t \leq \delta_0$, which contradicts the definition of δ_0 again. □

Proof of Theorem 7.11 Let $T_0 > 0$ be such that

$$\mathbf{I}(t) = 0 \quad \text{for} \quad x \in \Gamma_0, \ t \geq T_0. \tag{7.195}$$

By (7.192), we have

$$\mathcal{Q}_\lambda(t) + \omega(\eta) \int_s^t \mathcal{Q}_\lambda(\tau)d\tau \leq \mathcal{Q}_\lambda(s) \tag{7.196}$$

for $\max\{T_1, T_0\} \leq s < t < \infty$ with $t - s > T_1$. Let $\omega_0 = \max\{T_1, T_0\}$ and $t > \omega_0$. We multiply (7.196) by s^k, integrate it from 0 to $t - \omega_0$, and obtain

$$\int_0^{t-\omega_0} s^k \mathcal{Q}_\lambda(s)ds \geq \frac{(t-\omega_0)^{k+1}}{k+1} \mathcal{Q}_\lambda(t) + \frac{\omega(\eta)}{k+1} \int_0^{t-\omega_0} s^{k+1} \mathcal{Q}_\lambda(s)ds$$

for all $k \geq 0$ which yields

$$Q(t) \leq Q(0)e^{-\omega(\eta)(t-\omega_0)} \quad \text{for} \quad t \geq \omega_0. \tag{7.197}$$

The estimate (7.124) follows from (7.179) and (7.197) since the condition (7.195) implies $\mathcal{E}_{\Gamma_0,R}(t) = 0$ for $t \geq T_0$. □

7.4　Structure of Control Regions for Internal Feedbacks

After an internal local damping act on *an escape region for the metric*, we study the existence of global smooth solutions and the exponential decay of the energy for the quasi-linear wave equation when initial data are close to a given equilibrium.

Introduction and Main Results　Let $n \geq 2$ be an integer, $\Omega \subset R^n$ be a bounded open set with smooth boundary Γ, and

$$\mathbf{a}(x,y) = (a_1(x,y), a_2(x,y), \ldots, a_n(x,y))$$

be a smooth mapping from $\overline{\Omega} \times R^n$ to R^n with

$$\mathbf{a}(x,0) = 0 \quad \text{for} \quad x \in \overline{\Omega} \tag{7.198}$$

such that $(a_{ij}(x,y))$ is symmetrical and

$$(a_{ij}(x,y)) > 0 \quad \text{for} \quad (x,y) \in \overline{\Omega} \times R^n \tag{7.199}$$

where $a_{ij} = a_{i_{y_j}}$ are the partial derivatives of a_i with respect to the variable y. Let $f(x, s): \overline{\Omega} \times R \to R$ be a smooth function such that

$$f(x, 0) = 0 \quad \text{for} \quad x \in \overline{\Omega}. \tag{7.200}$$

We consider the problem:

$$\begin{cases} \ddot{\phi}(t, x) - \operatorname{div} \mathbf{a}(x, \nabla w + \nabla \phi) + f(x, \dot{\phi}) = 0 & \text{in} \quad (0, \infty) \times \Omega, \\ \phi = 0 \quad \text{on} \quad (0, \infty) \times \Gamma, \\ \phi(0, x) = \phi_0(x), \quad \dot{\phi}(0, x) = \phi_1(x) \quad \text{on} \quad \Omega \end{cases} \tag{7.201}$$

where w is an equilibrium solution, which satisfies

$$\operatorname{div} \mathbf{a}(x, \nabla w) = 0 \quad \text{for} \quad x \in \Omega. \tag{7.202}$$

It is well known that solutions to the problem (7.201) usually develop singularities after some time even if initial data (ϕ_0, ϕ_1) are small ([93]). Here we study what condition on f can guarantee that the problem (7.201) admits global smooth solutions when initial data (ϕ_0, ϕ_1) are small. For this purpose, we assume that f satisfies

$$f_s(x, s) \geq 0 \quad \text{for} \quad x \in \overline{\Omega}, \ |s| \leq 1 \tag{7.203}$$

where $f_s(x, s) = \dfrac{\partial f(x, s)}{\partial s}$.

Let

$$G = \{\, x \in \overline{\Omega} \mid f_s(x, 0) > 0 \}. \tag{7.204}$$

Definition 7.5 *G is called the damping region of the system (7.201).*

The structure of G reflects the effect of the internal dissipation f. We seek geometric conditions on G such that the problem (7.201) has global smooth solutions in time and its energy decays. We note that $G = \overline{\Omega}$ is one of the choices for such purposes (see Theorem 7.13). However, we are particularly interested in the case where G is not the whole domain Ω and is as small as possible in a technical sense. One of such conditions is *an escape region for a metric*, given by (7.33) below, see Corollaries (2.7)-(2.8).

Remark 7.13 *Problems like (7.201) have been extensively studied, and a wealth of results on this subject is available in the literature.*

Let $m \geq [n/2] + 3$ be a given positive integer. We say that $(\phi_0, \phi_1) \in H^m(\Omega) \times H^{m-1}(\Omega)$ satisfies the compatibility conditions of order m if

$$\phi_0|_\Gamma = 0, \quad \phi_k|_\Gamma = 0, \quad 1 \leq k \leq m - 1$$

where for $k \geq 2$, $\phi_k = \phi^{(k)}(0)$ as computed formally (and recursively) in terms of ϕ_0 and ϕ_1, using (7.201).

We seek geometric conditions on G such that solutions $\phi(t, x)$ to the problem (7.201) are in

$$\bigcap_{k=0}^{m} C^k((0, +\infty), H^{m-k}(\Omega)),$$

if

$$(u_0, u_1) \in H^m(\Omega) \times H^{m-1}(\Omega)$$

are small in $(w, 0) \in H^m(\Omega) \times H^{m-1}(\Omega)$, and satisfies the compatibility conditions of order m with w.

Let $A(x, y) = (a_{ij}(x, y))$ be the $n \times n$ matrix, given in (7.199), for each $(x, y) \in \overline{\Omega} \times R^n$. We define

$$g = A^{-1}(x, \nabla w(x)) \quad \text{for} \quad x \in \Omega \tag{7.205}$$

as a Riemannian metric on Ω and consider the couple (Ω, g) as a Riemannian manifold with the boundary Γ. Here the metric g depends not only on the coefficients $a_{ij}(\cdot, \cdot)$ but also on the equilibrium w. We denote the covariant differential of the metric g by D. If H is a vector field on $\overline{\Omega}$, then the covariant differential DH of H is a tensor field of rank 2 on Ω.

We recall that $G \subset \overline{\Omega}$ is said to be *an escape region for the metric g* if the following things are true: There exist $\varepsilon > 0$, $\varrho_0 > 0$, $\Omega_i \subseteq \Omega$ with C^∞ boundary $\partial \Omega_i$ and vector fields H^i, $i = 1, 2, \ldots, J$, such that $\Omega_i \cap \Omega_j = \emptyset$ for $1 \leq i < j \leq J$ and

$$DH^i(X, X) \geq \varrho_0 |X|_g^2 \quad \text{for} \quad X \in R_x^n, \ x \in \Omega_i, \tag{7.206}$$

$$G \supseteq \overline{\Omega} \cap \mathcal{N}_\varepsilon \big[\cup_{i=1}^{J} \Gamma_0^i \cup (\Omega \backslash \cup_{i=1}^{J} \Omega_i) \big] \tag{7.207}$$

where

$$\mathcal{N}_\varepsilon(S) = \cup_{x \in S} \{ y \in R^n \big| \ |y - x| < \varepsilon \}, \quad S \subset R^n,$$

$$\Gamma_0^i = \{ x \in \partial \Omega_i \,|\, H^i(x) \cdot \nu^i(x) > 0 \}, \tag{7.208}$$

and $\nu^i(x)$ is the unit normal of $\partial \Omega_i$ at x in the Euclidean metric of R^n, pointing towards the exterior of Ω_i. For the structure of an escape region, see Section 2.3.

We assume that short time solutions to the system (7.201) exist to consider the global smooth solutions in time and the decay of the energy.

Theorem 7.13 *Let $w \in H^m(\Omega)$ be an equilibrium and let G be an escape region for the metric g. Suppose that $(\phi_0, \phi_1) \in H^m(\Omega) \times H^{m-1}(\Omega)$ satisfies the compatibility conditions of order m. Then for (ϕ_0, ϕ_1) small in $H^m(\Omega) \times H^{m-1}(\Omega)$, the system (7.201) has a global solution ϕ in*

$$\bigcap_{k=0}^{m} C^k((0, +\infty), H^{m-k}(\Omega)).$$

Moreover, there exist positive constants $C > 0$, $\sigma > 0$ such that

$$\mathcal{E}(t) \leq Ce^{-\sigma t} \quad for \quad t \geq 0 \tag{7.209}$$

where $\mathcal{E}(t) = \sum_{k=0}^{m} \|\phi^{(k)}\|_{H^{m-k}(\Omega)}^2$.

The proof of Theorem 7.13 will be given at the end of the following Subsection.

Let us see an example.

Example 7.5 *Let $\mathbf{a}$, w, and g be given by Example 7.4. Then the Gauss curvature $\kappa \leq 0$. Consider the problem*

$$\begin{cases} \ddot{\phi} - \dfrac{\phi_{x_1 x_1}}{1 + (x_2 + \phi_{x_1})^2} - \dfrac{\phi_{x_2 x_2}}{1 + (x_1 + \phi_{x_2})^2} + f(x, \dot{\phi}) = 0 \quad in \quad (0, \infty) \times \Omega, \\ \phi = 0 \quad on \quad (0, \infty) \times \Gamma, \\ \phi(0) = \phi_0, \quad \dot{\phi}(0) = \phi_1 \quad on \quad \Omega. \end{cases}$$

Let H be an escape vector field on $\overline{\Omega}$. By Theorem 2.7, the damping region of the above problem can be supported on a neighborhood of

$$\Gamma_0 = \{ x \in \Gamma \,|\, \langle H, \nu \rangle > 0 \}.$$

Proofs of the Main Results We assume that solution ϕ to the system (7.201) exists for some $T > 0$ where initial data (ϕ_0, ϕ_1) are in $H^m(\Omega) \times H^{m-1}(\Omega)$ satisfying the compatibility conditions of order m.

Denote

$$B_\phi(t) = \Big(a_{ij}(x, \nabla w + \nabla \phi)\Big) \quad for \quad (t, x) \in Q; \tag{7.210}$$

then

$$[\mathbf{a}(x, \nabla w + \nabla \phi)]' = B_\phi(t) \nabla \dot{\phi}, \tag{7.211}$$

and for $j \geq 2$,

$$[\mathbf{a}(x, \nabla w + \nabla \phi)]^{(j)} = B_\phi(t) \nabla \phi^{(j)} + \sum_{k=1}^{j-1} C_k B_\phi^{(k)}(t) \nabla \phi^{(j-k)}. \tag{7.212}$$

We compute the j-th derivatives of $f(x, \dot{\phi})$ with respect to t and obtain

$$(f(x, \dot{\phi}))^{(j)} = \sum_{i=1}^{j} \sum_{l_1 + \cdots + l_i = j} f_s^{(i)}(x, \dot{\phi}) \dot{\phi}^{(l_1)} \dot{\phi}^{(l_2)} \cdots \dot{\phi}^{(l_i)} \tag{7.213}$$

$$= f_s(x, \dot{\phi}) \dot{\phi}^{(j)} + \sum_{i=2}^{j} \sum_{l_1 + \cdots + l_i = j} f_y^{(i)}(x, \dot{\phi}) \dot{\phi}^{(l_1)} \dot{\phi}^{(l_2)} \cdots \dot{\phi}^{(l_i)}.$$

We define

$$\mathcal{B}_\phi(t) v = \operatorname{div} B_\phi(t) \nabla v \quad for \quad v \in H^2(\Omega) \tag{7.214}$$

and
$$v_{\nu_B} = \langle B_\phi(t)\nabla v, \nu\rangle \quad \text{for} \quad v \in H^2(\Omega), \ x \in \Gamma. \tag{7.215}$$

Then
$$(\mathcal{B}_\phi(t)v, \phi) = -(B_\phi(t)\nabla v, \nabla\phi) \quad \text{for} \quad v, \ \phi \in H^2(\Omega) \cap H_0^1(\Omega). \tag{7.216}$$

For $T > 0$, differentiating the system (7.201) j times with respect to t, we get

$$\begin{cases} \phi^{(j)} - \mathcal{B}_\phi(t)\phi^{(j)} + f_s(x, \dot\phi)\phi^{(j)} + r_j(t) = 0 & \text{in} \quad Q^\infty, \\ \phi^{(j)} = 0 & \text{on} \quad \Sigma^\infty \end{cases} \tag{7.217}$$

where $r_j(t)$ is defined by

$$\begin{aligned} r_j(t) &= \sum_{i=2}^{j} \sum_{l_1+\cdots+l_i=j} f_y^{(i)}(x, \dot\phi)\dot\phi^{(l_1)}\dot\phi^{(l_2)}\cdots\dot\phi^{(l_i)} \\ &\quad - \sum_{k=1}^{j-1} \operatorname{div} B_\phi^{(k)}(t)\nabla\phi^{(j-k)} \end{aligned} \tag{7.218}$$

for $2 \le j \le m - 1$ and $r_1(t) = 0$.

Similar arguments as in the proof of Lemma 7.7 yield

Lemma 7.13 *Let $\gamma > 0$ be given and let ϕ be a solution to the problem (7.201) on $[0, T)$ such that*

$$\sup_{0 \le t \le T} \|\phi(t)\|_{H^m(\Omega)} \le \gamma, \quad \sup_{0 \le t \le T} \|\dot\phi(t)\|_{H^{m-1}(\Omega)} \le \gamma. \tag{7.219}$$

Then for $2 \le j \le m - 1$, there exists a constant $c_\gamma > 0$, depending on γ, such that

$$\|r_j(t)\|_{H^{m-1-j}(\Omega)}^2 \le c_\gamma \sum_{k=2}^{m-1} \mathcal{E}^k(t) \tag{7.220}$$

where

$$\mathcal{E}(t) = \sum_{k=0}^{m} \|\phi^{(k)}(t)\|_{H^{m-k}(\Omega)}^2. \tag{7.221}$$

For the system (7.217) with $1 \le j \le m - 1$, we define the corresponding energy as
$$\vartheta_j(t) = \|\phi^{(j)}(t)\|^2 + (B_\phi(t)\nabla\phi^{(j)}(t), \nabla\phi^{(j)}(t)). \tag{7.222}$$

Moreover, we introduce an operator
$$\mathcal{N}_\phi(t)v = \operatorname{div} N_\phi(t)\nabla v \quad \text{for} \quad v \in H^2(\Omega) \tag{7.223}$$

where
$$N_\phi(t) = \int_0^1 \Big(a_{ij}(x, \nabla w + s\nabla\phi)\Big)ds.$$

Then the system (7.201) can be rewritten as

$$\begin{cases} \ddot{\phi} - \mathcal{N}_\phi(t)\phi + f(x,\dot{\phi}) = 0 & \text{in} \quad Q^\infty, \\ \phi = 0 & \text{on} \quad \Sigma^\infty \end{cases} \tag{7.224}$$

with the energy defined as

$$\vartheta_0(t) = \|\dot{\phi}(t)\|^2 + (N_\phi(t)\nabla\phi(t), \nabla\phi(t)). \tag{7.225}$$

Next, we define

$$\vartheta(t) = \sum_{j=0}^{m-1} \vartheta_j(t), \quad \mathcal{L}(t) = \sum_{k=3}^{2m} \mathcal{E}^{k/2}(t),$$

$$\mathcal{U}(f,\phi) = \sum_{j=1}^{m-1} (f_s(x,\dot{\phi})\phi^{(j+1)}, \phi^{(j+1)}) + (f(x,\dot{\phi}), \dot{\phi}), \tag{7.226}$$

$$\mathcal{P}(t) = \|\phi(t)\|^2 + \|\dot{\phi}(t)\|^2 + (N_\phi(t)\nabla\phi(t), \nabla\phi(t)).$$

Theorem 7.14 *Let $\gamma > 0$ be given and let ϕ be a solution to the problem (7.201) on the interval $[0,T]$ for some $T > 0$ such that the conditions (7.219) hold true. Then there are constants $c_{0,\gamma} > 0$ and $c_\gamma > 0$, which depend only on γ, such that for $0 \le t \le T$,*

$$c_{0,\gamma}\vartheta(t) \le \mathcal{E}(t) \le c_\gamma\vartheta(t) + c_\gamma\mathcal{L}(t), \tag{7.227}$$

$$-\dot{\vartheta}(t) \le c_\gamma\mathcal{L}(t) + 2\mathcal{U}(f,\phi), \tag{7.228}$$

$$-\dot{\vartheta}(t) + c_\gamma\mathcal{L}(t) \ge 2c_{0,\gamma}\mathcal{U}(f,\phi), \tag{7.229}$$

$$\dot{\vartheta}(t) \le c_\gamma\mathcal{L}(t). \tag{7.230}$$

Proof. Clearly, the conditions (7.219) imply that $c_{0,\gamma}Q(t) \le \vartheta(t) \le c_{0,\gamma}Q(t)$ for $0 \le t \le T$ where $Q(t)$ is defined in (7.9). Then the estimate (7.227) follows from the inequality (7.12) in Theorem 7.1 since $L(t) \le \mathcal{L}(t)$ and $\mathcal{E}_{\Gamma_0,D}(t) = 0$ for the problem (7.201).

We differentiate $\vartheta_j(t)$ and have

$$\dot{\vartheta}_j(t) = 2(\phi^{(j+2)}, \phi^{(j+1)}) + 2(B_\phi(t)\nabla\phi^{(j)}, \nabla\phi^{(j+1)}) + (\dot{B}_\phi(t)\nabla\phi^{(j)}, \nabla\phi^{(j)}). \tag{7.231}$$

Using the formulas (7.216) and (7.217) in (7.231), and by Lemma 7.13, we obtain

$$\begin{aligned} -\dot{\vartheta}_j(t) &= -(\dot{B}_\phi(t)\nabla\phi^{(j)}, \nabla\phi^{(j)}) + 2(r_j(t), \phi^{(j+1)}(t)) \\ &\quad + 2(f_s(x,\dot{\phi})\phi^{(j+1)}, \phi^{(j+1)}) \\ &\le c_\gamma\mathcal{L}(t) + 2(f_s(x,\dot{\phi})\phi^{(j+1)}, \phi^{(j+1)}) \end{aligned} \tag{7.232} \\ \tag{7.233}$$

for $1 \leq j \leq m - 1$. In addition, the identity (7.232) also gives

$$-\dot{\vartheta}_j(t) \geq -c_{\varepsilon,\gamma}\mathcal{L}(t) - \varepsilon\|\phi^{(j)}\|^2 + 2(f_s(x,\dot{\phi})\phi^{(j+1)}, \phi^{(j+1)}) \qquad (7.234)$$

for $\varepsilon > 0$ small.

We have, by (7.224),

$$-\dot{\vartheta}_0(t) = -(\dot{N}_\phi(t)\nabla\phi, \nabla\phi) + 2(f(x,\dot{\phi}), \dot{\phi}). \qquad (7.235)$$

The inequalities (7.228)-(7.230) follow from (7.233)-(7.235). $\qquad\square$

Let ϕ satisfy the problem (7.217) on the interval $[0, T]$ for some $T > 0$. For $t \in [0, T]$, let g_ϕ be the metric on $\overline{\Omega}$ given by

$$g_\phi = B_\phi^{-1}(t) \qquad (7.236)$$

where the matrix $B_\phi(t)$ is defined by (7.210). Consider the pair $(\overline{\Omega}, g_\phi)$ as a Riemannian manifold for fixed $t \in [0, T)$.

Let $\hat{\Omega} \subset \Omega$ be given. We apply the identities (2.13) and (2.14) to the problem (7.217) and integrate them over $(s, t) \times \hat{\Omega}$ to have

Lemma 7.14 *Let $\phi^{(j)}$ be a solution to (7.217) for $2 \leq j \leq m - 1$ and let $\hat{\Omega} \subseteq \Omega$ be a subset. Suppose that H and h are a vector field and a function on $\hat{\Omega}$, respectively. Then*

$$\Phi(H, h, \partial\hat{\Omega}) = 2\int_s^t \int_{\hat{\Omega}} D_{g_\phi}H(\nabla_{g_\phi}\phi^{(j)}, \nabla_{g_\phi}\phi^{(j)})dxd\tau$$

$$+\Phi(H, h, \hat{\Omega}) \qquad (7.237)$$

where

$$\Phi(H, h, \partial\hat{\Omega}) = \int_s^t \int_{\partial\hat{\Omega}} \left[2H(\phi^{(j)})\phi_{\nu_B}^{(j)} + ((\dot{\phi}^{(j)})^2 - |\nabla_{g_\phi}\phi^{(j)}|_{g_\phi}^2)\langle H, \nu\rangle\right]dxd\tau$$

$$-\int_s^t \int_{\partial\hat{\Omega}} \left((\phi^{(j)})^2 h_{\nu_B} - 2h\phi^{(j)}\phi_{\nu_B}^{(j)}\right)dxd\tau, \qquad (7.238)$$

$$\Phi(H, h, \hat{\Omega}) = 2\left(\dot{\phi}^{(j)}, \mathcal{M}(\phi^{(j)})\right)\big|_s^t - \int_s^t \int_{\hat{\Omega}} (\dot{\phi}^{(j)})^2 \mathcal{B}_\phi(\tau)h\, dxd\tau$$

$$+2\int_s^t \int_{\hat{\Omega}} \left[r_j(t) + f_s(x,\dot{\phi})\phi^{(j)}\right]\mathcal{M}(\phi^{(j)})dxd\tau$$

$$+\int_s^t \int_{\hat{\Omega}} (\dot{\phi}^{(j)})^2 - |\nabla_{g_\phi}\phi^{(j)}|_{g_\phi}^2)p(\mathcal{H}, h)dxd\tau, \qquad (7.239)$$

$$\mathcal{M}(\phi^{(j)}) = H(\phi^{(j)}) + h\phi^{(j)}, \quad p(H, h) = (\text{div } H - 2h).$$

Lemma 7.15 *Let G, given by (7.204), be an escape region for the metric g. Let $\gamma > 0$ be given and let ϕ be a solution to the problem (7.201) on the interval $[0, T]$ for some $T > 0$ such that the conditions (7.219) are true. Then there exist constants $c_\gamma > 0$ and $T_\gamma > 3 \sup_{x \in \bar{\Omega}} |H|_g / \varrho_0$ such that, if $0 \le s \le t \le T$ with $t - s \ge T_\gamma$, then*

$$\int_s^t \vartheta(\tau) d\tau \le c_\gamma \int_s^t \mathcal{L}(\tau) d\tau + c_\gamma \int_s^t R(\tau) d\tau + c_\gamma \int_s^t \mathcal{U}(f, \phi) dx d\tau \quad (7.240)$$

where $R(t) = \sum_{j=0}^{m-1} \|\phi^{(j)}(t)\|^2$ and $\mathcal{U}(f, \phi)$ is given by (7.226).

Proof. For $0 < \varepsilon_2 < \varepsilon_1 < \varepsilon_0 < \varepsilon$, set

$$Q_k = \mathcal{N}_{\varepsilon_k}[\cup_{i=1}^J \Gamma_0^i \cup (\Omega \setminus \cup_{j=1}^J \Omega_i)] \quad \text{for} \quad k = 0, 1, 2. \quad (7.241)$$

Obviously we have

$$Q_2 \subset Q_1 \subset \overline{Q_0} \subset G. \quad (7.242)$$

Let $\beta^i, i = 1, \ldots, J$ satisfy

$$\beta^i \in C_0^\infty(R^n) \quad \text{for} \quad 0 \le \beta^i \le 1, \quad \beta^i = \begin{cases} 1 & \text{on} \quad \overline{\Omega_i} \setminus Q_1, \\ 0 & \text{on} \quad Q_2. \end{cases} \quad (7.243)$$

For each i, $1 \le i \le J$, set

$$\hat{\Omega} := \Omega_i, \quad H := \beta^i H^i, \quad h := \frac{1}{2} \text{div}\, (\beta^i H^i) = h^i. \quad (7.244)$$

Step 1 For $1 \le j \le m - 1$, we estimate $\vartheta_j(t)$.
(a) We estimate the integral

$$\int_s^t \int_{\Omega \setminus Q_1} |\nabla_{g_\phi} \phi^{(j)}|_{g_\phi}^2 \, dx d\tau$$

where $|\nabla_{g_\phi} \phi^{(j)}|_{g_\phi}^2 = \langle B_\phi(t) \nabla \phi^{(j)}, \nabla \phi^{(j)} \rangle$.
It follows from (7.237) that

$$2 \int_s^t \int_{\Omega_i} D_{g_\phi}(\beta^i H^i)(\nabla_{g_\phi} \phi^{(j)}, \nabla_{g_\phi} \phi^{(j)}) dx d\tau$$
$$= \Phi(\beta^i H^i, h^i, \partial \Omega_i) - \Phi(\beta^i H^i, h^i, \Omega_i). \quad (7.245)$$

We show that

$$\Phi(\beta^i H^i, h^i, \partial \Omega_i) \le 0. \quad (7.246)$$

In fact, we notice that

$$\partial \Omega_i = [\Gamma_0^i \cup (\partial \Omega_i \setminus \Gamma)] \cup ((\partial \Omega_i \setminus \Gamma_0^i) \cap \Gamma) \equiv I_1 \cup I_2.$$

Since $\partial\Omega_i \setminus \Gamma \subseteq \Omega \setminus \cup_{i=1}^{J}\Omega_i$, we have $\beta^i = 0$ in $I_1 \subseteq Q_2$. Since $I_2 \subseteq \Gamma$ and $\phi^{(j)} = 0$ on I_2,

$$\beta^i H^i(\phi^{(j)})\phi_{\nu_B}^{(j)} = |\nabla_{g_\phi}\phi^{(j)}|_{g_\phi}^2 \langle \beta^i H^i, \nu^i \rangle \quad \text{for} \quad x \in I_2.$$

We get

$$\int_s^t \int_{\partial\Omega_i} \left[2\beta^i H^i(\phi^{(j)})\phi_{\nu_B}^{(j)} + (\dot{\phi}^{(j)})^2 - |\nabla_{g_\phi}\phi^{(j)}|_{g_\phi}^2)\langle \beta^i H^i, \nu \rangle \right] dx d\tau$$

$$= \int_s^t \int_{(\partial\Omega_i \setminus \Gamma_0^i) \cap \Gamma} |\nabla_{g_\phi}\phi^{(j)}|_{g_\phi}^2 \langle \beta^i H^i, \nu \rangle dx d\tau \le 0 \qquad (7.247)$$

since, by (7.208), $\langle \beta^i H^i, \nu \rangle \le 0$ for $x \in \partial\Omega_i \setminus \Gamma_0^i$. Furthermore, we decompose $\partial\Omega_i = (\partial\Omega_i \setminus \Omega) \bigcup (\partial\Omega_i \cap \Omega) \equiv J_1 \bigcup J_2$. Since $J_1 \subseteq \Gamma$, by the boundary condition, we get

$$\int_s^t \int_{J_1} \left[\phi^{(j)2}(\operatorname{div}\beta^i H^i)_{\nu_B} - 2\operatorname{div}(\beta^i H^i)\phi^{(j)}\phi_{\nu_B}^{(j)} \right] dx d\tau = 0. \qquad (7.248)$$

It is easy to verify that $J_2 = \partial\Omega_i \cap \Omega \subseteq \Omega \setminus \cup_{i=1}^{I}\Omega_i \subseteq Q_2$ and

$$\int_s^t \int_{J_2} \left[\phi^{(j)2}(\operatorname{div}\beta^i H^i)_{\nu_B} - 2\operatorname{div}(\beta^i H^i)\phi^{(j)}\phi_{\nu_B}^{(j)} \right] dx d\tau = 0. \qquad (7.249)$$

The estimate (7.246) is from (7.247), (7.248), (7.249) and (7.238).

Next, noting that $p(H^i, h^i) = 0$ and $f_s(x, \dot{\phi}) \ge 0$, we obtain, from Lemma 7.13 and the formula (7.239), that

$$|\Phi(\beta^i H^i, h^i, \Omega_i)| \le c_\gamma[\vartheta_j(s) + \vartheta_j(t)]$$

$$+ \varepsilon \int_s^t \vartheta_j(\tau)d\tau + c_\gamma \int_s^t \|\phi^{(j)}\|^2 d\tau + c_{\varepsilon,\gamma} \int_s^t \mathcal{L}(\tau)d\tau$$

$$+ c_{\varepsilon,\gamma} \int_s^t \int_{\Omega_i} f_s(x, \dot{\phi})[\dot{\phi}^{(j)}]^2 dx d\tau \qquad (7.250)$$

for $\varepsilon > 0$ small and $1 \le j \le m - 1$ where the estimate $\|\phi^{(j)}\|^2 \le c_\gamma\|\nabla_{g_\phi}\phi^{(j)}\|^2 \le c_\gamma\vartheta_j$ has been used.

Moreover, it follows from Lemma 7.10, the relations (7.245), (7.246), and (7.250) that

$$2\varrho_0 \int_s^t \int_{\Omega \setminus Q_1} |\nabla_{g_\phi}\phi^{(j)}|_{g_\phi}^2 dx d\tau$$

$$\le \sum_{i=1}^{J} \int_s^t \int_{\Omega_i} 2D_{g_\phi}(\beta^i H^i)(\nabla_{g_\phi}\phi^{(j)}, \nabla_{g_\phi}\phi^{(j)})dx d\tau + c_\gamma \int_s^t \mathcal{L}(\tau)d\tau$$

$$\le c_\gamma[\vartheta_j(s) + \vartheta_j(t)] + \varepsilon \int_s^t \vartheta_j(\tau)d\tau + c_\gamma \int_s^t \|\phi^{(j)}\|^2 d\tau$$

$$+ c_{\varepsilon,\gamma} \int_s^t \int_\Omega f_s(x, \dot{\phi})[\dot{\phi}^{(j)}]^2 dx d\tau + c_{\varepsilon,\gamma} \int_s^t \mathcal{L}(\tau)d\tau \qquad (7.251)$$

since $\Omega/Q_1 \subset \cup_{i=1}^{J}\Omega_i$.

(b) We estimate

$$\int_s^t \int_{\Omega \cap Q_1} |\nabla_{g_\phi}\phi^{(j)}|^2_{g_\phi}\,dxd\tau.$$

Let $\xi \in C_0^\infty(R^n)$ such that $0 \le \xi \le 1$ and

$$\xi = 0 \quad \text{for} \quad x \in R^n \setminus Q_0; \quad \xi = 1 \quad \text{for} \quad x \in Q_1.$$

We multiply the equation in (7.217) by $\xi\phi^{(j)}$, integrate it by parts, and obtain

$$\int_s^t \int_\Omega \xi|\nabla_{g_\phi}\phi^{(j)}|^2_{g_\phi}\,dxd\tau = -(\dot\phi^{(j)}, \xi\phi^{(j)})|_s^t + \int_s^t \int_\Omega \xi[\dot\phi^{(j)}]^2\,dxd\tau$$

$$-\int_s^t \int_\Omega \phi^{(j)}\langle\nabla_{g_\phi}\phi^{(j)}, \nabla_{g_\phi}\xi\rangle_{g_\phi}]\,dxd\tau - \int_s^t \int_\Omega f_s(x,\dot\phi)\phi^{(j)}\xi\phi^{(j)}\,dxd\tau$$

$$-\int_s^t \int_\Omega r_j(\tau)\xi\phi^{(j)}\,dxd\tau$$

which yields

$$\int_s^t \int_{\Omega \cap Q_1} |\nabla_{g_\phi}\phi^{(j)}|^2_{g_\phi}\,dxd\tau \le c_\gamma[\vartheta_j(s) + \vartheta_j(t)]$$

$$+c_\gamma \int_s^t \int_{Q_0} [\dot\phi^{(j)}]^2\,dxd\tau + \varepsilon \int_s^t \vartheta_j(\tau)\,d\tau + c_{\varepsilon,\gamma}\int_s^t \|\phi^{(j)}\|^2\,d\tau$$

$$+c_\gamma \int_s^t \int_\Omega f_s(x,\dot\phi)[\dot\phi^{(j)}]^2\,dxd\tau. \tag{7.252}$$

(c) Furthermore, we need to estimate

$$\int_s^t \int_{Q_0} \left(\phi^{(\dot j)}\right)^2\,dxd\tau.$$

By (7.204) and (7.242), we see that there exists $c_0 > 0$ such that

$$f_s(x,0) \ge c_0 > 0 \quad \text{for} \quad x \in \overline{Q_0}. \tag{7.253}$$

We can take γ small in (7.219), which can guarantee that

$$2f_s(x,\dot\phi) \ge c_0 \quad \text{for} \quad x \in \overline{Q_0}, \tag{7.254}$$

so that we have

$$c_0 \int_s^t \int_{Q_0} \left(\phi^{(\dot j)}\right)^2\,dxd\tau \le 2\int_s^t \int_\Omega f_s(x,\dot\phi)\left(\phi^{(\dot j)}\right)^2\,dxd\tau. \tag{7.255}$$

(d) From the relations (7.251), (7.252), and (7.255), we obtain

$$2\varrho_0 \sum_{j=1}^{m-1} \int_s^t \int_\Omega |\nabla_{g_\phi} \phi^{(j)}|_{g_\phi}^2 dx d\tau$$

$$\leq c_\gamma [\vartheta(s) + \vartheta(t)] + \varepsilon \int_s^t \vartheta(\tau) d\tau + c_\gamma \int_s^t R(\tau) d\tau$$

$$+ c_{\varepsilon,\gamma} \sum_{i=1}^{m-1} \int_s^t \int_\Omega f_s(x, \dot\phi) [\dot\phi^{(j)}]^2 dx d\tau + c_{\varepsilon,\gamma} \int_s^t \mathcal{L}(\tau). \qquad (7.256)$$

Step 2 We shall estimate $\vartheta_0(t)$. We define $\hat g_\phi = N_\phi^{-1}(t)$ and replace g_ϕ in Step 1 by $\hat g_\phi$. By a similar process, we obtain

$$\int_s^t \int_\Omega |\nabla_{\hat g_\phi} \phi|_{\hat g_\phi}^2 dx d\tau \leq c_\gamma [\vartheta_0(s) + \vartheta_0(t)] + \varepsilon \int_s^t \vartheta_0(\tau) d\tau + c_\gamma \int_s^t \|\phi\|^2 d\tau$$

$$+ c_{\varepsilon,\gamma} \int_s^t \int_\Omega f(x, \dot\phi) \dot\phi dx d\tau + c_{\varepsilon,\gamma} \int_s^t \mathcal{L}(\tau). \qquad (7.257)$$

Step 3 We multiply the equation in (7.217) by $\phi^{(j)}$, integrate it over $(s,t) \times \Omega$, and obtain for $1 \leq j \leq m-1$

$$\int_s^t \|\dot\phi^{(j)}\|^2 d\tau \leq \int_s^t \int_\Omega |\nabla_{g_\phi} \phi^{(j)}|_{g_\phi}^2 dx d\tau + \vartheta_j(s) + \vartheta_j(t)$$

$$+ \int_s^t \int_\Omega f_s(x, \dot\phi) [\phi^{(j)}]^2 dx d\tau + c_\gamma \int_s^t [\mathcal{L}(\tau) + \|\phi^{(j)}\|^2] d\tau. \qquad (7.258)$$

Similarly, we have

$$\int_s^t \|\dot\phi\|^2 d\tau \leq \int_s^t \int_\Omega |\nabla_{\hat g_\phi} \phi|_{\hat g_\phi}^2 dx d\tau + \vartheta_0(s) + \vartheta_0(t)$$

$$+ \int_s^t \int_\Omega f(x, \dot\phi) \dot\phi dx d\tau + c_\gamma \int_s^t [\mathcal{L}(\tau) + \|\phi\|^2] d\tau. \qquad (7.259)$$

Now combining the relations (7.256)-(7.259) yields

$$\int_s^t \vartheta(\tau) d\tau \leq c_\gamma [\vartheta(s) + \vartheta(t)] + c_\gamma \int_s^t [R(\tau) + \mathcal{L}(\tau)] d\tau$$

$$+ c_\gamma \int_s^t \mathcal{U}(f, \phi) d\tau. \qquad (7.260)$$

Step 4 From the inequalities (7.228) and (7.230), we have

$$\max\{\vartheta(t), \vartheta(s)\} \leq \frac{1}{t-s} \int_s^t \vartheta(\tau) d\tau + c_\gamma \int_s^t \mathcal{L}(\tau) d\tau \qquad (7.261)$$

$$+ 2 \int_s^t \mathcal{U}(f, \phi)) d\tau.$$

Let $T_\gamma > \max\{2c_\gamma, 3\sup_{x\in\bar\Omega}|H|_g/\varrho_0\}$ be given. Inserting the inequality (7.261) into the inequality (7.260), we obtain the inequality (7.240). $\square$

An argument as in the proof of Lemma 7.11 yields

Lemma 7.16 *Let all assumptions in Lemma 7.15 hold. Then we have*

$$\int_s^t R(\tau)d\tau \le c_\gamma \int_s^t \mathcal{L}(\tau)d\tau + c_\gamma \int_s^t \mathcal{U}(f,\phi)d\tau \qquad (7.262)$$

for $t - s \ge T_0$, where $T_0 \ge 3\sup_{x\in\bar\Omega}|H|_g/\varrho_0$ and $\mathcal{U}(f,\phi)$ is given in (7.226).

Next, we shall establish the following

Theorem 7.15 *Let G be an escape region for the metric g. Let $\gamma > 0$ be given and let ϕ be a solution to the problem (7.201) on the interval $[0,T]$ for some $T > 0$ such that the conditions (7.219) are true. Then there are constants $c_\gamma > 0$ and $T_\gamma > 3\sup_{x\in\bar\Omega}|H|_g/\varrho_0$ such that for $0 \le s \le t \le T$ with $t - s \ge T_\gamma$,*

$$\int_s^t \vartheta(\tau)d\tau + c_\gamma\vartheta(t) \le c_\gamma\vartheta(s) + c_\gamma \int_s^t \mathcal{L}(\tau)d\tau \qquad (7.263)$$

where $\mathcal{L}(t)$ and ϱ_0 are as given in Theorem 7.14 and (7.206), respectively.

Proof. Using the inequality (7.262) in the inequality (7.240) gives

$$\int_s^t \vartheta(\tau)d\tau \le c_\gamma \int_s^t \mathcal{L}(\tau)d\tau + c_\gamma \int_s^t \mathcal{U}(f,\phi)d\tau. \qquad (7.264)$$

On the other hand, from the inequality (7.229), we obtain

$$\int_s^t \mathcal{U}(f,\phi)dxd\tau \le c_\gamma[\vartheta(s) - \vartheta(t)] + c_\gamma \int_s^t \mathcal{L}(\tau)d\tau. \qquad (7.265)$$

Finally inserting the inequality (7.265) into the inequality (7.264), we obtain the inequality (7.263). $\square$

Proof of Theorem 7.13 We assume that all the notation remains the same as before. Let $(\phi_0,\phi_1) \in H^m(\Omega) \times H^{m-1}(\Omega)$ be given such that the problem (7.201) has a short time solution. We look for a constant η_0 such that, if

$$\mathcal{E}(0) \le \eta_0, \qquad (7.266)$$

then solutions of the system (7.201) are global.

Throughout this section we take $\gamma = 1$. Let $0 < c_{0,1} \le 1$, $c_1 \ge 1$, and $T_1 > 3\sup_{x\in\bar\Omega}|H|_g/\varrho_0$ be given according to $\gamma = 1$ in Theorems 7.14 and 7.15. Let

$$\Xi(\eta) = \sum_{k=0}^{2m-3} \eta^{k/2}, \qquad (7.267)$$

$$\Theta(\eta) = 2\max\{c_1 c_{0,1}^{-1}, 2c_1\Xi(1), c_1^2\Xi(1)\eta\} \tag{7.268}$$

for $\eta \in (0, \infty)$.

Let

$$0 < \eta < \min\{1, \Theta^{-2}(2T_1)/4\} \tag{7.269}$$

be given. We assume that

$$\mathcal{E}(0) \le \eta^{3/2} < \eta. \tag{7.270}$$

Then there is some $\delta > 0$ such that

$$\mathcal{E}(t) < \eta \tag{7.271}$$

for $t \in [0, \delta)$. Let $\delta_0 > 0$ be the largest number such that inequality (7.271) holds for $t \in [0, \delta_0)$. We shall show that $\delta_0 = +\infty$.

Step 1   We claim $\delta_0 > 2T_1$. Suppose that this is not true, i.e., $\delta_0 \le 2T_1$. We have a contradiction as follows.

Using the inequalities (7.227) and (7.230), we obtain

$$\vartheta(0) \le c_{0,1}^{-1}\mathcal{E}(0), \quad \vartheta(t) \le c_{0,1}^{-1}\mathcal{E}(0) + 2c_1 T_1 \eta^{3/2}\Xi(1),$$

for $0 \le t \le \delta_0 \le 2T_1$, where the relations (7.270) and

$$\mathcal{L}(t) = \sum_{k=3}^{2m} \mathcal{E}^{k/2}(t) \le \eta^{3/2}\Xi(\eta) \le \eta^{3/2}\Xi(1) \tag{7.272}$$

have been used.

From (7.227) and (7.270)-(7.272), we obtain

$$\begin{aligned}
\mathcal{E}(t) &\le c_1 c_{0,1}^{-1}\mathcal{E}(0) + c_1(2c_1 T_1 + 1)\Xi(1)\eta^{3/2} \\
&\le 2\max\{c_1 c_{0,1}^{-1}, 2c_1^2 T_1\Xi(1), 2c_1\Xi(1)\}[\mathcal{E}(0) + \eta^{3/2}] \\
&\le \Theta(2T_1)[\mathcal{E}(0) + \eta^{3/2}] \le 2\Theta(2T_1)\eta^{1/2}\eta < \eta \tag{7.273}
\end{aligned}$$

for all $t \in [0, \delta_0]$. This contradicts the definition of δ_0.

Step 2   Next, we prove that $\delta_0 = +\infty$ by contradiction again. Let $T_1 \le s < t < \delta_0$ with $t - s \ge T_1$. Integrating the inequality (7.227) over interval (s, t) yields

$$\int_s^t \mathcal{E}(\tau)d\tau \le c_1 \int_s^t \vartheta(\tau)d\tau + c_1\Xi(1)\eta^{1/2}\int_s^t \mathcal{E}(\tau)d\tau,$$

that is,

$$\int_s^t \mathcal{E}(\tau)d\tau \le c_1[1 - c_1\Xi(1)\eta^{1/2}]^{-1}\int_s^t \vartheta(\tau)d\tau,$$

which, in turn, shows that

$$\int_s^t \mathcal{L}(\tau)d\tau \le \Xi(1)\eta^{1/2}\int_s^t \mathcal{E}(\tau)d\tau \le h(\eta^{1/2})\int_s^t \vartheta(\tau)d\tau \tag{7.274}$$

where
$$h(x) = c_1 \Xi(1)x[1 - c_1 \Xi(1)x]^{-1} \quad \text{for} \quad 0 \le x \le 1.$$

Furthermore, combining the inequalities (7.263) and (7.274), we obtain

$$c_1 \vartheta(t) + \int_s^t \vartheta(\tau)d\tau \le c_1 \vartheta(s) + c_1 h(\eta^{1/2}) \int_s^t \vartheta(\tau)d\tau \,,$$

that is,

$$\vartheta(t) + w(\eta) \int_s^t \vartheta(\tau)d\tau \le \vartheta(s) \tag{7.275}$$

where $w(\eta) = c_1^{-1} - h(\eta^{1/2})$.

We fix $\eta > 0$ small such that

$$\eta^{1/2} \le [2c_1 c_{0,1}^{-1}\Theta(2T_1) + c_1\Xi(1)]^{-1},$$
$$w(\eta) > 0.$$

If $\delta_0 < \infty$, we find a contradiction as follows. First, the inequality (7.275) implies that

$$\vartheta(t) \le \vartheta(T_1) \quad \text{for} \quad 2T_1 \le t \le \delta_0. \tag{7.276}$$

From (7.227), (7.276) and (7.273), we obtain

$$\begin{aligned}
\mathcal{E}(t) &\le c_1 \vartheta(T_1) + c_1\Xi(1)\eta^{3/2} \le c_1 c_{0,1}^{-1}\mathcal{E}(T_1) + c_1\Xi(1)\eta^{3/2} \\
&\le [2c_1 c_{0,1}^{-1}\Theta(2T_1) + c_1\Xi(1)]\eta^{1/2}\eta < \eta
\end{aligned}$$

for $2T_1 \le t \le \delta_0$, which contradicts the definition of δ_0 again. Then $\delta_0 = \infty$. Finally, the inequality (7.275) yields

$$\int_0^{t-T_1} s^k \vartheta(s)ds \ge \frac{(t-T_1)^{k+1}}{k+1}\vartheta(t) + \frac{w(\eta)}{k+1}\int_0^{t-T_1} \tau^{k+1}\vartheta(\tau)d\tau$$

for all $k \ge 0$, which gives

$$\vartheta(t) \le \vartheta(0)e^{-w(\eta)(t-T_1)} \quad \text{for} \quad t \ge T_1. \tag{7.277}$$

The estimate (7.209) follows from (7.277).   □

7.5   Notes and References

Section 7.2 is from [216]; Section 7.3 is from [215]; Section 7.4 is from [218].

Boundary Exact Controllability For the quasilinear string, the earlier works on boundary controllability were [45], [46] where the author changed a control problem into one of solving boundary-initial problems by exchanging the time variable for the material variable. The ideas from [45], [46] have been developed for the quasilinear systems in [135]-[145]. Very recently, this method was applied to the network systems by [73], [74]. One of the advantages of the works [45], [46], [73], [74], [135]-[145], is that their controls are constructive but this method does not work in higher dimensions.

Some excellent ideas were given by [184]-[186] which transformed the local exact controllability of the quasilinear string to that of a linear string. One of the advantages of [184]-[186] is that this approach opens up the vast reserves of the control theory of the linear systems. Moreover, the methods here work for the quasilinear systems in higher dimensions. The disadvantage is that their controls were not constructive.

In the case of higher dimensions a special quasilinear model was considered by [220] and some semilinear abstract systems were studied by [121].

Global Solutions of Boundary Feedback A string with boundary dissipation was studied by [4], [70], [90], [214], [217], and many other authors.

The first paper in higher dimensions was [173] where the constant coefficient case was handled: The coefficients $a_i(x, y) = a_i(y)$ do not depend on the variable x and the given equilibrium is zero. In this case, Riemann geometry was not needed because the metric, given by (7.119), is

$$g = \left(a_{iy_j}(0) \right) \quad \text{for} \quad x \in \Omega$$

which has the zero curvature. However, for a general case the metric (7.119) is $g = \left(a_{iy_j}(x, \nabla w(x)) \right)$ which depends on the variable $x \in \Omega$. Then the Riemannian geometrical approach is a necessary tool to provide the quasilinear wave equation (7.114) with the geometrical assumption (7.120) to guarantee the boundary stabilization.

Global Solution in Internal Damping Internal damping has been extensively studied and a wealth of results on this subject is available in the literature, in particularly, for the linear wave equation. For bounded domains, see [7], [36], [66], [82], [99], [103], [114], [122], [151], [159], [166], [176], [189], [196], [221]. For the Cauchy problem, see [197], [198], [199], [222]. For exterior domains, see [161] and the references therein.

The conditions (7.206) and (7.207) of an *escape region* represent a basic structure of localized damping. They were given in the present form, for the first time, by [151] where the piecewise multiplier method was introduced. [66] and [67] generalized them to the wave equation with variable coefficients and the shallow shell by the Riemann geometrical approach, respectively. Recently, [218] established global solutions to the quasilinear wave equation under those conditions.

Bibliography

[1] R. A. Adams, Sobolev Spaces, Academic Press, New York, 1975.

[2] S. Agmon, Elliptic Boundary Value Problems, Van Nostrand, Princeton, NJ, 1965.

[3] S. Agmon, A. Douglis, and L. Nirenberg, Estimates near the boundary for the solutions of elliptic partial equations satisfying general boundary conditions II, Comm. Pure Appl. Math. 17 (1964), 35-92.

[4] H. D. Alber and J. Cooper, Quasilinear hyperbolic 2×2 systems with a free, damping boundary condition, J. Reine Angew. Math. 406 (1990), 10-43.

[5] N. Aronszajn, A unique continuation theorem for solutions of elliptic partial differential equations or inequalities of second order, Journ. de Math. Vol. XXXVI. - Fasc. 3 (1957), 235-249.

[6] V. Barbu, Nonlinear Semigroups and Differential Equations in Banach Space, Noordhoff, 1976.

[7] V. Barbu, I. Lasiecka, and M. A. Rammaha, Blow-up of generalized solutions to wave equations with nonlinear degenerate damping and source terms, Indiana Univ. Math. J. 56 (2007), 995-1021.

[8] C. Bardos, G. Lebeau, and J. Rauch, Sharp suffcient conditions for the observation, control and stabilization of wave from the boundary, SIAM J. Control Optim. 30 (1992), 1024-1065.

[9] J. Bartolomeo, I. Lasiecka, and R. Triggiani, Uniform exponential energy decay of Euler-Bernoulli equations by suitable boundary feedback operators, Atti Accad. Naz. Lincei Rend. Cl. Sci. Fis. Mat. Natur. 83(8) (1989), 121-128.

[10] J. Bartolomeo and R. Triggiani, Uniform energy decay rates for Euler-Bernoulli equations with feedback operators in the Dirichlet/Neumann boundary conditions, SIAM J. Math. Anal. 22(1) (1991), 46-71.

[11] M. Berger, Nonlinearity and Functional Analysis, Academic Press, 1977.

[12] M. Bernadou and J. M. Boisserie, The Finite Element Method in Thin Shell Theory: Applications to Arch Dam Simulation, Progress in Scientific Computing, Vol. 1, Birkhäuser, Boston, 1982.

[13] M. Bernadou, P. G. Ciarlet, and B. Miara, Existence theorems for two-dimensional linear shell theories, J. Elasticity 34(2) (1994), 111-138.

[14] M. Bernadou and J. T. Oden, An existence theorem for a class of non-linear shallow shell problems, J. Math. Pures Appl. 60 (1981), 285-308.

[15] I. I. Blekhman, Vibratsionnaya mekhanika. (Russian) [Vibration mechanics] VO "Nauka", Moscow, 1994. 398 pp. ISBN: 5-02-014283-2.

[16] S. Bochner, Vector fields and Ricci curvature, Bull. Amer. Math. Soc. 52(1946), 776-797.

[17] —, Curvature in Hermitian metric, Bull. Amer. Math. Soc. 53 (1947), 179-195.

[18] W. Borchers and H. Sohr, On the equation $rotv = g$ and $divu = f$ with zero boundary conditions, Hokkaido Math. J. 19 (1990), 67-87.

[19] J. Cagnol and C. Lebiedzik, On the free boundary conditions for a dynamic shell model based on intrinsic differential geometry, Appl. Anal. 83(6) (2004), 607-633.

[20] M. M. Cavalcanti, A. Khemmoudj, and M. Medjden, Uniform stabilization of the damped Cauchy-Ventcel problem with variable coefficients and dynamic boundary conditions, J. Math. Anal. Appl. 328(2) (2007), 900-930.

[21] M. M. Cavalcanti, V. N. Domingos Cavalcanti, R. Fukuoka, and J. A. Soriano, Asymptotic stability of the wave equation on compact surfaces and locally distributed damping–a sharp result, Trans. Amer. Math. Soc. 361(9) (2009), 4561-4580.

[22] —, Uniform stabilization of the wave equation on compact manifolds and locally distributed damping–a sharp result, J. Math. Anal. Appl. 351(2) (2009), 661-674.

[23] —, Uniform stabilization of the wave equation on compact surfaces and locally distributed damping, Methods Appl. Anal. 15(4) (2008), 405-425.

[24] S. G. Chai, Stabilization of thermoelastic plates with variable coefficients and dynamical boundary control, Indian J. Pure Appl. Math. 36(5) (2005), 227-249.

[25] —, Boundary feedback stabilization of Naghdi's model, Acta Math. Sin. (Engl. Ser.) 21(1) (2005), 169-184.

[26] —, Uniqueness in the Cauchy problem for the Koiter shell, J. Math. Anal. Appl. 369(1) (2010), 43-52.

[27] —, Uniform decay rate for the transmission wave equations with variable coefficients, to appear in J. Syst Sci and Complexity.

[28] S. G. Chai and B. Z. Guo, Analyticity of a thermoelastic plate with variable coefficients, J. Math. Anal. Appl. 354 (2009), 330-338.

[29] —, Feedthrough operator for linear elasticity system with boundary control and observation, SIAM J. Control Optim. 48(6)(2010), 3708-3734.

[30] —,Well-posedness and regularity of Naghdi's shell equations under boundary control and observation, J. Differential Equations 249 (2010), 3174-3214.

[31] S. G. Chai and Y. X. Guo, Boundary stabilization of wave equations with variable coefficients and memory, Differential Integral Equations 17(5-6) (2004), 669-680.

[32] S. G. Chai, Y. X. Guo and P. F. Yao, Boundary feedback stabilization of shallow shells, SIAM J. Control Optim. 42(1)(2004), 239-259.

[33] S. G. Chai and K. Liu, Boundary stabilization of the transmission of wave equations with variable coefficients (in Chinese). Ann. Math. Ser. A 26(5) (2005), 605-612; translation in Chinese J. Contemp. Math. 26(4) (2005) 337-346 (2006).

[34] —, Observability inequalities for transmission of shallow shells, Systems Control Letters 55(9) (2006), 726-735.

[35] —, Boundary stabilization of the transmission problem of Naghdi's models, J. Math. Anal. Appl. 319(1) (2006), 199-214.

[36] —, Boundary stabilization of the transmission of wave equations with variable coefficients, Chinese Ann. Math. Ser. A, 26 (2005), 605-612 (in Chinese).

[37] S. G. Chai and P. F. Yao, Observability inequality for thin shell, Science in China Ser. A 44(3) (2003), 300-311.

[38] J. Cheeger and D. Ebin, Comparison Theorem in Riemannian Geometry, North-Holland, Amsterdam, 1975.

[39] G. Chen, Energy decay estimates and exact boundary value controllability for the wave equation in a bounded domain, J. Math. Pures. (9) 58 (1979), no. 3, 249-273.

[40] —, Control and stabilization for wave equation in a bounded domain I, SIAM J. Control and Opt. 17 (1979), 66-81.

[41] —, Control and stabilization for wave equation in a bounded domain II, SIAM J. Control and Opt. 19 (1981), 114-122.

[42] G. Chen, M. P. Coleman, and K. Liu, Boundary stabilization of Donnell's shallow circular cylindrical shell, J. Sound Vibration 209(2) (1998), 265-298.

[43] I. Chueshov and I. Lasiecka, Attractors and long time behavior of von Karman thermoelastic plates, Appl. Math. Optim. 58 (2008), no. 2, 195-241.

[44] P. G. Ciarlet, An introduction to differential geometry with applications to elasticity. Reprinted from J. Elasticity 78/79 (2005), no. 1-3. Springer, Dordrecht, 2005. iv+209 pp. ISBN: 978-1-4020-4247-8; 1-4020-4247-7, 74-02.

[45] M. Cirina, Boundary controllability of nonlinear hyperbolic systems, SIAM J. Control 7 (11) (1969), 198-212.

[46] ──, Nonlinear hyperbolic problems with solutions on preassigned sets, Mich. Math. J. 17 (1970), 193-209.

[47] P. Cornilleau, J.-P. Lohac, and A. Osses, Nonlinear Neumann boundary stabilization of the wave equation using rotated multipliers, J. Dyn. Control Syst. 16(2) (2010), 163-188.

[48] C. M. Dafermos and W. J. Hrusa, Energy methods for quasilinear hyperbolic initial-boundary value problems. Applications to elastodynamics, Arch. Rational Mech. Anal. 87(2) (1985), 267-292.

[49] M. C. Delfour, Intrinsic differential geometry methods in the asymptotic analysis of linear thin shells, in "Boundaries, interfaces and transitions", M. Delfour, ed., 19-90, CRM Proc. Lect. Notes Ser., AMS Publications, Providence, R.I. 1998.

[50] M. C. Delfour and J.P. Zolésio, Differential equations for linear shells: comparison between intrinsic and classical models, Centre de Recherches Mathématiques, CRM Proceedings and Lecture Notes, 1997, 42-124.

[51] ──, Convergence of the linear $P(1,1)$ and $P(2,1)$ thin shells, CRM Proceeding and Lecture Notes, AMS Publications, Providuce, RI, 1997.

[52] ──, A boundary differential equations for thin shells, J. Differential Equations 119 (1995), 426-445.

[53] ──, Tangential differential equations for dynamical thin/shallow shells, J. Differential Equations 128 (1996), 125-167.

[54] ──Convergence of the linear $P(1,1)$ and $P(2,1)$ thin shells to asymptotic shells, in "Proc. Plates and Shells: from theory to practice", M. Fortin, ed., CRM Proc. Lect. Notes ser., AMS Publications, Providence, R.I. 1998.

[55] L. Deng and P. F. Yao, Global smooth solutions for semilinear Schrodinger equations with boundary feedback on 2-dimensional Riemannian manifolds, J. Syst. Sci. Complex. 22(4) (2009), 749-776.

[56] —, Boundary controllability for the semilinear Schrodinger equations on Riemannian manifolds, J. Math. Anal. Appl. 372(1) (2010), 19-44.

[57] —, Boundary Controllability for the Quasi-linear Wave Equations Coupled in Parallel, to appear in Nonlinear Analysis Series A: Theory, Methods, and Applications.

[58] S. Dolecki and D. L. Russell, A general theory of observation and control, SIAM J. Control Optizm., 15 (1977), 185-220.

[59] L.H. Donnell, Beams, Plates, and Shells, McGraw-Hill, New York, 1976.

[60] T. Duyckaerts, X. Zhang, and E. Zuazua, On the optimality of the observability inequalities for parabolic and hyperbolic systems with potentials, Ann. Inst. H. Poincar Anal. Non Linaire 25(1) (2008), 1-41.

[61] D. Gromoll, W. Klingenberg and W. Meyer, Riemannsche Geometrie in Grossen, Springer-Verlag, Berlin, Heidelberg and New York, 1966.

[62] S. Feng and D. X. Feng, Locally distributed control of wave equations with variable coefficients, Sci. China Ser. F 44(4) (2001), 309-315.

[63] —, Boundary stabilization of wave equations with variable coefficients, Sci. China Ser. A 44 (3) (2001), 345-350.

[64] —, A note on geometric conditions for boundary control of wave equations with variable coefficients, J. Math. Anal. Appl. 271 (2002), 59-65.

[65] —, Nonlinear boundary stabilization of wave equations with variable coefficients, Chinese Ann. Math. Ser. B 24(2)(2003), 239-248.

[66] —, Nonlinear internal damping of wave equations with variable coefficients, Acta Math. Sin. (Engl. Ser.) 20 (2004), 1057-1072.

[67] —, Exact internal controllability for shallow shells, Sciences in China 49(4) (2006), 1-12.

[68] D. Gilbarg and N. S. Trudinger, Elliptic Partial Differential Equations of Second Order, Second Edition and Revised Third Printing, Springer-Verlag, 1998.

[69] S. Goldberg, Unbounded Linear Operators, McGraw-Hill, New York, 1966.

[70] J. M. Greenberg and T. T. Li, The effect of boundary damping for the quasilinear wave equation, Differential Equations 52(1) (1984), 66-75.

[71] R. E. Greene and H. Wu, C^∞ convex functions and manifolds of positive curvature, Acta Math. 137 (1976), 209-245.

[72] D. Gromoll, W. Klingenberg and W. Meyer, Riemannsche Geometrie in Grossen, Springer-Verlag, Berlin, Heidelberg and New York, 1966.

[73] Q. Gu and T. T. Li, Exact boundary observability of unsteady flows in a tree-like network of open canals, Math. Methods Appl. Sci. 32(4) (2009), 395-418.

[74] —, Exact boundary controllability for quasilinear wave equations in a planar tree-like network of strings, Ann. Inst. H. Poincar Anal. Non Linaire 26(6) (2009), 2373-2384.

[75] R. Gulliver, I. Lasiecka, W. Littman, and R. Triggiani, The case for differential geometry in the control of single and coupled PDEs: the structural acoustic chamber, Geometric methods in inverse problems and PDE control, 73-181, IMA Vol. Math. Appl. Vol. 137, Springer, New York, 2004.

[76] R. Gulliver and W. Littman, Chord uniqueness and controllability: the view from the boundary. I. Differential geometric methods in the control of partial differential equations (Boulder, CO, 1999), 145-175, Contemp. Math., 268, Amer. Math. Soc., Providence, RI, 2000.

[77] Y. X. Guo, S. G. Chai and P. F. Yao, Stabilization of elastic plates with variable coefficients and dynamical boundary control, Quart. of Appl. Math. Vol. 61(2)(2002), 383-400.

[78] B. Z. Guo and Z. C. Shao, On well-posedness, regularity and exact controllability for problems of transmission of plate equation with variable coefficients, Quart. Appl. Math. 65(4) (2007), 705-736.

[79] —, On exponential stability of a semilinear wave equation with variable coefficients under the nonlinear boundary feedback, Nonlinear Anal. 71(12) (2009), 5961-5978.

[80] Y. X. Guo and P. F. Yao, On Boundary Stability of Wave Equations with Variable Coefficients, Acta Mathematicae Applicatae Sinica, English Series Vol. 18(4) (2002), 589-598.

[81] —, Stabilization of Euler-Bernoulli plate equation with variable coefficients by nonlinear boundary feedback, J. Math. Anal. Appl. 317(1) (2006), 50-70.

[82] A. Haraux, Stabilization of trajectories for some weakly damped hyperbolic equations, J. Differential Equations 59 (1985), 145-154.

[83] I. Hamchi, Uniform decay rates for second-order hyperbolic equations with variable coefficients, Asymptot. Anal. 57(1-2) (2008), 71-82.

[84] I. Hamchi and S. E. Rebiai, Indirect boundary stabilization of a system of Schrodinger equations with variable coefficients, NoDEA Nonlinear Differential Equations Appl. 15(4-5) (2008), 639-653.

[85] E. Hebey, Sobolev Spaces on Riemannian Manifolds, Lecture Notes in Mathematics 1635, Springer-Verlag, Berlin, Heidelberg, 1996.

[86] L. F. Ho, Observabilite frontiere de lequation des ondes, C. R. Acad. Sci. Paris, Ser. I Math. 302 (1986), 443-446.

[87] L. Hörmander, The Analysis of Linear Partial Differential Operators, Vol. III, Springer-Verlag, Berlin, New York, 1985.

[88] M. A. Horn, Sharp trace regularity of the traces to solution of dynamic elasticity, J. Math. Systems Estim. Control. 8(2) (1998), 1-11.

[89] —, Exact controllability of the Euler-Bernoulli plate via bending moments only on the space of optimal regularity, J. Math. Anal. Appl. 167(2) (1992), 557-581.

[90] L. Hsiao and R. Pan, Initial boundary value problem for the system of compressible adiabatic flow through porous media, J. Differential Equations 159 (1999), 280-305.

[91] J. R. Hughes, Thomas, T, Kato, and J. E. Marsden, Well-posed quasilinear second-order hyperbolic systems with applications to nonlinear elastodynamics and general relativity, Arch. Rational Mech. Anal. 63(3) (1976), 273-294.

[92] R. E. Kalman, Mathematical description of linear dynamical systems, SIAM J. Control 1 (1963), 152-192.

[93] S. Klainerman, A. Majda, Formation of singularities for wave equations including the nonlinear string, Comm. Pure Appl. Math. 33 (1980), 241-263.

[94] W. T. Koiter, A consistent first approximation in the general theory of thin elastic shells, Proc. IUTAM Symposium on the Theory of Thin Shells, Delft (August 1959), 12-33, North-Holland Publishing Cy., Amsterdam (1960).

[95] —, On the nonlinear theory of thin elastic shells I, Proc. Kon. Ned. Akad. Wetensch. B69 (1966), 1-17.

[96] —, On the nonlinear theory of thin elastic shells II, Proc. Kon. Ned. Akad. Wetensch. B69 (1966), 18-32.

[97] —, On the nonlinear theory of thin elastic shells III, Proc. Kon. Ned. Akad. Wetensch. B69 (1966), 33-54.

[98] —, On the foundations of the linear theory of thin elastic shells, Proc. Kon. Ned. Akad. Wetensch. B73 (1970), 169-195.

[99] V. Komornik, Exact Controllability and Stabilization. The multiplier method, RAM: Research in Applied Mathematics. Masson, Paris; John Wiley and Sons, Ltd., Chichester, 1994.

[100] V. Kormornik and E. Zuazua, A directed method for the boundary stabilization of the wave equation, J. Math. Pures et Appl. 69 (1990), 33-54.

[101] S. Kobayashi and K. Nomizu, Foundations of Differential Geometry, New York Interscience, Vol. I (1963), Vol. II (1969).

[102] M. Jamshidi and M. Malek-Zavarei, Linear Control Systems: A Compter-Aided Approach, International Series on Systems and Control, Vol. 7, Pergamon Press, 1986.

[103] J. E. Lagnese, Control of wave processes with distributed controls supported on a subregion, SIAM J. Control Optim. 21 (1983), 68-85.

[104] —, Decay of solutions of wave equations in a bounded region with boundary dissipation, J. Differential Equations, 50 (1983), no. 2, 163-182.

[105] —, Boundary Stabilization of Thin Plates, SIAM Studies in Applied Mathematics, 1989.

[106] —, Modeling and stabilization of nonlinear plates, Internat. Ser. Numer. Math. 100 (1991), 247-264.

[107] —, Uniform asymptotic energy estimates for solutions of the equations of dynamic plane elasticity with nonlinear dissipation of the boundary, Nonlinear Anal. 16 (1991), 35-54.

[108] —, Boundary controllability in problems of transmission for a class of second order hyperbolic systems, ESAIM Control Optim. Calc. 2 (1997), 343-357.

[109] J. E. Lagnese and J. L. Lions, Modelling Analysis and Control of Thin Plates. Recherches en Mathmatiques Appliques, 6. Masson, Paris, 1988.

[110] I. Lasiecka, Mathematical Control Theory of Coupled Systems, Lecture Notes CMBS-NSF, SIAM, 2001.

[111] —, Uniform stabilizability of a full von Karman system with nonlinear boundary feedback, SIAM J. Control 36 (1998), 1376-1422.

[112] —, Uniform decay rates for the full von Karman system of dynamic thermo-elasticity with free boundary conditions and partial dissipation, Comm. in PDEs 24 (1999), 1801-1849.

[113] I. Lasiecka, R. Marchand, Uniform decay for solutions to nonlinear shells with nonlinear dissipation, Nonlinear Anal. 30 (1997), 5409-5418.

[114] I. Lasiecka and J. Ong, Global solvability and uniform decays of solutions to quasilinear equations with nonlinear boundary dissipation, Comm. in PDEs 24 (1999), 2069-2107.

[115] I. Lasiecka, D. Tataru, Uniform boundary stabilization of semilinear wave equations with nonlinear boundary damping, Differential Integral Equations 6 (1993), 507-533.

[116] I. Lasiecka and R. Triggiani, Regularity of hyperbolic equations under $L_2(0, T); L_2(\Gamma))$ boundary terms, Appl. Math. Optimiz. 10 (1983), 275-286.

[117] —, Exact controllability of the wave equation with Neumann boundary control, Appl. Math. Optimiz. 19 (1989), 243-209.

[118] —, Regularity theory for a class of a Euler-Bernoulli equations: a cosine operator approach, Boll. Un. Mat. Ital. (7), 3-B (1989), 199-228.

[119] —, Exact controllability of the Euler-Bernoulli equation with controls in Dirichlet Neumann boundary conditions: A nonconservative case, SIAM J. Contr. and Optim. 27(2)(1989), 330-373.

[120] —, Exact controllability of the Euler-Bernoulli equation with boundary controls for displacement and moment, J. Math. Anal. Appl. 146 (1990), 1-33.

[121] —, Exact controllability of semilinear abstract systems with applications to waves and plates boundary control problems, Appl. Math. Optim. 23 (1991), 109-145.

[122] —, Uniform stabilization of the wave equation with Dirichlet and Neumann feedback control without geometrical conditions, Appl. Math. Optim. 25 (1992), 189-224.

[123] —, Sharp trace estimate of solutions to Kirchhoff and Euler-Bernoulli equations, Appl. Math. Optim 28 (1993), 277-306.

[124] —, Control Theory for Partial Differential Equations, Vol. 1, Abstract Parabolic Systems: Continuous and Approximation Theories (Encyclopedia of Mathematics and its Applications), Cambridge University Press, 2000.

[125] —, Control Theory for Partial Differential Equations: Volume 2, Abstract Hyperbolic-like Systems over a Finite Time Horizon: Continuous and Approximation ... of Mathematics and its Applications), Cambridge University Press, 2000.

[126] —, Uniform stabilization of a shallow shell model with nonlinear boundary feedbacks, J. Math. Anal. Appl. 269(2) (2002), 642-688.

[127] —, Uniform energy decay rates of hyperbolic equations with nonlinear boundary and interior dissipation. Control Cybernet. 37 (2008), no. 4, 935C969.

[128] —, Linear hyperbolic and Petrowski type PDEs with continuous boundary control → boundary observation open loop map: implication on nonlinear boundary stabilization with optimal decay rates, Sobolev spaces in mathematics, III, 187-276, Int. Math. Ser. (N.Y.), 10, Springer, New York, 2009.

[129] I. Lasiecka, R. Triggiani, and W. Valente, Uniform stabilization of spherical shells by boundary dissipation, Adv. Differential Equations 1 (1996), 635-674.

[130] I. Lasiecka, R. Triggiani, and P. F. Yao, Inverse/observability estimates for second-order hyperbolic equations with variable systems. J. Math. Anal. Appl. 235 (1999), 13-57.

[131] —, Carleman estimates for a plate equation on a Riemann manifold with energy level terms. Analysis and applications—ISAAC 2001 (Berlin), 199-236, Int. Soc. Anal. Appl. Comput., 10, Kluwer Acad. Publ., Dordrecht, 2003.

[132] I. Lasiecka, W. Valente, Uniform boundary stabilization of a nonlinear shallow and tin spherical cap, J. Math. Anal. Appl. 22 (1996), 951-994.

[133] P. D. Lax, Development of singularities of solutions of nonlinear hyperbolic partial differential equations, J. Math. Phys. 5 (1964), 611-613.

[134] K. C. Le, Vibrations of Shells and Rods, Springer-Verlag, Berlin, Heidelberg, 1999.

[135] T. T. Li, Global Classical Solutions for Quasilinear Hyperbolic Systems, Masson/Wiley, Paris/New York, 1994.

[136] —, Exact boundary controllability for quasilinear hyperbolic systems and its application to unsteady flows in a network of open canals, Math. Meth. Appl. Sci. 27, 1089-1114 (2004), 26.

[137] —, Exact boundary controllability of unsteady flows in a network of open canals, Math. Nachr. 278(27)(2005), 278-289.

[138] —, Controllability and Observability for Quasilinear Hyperbolic Systems, AIMS on Applied Mathematics Vol. 3, American Institute of Mathematical Sciences and Higher Education Press, 2010.

[139] —, Controllability and observability: from ODEs to quasilinear hyperbolic systems, ICIAM 07-6th International Congress on Industrial and Applied Mathematics, 251-278, Eur. Math. Soc., Zurich, 2009.

[140] T. T. Li and B. P. Rao, Exact boundary controllability for quasilinear hyperbolic systems, SIAM J. Control Optim. 41(6)(2003), 1748-1755.

[141] —, Exact controllability for first order quasilinear hyperbolic systems with vertical characteristics, Acta Math. Sci. Ser. B Engl. Ed. 29(4) (2009), 980-990.

[142] T. T. Li, B. P. Rao, and Z. Q. Wang, A note on the one-side exact boundary controllability for quasilinear hyperbolic systems, Commun. Pure Appl. Anal. 8(1) (2009), 405-418.

[143] —, A note on the one-side exact boundary observability for quasilinear hyperbolic systems, Georgian Math. J. 15(3) (2008), 571-580.

[144] T. T. Li and L. Yu, Exact boundary controllability for 1-D quasilinear waves equations, SIAM J. Control Optim. 45 (2006), 1074-1083.

[145] T. T. Li and B. Y. Zhang, Global exact boundary controllability of a class of quasilinear hyperbolic systems, J. Math. Anal. Appl. 225 (1998), 289-311.

[146] J. L. Lions, Exact controllability, stabilization and perturbations for distributed system, SIAM Review 30 (1988), 1-68.

[147] —, Contrôlabilite exacte des systemes distribues. (French) [Exact controllability of distributed systems], C. R. Acad. Sci. Paris Sr. I Math. 302(13) (1986), 471-475.

[148] W. Littman, Boundary control theory for hyperbolic and parabolic partial differential equations with constant coefficients, Ann. Scuola Norm. Sup. Pisa Cl. Sci. (4) 5(3) (1978), 567-580.

[149] —, Near optimal time boundary controllability for a class of hyperbolic equations, Lecture Notes in Control and Information, Vol. 97, pp. 307-312, Springer-Verlag, Berlin, New York, 1987.

[150] H. W. Liu, The Theory of Plates and Shells (in Chinese), University of Zhejiang, 1987.

[151] K. Liu, Locally distributed control and damping for the conservative system, SIAM Control Optim. 35 (1997), 1574-1590.

[152] W. Liu, Stabilization and controllability for the transmission wave equation, IEEE Trans. Automat. Control 46 (2001), 1900-1907.

[153] W. Liu and G. Williams, The exponential stability of the problem of transmission of the wave equation, Bull. Austral. Math. Soc. 57(2) (1998), 305-327.

[154] L. Miller, Escape function conditions for the observation, control, and stabilization of the wave equation, SIAM J. Control Optim. 41(5) (2002), 1554-1566.

[155] C. S. Morawetz, Time decay for nonlinear Klein-Gordon equations, Proc. Royal. Soc. London A 306 (1968), 291-296.

[156] —, Decay for solutions of the exterior problem for the wave equation, Comm. Pure Appl. Math. 28 (1975), 229-246.

[157] P. M. Naghdi, Foundations of elastic shell theory. 1963 Progress in Solid Mechanics, Vol. IV, pp. 1-90, North-Holland, Amsterdam.

[158] —, Theorey of shell and plates, Handbuch der Physik, Vol. VI a-2, Springer-Verlag, Berlin, pp. 425-460, 1972.

[159] M. Nakao, Decay of solutions of the wave equation with a local nonlinear dissipation, Math. Ann. 305 (1996), 403-417.

[160] —, Energy decay for the linear and semilinear wave equations in exterior domains with some localized dissipations, Math. Z. 238 (2001), 781-797.

[161] —, Existence of global solutions for the Kirchhoff-type quasilinear wave equation in exterior domains with a half-linear dissipation, Kyushu J. Math. 58 (2004), 373-391.

[162] S. Nicaise and C. Pignotti, Stabilization of the wave equation with variable coefficients and boundary condition of memory type, Asymptot. Anal. 50(1-2) (2006), 31-67.

[163] —, Internal and boundary observability estimates for the heterogeneous Maxwell's system, Appl. Math. Optim. 54 (2006), 47-70.

[164] —, Exponential and polynomial stability estimates for the wave equation and Maxwell's system with memory boundary conditions, Functional analysis and evolution equations, 515-530, Birkhäuser, Basel, 2008.

[165] F. I. Niordson, Shell Theory, North-Holland Series in Applied Mathematics and Mechanics, 1985.

[166] K. Ono, On global solutions and blow-up solutions of nonlinear Kirchhoff strings with nonlinear dissipation, J. Math. Anal. Appl. 216 (1997), 321-342.

[167] H. P. Oquendo, Nonlinear boundary stabilization for a tranmission problem in elasticity, Nonlinear Analysis 52(4)(2003), 1331-1354.

[168] A. Osses, A rotated multiplier applied to the controllability of waves, elasticity, and tangential Stokes control, SIAM J. Control Optim. 40(3) (2001), 777-800.

[169] A. Pazy, Semigroup of Linear Operator and Application to Partial Equations, Springer-Verlag, New York, Berlin, 1983.

[170] R. N. Pederson, On the unique continuation theorem for certain second and fourth order elliptic equations, Comm. Pure Appl. Math. XI (1958), 67-80.

[171] P. Petersen, Riemannian Geometry. Second edition. Graduate Texts in Mathematics, 171, Springer, New York, 2006.

[172] J. Puel and M. Tucsnak, Boundary stabilization for the von Karman equations, SIAM J. Control Optim. 33 (1995), no. 1, 255-273.

[173] T. Qin, The global smooth solutions of second order quasilinear hyperbolic equations with dissipative boundary conditions, Chin. Ann. of Math. 9B(3) (1988), 251-269.

[174] L. Ralston, Gaussian beams and the propagation of singularities, in: W. Littman (Ed.), Studies in Partial Differential Equations, in: MAA Stud. Math., Mathematical Association of America, Washington, DC, 23 (1982), 204-248.

[175] —, Solution of the wave equation with localized energy, Comm. Pure Appl. Math. 22 (1969), 807-823.

[176] M. A. Rammaha and T. A. Strei, Global existence and nonexistence for nonlinear wave equations with damping and source terms, Trans. Amer. Math. Soc. 354 (2002), 3621-3637.

[177] B. Rao, Stabilization of elastic plate with dynamical boundary control, SIAM J. Control Optim., 36 (1998), 148-163.

[178] S. E. Rebiai, Boundary stabilization of Schrodinger equations with variable coefficients, Afr. Diaspora J. Math. 1(1) (2004), 33-41.

[179] —, Uniform stabilization of the Euler-Bernoulli equation with variable coefficients, Proceedings (CD-ROM) of the 16th International Symposium on Mathematical Theory of Networks and Systems, Leuven, Belgium (2004).

[180] D. L. Russell, A unified boundary controllability theory for hyperbolic and parabolic partial differential equations, Studies in Appl. Math. 52 (1973), 189-211.

[181] —, Exact boundary value controllability theorems for wave and heat processes in star-complemented regions. Differential games and control theory, Proc. NSFCBMS Regional Res. Conf., Univ. Rhode Island, Kingston, R.I., 1973, pp. 291-319, Lecture Notes in Pure Appl. Math., Vol. 10, Dekker, New York, 1974.

[182] —, Controllability and stabilizability theory for linear partial differential equations: recent progress and open questions, SIAM Review 20(4) (1978), 639-739.

[183] H. S. Rutten, Theory and design of shells on the basis of asymptotic analysis, Rutten + Kruisman, Consulting engineers, Rijswijk, Holland, 1973.

[184] E. J. P. G. Schmidt, On the modelling and exact controllability of networks of vibrating strings, SIAM J. Control Optim. 31 (1992), 230-245.

[185] —, On the control of mechanical systems from one equilibrium location to another, J. Differential Equations 175 (2001), 189-208.

[186] —, On a non-linear wave equation and the control of an elastic string from one equilibrium location to another, J. Math. Anal. Appl. 272 (2002), 536-554.

[187] R. Schoen and S.-T. Yau, Lectures on Differential Geometry, Conference Proceedings and Lecture Notes in Geometry and Topology, International Press, 1994.

[188] T. Shirota, A remark on the unique continuation theorem for certain fourth order equations, Proc. Japan Acad. Ser. A Math. Sci. 36 (1960), 571-573.

[189] M. Slemrod, Weak asymptotic decay via a "related invariance principle" for a wave equation with nonlinear, nonmonotone damping, Proc. Roy. Soc. Edinburgh Sect. A 113 (1989), 87-97.

[190] M. Spivak, A Comprehensive Introduction to Differential Geometry I, II, IV, Boston Publish or Perish, 1970, 1975.

[191] M. Strauss and F. Treves, First order linear PDEs and uniqueness in the Cauchy problem, J. of Differential Equations 15 (1974), 195-209.

[192] D. Tataru, Boundary controllability for conservative PDEs, Appl. Math. Optim. 31 (1995), 257-295.

[193] —, A priori estimate of Carleman's type in domains with boundary, J. Math. Pure Appl. 73 (1994), 355-357.

[194] M. E. Taylor, Partial Differential Equations I, Springer-Verlag, New York, 1996.

[195] —, Partial Differential Equations III, Springer-Verlag, New York, 1996.

[196] L. R. T. Tebou, Stabilization of the wave equation with localized nonlinear damping, J. Differential Equations 145 (1998), 502-524.

[197] G. Todorova, Cauchy problem for a nonlinear wave equation with nonlinear damping and source terms, Nonlinear Anal. Ser. A, Theory Methods 41 (2000), 891-905.

[198] G. Todorova and B. Yordanov, The energy decay problem for wave equations with nonlinear dissipative terms in R^n, Indiana Univ. Math. J. 56 (2007), 389-416.

[199] —, Critical exponent for a nonlinear wave equation with damping, J. Differential Equations, 174 (2001), 464-489.

[200] R. Triggiani, Regularity theory, exact controllability and optimal quadratic cost problem for spherical shells with physical boundary controls, Special Issue of Control and Cybernetics 25(3) (1996), 553-568.

[201] R. Triggiani and P. F. Yao, Carleman estimates with no lower-order terms for general Riemann wave equations. Global uniqueness and observability in one shot, Appl. Math. Optim. 46 (2002), 331-375.

[202] M. Tucsnak and G. Weiss, Observation and control for operator semigroups, Birkhauser Advanced Texts: Basler Lehrbcher. (Birkhauser Advanced Texts: Basel Textbooks) Birkhauser Verlag, Basel, 2009, xii+483 pp. ISBN: 978-3-7643-8993-2.

[203] G. Weiss, Transfer functions of regular linear systems. I. Characterizations of regularity, Trans. Amer. Math. Soc. 342 (1994), no. 2, 827-854.

[204] H. Wu, Selected Lectures in Riemannian Geometry (in Chinese), Univ. of Beijing, 1981.

[205] —, The Bochner Technique in Differential Geometry, Mathematical Reports, Vol. 3, part 2, Harwood Academic Publishers, London-Paris, 1988.

[206] H. Wu, C. L. Shen, and Y. L. Yu, A Introduction to Riemannian Geometry (in Chinese), Univ. of Beijing, 1989.

[207] A. Wyler, Stability of wave equations with dissipative boundary conditions in a bounded domain, Differential Integral Equations 7 (1994), 345-366.

[208] P. F. Yao, On the observability inequalities for the exact controllability of the wave equation with variable coefficients, SIAM J. Control Optim. 37(6) (1999), 1568-1599.

[209] —, Observability inequalities for the Euler-Bernoulli plate with variable coefficients, Differential Geometric Methods in the Control of Partial Differential Equations, Contemp. Math. 268, Amer. Math. Soc., Providence, RI, 2000, pp. 383-406.

[210] —, On shallow shell equations, Discrete and Continuous Dynamical Systems Series 2(3) (2009), 697-722.

[211] —, Observability inequalities for shallow shells, SIAM J. Contr. and Optim. 38(6) (2000), 1729-1756.

[212] —, The ellipticity of the elliptic membrane, Acta Anal. Funct. Appl. 3(4) (2001), 322-333.

[213] —, Observability inequalities for Naghdi's model, MMAR 2000, the 6th International Conference on Methods and Models in Automation and Robotics, Miedzyzdroje, Poland, 2000, pp. 133-138.

[214] —, An energy-preserving nonlinear system modeling a string with input and output on the boundary, Nonlinear Anal. 66(10) (2007), 2128-2139.

[215] —, Global smooth solutions for the quasilinear wave equation with boundary dissipation, J. Differential Equations 241(1) (2007), 62-93.

[216] —, Boundary controllability for the quasilinear wave equation, Appl. Math. Optim. (61)(2010), No. 2, 191-233.

[217] P. F. Yao and G. Weiss, Global smooth solutions for a nonlinear beam with boundary input and output, SIAM J. Control Optim. 45(6) (2007), 1931-1964.

[218] Z. F. Zhang and P. F. Yao, Global smooth solutions of the quasi-linear wave equation with internal velocity feedback, SIAM J. Control Optim. 47(4) (2008), 2044-2077.

[219] —, Global smooth solutions and stabilization of nonlinear elastodynamic systems with locally distributed dissipation. Systems Control Lett. 58 (2009), no. 7, 491-498.

[220] Y. Zhou and Z. Lei, Local exact boundary controllability for nonlinear wave equations, SIAM J. Control Optim. 46(3) (2007), 1022-1051.

[221] E. Zuazua, Exponential decay for the semilinear wave equation with locally distributed damping, Comm. Partial Differential Equations 15 (1990), 205-235.

[222] —, Exponential decay for the semilinear wave equation with localized damping in unbounded domains, J. Math. Pures Appl. 70(9) 70 (1991), 513-529.

[223] C. Zuily, Uniqueness and Non-uniqueness in the Cauchy Problem, Progress in Mathematics, 1983.

Index